COSMO-98

COSMO-98

Second International Workshop on Particle Physics and the Early Universe

Asilomar, CA November 1998

EDITOR
David O. Caldwell
Institute for Nuclear and Particle Astrophysics and Cosmology, and Department of Physics, University of California, Santa Barbara

AIP

AIP CONFERENCE PROCEEDINGS 478

American Institute of Physics Woodbury, New York

Editor:

David O. Caldwell
Institute for Nuclear and Particle Astrophysics and Cosmology
and Department of Physics
University of California, Santa Barbara
Santa Barbara, CA 93106

E-mail: caldwell@slac.stanford.edu

L.C. Catalog Card No. 99-62726
ISBN 1-56396-853-3
ISSN 0094-243X
DOE CONF- 981150

Printed in the United States of America

Contents

*Italicized name indicates author who presented the paper.

OBSERVATIONAL COSMOLOGY

*Italicized name indicates author who presented the paper.

PARTICLE PHYSICS AND COSMOLOGY

THE MISSING ENERGY/MATTER OF THE UNIVERSE

*Italicized name indicates author who presented the paper.

NEUTRINOS

*Italicized name indicates author who presented the paper.

*Italicized name indicates author who presented the paper.

Preface

The confluence of work on particle physics, astrophysics, and cosmology has resulted in impressive advances in all three areas. The rapidity of progress via this interaction resulted in a decision during the first of these meetings, COSMO-97 in the English Lake District, to have such conferences yearly for at least the immediate future. As a result, the second was held at Asilomar, California, with subsequent ones to be held in 1999 and 2000 in Trieste, Italy and in Korea, respectively. This site in the lovely Carmel-Monterey area on the Pacific Ocean was especially suitable for the meeting, since shared meals and living quarters on the pinetree-covered grounds enhanced interactions among practitioners of the different disciplines. As breaks from the intensive program of 38 plenary talks of about a half-hour each and 50 generally 20-minute talks in 5 parallel sessions during two half days, there was a half-day tour of the Monterey peninsula and a banquet at the Monterey Bay Aquarium. Two evening sessions were required, one of which was an unusual debate on quantum cosmology among Stephen Hawking, Andrei Linde, and Alexander Vilenkin, moderated by Rocky Kolb.

I wish to thank the International Advisory Committee for initial help in arranging a program and speakers, but I am especially grateful to members of the Local Organizing Committee who put together much of the program. Aid with the quantum cosmology sessions came from Andrei Linde, on observational cosmology from Joel Primack and Bernard Sadoulet. Phil Lubin helped arrange microwave background talks and saw to getting conference transparencies onto the web. Steve Yellin provided valuable assistance in making computers function, and Dan Bauer helped in many ways with e-mail communications, participant lists, and generally in keeping the conference running smoothly.

Funds from the University of California Office of the President and from the Santa Barbara campus to our Institute for Nuclear and Particle Astrophysics and Cosmology were essential to COSMO-98. Additional funds were provided by France Córdova, the Vice Chancellor for Research at UCSB, who also set the tone for the conference with her introductory remarks, and by the National Science Foundation, which helped support students.

Even with all this assistance, the ultimate success of a conference depends on the participants. The reviews and recent advances presented were exciting and thought-provoking, as hopefully the reader will agree. Unfortunately, some of the talks, especially by astronomers, are not presented here, and some last-minute contributions did not conform to length criteria, but there is a wealth of information and ideas. Apparently widely judged as a success, COSMO-98 should provide increased interest in the next conferences in the series.

David O. Caldwell

INTERNATIONAL ADVISORY COMMITTEE

R. Arnowitt (Texas A&M)
A. Bottino (Torino)
J. Ellis (CERN)
F. Halzen (Wisconsin)
G. Kane (Michigan)
H. Klapdor-Kleingrothaus (MPI Heidelberg)
D. Lyth (Lancaster)
A. Morales (Zaragoza)
G. Ross (Oxford)
Y. Totsuka (Tokyo)
V. Barger (Wisconsin)
R. Cowsik (Bangalore)
J. Frieman (Fermilab)
S. Hawking (Cambridge)
J. Kim (Seoul)
E. Kolb (Fermilab)
P. Langacker (Pennsylvania)
R. Mohapatra (Maryland)
L. Roszkowski (Lancaster)
G. Senjanovic (ICTP Trieste)
S. Weinberg (Texas)

LOCAL ORGANIZING COMMITTEE

D. Bauer (Santa Barbara)
M. Dine (Santa Cruz)
G. Gelmini (Los Angeles)
P. Lubin (Santa Barbara)
B. Sadoulet (Berkeley)
S. Yellin (Santa Barbara)
D. Caldwell (Santa Barbara), Chair
G. Fuller (San Diego)
A. Linde (Stanford)
J. Primack (Santa Cruz)
J. Silk (Berkeley)
G. Yodh (Irvine)

QUANTUM COSMOLOGY AND INFLATION

A View of the Early Universe

Edward W. Kolb

NASA/Fermilab Theoretical Astrophysics Group
Fermi National Accelerator Laboratory
and
Department of Astronomy and Astrophysics, Enrico Fermi Institute
The University of Chicago

Abstract. During this conference we have heard a great deal about recent progress in understanding the development of cosmic structures from small initial seed perturbations. In these concluding remarks, I would like to comment on the two parts of the picture involving elementary-particle physics: 1) the generation of the seed perturbations in the early universe during inflation, and 2) the nature of the dark matter.

INTRODUCTION

In my opinion, the most striking illustration of the unity of science is the interdisciplinary field of "particle cosmology," the subject of the COSMO98 meeting. High-energy physics examines nature on the smallest scales, while cosmology studies the universe on the largest scales. Although the two fields examine the physical world at the opposite end of the size spectrum, they are united in the sense that it is impossible to understand the origin and evolution of large-scale structures in the universe without understanding the "initial conditions" that led to the their development. It is also possible to learn something about fundamental physics from cosmological observations.

The initial data for structure formation was set in the very early universe when the fundamental particles and forces acted to produce perturbations in the cosmic density field. To understand structure formation we must understand the origin of the perturbations. A complete understanding of the present structure of the universe will also be impossible without accounting for the dark component in the density field. The most likely possibility is that this ubiquitous dark component is an elementary particle relic from the early universe.

At present we have a *theory* for the origin of the initial seed perturbations during the era of cosmic inflation. However we have no *'standard model'* of inflation. The situation is no better with respect to dark matter. We have no standard model of dark matter, but we do have many candidates.

CP478 , *COSMO—98*, edited by David O. Caldwell

Much of the progress we heard about during the conference was driven by experimental programs of sufficient scope and accuracy for the confrontation of models with observation. We can look forward to an even richer environment for phenomenological cosmology.

Two of the issues that could be cleared up in the new era of phenomenological cosmology have to do with the origin of the seed perturbations necessary for structure formation, and the nature of the ubiquitous dark matter.

PERTURBATIONS

Just two years ago there were two theories for the generation of the seed perturbations required for structure formation: cosmic defects and inflation. But heroic calculations of the angular power spectrum of temperature fluctuations in the cosmic background radiation (CBR) by several groups [1–3] ruled out cosmic defects. The only theory left standing is cosmic inflation.

Inflation

One of the striking features of the CBR temperature fluctuations is that they *appear* to be noncausal. The CBR fluctuations were largely imprinted at the time of last-scattering, about 300,000 years after the bang. However, there are fluctuations on length scales much larger than 300,000 light years! How could a causal process imprint fluctuations on scales larger than the light-travel distance since the time of the bang? The answer is inflation.

Sometime during the early evolution of the universe the expansion must have been such that $\ddot{a} > 0$. From Einstein's equations, the "acceleration" is related to the energy density and the pressure as $\ddot{a} \propto (\rho + 3p)$. Therefore, an acceleration (positive $\ddot{a}$) requires an unusual equation of state with $\rho + 3p < 0$. This is referred to as "accelerated expansion" or "inflation."

A universe dominated by vacuum energy has the equation of state $p = -\rho$, so a vacuum-dominated phase satisfies the criterion for inflation, $\rho + 3p < 0$. How can one imagine a universe dominated by vacuum energy making a transition to a matter-dominated or radiation-dominated universe? A simple way to picture this is by the action of a scalar field ϕ with potential $V(\phi)$. Let's imagine the scalar field is displaced from the minimum of its potential as illustrated in Fig. 1. If the energy density of the universe is dominated by the potential energy of the scalar field ϕ, known as the *inflaton*, then $\rho + 3p$ will be negative. The vacuum energy disappears when the scalar field evolves to its minimum. The amount of time required for the scalar field to evolve to its minimum and inflation to end (or even more useful, the number of e-folds of growth of the scale factor) can be found by solving the classical field equation for the evolution of the inflaton field $\ddot{\phi} + 3H\dot{\phi} + dV/d\phi = 0$.

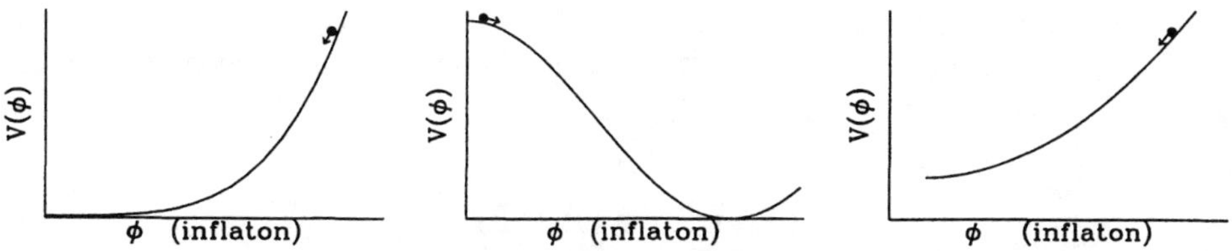

FIGURE 1. Schematic illustrations of the inflaton potential energy. The potential on the left is a "large-field" (Type 1-a) model. The middle figure illustrates a "small-field" (Type-Ib) model. The figure on the right illustrates an example of hybrid (Type-Ic) inflation.

Of course there are more complicated ways inflation could be realized, but even within the scope of this simple single-field picture, there are many models of inflation. One of the immediate goals is to sort through the models. Of course, in some sense the best we can hope for is that large numbers of models will be excluded by experiment, and somehow this thinning of the model space will point the way for construction of the most promising models. So we must find a way to kill models (or in the colorful language of Chicago, "whack" some models).

One possible way to sort through models is by determining from CBR observations the slope of the scalar spectrum (usually denoted as n), and the ratio of the

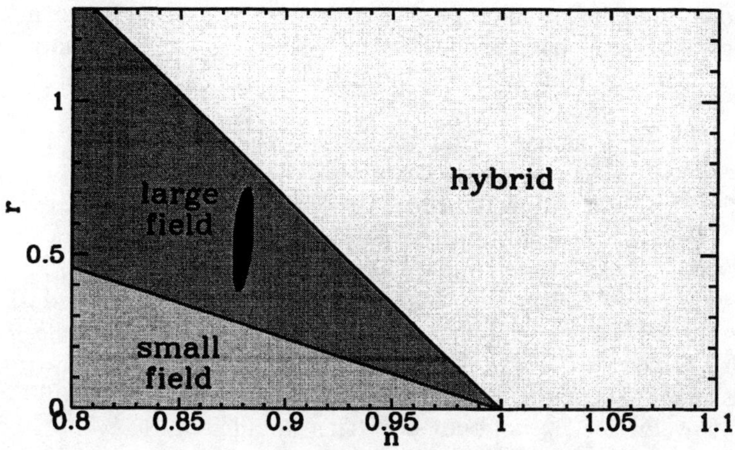

FIGURE 2. Large-field (Type 1-a) inflation models, small-field (Type-Ib) inflation models, and hybrid (Type-Ic) inflation models seem to populate different regions of the n–r plane. Observational determination of n and r could rule out entire classes of models. This figure is from Dodelson, Kinney and Kolb [4]. The dark ellipse illustrates the size (but not necessarily the location) of the expected 2σ region after the MAP and Planck satellites.

tensor-to-scaler contribution to C_2 (usually denoted as r). Perhaps the data will be good enough to rule out one or more types of models and provide some guidance to model builders.

DARK MATTER

Just a few years ago it was thought that we live in a simple universe where the only dynamically important matter components in the universe were dissapationless cold dark matter and baryons. I now feel a lot of nostalgia for the good old simple days of CDM.

The Inventory

In the last year, our inventory of the mass content of the universe seemed to shift. The shift was driven by observation of atmospheric neutrino oscillations, and evidence for a cosmological constant.

Evidence for an atmospheric-neutrino anomaly becomes stronger and stronger [5]. The most likely explanation is $\nu_\mu \leftrightarrow \nu_\tau$ oscillation with $10^{-3} < \Delta m^2 < 10^{-1}\,\mathrm{eV}^2$. Of course oscillation experiments only measure Δm^2, but barring any near degeneracy in masses, it is reasonable to assume we are seeing an individual mass between 0.03 and 0.3 eV. We have no theory for any mass hierarchy among neutrinos, but the traditional prejudice is that the tau neutrino should be the most massive. The most conservative explanation of the atmospheric neutrino anomaly is that the tau neutrino has a mass in the 0.03 and 0.3 eV range, and it is the most massive neutrino.[1]

Let's pretend for a moment that the most massive neutrino is a tau neutrino with $m_\nu = 0.3\mathrm{eV}$. Using the well know result $\Omega_\nu h^2 = m_\nu/92\,\mathrm{eV}$, we find $\Omega_\nu \simeq 3\times 10^{-3}h^{-2}$. Using the currently fashionable value of $h = 0.65$, $\Omega_\nu \simeq 0.008$. This is too small to be dynamically important or useful, but perhaps just a little to large to ignore completely in precision calculations.

The latest exciting piece of experimental information regarding dark matter is the announcement of evidence by two groups, *The Supernova Cosmology Project* [6] and the *Hi-z Supernova Search Team* [7], that we live in an accelerating universe. (The evidence was discussed in this workshop by Adam Reiss and Saul Perlmutter.) Their best-fit value for Ω_Λ is about 0.6. This would mean that most of the mass density of the universe is in some sort of vacuum energy.

While there is still a fair amount of skepticism about the results, perhaps particle physicists will be given the job of explaining why the vacuum energy is about 10^{-30}g cm^{-3}, or in more useful units, about $(10^{-3}\mathrm{eV})^4$. If this result holds, then in a couple of years future COSMO meetings may be dominated by his issue.

---

1) This conclusion would prove to be completely wrong if the LSND results are confirmed.

With neutrinos of a mass suggested by the atmospheric anomaly, or with vacuum energy of the magnitude suggested by the supernova results, or even with neutrinos of a mass suggested by the LSND results, there is still a need for a cold component of dark matter.

Cold Dark Matter

Thermal Relics

It is usually assumed that the dark matter consists of a species of a new, yet undiscovered, massive particle, traditionally denoted by X. It is also often assumed that the dark matter is a thermal relic, *i.e.*, it was in chemical equilibrium in the early universe.

A thermal relic is assumed to be in local thermodynamic equilibrium (LTE) at early times. The *equilibrium* abundance of a particle, say relative to the entropy density, depends upon the ratio of the mass of the particle to the temperature. Define the variable $Y \equiv n_X/s$, where n_X is the number density of WIMP X with mass M_X, and $s \sim T^3$ is the entropy density. The equilibrium value of Y, Y_{EQ}, is proportional to $\exp(-x)$ for $x \gg 1$, while $Y_{EQ} \sim$ constant for $x \ll 1$, where $x = M_X/T$.

A particle will track its equilibrium abundance as long as reactions which keep the particle in chemical equilibrium can proceed rapidly enough. Here, rapidly enough means on a timescale more rapid than the expansion rate of the universe. When the reaction rate becomes smaller than the expansion rate, then the particle can no longer track its equilibrium value, and thereafter Y is constant. When this occurs the particle is said to be "frozen out." A schematic illustration of this is given in Fig. 3.

The more strongly interacting the particle, the longer it stays in LTE, and the smaller its eventual freeze-out abundance. Conversely, the more weakly interacting the particle, the larger its present abundance. The freeze-out value of Y is related to the mass of the particle and its annihilation cross section (here characterized by σ_0) by [8] $Y \propto (M_X m_{Pl} \sigma_0)^{-1}$. Since the contribution to Ω is proportional to $M_X n_X$, which in turn is proportional to $M_X Y$, the present contribution to Ω from a thermal relic roughly is *independent* of its mass,[2] and depends only upon the annihilation cross section. The cross section that results in $\Omega_X h^2 \sim 1$ is of order $10^{-37}\mathrm{cm}^2$, which is the order of the weak scale. This is one of the attractions of thermal relics. The scale of the annihilation cross section is related to a known mass scale.

The simple assumption that dark matter is a thermal relic is surprisingly restrictive. The largest the annihilation cross section possible is roughly M_X^{-2}. This implies that large-mass WIMPS would have such a small annihilation cross section

---

[2] To first approximation the relic dependence depends upon the mass only indirectly through the dependence of the annihilation cross section on the mass.

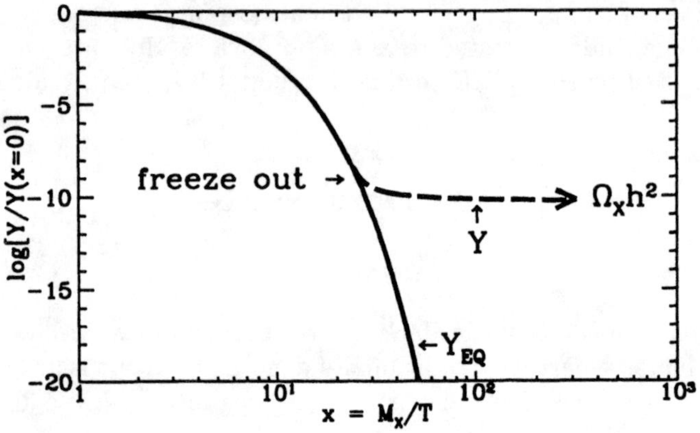

FIGURE 3. A thermal relic starts in LTE at $T \gg M_X$. When the rates keeping the relic in chemical equilibrium become smaller than the expansion rate, the density of the relic relative to the entropy density freezes out.

that their present abundance would be too large. Thus one expects a maximum mass for a thermal WIMP, which turns out to be a few hundred TeV [9].

The standard lore is that the hunt for dark matter should concentrate on particles with mass of the order of the weak scale and with interaction with ordinary matter on the scale of the weak force. This has been the driving force behind the vast effort in dark matter direct detection described in this meeting by Sadoulet.

Nonthermal relics—WIMPZILLAS

In this meeting we heard of scenarios for production of *nonthermal* dark matter. A example of nonthermal WIMP productions is illustrated in Fig. 4.

Several ways to produce nonthermal relics have been explored over the past year [10]. The WIMPZILLA, again denoted as X, may have a mass many orders of magnitude larger than the weak scale, possibly as large as the Grand Unified Theory (GUT) scale. It is very intriguing that these considerations resurrect the possibility that the dark matter might be charged or even strongly interacting!

An attractive origin for WIMPZILLAS is during the defrosting phase after inflation. It is important to realize that it is not necessary to convert a significant fraction of the available energy into massive particles; in fact, it must be an infinitesimal amount. If a fraction ϵ of the available energy density is in the form of a massive, stable X particle, then $\Omega_X = \epsilon\Omega_\gamma(T_{RH}/T_0)$, where T_{RH} is the "reheat" temperature. For $\Omega_X = 1$, this leads to the limit $\epsilon \lesssim 10^{-17}(10^9\,\text{GeV}/T_{RH})$.

Since the masses of the nonthermal WIMPS are in the range 10^{12} to 10^{16} GeV, much larger than the mass of thermal wimpy WIMPS, they may be referred to as

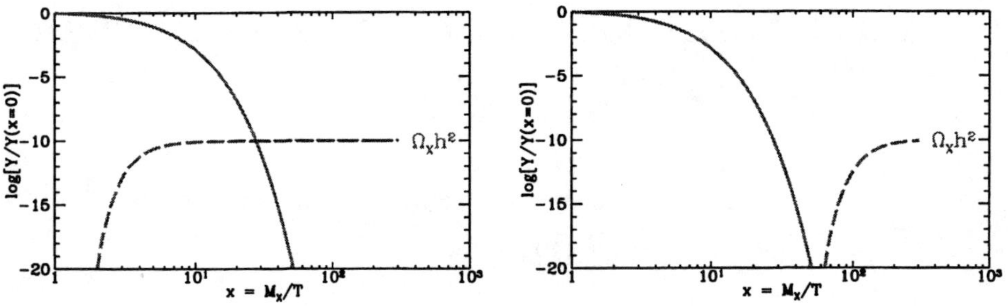

FIGURE 4. A nonthermal relic typically has abundance much less than LTE at $T \gg M_X$. Here are two examples of the evolution of a nonthermal relic. In the left panel $Y \ll Y_{EQ}$ at freezeout, while in the right panel $Y \gg Y_{EQ}$ at freezeout. Again, the solid curve is the LTE abundance while the dashed curve is the actual abundance.

WIMPZILLAS. In searches for dark matter it may be well to remember that "size does matter."

ACKNOWLEDGEMENTS

This work was supported in part by the Department of Energy and NASA (NAG5-7092).

REFERENCES

1. U.-L. Pen, U. Seljak, N. Turok, Phys. Rev. Lett. 79, 1611 (1997).
2. B. Allen, R. R. Caldwell, S. Dodelson, L. Knox, E. P. S. Shellard, A. Stebbins, Phys. Rev. Lett. 79, 2624 (1997).
3. A. Albrecht, R. A. Battye, J. Robinson, Phys. Rev. Lett. 79, 4736 (1997).
4. S. Dodelson, W. H. Kinney, and E. W. Kolb, Phys. Rev. D56, 3207 (1997) .
5. See, e.g., Takaaki Kajita, for the Super-Kamiokande and Kamiokande collaborations, to appear in the proceedings of XVIII International Conference on Neutrino Physics and Astrophysics (Neutrino'98), Takayama, Japan, June 1998.
6. S. Perlmutter, et al., astro-ph/9812133.
7. A. G. Riess, et al., Astron. J., **116**, 1009 (1998).
8. E. W. Kolb and M. S. Turner, *The Early Universe*, (Addison-Wesley, Menlo Park, Ca., 1990).
9. K. Griest and M. Kamionkowski, Phys. Rev. Lett. **64**, 615 (1990).
10. D. J. H. Chung, E. W. Kolb, and A. Riotto, Phys. Rev. D 59, 023501 (1999); D. J. H. Chung, E. W. Kolb, and A. Riotto, hep-ph/9809453; D. J. H. Chung, E. W. Kolb and A. Riotto, Phys. Rev. Lett. 81 4048 (1998); V. Kuzmin and I. I. Tkachev, JETP Lett. 68 271 (1998); E. W. Kolb, A. Riotto and I. I. Tkachev, Phys. Lett. B423, 348 (1998); D. J. Chung, hep-ph/9809489.

Models of inflation and their predictions

David H. Lyth

Department of Physics,
Lancaster University,
Lancaster LA1 4YB. U. K.

Abstract. Taking field theory seriously, inflation model-building is difficult but not impossible. The observed value of the spectral index of the adiabatic density perturbation is starting to discriminate between models, and may well pick out a unique one in the forseeable future.

I shall summarise the present status of inflation model-building, and its comparison with observation. This is already a substantial area of research, and with the advent of new observations in the next few years it will become a major industry. For an extensive review with references, see [1].

I focus on the simplest paradigm, which following the usual scientific practice should be tested to destruction before complications are entertained. There is a *slowly-rolling, single-component* inflaton field, which experiences *Einstein gravity* and drives the observable Universe into a *spatially flat* condition. The *gaussian adiabatic* density perturbation, generated by the vacuum fluctuation of the inflaton field ϕ, is solely responsible for the origin of structure.

During inflation, the potential $V(\phi)$ of the inflaton field ϕ satisfies the flatness conditions

$$\epsilon \ll 1, \qquad |\eta| \ll 1, \tag{1}$$

where $\epsilon \equiv \frac{1}{2}M_{\rm P}^2(V'/V)^2$ and $\eta \equiv M_{\rm P}^2 V''/V$. The inflaton field satisfies the slow-roll approximation $3H\dot\phi = -V'$ where H is the Hubble parameter given by $3H^2 = V/M_{\rm P}^2$. To work out the predictions one needs the number of e-folds N between a given epoch and the end of slow-roll inflation (with the inflaton ϕ). Its small change is defined by $dN \equiv -H\,dt (= -d\ln a)$, which with the slow-roll approximation leads to

$$N(\phi) = \int_{\phi_{\rm end}}^{\phi} M_{\rm P}^{-2}\frac{V}{V'}d\phi\,. \tag{2}$$

Here $\phi_{\rm end}$ marks the end of slow-roll inflation, caused by the failure of the flatness conditions or by the destabilization of a non-inflaton field. Often, the integral is

CP478, *COSMO—98*, edited by David O. Caldwell

dominated by the other limit ϕ in which case the predictions are independent of $\phi_{\rm end}$.

The vacuum fluctuation of the inflaton field generates a gaussian adiabatic primordial density perturbation, whose conventionally-defined spectrum is given by

$$\delta_H^2(k) = \frac{1}{75\pi^2 M_{\rm P}^6} \frac{V^3}{V'^2} . \tag{3}$$

The right hand side is evaluated at the value $\phi(k)$ which corresponds to horizon exit $k = aH$. It satisfies $d\ln k = -dN(\phi)$ and therefore $\ln(k_{\rm end}/k) = N(\phi)$, where $k_{\rm end}$ is the scale leaving the horizon at the end of slow-roll inflation. With Eq. (2), this determines $\phi(k)$ provided that we know the value of $N(\phi)$ when some reference scale leaves the horizon. This scale is conveniently taken to be the central scale probed by COBE, $k_{\rm COBE} \simeq 7.5 H_0$ where H_0 is the present Hubble parameter. Depending on the history of the Universe, one has

$$N_{\rm COBE} \simeq 60 - \ln(10^{16}\,{\rm GeV}/V^{1/4}) - \frac{1}{3}\ln(V^{1/4}/T_{\rm reh}) - \Delta N , \tag{4}$$

where $T_{\rm reh}$ is the reheat temperature and $\Delta N > 0$ allows for matter domination and thermal inflation between reheating and nucleosynthesis (and any continuation of inflation after the epoch $\phi_{\rm end}$).

Differentiating Eq. (3) with the aid of Eq. (2), the spectral index is

$$\frac{n(k)-1}{2} \equiv \frac{d\delta_H}{d\ln k} = \eta - 3\epsilon . \tag{5}$$

If n is constant then $\delta_H^2 \propto k^{n-1}$.

Inflation also generates gravitational waves with primordial spectrum $\mathcal{P}_{\rm grav}(k) = \frac{2}{M_{\rm P}^2}\left(\frac{H}{2\pi}\right)^2$. No gravitational wave signal is seen in the cmb anisotropy, which translates into a bound $\epsilon \lesssim 0.1$. The signal will probably never be seen unless $\epsilon \gtrsim 10^{-3}$.

At $k_{\rm COBE}$, the COBE observations give the accurate normalization (ignoring gravitational waves) $\delta_H = 1.91 \times 10^{-5}$, which corresponds to

$$V^{1/4}/\epsilon^{1/4} = .027 M_{\rm P} = 6.7 \times 10^{16}\,{\rm GeV}. \tag{6}$$

The present bound $\epsilon \lesssim 0.1$ on gravitational waves implies $V^{1/4} \lesssim 3.6 \times 10^{16}$ GeV. Gravitation waves will never be detectable if $V^{1/4} \lesssim 1 \times 10^{16}$ GeV, and *most inflation models give a lower value* when normalized to satisfy Eq. (6).

Over the range of cosmological scales, say $H_0 < k < 10^4 H_0$, there is an observational bound on the scale-dependence of δ_H. Until recently uncertainties in the cosmological parameters allowed only the weak result $|n-1| \lesssim 0.2$, but new data give a preliminary result $|n-1| < 0.05$ [2]. After Planck flies we shall probably know $n(k)$ with an uncertainty of ± 0.01.

At the most primitive level, a model of inflation consists of a form for $V(\phi)$, plus a prescription for $\phi_{\rm end}$ if the latter is not determined by V as happens in some hybrid inflation models. From a field theory viewpoint, one expects $V(\phi)$ to be schematically of the following form[1]

$$\begin{aligned} V = V_0 &+ \frac{1}{2}m^2\phi^2 + M\phi^3 + \frac{1}{4}\lambda\phi^4 \\ &+ (\tilde{m}^4 + 2g\phi^2 + g^2\phi^4)\ln(g\phi/Q) \\ &+ \sum_{d=5}^{\infty} \lambda_d M_{\rm P}^{4-d}\phi^d \\ &+ [\Lambda^{4+\alpha}\phi^{-\alpha} - \tilde{\Lambda}^{4\pm\beta}\phi^{\mp\beta}] . \end{aligned} \tag{7}$$

In the first line are *renormalizable tree-level* terms, with the origin is chosen so that $V' = 0$; the coefficients can have either sign. In the second line is the *one-loop correction* due to a particle with mass $\tilde{m}$ and coupling g, valid if $g\phi \gtrsim \tilde{m}$. (It is suppressed at smaller ϕ.) The renormalization scale Q should be fixed at a typical relevant value of $g\phi$ to minimize higher loop contributions. One sums over particles with a plus/minus sign for bosons/fermions, and unbroken global supersymmetry would make the total vanish. During inflation susy is broken, but the ϕ^4 term still vanishes, and the ϕ^2 term *may* vanish, but one expects no cancellations between the contributions to the the constant term. The third line contains the *non-renormalizable* terms which summarise unknown Planck scale physics; the coefficients λ_d are generically of order 1, but supersymmetry can make a finite number of them tiny. (By appealing to a continuous global symmetry it can make them all tiny, but no such symmetry comes out of string theory.) The fourth line contains a $\phi^{-\alpha}$ term that might come from dynamical symmetry breaking, and a $-\phi^{\mp\beta}$ term that might come from mutated hybrid inflation. These terms will be present only in exceptional cases, unlike the others which are generic.

One who presumes to use field theory ought to take this expression seriously, and when that is done the flatness conditions Eq. (1) turn out to be extraordinarily difficult to satisfy. The non-renormalizable terms are obviously dangerous. So are the loop corrections, especially in the context of hybrid inflation where some coupling has to be substantial. Less obviously, a generic supergravity theory gives a prediction of the form $M_{\rm P}^2 V''/V = 1 + \cdots$, in which case there has to be some cancellation whose origin is at present obscure.

The simplest proposed model, usually called chaotic inflation, is a monomial $V \propto \phi^p$ with p usually 2 or 4. Inflation takes place at $\phi > \phi_{\rm end} \sim pM_{\rm P}$, giving $n - 1 = -(2+p)/(2N)$ and significant gravitational waves ($\epsilon = p/(5N)$).

If non-renormalizable terms are there, they kill the above model. To live with them one needs $\phi \lesssim M_{\rm P}$ or $\phi \ll M_{\rm P}$. (The latter case is preferable, but one has to watch the loop correction which generates a term $V'' \propto \phi^{-2}$.) Making the reasonable assumption that only one term of Eq. (7) is relevant, Eq. (1) then requires $V \simeq V_0$.

[1] The form is more restrictive if ϕ is a pseudo-Goldstone boson, but that hypothesis has not so far lead to an attractive model of inflation.

TABLE 1. Predictions for the spectral index $n(k)$, using $N = \ln(k_{\rm end}/k)$ where $k_{\rm end} = aH$ at the end of slow-roll inflation. Constants c, q and A are positive while σ and p can have either sign. In the first two cases one expects $|c| \sim 10^{-1}$ to 10^{-2}, and in the second case, one expects $|\sigma| \gtrsim |c|$.

Comments	$V(\phi)/V_0 \simeq 1$	$\frac{1}{2}(n-1)$	$\frac{1}{2}\frac{dn}{d\ln k}$
Mass term	$1 \pm \frac{1}{2}c\frac{\phi^2}{M_P^2}$	$\pm c$	0
Softly broken susy	$1 \pm \frac{1}{2}c\frac{\phi^2}{M_P^2}\ln\frac{\phi}{A}$	$\pm c + \sigma e^{\pm cN}$	$\mp c\sigma e^{\pm cN}$
Spont. broken susy	$1 + c\ln\frac{\phi}{A}$	$-\frac{1}{2N}$	$-\frac{1}{2}\frac{1}{N^2}$
$p > 2$ or $-\infty < p < 1$	$1 - c\phi^p$	$-\left(\frac{p-1}{p-2}\right)\frac{1}{N}$	$-\left(\frac{p-2}{p-1}\right)\left(\frac{n-1}{2}\right)^2$
p integer ≤ -1 or ≥ 3	$1 + c\phi^p$	$\frac{p-1}{p-2}\frac{1}{N_{\max}-N}$	$-\left(\frac{p-2}{p-1}\right)\left(\frac{n-1}{2}\right)^2$

TABLE 2. Some predicted values.

$V(\phi)/V_0$	$1-n$		$-10^3 dn/d\ln k$	
	$N=50$	$N=20$	$N=50$	$N=20$
$1 + c\ln(\phi/Q)$	0.02	0.05	(0.4)	2.6
$1 - c\phi^{-2}$	0.03	0.075	(0.6)	3.8
$1 - c\phi^4$	0.06	0.15	(1.2)	5.4
$1 - c\phi^3$	0.08	0.20	(1.6)	10.0

Predictions for the spectral index are given in the Tables. An inflaton field with negligible interaction ($V = V_0 \pm \frac{1}{2}m^2\phi^2$) gives a constant $n-1$, which can be positive or negative and is typically not extremely small or interactions *would* be significant. One significant interaction term typically gives n close to 1, with weak scale-dependence.

A dramatic exception is the case [3] where a $\phi^2\ln\phi$ loop correction dominates the mass term, as shown in the second line of Table 1. The correct COBE normalization Eq. (6) is obtained with a reasonable value $c \sim 10^{-1}$ to 10^{-2} of the coupling c.[2] Furthermore, such a coupling allows n to pass through 1 on cosmologically interesting scales!

The observed value of $n(k)$ will become an increasingly powerful discriminator in the future. If one were to take it seriously, the preliminary result $|n-1| < 0.05$ would already rule out the cubic self-interaction in Table 2. It would also strongly constrain the parameters c and σ in the case just mentioned, perhaps demanding physically unreasonable values for them.

REFERENCES

1. D. H. Lyth and A. Riotto, hep-ph/9807278, to appear in Phys. Reports.

[2] To calculate this normalization one has to take into account higher loop corrections by using a renormalization-group-improved potential, but the order of magnitude is unaffected.

2. R. Bond, Pritzker Symposium,
http://www-astro-theory.fnal.gov/Personal/psw/talks/bond/bond.03.gif.
3. E. D. Stewart, Phys. Lett. **B391**, 34 (1997); Phys. Rev. **D56**, 2019 (1997).

A Debate on Open Inflation

Professor S.W. Hawking

Department of Applied Mathematics and Theoretical Physics,
University of Cambridge, Silver Street, Cambridge, CB3 9EW

Abstract. This is a reproduction of Professor Stephen Hawking's part in a debate, which took place at the COSMO 98 Coference, in Monterey, California. Two other physicists, Andrei Linde and Alexander Villenkin, also took part. Professor Hawking is the Lucasian Professor of Mathematics at the University of Cambridge, in England.

OPEN INFLATION

In February this year, Neil Turok and I proposed a new model for inflation. The distinctive feature was that it produced an infinite, open universe, yet it satisfied the no boundary condition and came from an instanton of finite size, with a mass of the order of one gram. Our paper aroused a lot of interest, as shown by 37 citations on HEP-TH, but it brought a lot of opposition. This centered on three features of our model.

First, we were attacked for using the no boundary proposal. People like Linde and Vilenkin, claimed that one should use the quantum tunneling wave function instead.

Second, we were criticized because our instanton contained a singularity. It was said this was contrary to the spirit of the no boundary proposal; that the singularity would be naked and would make the universe non predictable.

Third, we invoked the anthropic principle, to avoid the model predicting a totally empty universe. We were attacked both for using anthropic arguments, and for the very low value for the density of the universe, that they seemed to lead to.

In this talk, I will describe the open inflation model that Neil and I proposed, and answer some of the objections that have been raised.

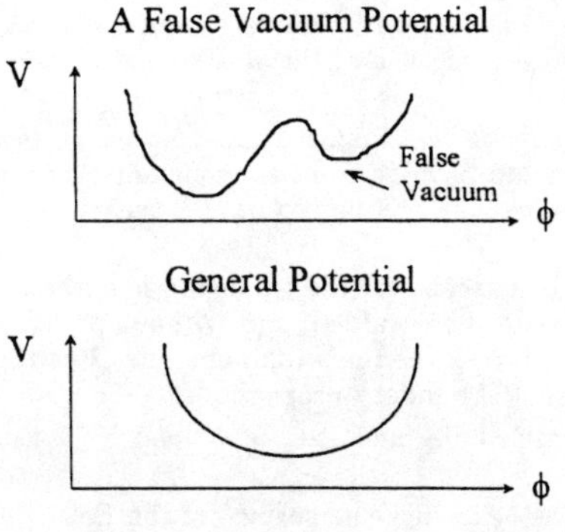

Fig. 1: False vacuum potential and general potential.

CP478 , *COSMO—98*, edited by David O. Caldwell

The original idea for inflation, was that in some way, the universe got trapped in what was called, a false vacuum state. A false vacuum state corresponds to some scalar field being in a local minimum of the potential, that has more energy than the true vacuum, which is taken to have zero energy density.

$$T_{\mu\nu} = -\tfrac{1}{2} g_{\mu\nu} V(\phi) \tag{1}$$

False vacuum energy momentum tensor.

Because a false vacuum is Lorentz invariant, its energy momentum tensor must be proportional to the metric. Since the false vacuum has positive energy density, the coefficient of proportionality must be negative. The Einstein equations then imply that the scale factor increases exponentially with time.

In such a universe, the integral of one over the scale factor, diverges as one goes back in time. This means that different regions in the early universe could have communicated with each other, and come to equilibrium at a common state. So explaining why the microwaves, look the same in different directions.

The original model of inflation, which came to be known as old inflation, had various problems. How did the universe get into a false vacuum state in the first place, and how did it get out again? Various modifications were proposed, that went under the names of new inflation, or extended inflation. I won't describe them, because I have got into trouble in the past, about who should have credit for what, and because I now consider them irrelevant.

As Linde first pointed out, it is not necessary for the universe to be in a false vacuum, to get inflation. A scalar field with a potential V, will have the energy momentum tensor shown below.

$$T_{\mu\nu} = \phi_{,\mu}\phi_{,\nu} - \tfrac{1}{2} g_{\mu\nu}[\phi_{,\Lambda}\phi^{,\Lambda} + V(\phi)] \tag{2}$$

Energy momentum tensor.

If the field is nearly constant in a region, the gradient terms will be small, and the energy momentum tensor, will be minus half V, times the metric. This is just what one needs for inflation.

In the false vacuum case, the scalar field sits in a local minimum of the potential, V. In that case, the field equation allows the scalar field, to remain constant in space and time. If the scalar field is not at a local minimum, it can not remain constant in time, even if it is initially constant in space. However, Linde pointed out that if the potential is not too steep, the expansion of the universe, will slow down the rate at which the field rolls down the potential, to the minimum. The gradient terms in the energy momentum tensor, will remain small, and the scale factor will increase almost exponentially. One can get inflation with any reasonable potential V, even if it didn't have local minima, corresponding to false vacua. The work that Neil and I have done, is a logical extension of Andrei's idea. But I'm not sure if Andrei agrees with it, though I think he's coming round.

Andrei's idea removed the need to believe that the universe began in a false vacuum. However, one still needed to explain, why the field should have been nearly constant over a region, with a value that was not at the minimum of the potential. To do this, one has to have a theory of the initial conditions of the universe. There are three main candidates.

They are, the so called pre-big bang scenario, the tunneling hypothesis, and the no boundary proposal. In my opinion, the pre-big bang scenario is misguided, and without predictive power. And I feel the tunneling hypothesis, is either not well defined, or gives the wrong answers. But then I'm biased, for it was Jim Hartle and I, that were responsible for the no boundary proposal.

This says that the quantum state of the universe, is defined by a Euclidean path integral over compact metrics, without boundary.

One can picture these metrics, as being like the surface of the Earth, with degrees of latitude, playing the role of imaginary time. One starts at the north pole, with the universe as a single point. As one goes south, the spatial size of the universe, increases like the lengths of the circles of latitude. The spatial size of the universe, reaches a maximum size at the equator, and then shrinks again to a point at the south pole.

Of course, spacetime is four dimensional, not two dimensional, like the surface of the Earth, but the idea is much the same. I shall go through it in detail, because it is basic to the work I'm going to describe.

The simplest compact four dimensional metric that might represent the universe, is the four sphere.

$$ds^2 = d\sigma^2 + \frac{1}{H^2} \sin^2 H\sigma \,(d\chi^2 + \sin^2\chi \; d\Omega^2)$$

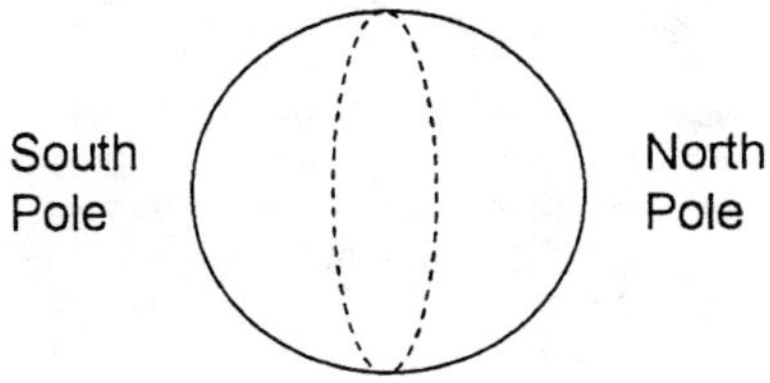

Fig. 2: The simplest compact four dimensional metric that might represent the universe; the four sphere.

One can give its metric in terms of a coordinate, σ, that measures the distance from the pole, and three coordinates, χ, θ and ϕ, on a three sphere, that represents the spatial size of the universe. Again, one starts at the north pole, σ =0, with a universe of zero spatial size, and expands up to a maximum size at the equator, $\sigma = \pi$, over 2H. But we live in a universe with a Lorentzian metric, like Minkowski space, not a Euclidean, positive definite metric. One therefore has to analytically continue, the Euclidean metrics used in the path integral, for the no boundary proposal. There are several ways one can analytically continue, the metric of the four sphere, to a Lorentzian spacetime metric.

Next analytically continue $\sigma = \frac{\pi}{2H} + it$

$$ds^2 = - dt^2 + \frac{1}{H^2} \cosh^2 Ht \;(d\chi^2 + \sin^2\chi \; d\Omega^2)$$

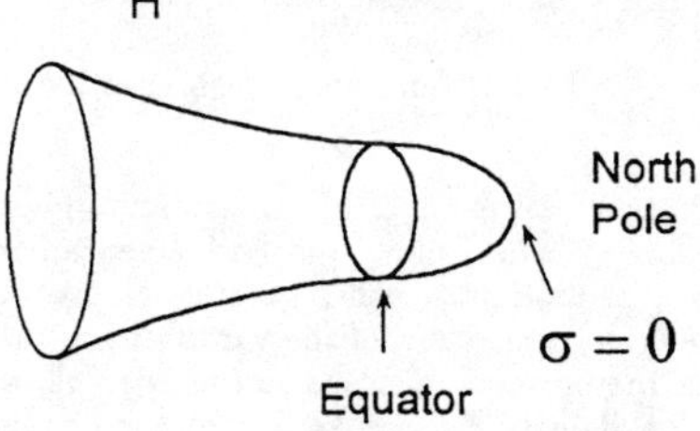

Fig. 3

The most obvious is to follow the Euclidean time variable, σ, from the north pole to the equator, and then go in the imaginary σ direction, and call that real Lorentzian time, t.

Instead of the size of the three spheres going as the $\sin(H\sigma)$, they now go as the $\cosh(Ht)$. This gives a closed universe, that expands exponentially with real time. At late times, the expansion will change from being exponential, to being slowed down by matter in the normal way. This departure of the scale factor from a cosh behavior, will occur because the original Euclidean four sphere, was not perfectly round. But the universe would still be closed, however deformed the four sphere.

For nearly 15 years, I believed that the no boundary proposal, predicted that the universe was spatially closed.

$$\Omega_{matter} + \Omega_\Lambda = 1 + \frac{\kappa}{3\dot{s}^2} \tag{3}$$
Einstein equations.

But the Einstein equations, relate the energy density in the universe, plus lambda, to the rate of expansion, and the curvature, k, of the surfaces of constant time. Define Ω_{matter} and Ω^Λ, to be the density and Λ, divided by the critical value. If the universe is closed, that is, k=+1, Ω_{matter} plus Ω^Λ, must be greater than one. Observations seemed to show that omega matter at least, was significantly less than one. Still, Eddington once said, if your theory doesn't agree with the observations, don't worry. The observations are probably wrong. But if your theory doesn't agree with the second law of thermodynamics, forget it. I firmly believed in the no boundary proposal, and I thought it implied that the universe had to be closed. Since a closed universe, is not incompatible with the second law of thermodynamics, I was sure the observers had missed something, and there really was enough matter to close the universe. At that time, I didn't take seriously the possibility of a small cosmological constant.

You will hear from other people at this conference, about observations of Ω. In my opinion, they do not yet show that the universe is definitely open, or that Λ is non zero, but it is beginning to look like one or the other, if not both. One can summarize the observations in this diagram.

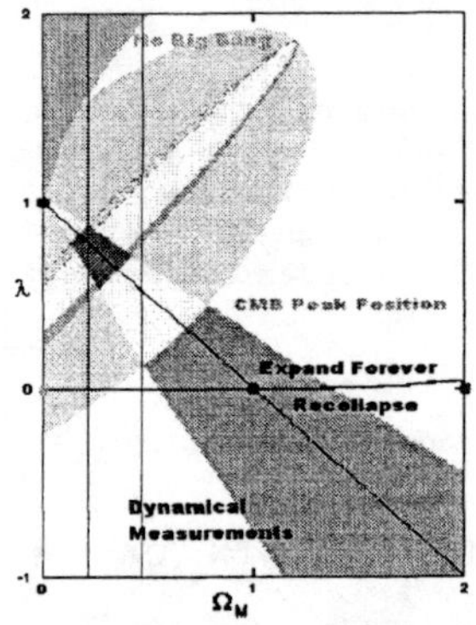

Fig. 4: Summary of observations.

The vertical strip, corresponds to measurements of gravitational clustering, which suggests that Ω_{matter}, is in the range of point 2, to point 4. The yellow, red and green areas, represent the formal errors of the supernova observations, and the large pink area, other possible errors. Also shown in blue, are the limits set by the position of a peak in the angular spectrum, of the variations of the microwave background. As you can see, the observations suggest that the universe is closed to the open close divide, but with a non zero Λ.

Despite these indications of a low density Λ universe, I continued to believe that the cosmological constant was zero, and the no boundary proposal, implied that the universe must be closed. Then in conversations with Neil Turok, I realized there was another way of looking at the no boundary universe, that made it appear open.

One starts with the point that Andrei Linde made, that inflation doesn't need a false vacuum, a local minimum of the potential. But if the scalar field is not at a stationary point of the potential, then it can not be constant on an instanton, a Euclidean solution of the field equations. In turn, this implies that the instanton can't be a perfectly round four sphere. A perfectly round four sphere, would have the symmetry group, O(5). But with a non constant scalar field, the largest symmetry group that an instanton can have, is O(4). In other words, the instanton is a deformed four sphere.

$$ds^2 = d\sigma^2 + b^2(\sigma)(d\chi^2 + \sin^2\chi d\Omega^2) \tag{4}$$
O(4) Instanton.

One can write the metric of an O(4) instanton, in terms of a function, b of σ. Here b is the radius of a three sphere of constant distance, σ, from the north pole of the instanton. If the instanton were a perfectly round four sphere, b would be a sin function of σ. It would have one zero at the north pole, and a second at the south pole, which would also be a regular point of the geometry. However, if the scalar field at the north pole, is not at a stationary point of the potential, it will be almost constant over most of the four sphere, but will diverge near the south pole. This behavior is independent of he precise shape of the potential. The non constant scalar field, will cause the instanton not to be a perfectly round four sphere, and in fact there will be a singularity at the south pole. But it will be a very mild singularity, and the Euclidean action of the instanton will be finite.

This Euclidean instanton, has been described as the universe begining as a pea. In fact, a pea is quite a good image for a deformed sphere. Its size of a few thousand Planck lengths, makes it a very petty pwa. But the mass of the matter it contains, is about half a gram, which is about right for a pea.

I actually discovered this pea instanton in 1983, but I thought it could describe the birth of closed universes only.

$$ds^2 = -dt^2 + \frac{1}{H^2}\cosh^2 Ht(d\chi^2 + \sin^2\chi d\Omega^2) \tag{5}$$

Closed universe from the P-instanton.

To get a closed universe, one starts with $\sigma = 0$ at the north pole, and proceeds to the equator, or rather the value of σ at which the radius, b, of the three sphere is maximum. One then analytically continues sigma in the imaginary direction, as Lorentzian time. As I described earlier, this gives a closed universe with a scale factor that initially goes like $\cosh(t)$. The scalar field, will have a small imaginary part, but that can be corrected by giving the initial value of the scalar field at the north pole, a small imaginary part.

According to the no boundary proposal, the relative probability of such a closed universe, is e to minus twice the action of the part of the pea instanton, between the north pole, and the equator. Notice that as this part, doesn't contain the singularity at the south pole, there is no ambiguity about the action of a singular metric. The action of this part of the instanton, is negative, and is more negative, the bigger the pea. Thus the probability of the pea, is bigger, the bigger the pea. The negative sign of the action, may look counter intuitive, but it leads to physically reasonable consequences.

As I said, I thought the no boundary proposal, implied that the universe had to be spatially closed, and finite in size. But Neil Turok and I, realized his ideas on open inflation, could be fitted in with the no boundary proposal. The universe would still be closed and finite, in one way of looking at it. But in another, it would appear open and infinite.

$$ds^2 = -dt^2 + a^2(d\psi^2 + \sinh^2\psi d\Omega^2) \tag{6}$$

Open universe from the P-instanton.

Let's go back to the metric for the pea instanton, and analytically continue it in a different way. As before, one analytically continues the Euclidean latitude coordinate, in the imaginary direction, to become a Lorentzian time, t. The difference is that one goes in the imaginary sigma direction at the north pole, rather than the equator. One also continues the coordinate, χ, in the imaginary direction, as a coordinate, ψ. This changes the three sphere, into a hyperbolic space. One therefore gets an exponentially expanding open universe.

One can think of this open universe, as a bubble in a closed, de Sitter like universe. In this way, it is similar to the single bubble inflationary universes, that have been proposed by a number of authors.

The difference is, the previous models all required carefully adjusted potentials, with false vacuum local minima. But the pea instanton, will work for any reasonable potential. The price one pays for a general potential, is a singularity at the south pole. In the analytically continued Lorentzian spacetime, this singularity would be time like, and naked.

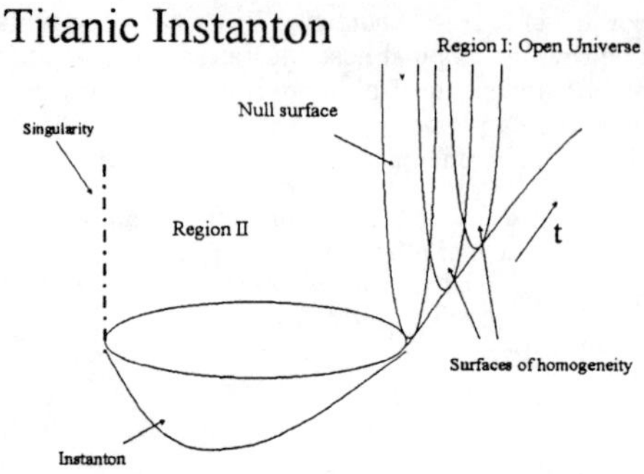

Fig. 5: The Titanic Instanton.

One might think that this naked singularity, would mean one couldn't evaluate the action of the instanton, or of perturbations about it. This would mean that one couldn't predict the quantum fluctuations, or what would happen in the universe. However, the singularity at the south pole, the stalk of the pea, is so mild, that the actions of the instanton, and of perturbations around it, are well defined. At first, it seemed just a lucky accident, that the singularity had these properties, but it now appears that there may be a deep reason, that I will come on to later. This behavior of the singularity, means one can determine the relative probabilities of the instanton, and of perturbations around it. The action of the instanton itself, is negative, but the effect of perturbations around the instanton, is to increase the action, that is, to make the action less negative.

$$Probability = e^{-2S_E} \tag{7}$$

No boundary proposal (Hawking Hartle)

$$Probability = e^{+2S_E} \tag{8}$$

Tunnelling hypothesis (Linde Vilenkin)

According to the no boundary proposal, the probability of a field configuration, is e to minus its action. Thus perturbations around the instanton, have a lower probability, than the unperturbed background. This means that quantum fluctuation are suppressed, the bigger the fluctuation, as one would hope. On the other hand, according to the tunneling hypothesis, favored by Vilenkin and Lindeh , probabilities are proportional to e to the plus action. This would mean that quantum fluctuation would be enhanced, the bigger the fluctuation. There is no way this could lead to a sensible description of the universe. Lindeh therefore proposes to take e to the plus action, for the probability of the background universe, but e to the minus action, for the perturbations. However, there is no invariant way, in which one can divide the action, into a background part, and a part due to fluctuations. So Linde's proposal, does not seem well defined in general. By contrast, the no boundary proposal, is well defined. Its predictions may be surprising, but they are not obviously wrong.

So we see that a general potential, without false vacuums, or local minima, leads to the P-instanton. This can be analytically continued, to either an open, or a closed universe. The no boundary proposal, then allows one to calculate the relative probabilities of different backgrounds, and the quantum fluctuations about them.

There isn't just a single P-instanton, but a whole family of them, labelled by different values of the scalar field at the north pole. The higher the value of the potential at the north pole, the smaller the instanton, and the less negative the value of the action. Thus the no boundary proposal, predicts that large instantons, are more probable than small ones. This is a problem, because large instantons, will lead to a shorter period of exponential expansion or inflation, than small ones. In the closed universe case, a short period of inflation, would mean the universe would recollapse before it reached the present size and density. On the other hand, an open universe with a short period of inflation, would become almost empty early on.

Clearly, the universe we live in, didn't collapse early on, or become almost empty. So we have to take account of the anthropic principle, that if the universe hadn't been suitable for our existence, we wouldn't be asking why it is, the way it is. Many physicists don't like the anthropic principle, but I think some version of it is essential, in quantum cosmology. M theory, or whatever the ultimate theory is, seems to allow a very large number of possible solutions, and compactifications. One has to have some criterion, for discarding most of them. Otherwise, why isn't the universe, eleven dimensional Minkowski space. According to the no boundary proposal, there should be an amplitude for any kind of spacetime, but most of them won't be of much interest, because they won't contain life.

The approach Neil Turok and I took, was to invoke the weakest version of the anthropic principle. We adopted Bayes statistics.

$$P(\Omega_{matter}, \Omega_\Lambda | Galaxy) \text{ is proportional to } P(Galaxy | \Omega_{matter}, \Omega_\Lambda) \times P(\Omega_{matter}, \Omega_\Lambda) \tag{9}$$

Bayesian statistics

In this, one starts with an a-priori probability distribution, and then modifies it in light of ones knowledge of the system. In this case, we took the a-priori distribution, to be the e to the minus action, predicted by the no boundary proposal. We then modified it, by the probability that the model contained galaxies, which are presumably a necessary condition, for the existence of intelligent observers. An open universe, has an infinite spatial volume. Thus the total number of galaxies in an open universe, would always be infinite, no matter how low the probability of finding a galaxy, in a given comoving volume. One therefore can not weight the a-priori probability, given by the no boundary proposal, by the total number of galaxies in the universe. Instead, we weighted by the comoving density of galaxies, predicted from the growth of quantum fluctuations, about the pea instanton. This gives a modified probability distribution for omega, the present density, divided by the critical density. For the open models, this probability distribution, is sharply peaked at an omega of about zero point zero one. This is lower than is compatible with the observations, but it is not such a bad miss. As far as I'm aware, this is the first attempt to predict a value of Ω for an open universe, rather than fine tune a false vacuum potential, to obtain a value in the range indicated by observation.

The reason we obtained such a low value for Ω, was that the a-priori probability distribution, given by the no boundary proposal, depended so steeply on omega. This meant that the modified probability distribution, was sharply peaked at the minimum value of omega, that would allow a single galaxy in our Hubble volume. Clearly this is not the universe we live in, which is just as well, because we cosmologists would all be out of a job, if there were no other no other galaxy visible. Refining the anthropic arguments, will not solve this problem. We just don't need all those other galaxies, for our own existence. The only way I can see of explaining omega without fine tuning, is if there is not a continuous family of instantons, with different values of omega, but only a discrete set. If the spacing in omega is large, there may be only one value of omega in the allowed range. Other values of Ω might have much higher a-priori probabilities, but they would correspond to uninhabited universes.

It seems there may be a way of geting such a discrete set, from eleven dimensional super gravity, which is the best candidate we have, for a theory of everything. It has a three form potential, with a four form field strength. When dimensionally reduced to four dimensions, this can act as a cosmological constant. For a real four form in four dimensions, the contribution to the cosmological constant is negative, which isn't what we want. However, super symmetry is broken, in the universe we live in.

Indeed, super symmetry breaking, is a necessary condition for life. Life could not develop in a super symmetric universe, filled with mass less particles. The breaking of super symmetry, will give rise to a positive contribution to the cosmological constant. This has been seen as a great problem, and there have been elaborate fine tuning schemes proposed, to get rid of it. But maybe the positive contribution from symmetry breaking, is canceled by the negative contribution from the four form. The fine tuning would be provided by the anthropic principle. Galaxies would not form, unless the total cosmological constant, was almost zero.

But symmetry breaking and the four form, need not cancel each other exactly. The anthropic requirement, can probably be satisfied by any omega lambda between about minus one point five, and plus one point five. This is consistent with the observations of omega lambda.

When I gave a talk on this in Santa Barbara, Joe Polchinski pointed out that the four form was not continuously variable, but that the integral of its dual over a seven cycle, was quantized. Unless the compact dimensions were unacceptably large, the spacing of the contribution to the cosmological constant, from different levels of the four form, was larger than the observational limit on lambda. Thus one could not adjust the four form, to cancel symmetry breaking, to the accuracy required. At the time, I saw this as a setback, but now I see it as a positive advantage. Coupled with another idea I shall describe, it could mean that there is not a continuous variation in the size of the instantons, with all the problems that causes with low omega, but only a discrete set.

The other idea I mentioned, is due to Garriga. He took a regular higher dimensional space, and dimensionally reduced it with respect to a U1 isometry group, that had a fixed point. He obtained a four dimensional geometry, with a scalar field that was the log of the length of the U1 fibers. Near the fixed point, the scalar field and four geometry behaved just like those in the pea instanton, near the south pole. So the apparent singularity in the pea instanton, may be resolved by going to higher dimensions. This could be the deep reason, why the singularity is so mild, that the solution, and perturbations around it, have well defined actions, so the quantum fluctuations, are well defined. Having a discrete set of non singular instantons, would be more in keeping with the spirit of the no boundary proposal. As I said earlier, it might also offer a way out of the problem of low omega. However, a lot more work needs to be done, to get a concrete model. The aim is to find a description of the origin of the universe, on the basis of fundamental theory.

Assuming that one can find a model that predicts a reasonable omega, how can we test it by observation. The best way is by observing the spectrum of fluctuations, in the microwave background. This is a very clean measurement of the quantum fluctuations, about the initial instanton. However, there is an important difference between our instanton, and previous proposals for open inflation. They have all assumed false vacuum potentials, and have used the Coleman De Lucia instanton, which is non singular. On the other hand, our instanton has a singularity at the south pole. As I said, quantum fluctuations around the instanton are well defined, despite the singularity. Perturbations of the Euclidean instanton, have finite action if and only, they obey a Dirichelet boundary condition at the singularity. Perturbation modes that don't obey this boundary condition, will have infinite action, and will be suppressed. The Dirichelet boundary condition also arises, if the singularity is resolved in higher dimensions.

When one analytically continues to Lorentzian spacetime, the Dirichelet boundary condition, implies that perturbations reflect at the time like singularity. This has an effect on the two point correlation function of the perturbations, but it seems to be quite small. The present observations of the microwave fluctuations, are certainly not sensitive enough to detect this effect. But it may be possible with the new observations that will be coming in, from the map satellite in two thousand and one, and the Planck satellite in two thousand and six. Thus the no boundary proposal, and the pea instanton, are real science. They can be falsified by observation.

I will finish on that note.

The quantum cosmology debate

Alexander Vilenkin

Institute of Cosmology,
Department of Physics, Tufts University, Medford MA 02155

Abstract.
I review the tunneling, Hartle-Hawking and Linde proposals for the wave function of the universe and comment on the recent work on quantum creation of open universes.

INTRODUCTION

In quantum cosmology the whole universe is treated quantum-mechanically and is described by a wave function rather than by a classical space-time. This quantum approach to cosmology may help us avoid the cosmological singularity problem and understand what determined the initial state of the universe.

The wave function of the universe ψ satisfies the Wheeler-DeWitt equation

$$\mathcal{H}\psi = 0 \tag{1}$$

which is analogous to the Schrodinger equation in ordinary quantum mechanics. To solve this equation, one has to specify some boundary conditions for ψ. In quantum mechanics, the boundary conditions are determined by the physical setup external to the system. But since there is nothing external to the universe, it appears that boundary conditions for the wave function of the universe should be postulated as an independent physical law. The possible form of this law has been debated for some 15 years and is one of the central points of our debate here.

There are at least three proposals on the table: the Hartle-Hawking wave function [1], the Linde wave function [2], and the tunneling wave function [3,4]. In this talk I will first review the tunneling wave function. Then, since this is supposed to be a debate, I will attack my opponents. And finally, I will comment on the recent work on quantum creation of open universes.

THE TUNNELING WAVE FUNCTION

To introduce the tunneling wave function, let us consider a very simple model of a closed Robertson-Walker universe filled with a vacuum of constant energy

CP478 , *COSMO—98*, edited by David O. Caldwell

density ρ_v and some radiation. The total energy density of the universe is given by $\rho = \rho_v + \epsilon/a^4$, where a is the scale factor and ϵ is a constant characterizing the amount of radiation. The evolution equation for a can be written as

$$p^2 + a^2 - a^4/a_0^2 = \epsilon. \tag{2}$$

Here, $p = -a\dot{a}$ is the momentum conjugate to a and $a_0 = (3/4)\rho_v^{-1/2}$.

Eq.(2) is identical to that for a "particle" of energy ϵ moving in a potential $U(a) = a^2 - a^4/a_0^2$. For sufficiently small ϵ, there are two types of classical trajectories. The universe can start at $a = 0$, expand to a maximum radius a_1 and then recollapse. Alternatively, it can contract from infinite size, bounce at a minimum radius a_2 and then re-expand (see Fig. 1). But in quantum cosmology there is yet another possibility. Instead of recollapsing, the universe can tunnel through the potential barrier to the regime of unbounded expansion. The semiclassical tunneling probability can be estimated as

$$\mathcal{P} \sim \exp\left(-2\int_{a_1}^{a_2} |p(a)|da\right). \tag{3}$$

It is interesting that this probability does not vanish in the limit of $\epsilon \to 0$, when there is no radiation and the size of the initial universe shrinks to zero. We then have tunneling from *nothing* to a closed universe of a finite radius a_0; the corresponding probability is

$$\mathcal{P} \sim \exp\left(-2\int_0^{a_0} |p(a)|da\right) = \exp(-3/8\rho_v). \tag{4}$$

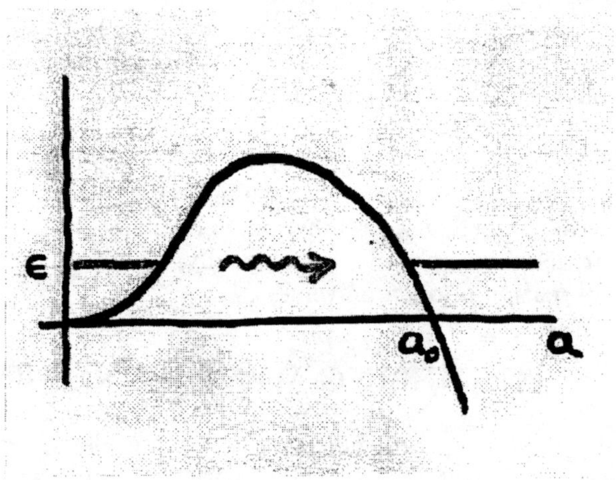

FIGURE 1. The potential for the scale factor in Eq.(2). Instead of recollapsing, the universe can tunnel through the potential barrier to the regime of unbounded expansion.

The tunneling approach to quantum cosmology assumes that our universe originated in a tunneling event of this kind. Once it nucleates, the universe immediately begins a de Sitter inflationary expansion.

The Wheeler-DeWitt equation for our simple model can be obtained by replacing the momentum p in (2) by a differential operator, $p \to -id/da$,

$$\left(\frac{d^2}{da^2} - a^2 + \frac{a^4}{a_0^2}\right)\psi(a) = 0. \tag{5}$$

This equation has outgoing and ingoing wave solutions corresponding to expanding and contracting universes in the classically allowed range $a > a_0$ and exponentially growing and decaying solutions in the classically forbidden range $0 < a < a_0$. The boundary condition that selects the tunneling wave function requires that ψ should include only an outgouing wave at $a \to \infty$ (see Fig.2). The under-barrier wave function is then a linear combination of the growing and decaying solutions. The two solutions have comparable magnitudes near the classical turning point, $a = a_0$, but the decaying solution dominates in the rest of the under-barrier region.

In a more realistic model, the constant vacuum energy density ρ_v is replaced by the potential $V(\phi)$ of some scalar field ϕ. If $V(\phi)$ is a sufficiently slowly-varying function of ϕ, one finds the same result as before, with the replacement $\rho_v \to V(\phi)$,

$$\mathcal{P} \sim \exp\left(-\frac{3}{8V(\phi)}\right). \tag{6}$$

Eq.(6) can be interpreted as the probability distribution for the initial values of ϕ in the ensemble of nucleated universes. The highest probability is obtained for the largest values of $V(\phi)$ (and smallest initial size a_0). Thus, the tunneling wave

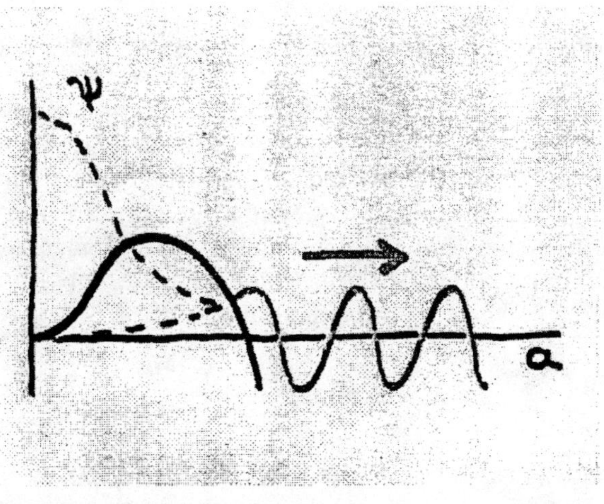

FIGURE 2. The tunneling wave function for the simple model (5).

function 'predicts' that the universe is most likely to nucleate with the largest possible vacuum energy. This is just the right initial condition for inflation.

In the general case, the wave function of the universe is defined on superspace, which is the space of all 3-dimensional geometries and matter field configurations, $\psi[g_{ij}(\mathbf{x}), \phi(\mathbf{x})]$, where g_{ij} is the 3-metric, and matter fields are represented by a single field ϕ. The tunneling boundary condition can be extended to full superspace by requiring that ψ should include only outgoing waves at the boundary of superspace, except the part of the boundary corresponding to vanishing 3-geometries.

Alternatively, the tunneling wave function can be defined as a path integral

$$\psi_T(g, \phi) = \int_{\emptyset}^{(g,\phi)} e^{iS}, \tag{7}$$

where the integration is over paths interpolating between a vanishing 3-geometry $\emptyset$ ('nothing') and (g, ϕ). In other words, the integration is over compact Lorentzian geometries bounded by the 3-geometry g with the field configuration ϕ.

At present these general definitions of the tunneling wave function remain largely formal since we do not know how to solve the Wheeler-DeWitt equation or how to calculate the path integral (7), except for simple models and small perturbations about them.

THE HARTLE-HAWKING WAVE FUNCTION

The Hartle-Hawking wave function is expressed as a path integral over compact Euclidean grometries bounded by a given 3-geometry g,

$$\psi_{HH}(g, \phi) = \int^{(g,\phi)} e^{-S_E}. \tag{8}$$

The Euclidean rotation of the time axis, $t \to i\tau$, is often used in quantum field theory because it improves the convergence of the path integrals. However, in quantum gravity the situation is the opposite. The gravitational part of the Euclidean action S_E is unbounded from below, and the integral (8) is badly divergent. One can hope that the problem will somehow be fixed in the future theory of quantum gravity, but at present we cannot meaningfully define an integral such as (8).

In practice, one assumes that the dominant contribution to the path integral is given by the stationary points of the action (the instantons) and evaluates ψ_{HH} simply as $\psi_{HH} \sim e^{-S_E}$. For our simple model, $S_E \approx -3/8V(\phi)$ and

$$\mathcal{P} \sim \exp\left(+\frac{3}{8V(\phi)}\right). \tag{9}$$

The wave function $\psi_{HH}(a)$ for this model has only the growing solution under the barrier and a superposition of ingoing and outgoing waves with equal amplitudes in the classically allowed region. This wave function appears to describe a contracting and re-expanding universe. The distribution (9) is similar to Eq.(6) for the tunneling wave function, but there is a crucial difference in sign. The distribution (9) is peaked at the smallest values of $V(\phi)$, and thus the Hartle-Hawking wave function tends to predict initial conditions that disfavor inflation.

THE LINDE WAVE FUNCTION

Linde suggested that the wave function of the universe is given by a Euclidean path integral like (8), but with the Euclidean time rotation performed in the opposite sense, $t \rightarrow +i\tau$, yielding

$$\psi_L = \int^{(g,\phi)} e^{+S_E}. \tag{10}$$

For our simple model, this wave function gives the same nucleation probability (6) as the tunneling wave function.

The problem with this proposal is that the Euclidean action is also unbounded from above and, once again, the path integral is divergent. This divergence is even more disastrous than in the Hartle-Hawking case, because now all integrations over matter fields and over inhomogeneous modes of the metric are divergent. It is not clear how Linde's proposal can be extended beyond the simplest model. This problem of Linde's wave function makes it an easy target, and I suspect it is for this reason that Stephen likes to confuse ψ_L and ψ_T and refer to both of them as "the tunneling wave function". In fact, the two wave functions are quite different, even in the simple model (5) [?]. The Linde wave function includes only the decaying solution under the barrier and a superposition of ingoing and outgoing modes with equal amplitudes outside the barrier.

QUANTUM CREATION OF OPEN UNIVERSES

Hawking and Turok (HT) have recently argued [5] that open universes can be spontaneously created from nothing and suggested an instanton to describe this

FIGURE 3. The Hawking-Turok instanton.

process. They considered a model with a very simple potential,

$$V(\phi) = \frac{1}{2}m^2\phi^2. \tag{11}$$

This model is known not to have regular instanton solutions, and indeed, the HT instanton is singular. Geometrically, it is like a sphere with a thorn, the tip of the thorn being the singularity where the curvature and the scalar field are infinite (see Fig. 3). HT point out, however, that the singularity is integrable and the instanton action is finite. Analytic continuation of this instanton gives a closed, singular spacetime. A part of this spacetime, is isometric to an open Robertson-Walker universe. The singularity has the form of an expanding singular bubble. However, it never hits an observer in the Robertson-Walker part of the universe, and HT argue that the singularity is therefore not a problem [7].

I think there are serious problems with HT approach. In a singular instanton, the field equations are not satisfied at the singularity, and such an instanton is not, therefore, a stationary point of the action. It is not clear why such instantons should dominate the path integral. Moreover, if HT instanton is allowed, we will then have to admit a host of other instantons with integrable singularities. I am going to give some examples of such instantons and argue that they lead to unacceptable consequences.

First, I will construct a singular instanton for nucleation of open universes which has a lower action than the Hawking-Turok one. Take two copies of HT instanton, cut off their thorns and match what remains of the instantons across the cut, as in Fig.4. The resulting instanton has an integrable domain-wall-type singularity at the matching surface. Its action is about twice as negative as that of the HT instanton. One can go further and use HT-type instantons with thorns on both sides[1] to construct singular instantons of arbitrarily large negative action (see Fig.5). In the HT approach, such instantons should completely dominate the path integral.

[1] Instantons with two singular thorns have been discussed by R. Bousso and A. Linde [8]

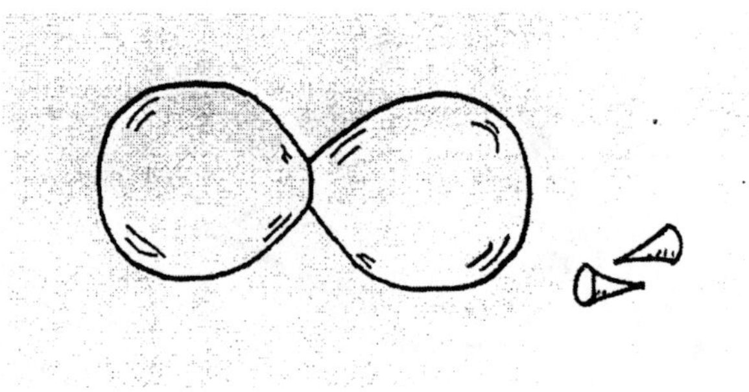

FIGURE 4. This singular instanton is made out of two HT instantons with their thorns cut off. It has a lower action than the HT instanton.

As a second example, for the same model (11) I have constructed an asymptotically-flat singular instanton [6]. Geometrically, it looks like a flat space with a thorn. The behavior of the fields near the singularity is identical to that in the HT instanton and the action is finite, so there is absolutely no reason to reject my instanton if HT instanton is legitimate. The analytic continuation of my instanton gives a flat space with a singular sphere which expands at a speed close to the speed of light. If this were indeed a legitimate instanton, then we would have to conclude that flat space is unstable with respect to nucleation of singular bubbles. The nucleation probability can be made very high by adjusting the strength of the singularity, and since this strength is a free parameter, the universe in this picture would have already been overrun by expanding singular bubbles. Since this is in a glaring contradiction with observations, we have to conclude that HT instanton, as it stands, cannot be used to describe the creation of open universes.

Let me conclude by pointing out one thing that I think is good about the HT instanton. It has invigorated the debate about the basic issues of quantum cosmology and will hopefully lead to a better understanding of some of these issues.

REFERENCES

1. Hartle J.B. and Hawking S.W., Phys. Rev. **D28**, 2960 (1983).
2. Linde A.D., Lett. Nuovo Cimento **39**, 401 (1984).
3. Vilenkin A., Phys. Rev. **D30**, 509 (1984).
4. Vilenkin A., Phys. Rev. **D33**, 3560 (1986).
5. Hawking S.W. and Turok N.G., Phys. Lett. **B425**, 25 (1998).
6. Vilenkin, A., Phys. Rev. **D58**, 067301 (1998).
7. Turok N.G. and Hawking S.W., Phys. Lett. **B 432**, 271 (1998).
8. Vilenkin A., Phys. Rev. **D57**, 7069 (1998).
9. R. Bousso and A. D. Linde, Phys. Rev. **D58**, 083503 (1998).

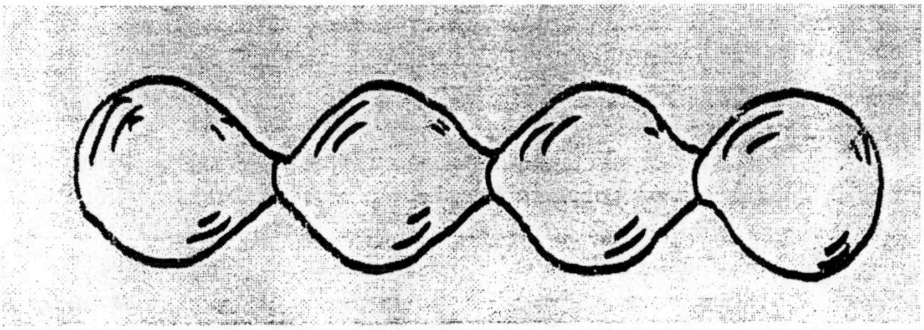

FIGURE 5. In this way one can construct singular instantons of arbitrarily large negative action.

Wave Function and Self-reproduction of the Universe

Andrei Linde

Department of Physics, Stanford University, Stanford, CA 94305, USA

Abstract.
I discuss the relation between various wave functions of the universe and argue that this issue becomes relatively unimportant in the context of the theory of a self-reproducing inflationary universe.

INTRODUCTION

For a long time it seemed almost meaningless to apply the notion of the wave function to the universe itself, since the universe is not a microscopic object. Only with the development of inflationary cosmology did it become clear that the whole universe could appear from a tiny part of space as small as the Planck length M_p^{-1} (at least in the chaotic inflation scenario [1]). Such a tiny region of space can appear as a result of quantum fluctuations of the metric, which should be studied in the context of quantum cosmology. Later it was found that the global structure of the universe in the chaotic inflation scenario is determined not by classical physics, but by quantum processes [2]. This makes quantum cosmology very important for understanding of the origin of the universe and of its global structure.

Unfortunately, quantum cosmology is not a well developed science. This theory is based on the Wheeler-DeWitt equation, which is the Schrödinger equation for the wave function of the universe. This equation has many solutions, and at the present time the best method for specifying preferable solutions of this equation, as well as for interpreting them, is based on the Euclidean approach to quantum gravity. This method is very powerful, but some of its applications are not well justified. In some cases this method may give incorrect answers, but rather paradoxically sometimes these answers appear to be correct when applied to some other questions. Therefore it becomes necessary not only to solve the problem in the Euclidean approach, but also to check, using one's best judgement (and sense of humor), whether the solution is related to the original problem or to something else. An alternative approach is based on the use of stochastic methods in inflationary cosmology [2–10]. These methods allow one to understand such effects as the creation of inflationary density

CP478 , *COSMO—98*, edited by David O. Caldwell

perturbations, the theory of tunneling, and even the theory of the self-reproduction of an inflationary universe. Both the Euclidean approach and stochastic approach to inflation have their limitations, and it is important to understand them.

CREATION OF THE UNIVERSE FROM NOTHING

According to classical cosmology, the universe appeared from a singularity in a state of infinite density. Of course, when the density was greater than the Planck density M_p^4 one could not trust the classical Einstein equations, but in many cases there is no demonstrated need to study the creation of the universe using the methods of quantum theory. For example, in the simplest versions of the chaotic inflation scenario [1], the process of inflation, at the classical level, could begin directly in the initial singularity. However, in certain models, such as the Starobinsky model [11] or the new inflationary universe scenario [12], inflation cannot start in a state of infinite density. In such cases one may speculate about the possibility that the inflationary universe appears due to quantum tunneling "from nothing."

The first idea as to how one can describe the creation of an inflationary universe "from nothing" was given in 1981 by Zeldovich [13] in application to the Starobinsky model [11]. His idea was qualitatively correct, but he did not propose any quantitative description of this process. A very important step in this direction was made in 1982 by Vilenkin [14]. He suggested that the probability of quantum creation of the universe is given by $\exp(-2S)$, where $S = -\frac{3M_p^4}{16V(\phi)}$ is the Euclidean action on de Sitter space with the energy density $V(\phi)$. Vilenkin suggested interpreting this Euclidean solution as the tunneling trajectory describing the creation of a universe with the scale factor $a = H^{-1} = \sqrt{\frac{3M_P^2}{8\pi V}}$ from the state with $a = 0$. This would imply that the probability of the quantum creation of the universe is given by

$$P_{HH} \sim \exp\left(\frac{3M_p^4}{8V(\phi)}\right). \tag{1}$$

A year later this result was reproduced by a different method by Hartle and Hawking [15]. They obtained this result by taking a square of the wave function of the ground state of the universe. $P_{HH} \sim |\Psi_0^{HH}|^2$. We will call this probability distribution P_{HH}.

There were many problems in both derivations of this result. Vilenkin obtained it by considering tunneling of a scale factor with negative kinetic energy through the negative potential barrier from the state containing no initial state. Meanwhile Hartle and Hawking obtained their result using an analogy between the universe and a harmonic oscillator. However, as explained in [2], their method of finding the wave function of the ground state Ψ_0 works only for the states with positive energy. This condition, which was crucial for the derivation of their result, may not seem restrictive. Indeed, it is satisfied for all usual physical systems. However, it is well known that the energy of the scale factor of the universe is negative.

Thus, in my opinion, there is no consistent mathematical derivation of Eq. (1). Moreover, the interpretation of this equation is also questionable. It was derived in [15] as an expression for the ground state of the universe, which should be its *final* state, but for whatever reason it was interpreted as a probability of creation of the universe from nothing, i.e. as a probability for the *initial* state of the universe. This ambiguity will be important for us later, when we will re-derive this equation using the stochastic approach to inflation. In this approach the interpretation of this equation will be quite unambiguous. This equation, in those few cases where it is valid, describes the probability distribution for the *final* state of the universe, rather than the probability of initial conditions [2].

Soon after the publication of the papers by Vilenkin and Hartle and Hawking, the problems mentioned above were revealed, and a different probability distribution was proposed:

$$P_T \sim |\Psi_0^T|^2 \sim \exp\left(-\frac{3M_p^4}{8V(\phi)}\right). \tag{2}$$

I obtained this probability distribution in 1983 by two different methods. The first method was to change the overall sign of the effective Lagrangian of the scale factor, in order to make its kinetic term canonically normalized (note that it does not change the Lagrange equations). Tunneling in this theory looks like a usual quantum mechanical motion through the potential barrier. The second derivation (the one which I decided to publish, because Zeldovich initially strongly objected to the first one) was based on a different way of performing the Wick rotation, which made it possible to find Ψ_0 despite the fact that the energy of the scale factor is negative [16].

Additional insight into this matter was given by Starobinsky, who obtained the same result by an extended version of the first of my methods, and then together with Zeldovich applied this method to the study of the creation of universes with nontrivial topology [17]. Rubakov used a similar method, but argued that particle production during the tunneling will lead to substantial modification of Eq. (2) [18]. A few months later Eq. (2) was also obtained by Vilenkin [19].

The two different derivations give the same expression for the probability of universe formation P_T, up to the subexponential factor. For a discussion of the subtle difference between these expressions one may look at the paper by Vilenkin in this volume. We will ignore this issue in our discussion, and concentrate on the exponents. In fact, the only thing which we want to know is whether it is more probable to create the universe with a large $V(\phi)$ but a small total mass, or with a small $V(\phi)$ and a large total mass. Indeed, the total energy of matter in a closed de Sitter space, at the moment when it is created having size H^{-1}, is proportional to its volume multiplied by its energy density, $E \sim H^{-3}V \sim M_p^3/\sqrt{V}$. Therefore the whole de Sitter space will contain matter with a total energy $O(M_p)$ if it is created at $V \sim M_p^4$. Such an event may easily occur within a time $\Delta t \sim E^{-1} \sim M_p^{-1}$. Meanwhile creation of the universe with $V \ll M_p^4$ would require an enormous fluctuation of energy, which does not seem probable. This is another way to understand the physical meaning of Eq. (2).

Thus the debate about the choice of different expressions is actually the debate between the people who believe that it is easier to create a small universe [16–19] and those who believe that it is easier to create a huge universe [15]. In my opinion, the answer to this question is obvious, but since we are discussing the "miracle of creation," one can always argue that a big miracle is easier than a small miracle:) As a result, this debate has continued for 15 years. Perhaps it is not surprising taking into account the rather esoteric character of the discussion.

Recently there was a renewal of interest in this subject when Hawking and Turok proposed a model of creation of an open universe. They used P_{HH} to calculate the most probable value of Ω. Not surprisingly, they obtained the maximum probability for a state with vanishingly small $V(\phi)$, no inflation, and $\Omega = 0$. By invoking the anthropic principle they were able to raise the predicted value of Ω to $\Omega = 10^{-2}$, which is inconsistent with observational data. One could consider this as an experimental proof that P_{HH} does not describe the quantum creation of the universe, but the situation is not that simple. First of all, in certain models of new inflation one may use P_{HH} and get $\Omega = 1$ [21]. This is not what Hawking and Turok wanted, but it is still better than $\Omega = 10^{-2}$. Another possibility is to use a theory of the scalar field which is strongly nonminimally coupled to gravity, take into account quantum corrections to the Euclidean action and the effects of boundary terms, and then fine-tune the parameters of this model to get $\Omega \sim 0.3$ [22]. While this construction defeats the main purpose of the Hawking-Turok model (no fine tuning in construction of inflationary models with $\Omega < 1$), it does not allow one to conclude that the Hartle-Hawking approach is already experimentally ruled out.

In what follows I will try to look at this subject from an alternative point of view, based on the stochastic approach to inflation. Within this approach Eqs. (1) and (2) can be derived in a much more clear and rigorous way, but they will have a somewhat different interpretation.

THE WAVE FUNCTION OF THE UNIVERSE AND THE STOCHASTIC APPROACH TO INFLATION

Consider the simplest model of chaotic inflation based on the theory of a scalar field ϕ minimally coupled to gravity, with the effective potential $V(\phi)$. If the classical field ϕ is sufficiently homogeneous in some domain of the universe, then its behavior inside this domain is governed by the equation $3H\dot{\phi} = -dV/d\phi$, where $H^2 = \frac{8\pi V(\phi)}{3M_p^2}$.

Inflation stretches all initial inhomogeneities. Therefore, if the evolution of the universe were governed solely by classical equations of motion, we would end up with an extremely smooth universe with no primordial fluctuations to initiate the growth of galaxies. Fortunately, new density perturbations are generated during inflation due to quantum effects. The wavelengths of all vacuum fluctuations of the scalar field ϕ grow exponentially in the expanding universe. When the wavelength of any particular fluctuation becomes greater than H^{-1}, this fluctuation stops oscillating, and its amplitude freezes at some nonzero value $\delta\phi(x)$ because of the large friction term $3H\dot{\phi}$ in the equation of motion of the field ϕ. The amplitude of this fluctuation then remains almost unchanged for a very long time, whereas its wavelength grows exponentially. Therefore, the appearance of such a frozen fluctuation is equivalent to the appearance of a classical field $\delta\phi(x)$ that does not vanish after averaging over macroscopic intervals of space and time.

Because the vacuum contains fluctuations of all wavelengths, inflation leads to the creation of more and more perturbations of the classical field with wavelengths greater than H^{-1}. The average amplitude of such perturbations generated during a time interval H^{-1} (in which the universe expands by a factor of e) is given by $|\delta\phi(x)| \approx \frac{H}{2\pi}$. The phase of each wave is random. Therefore, the sum of all waves at a given point fluctuates and experiences Brownian jumps in all directions in the field space.

One can describe the stochastic behavior of the inflaton field using diffusion equations for the probability distribution $P_c(\phi, t|\phi_0)$. This function describes the probability of finding the field ϕ at a given point at a time t, under the condition that at the time $t = 0$ the field ϕ at this point was equal to ϕ_0 [2–10]. The same function also describes the probability that the scalar field which at time t was equal to ϕ, at some earlier time $t = 0$ was equal to ϕ_0.

The probability distribution P_c is in fact the probability distribution per unit volume in *comoving coordinates* (hence the index c in P_c), which do not change during the expansion of the universe. This function satisfies two different equations, describing the Brownian motion (diffusion) of the field ϕ.

The first equation is called the backward Kolmogorov equation,

$$\frac{\partial P_c(\phi, t|\phi_0)}{\partial t} = \frac{H^{3/2}(\phi_0)}{8\pi^2}\frac{\partial}{\partial \phi_0}\left(H^{3/2}(\phi_0)\frac{\partial P_c(\phi, t|\phi_0)}{\partial \phi_0}\right) - \frac{V'(\phi_0)}{3H(\phi_0)}\frac{\partial P_c(\phi, t|\phi_0)}{\partial \phi_0} . \tag{3}$$

In this equation one considers the value of the field ϕ at the time t as a constant, and finds the time dependence of the probability that this value was reached during the time t as a result of the diffusion of the scalar field from different possible initial values $\phi_0 \equiv \phi(0)$.

The second equation is the adjoint of the first one; it is called the forward Kolmogorov equation, or the Fokker-Planck equation [4],

$$\frac{\partial P_c(\phi, t|\phi_0)}{\partial t} = \frac{\partial}{\partial \phi}\left(\frac{H^{3/2}(\phi)}{8\pi^2}\frac{\partial (H^{3/2}(\phi)P_c(\phi, t|\phi_0))}{\partial \phi} + \frac{V'(\phi)}{3H(\phi)}P_c(\phi, t|\phi_0)\right) . \tag{4}$$

For notational simplicity we took $M_p = 1$ in these equations.

These equations take into account classical motion of the scalar field, as well as its quantum jumps. Thus these equations contain much more information than the classical equations, but they are much easier to interpret than the Wheeler-DeWitt equation for the wave function of the universe.

One may try to find a stationary solution of Eqs. (3), (4), assuming that $\frac{\partial P_c(\phi,t|\phi_0)}{\partial t} = 0$. The idea to look for stationary solutions is very similar to the idea to find the wave function of the ground state of the universe proposed in [15]. The simplest stationary solution would be [4,8,10]

$$P_c(\phi, t|\phi_0) \sim N \ \exp\left(\frac{3M_p^4}{8V(\phi)}\right) \cdot \exp\left(-\frac{3M_p^4}{8V(\phi_0)}\right) \ . \tag{5}$$

This is a very interesting result. One could argue that Eq. (5) provides a direct confirmation and a simple physical interpretation of both the Hartle-Hawking wave function of the universe *and* the tunneling wave function. First of all, we see that the distribution of the probability to find the universe in a state with the field ϕ is proportional to $P_{HH}(\phi)$. Note that we are speaking here about the probability of a given state of the universe at the moment when the observation is performed, rather than the probability of its creation. Meanwhile, the probability that the universe emerged from the state with the field ϕ_0 is proportional to $P_T(\phi_0)$. Now we are speaking about the probability that a given part of the universe was created from the state with the field ϕ_0, and the result coincides with our expression for the probability of the quantum creation of the universe $P_T(\phi_0)$.

This would be a great peaceful resolution of the conflict between the two wave functions: they coexist in the same expression. However, interpretation of $P_{HH}(\phi)$ in this expression differs from the interpretation proposed in [15]. In fact, the situation is even more complicated. In all realistic cosmological theories, in which $V(\phi) = 0$ at its minimum, Eq. (5) is not a true solution of Eq. (4); all physically acceptable solutions for P_c are non-stationary [7]. The source of this difficulty can be easily understood: a distribution can be stationary only due to the compensation of the classical flow of the field ϕ downwards to the minimum of $V(\phi)$ by the diffusion motion upwards. But the diffusion of the field ϕ discussed above exists only during inflation. There is no diffusion upwards from the region near the minimum of the effective potential where inflation ends.

A stationary distribution of probability may still exist, though it will be a probability distribution of a different kind. In many realistic inflationary models quantum jumps of the scalar field $\delta\phi \sim H/2\pi$ may lead to an eternal process of self-reproduction of inflationary domains. This effect was known for many years in the new inflation scenario [23,24,3], but the real significance of this regime was not recognized until it was discovered in the context of chaotic inflation [7]. Indeed it was most unexpected that quantum fluctuations could lead to a regime where the inflaton field eternally climbs the wall. The probability of quantum jumps upwards is exponentially suppressed, but if the fields jumps high enough, the parts of the universe where it happens become rewarded by the exponentially fast growth of their volume.

This effect is very unusual. Its investigation still brings us new unexpected results. For example, for a long time it was believed that self-reproduction in the chaotic inflation scenario can occur only if the scalar field ϕ is extremely large [7]. For example, in the theory $\frac{m^2}{2}\phi^2$ it was shown that self reproduction begins in any domain of initial size H^{-1} if the field ϕ there is greater than $M_p\sqrt{\frac{M_p}{m}}$ (and the energy density is greater than mM_p^3, which is still much smaller than the Planck density M_p^4). However, it was shown in [10] that if the size of the initial inflationary domain is large enough, then the process of self-reproduction of the universe begins there even if the field ϕ is very small. This result is based on the investigation of the probability of quantum jumps with amplitude $\delta\phi \gg H/2\pi$. We have found, to our own surprise, that the probability of the beginning of the regime of self-reproduction starting from a small value of ϕ is suppressed by $P_T(\phi)$. We see here the same expression again, but now with a completely different interpretation. This result implies that self-reproduction begins in any inflationary domain having volume greater than $H^{-3}P_T^{-1}(\phi)$.

This is particularly important for the discussion of the quantum creation of an open inflationary universe. According to [20], such a universe is created in a state with a very small field ϕ (that is, if one uses the probability distribution P_{HH}). This leads to unacceptably small Ω. However, an open universe is infinitely large, so as soon as it appears, there will be infinitely many regions of this universe where the scalar field ϕ will jump upwards, and then the process of self-reproduction will create infinitely many domains of the universe with $\Omega = 1$. So why should one calculate P_{HH} even if it correctly describes the probability of initial conditions? We cannot live in the parts of the universe with extremely small Ω, but we can comfortably live in the parts with $\Omega \sim 1$ which are created in the process of self-reproduction of the universe.

In order to study this process using the stochastic approach to inflation, we have introduced the volume-weighted probability distribution $P_p(\phi, t|\phi_0)$ [7]. This probability distribution takes into account different speeds of the exponential growth of regions filled with different values of the field ϕ. The diffusion equations for $P_p(\phi, t|\phi_0)$ are similar to the equations for $P_c(\phi, t|\phi_0)$, but have some extra terms [10]. The investigation of this question shows [10] that under certain conditions $P_p(\phi, t|\phi_0)$ can be represented as a product of three independent functions,

$$P_p(\phi, t|\phi_0) \sim N(t)\ \pi(\phi) \cdot \psi(\phi_0)\ . \tag{6}$$

This equation is very similar to Eq. (5). The relative fraction of the volume occupied by the field ϕ is described by the function $\pi(\phi)$ which, however, is completely different from $P_{HH}(\phi)$. The relative fraction of the volume of the universe from which any given part of the universe could originate is given by a function $\psi(\phi_0)$, which in the limit $V(\phi_0) \ll M_p^4$ coincides with $P_T(\phi_0)$ [10]. This is additional evidence indicating that if one wants to find out where our part of the universe came from, one should use the probability distribution $P_T(\phi_0)$.

But the most surprising fact is that the probability distribution $\pi(\phi)$, which is most important for the description of the present state of the universe, does not depend on the probability of initial conditions. This suggests that the debate about the probability of initial conditions may not be relevant for the description of the present state of the universe in the context of the theory of eternal inflation.

REFERENCES

1. A.D. Linde, Phys. Lett. **129B**, 177 (1983).
2. A.D. Linde, *Particle Physics and Inflationary Cosmology* (Harwood, Chur, Switzerland, 1990).
3. A. Vilenkin, Phys. Rev. **D27**, 2848 (1983).
4. A.A. Starobinsky, in: *Fundamental Interactions* (MGPI Press, Moscow, 1984), p. 55.
5. A.S. Goncharov and A.D. Linde, Sov. J. Part. Nucl. **17**, 369 (1986).
6. A.A. Starobinsky, in: *Current Topics in Field Theory, Quantum Gravity and Strings*, Lecture Notes in Physics, eds. H.J. de Vega and N. Sanchez (Springer, Heidelberg 1986) **206**, p. 107.
7. A.D. Linde, Phys. Lett. **175B**, 395 (1986); Physica Scripta **T15**, 169 (1987); Physics Today **40**, 61 (1987); A.S. Goncharov, A. Linde and V.F. Mukhanov, Int. J. Mod. Phys. A2, 561 (1987).
8. M. Mijić, Phys. Rev. **D42** (1990) 2469; Int. J. Mod. Phys. **A6**, 2685 (1991).
9. A.D. Linde, Nucl. Phys. **B372**, 421 (1992).
10. A.D. Linde, D.A. Linde and A. Mezhlumian, Phys. Rev. D **49**, 1783 (1994).
11. A.A. Starobinsky, JETP Lett. **30**, 682 (1979); Phys. Lett. **91B**, 99 (1980).
12. A.D. Linde, Phys. Lett. **108B**, 389 (1982); **114B**, 431 (1982); **116B**, 335, 340 (1982); A. Albrecht and P.J. Steinhardt, Phys. Rev. Lett. **48**, 1220 (1982).
13. Ya. B. Zeldovich, Sov. Astron. Lett. **7**, 322 (1981); L.P. Grishchuk and Ya. B. Zeldovich, in: *Quantum Structure of Space-Time*, edited by M. Duff and C. J. Isham, Cambridge University Press, Cambridge (1983), p. 353.
14. A. Vilenkin, Phys. Lett. **B117**, 25 (1982).
15. J.B. Hartle and S.W. Hawking, Phys. Rev. **D28** (1983) 2960.
16. A.D. Linde, Sov. Phys. JETP **60**, 211 (1984); Lett. Nuovo Cimento **39**, 401 (1984).
17. Ya.B. Zeldovich and A.A. Starobinsky, Sov. Astron. Lett. **10**, 135 (1984).
18. V.A. Rubakov, Phys. Lett. **148B**, 280 (1984).
19. A. Vilenkin, Phys. Rev. D **30**, 549 (1984).
20. S.W. Hawking and N.Turok, Phys. Lett. **B425**, 25 (1998); S.W. Hawking and N.Turok, Phys. Lett. **B432** 271 (1998), hep-th/9803156.
21. A.D. Linde, Phys. Rev. D**58**, 083514 (1998), gr-qc/9802038.
22. A.O. Barvinsky, hep-th/9806093, gr-qc/9812058.
23. P.J. Steinhardt, in: *The Very Early Universe*, G.W. Gibbons, S.W. Hawking, S. Siklos, eds., Cambridge U.P. Cambridge, England (1982), p. 251.
24. A.D. Linde, *Nonsingular Regenerating Inflationary Universe*, Cambridge University preprint (1982).

Chaotic New Inflation

Jun'ichi Yokoyama

Yukawa Institute for Theoretical Physics, Kyoto University, Kyoto 606-8502, Japan

Abstract. In a number of scalar potentials with an unstable local maximum at the origin, chaotic inflation is followed by new inflation if model parameters of the potential are appropriately chosen. We calculate the spectrum of adiabatic fluctuations in such a double inflation model and discuss its cosmological implications.

It is now commonly believed that the large-scale homogeneity and isotropy observed in the Universe were realized as a result of a stage of accelerated expansion or inflation in the early universe [1]. Historically, the graceful exit problem of the original inflation model [2] was overcome by the slow roll-over new inflation model [3], and the initial condition problem of the latter was resolved by the chaotic inflation scenario [4], which assumes existence of an inflation-driving (inflaton) field, ϕ, and a potential, $V[\phi]$, with a gentle slope up to well above the Planck scale, M_{Pl}. Then inflationary expansion is realized in domains with $\phi \gtrsim M_{Pl}$ as ϕ gradually rolls down to a minimum. The root-mean-square amplitude of curvature perturbation on comoving scale $r = 2\pi/k$ at the time $\boldsymbol{k}$-mode reenters the Hubble radius in the post-inflationary universe is given by

$$\Phi\left(r = \frac{2\pi}{k}\right) = f\frac{H^2}{2\pi|\dot{\phi}|}\bigg|_{t_k} \equiv f\Delta(t_k). \tag{1}$$

Here $f = 3/5$ $(2/3)$ in the matter- (radiation-) dominated stage [5] and the right-hand-side should be evaluated at $t = t_k$ when $a(t_k)H(t_k) = k$ during inflation.

In this paper we argue that in a number of scalar potentials with an unstable local maximum at the origin chaotic inflation is followed by new inflation if model parameters are appropriately chosen [6,7]. We calculate the spectrum of adiabatic fluctuations in such a double inflation model and discuss its cosmological implications.

First we adopt the Coleman-Weinberg potential [8],

$$V[\phi] = \frac{\lambda}{4}\phi^4\left(\ln\left|\frac{\phi}{v}\right| - \frac{1}{4}\right) + \frac{\lambda}{16}v^4. \tag{2}$$

This potential can drive chaotic inflation for $|\phi| > \phi_e \equiv \max\left(\frac{M_{Pl}}{8\pi}, 4v\right)$, and new inflation around $\phi = 0$. Kung and Brandenberger [9] studied the initial distribution

CP478 , *COSMO—98*, edited by David O. Caldwell

of a scalar field with a double-well potential and concluded that chaotic inflation is much more likely than new inflation. Thus we also start with chaotic inflation with $\phi \gg M_{Pl}$. After chaotic inflation, the scalar field starts oscillation around $\phi = v$ if v is too large, or it overshoots the symmetric state $\phi = 0$ and approaches the other minimum $\phi = -v$ if v is too small. If v is appropriately chosen at an intermediate value, on the other hand, it can spend a long enough time near the origin and then slow roll-over new inflation can set in, which ends up with either $\phi = v$ or $\phi = -v$ depending on the sign of ϕ when classical slow roll-over regime begins.

We numerically solve equations of motion of the homogeneous part of the fields,

$$\ddot{\phi} + 3H\dot{\phi} + V'[\phi] = 0, \qquad H^2 = \left(\frac{\dot{a}}{a}\right)^2 = \frac{8\pi}{3M_{Pl}^2}\left(\frac{\dot{\phi}^2}{2} + V[\phi]\right), \tag{3}$$

with the above initial condition. As a result we find that new inflation can be realized after chaotic inflation if we take $v \simeq 0.22M_{Pl}$. More precisely, the number of e-folds of new inflationary expansion, N_{new}, with $|\dot{H}| < H^2$ is larger than 10 for $v = 0.2201M_{Pl} \sim 0.2259M_{Pl}$, and it satisfies $N_{new} \gtrsim 60$ for $v = 0.2223M_{Pl} \sim 0.2239M_{Pl}$. In the latter case, the comoving scales which leave the Hubble radius during chaotic inflation are stretched beyond the current Hubble horizon and hence they are unobservable today.

After new inflation ϕ settles to the positive potential minimum $\phi = v$ if $v \geq 0.2232M_{Pl}$ and to the negative potential minimum $\phi = -v$ if $v \leq 0.2231M_{Pl}$. If we take even smaller values of v, or $v \leq 0.1066M_{Pl}$, ϕ will return to the positive potential minimum after one oscillation around the negative potential minimum. These results are independent of the initial condition as long as ϕ has large enough initial value so that the system relaxes to the slow roll-over chaotic inflation solution [4]. Below we concentrate on the cases new inflation takes place when ϕ moves to the negative direction and relaxes to $\phi = -v$ without returning to $\phi = v$.

Figure 1 depicts the amplitude of fluctuation $\Delta\left(r = \frac{2\pi}{k}\right)$ as a function of the natural logarithm of the comoving wavenumber, $N(k) \equiv \ln(k/k_*)$, for various values of v with $\lambda = 10^{-14}$. Here k_* is the comoving wavenumber leaving the Hubble radius at the end of chaotic inflation. In this figure fluctuations on the left of the dip are due to chaotic inflation and those on the right are from new inflation. The spectrum is peaked on the comoving horizon scale at the onset of slow-roll new inflation. The value of $N(k)$ at the right end of each curve, N_f, where Δ sharply declines, is practically equal to N_{new} for each v. The amplitude of fluctuation $\Delta(r)$ on the current Hubble radius can be obtained from the value at $N(k) \cong N_f - 60$ and that on the comoving scale corresponding to $1h^{-1}$ Mpc today from $N(k) \cong N_f - 52$ for each v.

Let us summarize our main result shown in Figure 1. First for $v \leq 0.20M_{Pl}$, the number of e-folds during new inflation is less than unity and the density fluctuation is entirely due to chaotic inflation. For $0.2223M_{Pl} < v$ $(< 0.2239M_{Pl})$, on the contrary, all the currently observable scales have left the Hubble radius during new inflation regime. The case with $v = 0.22235M_{Pl}$ is particularly interesting because

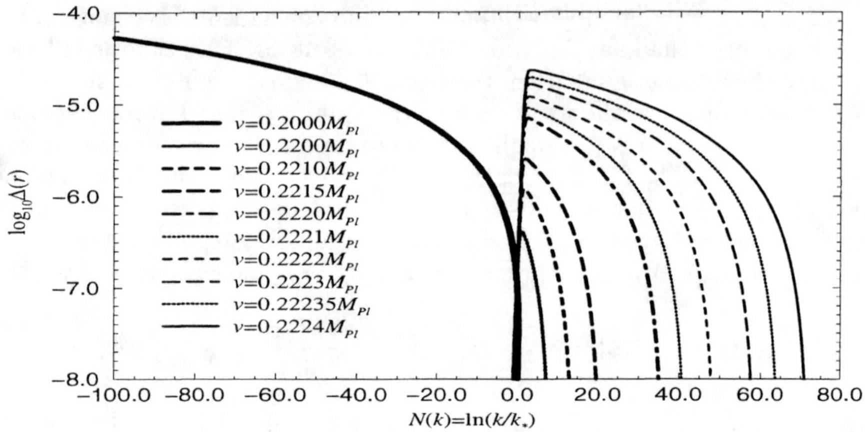

FIGURE 1. Spectrum of perturbation, $\Delta(r)$, generated in chaotic new inflation model for $\lambda = 10^{-14}$. Fluctuations generated during chaotic inflation are practically degenerate for all v's considered here.

there is a sharp decline in the spectrum of fluctuation just above the current Hubble radius. In fact, it has been argued in the literature [10] that such a spectrum fits the observation of the anisotropy of cosmic microwave background (CMB) by the COBE [11] better than the conventional scale-invariant spectrum.

For $0.20M_{Pl} < v \leq 0.2223M_{Pl}$, both chaotic inflation and new inflation contribute to density fluctuations within the current Hubble radius. In most cases, the spectrum of fluctuations on scales probed by the large-scale structures (LSS) and CMB, namely, those corresponding to the comoving scales between $N(k) = N_f - 52$ and $N_f - 60$ in the unit of the horizontal axis in Fig. 1, can be fit by a power-law. But models with $v = 0.2222M_{Pl} \sim 0.2223M_{Pl}$ should be ruled out, because they produce a large dip in the spectrum of fluctuations on scales probed by LSS and CMB. We can realize a tilted spectrum of adiabatic fluctuations with the spectral index $n_{\rm eff} = 0.7 \sim 0.95$ without resorting to an exponential potential.

Next we consider a different class of potential including another free parameter:

$$V[\phi] = -\frac{1}{2}m^2\phi^2 + \frac{\tilde{\lambda}}{4}\phi^4\left(\ln\left|\frac{\phi}{v}\right| - \frac{1}{4}\right) + V_0, \tag{4}$$

that is, typical one-loop effective potential with a nonvanishing mass term at the origin. The potential (4) has four parameters, but one of them, V_0, is fixed from the requirement that the vacuum energy density vanishes at the potential minima

$\phi \equiv \pm\phi_m$. Another parameter, say $\tilde{\lambda}$, can be fixed from the amplitude of large-scale CMB fluctuations using the COBE data [11] as in chaotic inflation. Hence we are essentially left with two free parameters, v and m. v mainly controls the speed of ϕ around the origin and its fate, *i.e.*, to which minimum it falls, and m mainly governs duration of new inflation. By choosing appropriate values of v and m we can realize a desired amplitude of peak fluctuations on a desired mass scale.

Let us now consider a specific example to explain MACHOs [12] in terms of primordial black holes. For this purpose we must realize a peak with $\Delta = 0.0801$ on the comoving scale leaving the Hubble radius 35 e-folds expansion before the end of new inflation, that is, we must have $N_{new} = 35$ [6,13]. After some iterative trials we find that the desired spectrum is obtained if we take $\tilde{\lambda} = 3 \times 10^{-12}$, $m = 6 \times 10^{-8} M_{Pl}$, and $v = 0.21384360 M_{Pl}$.

In summary, we have calculated the spectrum of adiabatic fluctuation generated in the chaotic new inflation model. As a result we have shown that, depending on the choice of the model parameters, this model can realize an almost scale-invariant spectrum with a sharp cut-off on a super horizon scale, which may be favored by COBE observation of CMB anisotropy, and a tilted spectrum with the spectral index $n_{\rm eff} < 0.95$ which may be desirable for successful structure formation. Furthermore it can also be applied to formation of primordial black holes.

REFERENCES

1. For a review of inflation see, *e.g.* A.D. Linde, Particle Physics and Inflationary Cosmology (Harwood, Chur, 1990); K.A. Olive, Phys. Rep. **190**, 181 (1990).
2. A.H. Guth, Phys. Rev. **D23**, 347 (1981); K. Sato, Mon. Not. R. astr. Soc. **195**, 467 (1981).
3. A.D. Linde, Phys. Lett. **108B**, 389 (1982); A. Albrecht and P.J. Steinhardt, Phys. Rev. Lett. **48**, 1220 (1982).
4. A.D. Linde, Phys. Lett. **129B**, 177 (1983).
5. S.W. Hawking, Phys. Lett. **115B**, 295 (1982); A.A. Starobinsky, Phys. Lett. **117B**, 175 (1982); A.H. Guth and S-Y. Pi, Phys. Rev. Lett. **49**, 1110 (1982).
6. J. Yokoyama, Phys. Rev. **D58**, 083510 (1998).
7. J. Yokoyama, preprint YITP-98-79.
8. S. Coleman and E. Weinberg, Phys. Rev. **D7**, 788 (1973).
9. J.H. Kung and R.H. Brandenberger, Phys. Rev. **D42**, 1008 (1990).
10. Y.-P. Jing and L.-Z. Fang, Phys. Rev. Lett. **73**, 1882 (1994); L.-Z. Fang and Y.-P. Jing, Mod. Phys. Lett. A **11**, 1531 (1996); A. Berera, L.-Z. Fang and G. Hinshaw, Phys. Rev. **D57**, 2207 (1998).
11. C.L. Bennet, *et al.* Astrophys. J. Lett. **464**, L1 (1996).
12. C. Alcock et al., Nature, **365**, 623 (1990); Phys. Rev. Lett. **74**, 2867 (1995); Astrophys. J. **486**, 697 (1997); E. Aubourg et al., Nature, **365**, 623 (1993); Astron. Astrophys. **301**, 1 (1995).
13. J. Yokoyama, Astron. Astrophys. **318**, 673 (1997).

A Next-to-Minimal model of Hybrid Inflation in Effective Supergravity Theories

M. Bastero-Gil and S. F. King

Department of Physics and Astronomy, University of Southampton, Southampton, SO17 1BJ, U. K.

Abstract. We propose a model of inflation based on a simple variant of the NMSSM, called ϕNMSSM where the additional singlet ϕ plays the role of the inflaton in hybrid (or inverted hybrid) type models. The ϕNMSSM solves the μ problem of the MSSM via the vacuum expectation value of the gauge singlet N, but also solves the strong CP problem through an approximate Peccei-Quinn symmetry. The potential energy which drives inflation originates from the F-term of the effective supergravity theory which result from a generic string theory. In the class of models considered the inflaton is protected from receiving mass during inflation by a Heisenberg symmetry, avoiding the η problem.

INTRODUCTION

Due to its intrinsic elegance, inflation [1] has become the almost universally accepted dogma for accounting for the flatness and homogeneity of the universe. One of the most popular versions of inflation these days is hybrid inflation, where there are (at least) two fields at work: the slowly rolling inflaton field ϕ, and a second field N whose role is to end inflation by developing a non-zero vacuum expectation value (VEV) when ϕ passes a certain critical value ϕ_c during its slow roll. During inflation $N = 0$, and the potential along the ϕ direction is approximately flat, with the flatness lifted by a ϕ mass which must be small enough to satisfy the slow-roll conditions. Given a particular model, the slow roll conditions and COBE constraints on the spectrum of density perturbations then determine the relationship between the height of the potential $V(0)^{1/4}$ and the inflaton mass during inflation. However, the origin of the vacuum energy $V(0)$ which drives inflation can only be properly understood within a framework which allows the possibility for the potential energy to settle to zero at the global minimum, and hence lead to an acceptable cosmological constant, and this implies supergravity (SUGRA). In SUGRA models a non-zero vacuum energy can be generated through the F-terms and/or the D-terms, but here we will assume the D-terms to be negligible or zero.

CP478 , *COSMO—98*, edited by David O. Caldwell

In F-term Hybrid inflation we have to face the so-called η problem. During inflation SUSY is broken by non-zero F-terms, and due to the exponential factor for the Kahler potential in front of the potential all the scalar fields including the inflaton will pick up masses of the order of the Hubble constant, $H \approx V(0)^{1/2}/\tilde{M}_P$. This will lead to a violation of the slow roll condition $|\eta| = \tilde{M}_P^2|V''/V|^2 \ll 1$. To overcome this problem different solutions can be found in the literature [2]. Here we would like to pursue the possibility that the inflaton mass remains zero at tree-level during inflation and its only contribution is given by very small radiative corrections, safely smaller than the Hubble constant. This can be achieved working in the context of no-scale SUGRA theories, where it is known that the soft scalar masses can be zero even in the presence of a non-zero gravitino mass. Moreover, the requirement of small inflaton mass, combined with the COBE constraint, imply that the height of the potential during inflation must be lower than the usual SUSY breaking scale, $M_{SUSY} \approx 10^{11}\, GeV$, and we will show how the no-scale structure allows this possibility [3]. We shall give an explicit example where the SUSY breaking sector, the height of the potential and the inflaton sector are all specified.

THE MODEL

The model is based on the superpotential [4]:

$$\tilde{W} = \lambda N H_1 H_2 - k\phi N^2 \,, \tag{1}$$

with the fields ϕ, N being gauge singlets, and H_1, H_2 the minimal supersymmetric standard model Higgs doublets. Since the VEV of N generates the effective μ mass term coupling the two Higgs doublets, we require that $\lambda < N > \sim 1$ TeV as in the well known particle physics next-to-minimal supersymmetric standard model (NMSSM) [5]. This superpotential is invariant under a global $U(1)_{PQ}$ symmetry, broken by the VEVs of ϕ and N, and which leads to a very light axion with its decay constant of order of the VEVs $< N > \sim < \phi >$. This implies $< N > \sim 10^{13}\, GeV$, in order to satisfy the cosmological axion bounds. The Higgs doublets develop electroweak VEVs, much smaller than $< N >$, and they may be ignored in the analysis.

The potential relevant for inflation then reads,

$$V(\phi, N) = V(0) + k^2 N^4 + (m_N^2 - 2kA_k\phi + 4k^2\phi^2)N^2 + m_\phi^2\phi^2 \,, \tag{2}$$

where the soft parameters above occur in the soft SUSY breaking potential, and they are typically of the order of $1\, TeV$, but m_ϕ^2 owes its origin to radiative corrections to the potential controlled by the small coupling k. We have added a constant vacuum energy $V(0)$ to the potential, whose origin we explain latter.

For large values of the field ϕ the effective N mass is positive and during inflation the field $N = 0$; ϕ slowly rolls until it reaches a critical value[1] $\phi_c \approx \frac{A_k}{4k}$. When the

[1] In fact the model has two different critical values, $\phi_c^{\pm}$, which allows the possibility of having

critical value is reached, inflation ends and the global minimum is achieved, with non zero VEVS $< \phi > \sim < N > \sim \phi_c$. After inflation ends $V(0)$ is assumed to remain unchanged, but be cancelled by a negative contribution from the remaining part of the potential at the global minimum, $V(< \phi >, < N >) = -k^2 < N >^4$. Due to the axion bound, we then have $k \approx 10^{-10}$, and $V(0)^{1/4} \approx 10^8\, GeV$. To satisfy the COBE constraint the model requires an inflaton mass in the range of a few eV, and this is consistent with this mass being generated by radiative corrections controlled by the small coupling[2] k.

We now wish to elevate this model to an effective no-scale SUGRA theory, where the tree-level inflaton mass is ensured to vanish. This kind of theory may be obtained from 4d effective string theories [7]. We place the inflaton and the N field in the untwisted sector (modular weight -1) along with the moduli fields, and assume the following conditions to hold during inflation:

(a) The superpotential is independent of the over-all modulus T and, together with Eq. (1), it includes a dilaton superpotential $W(S) = \Lambda^3 e^{-S/b_0}$. The dilaton will act as a source for SUSY breaking, with the gravitino mass given by:

$$m_{3/2}^2 \simeq e^K \frac{|W(S)|^2}{\tilde{M}_P^4} \simeq \frac{\Lambda^6}{\tilde{M}_P^4}\,. \tag{3}$$

The requirement of having a gravitino mass $m_{3/2} \approx 1\, TeV$ then fixes the effective scale Λ to be of the order of $10^{13} GeV$.

(b) The Kahler potential is given by,

$$K = -3\ln(\rho) - \ln(S + S^*) + \frac{\beta}{\rho^3} - \frac{2s_0}{S + S^*} + \frac{b + 4s_0^2}{6(S + S^*)^2}\,, \tag{4}$$

It depends only on the combination $\rho = T + T^* - \sum_i \phi_i \phi_i^*$, with ϕ_i any untwisted field of the theory, in particular ϕ and N. This condition can be formalised in terms of a Heisenberg symmetry [8]. The twisted fields are switched off during inflation, and they do not contribute. We remark that we only demand the theory to posses a Heisenberg symmetry *during inflation.* After inflation ends this symmetry may or may not be broken by the contribution from the twisted sector. The last three terms of Eq. (4) model non-perturbative terms for both the field ρ and the dilaton S [9], needed in order to stabilise them at a fixed value during inflation, $\rho_0 \approx (2\beta)^{1/3}$ and $2ReS_0 \approx s_0$.

The above conditions ensure that the inflaton remains massless at tree-level. The key point is that ρ is fixed at its minimum, for which the condition $dV/d\rho = 0$ is fulfilled. Then, computing the mass matrix for the fields (T, ϕ) during inflation

---

either standard hybrid inflation or inverted hybrid inflation [6] depending on the sign of the mass squared m_ϕ^2.

2) The smallness of k seems to indicate that it has a non-renormalisable origin; this may involve an additional sector which obeys a discrete $Z_3 \times Z_5$ symmetry from which the Peccei-Quinn symmetry emerges as an approximation (for details see [4]).

it can be shown that there is a zero eigenvalue which corresponds to the massless inflaton.

The non-zero potential $V(0)$ is given by the SUGRA potential:

$$V = |F_T|^2 + |F_S|^2 - 3m_{3/2}^2 \tilde{M}_P^2 \,, \tag{5}$$

evaluated at the minimum ρ_0 and S_0. If we now require that the potential vanish in the global minimum at the end of inflation, when the fields ϕ and N also contribute, then we obtain that during inflation the height of the potential is $V^{1/4} \sim \epsilon^{1/4}\sqrt{m_{3/2}\tilde{M}_P}$, with $\epsilon^{1/4} \sim 10^{-3}$. In our approach, the fact that the potential is much smaller than the typical SUSY breaking scale is a consequence of having a very small coupling k, otherwise needed to control the radiative corrections to the inflaton mass. Another good feature of such a small coupling is that the contribution of the hybrid superpotential at the end of inflation will be highly suppressed with respect to the dilaton contribution, and then the minima for the moduli and the dilaton (and the gravitino mass) are mainly the same during and after inflation, i. e., no cosmological moduli problem is present in our scenario. As discussed in [4] the small couplings λ, k may have a natural explanation in terms of non-renormalisable effective operators.

REFERENCES

1. See for example Lyth, D. H., and Riotto, A., hep-ph/9807278, and references therein.
2. See for example Covi, L., in this proceedings.
3. Bastero-Gil, M., and King, S. F., hep-ph/9806477.
4. Bastero-Gil, M., and King, S. F., *Phys. Lett.* **B423**, 27 (1998).
5. Fayet, P., *Nucl. Phys.* **B90**, 104 (1975); Nilles, H. -P., Srednicki, M., and Wyler, D., *Phys. Lett.* **B120**, 346(1983); Derendinger, J. -P., and Savoy, C. A., *Nucl. Phys.* **B237**, 307(1984); Ellis, J., *et al.*, *Phys. Rev.* **D39**, 844 (1989); Durand, L., and Lopez, J. L., *Phys. Lett.* **B217**, 463 (1989); Drees, M., *Int. J. Mod. Phys.* **A4**, 3635 (1989); King, S. F., and White, P. L., *Phys. Rev.* **D52**, 4183 (1995).
6. Lyth, D. H., and Stewart, E. D., *Phys. Rev.* **D54**, 7186 (1996); King, S. F., and Sanderson, J., *Phys. Lett.* **B412**, 19 (1997).
7. Ferrara, S., Kounnas, C., and Porrati, M., *Phys. Lett.* **B181**, 263 (1986); Cvetic, M., Louis, J., and Ovrut, B., *ibid* **B206**,227 (1988); Ibáñez, L. E., and Lüst, D., *Nucl. Phys.* **B382**,305 (1992).
8. Binetruy, P., and Gaillard, M. K., *Phys. Lett.* **B195**, 382 (1987); Gaillard, M. K., Murayama, H., and Olive, K., *ibid* **B355**, 71 (1995); Gaillard, M. K., Lyth, D. H., and Murayama, H., *Phys. Rev.* **D58**, 1235 (1998).
9. Banks, T., and Dine, M., *Phys. Rev.* **D50**, 7454 (1994); Casas, J. A., *Phys. Lett.* **B384**, 103 (1996); Binétruy, P., Gaillard, M. K. and Wu, Y.-Y, *Nucl. Phys.* **B481**, 109 (1996); Barreiro, T., de Carlos, B. and Copeland, E. J., *Phys. Rev.* **D57**, 7354 (1998); for the dilaton non-perturbative Kahler potential we follow the particular ansatz given in: Arkani-Hamed, N., Dine, M., and Martin, S., *Phys. Lett.* **B431**, 329 (1998).

Running mass models of inflation

Laura Covi

Department of Physics, Lancaster University
Lancaster LA1 4YB

Abstract. We study models of inflation where the inflaton corresponds to a flat direction and its mass-squared is generated by gravity mediated soft supersymmetry breaking at high scale. If the inflaton has non negligible couplings to other fields, its mass runs with the scale and can reach the small value required by slow roll inflation at lower scale. Slow roll inflation can therefore take place in such a regime, but the spectral index is then scale dependent. We explore the parameter space of this kind of models to find the region compatible with the present observations.

INTRODUCTION

Slow roll inflation[1] requires the flatness conditions $\epsilon \ll 1$ and $|\eta| \ll 1$ on the potential, where

$$\epsilon \equiv \frac{1}{2} M_{Pl}^2 \left(\frac{V'}{V}\right)^2 ; \ \eta \equiv M_{Pl}^2 \frac{V''}{V}, \tag{1}$$

and $M_{Pl} \equiv (8\pi G)^{-1/2} = 2.4 \times 10^{18}$GeV. The first condition is easily satisfied, but the second is problematic. In fact, a generic supergravity theory gives all scalar fields [2,3], and in particular the inflaton [4] a mass-squared with magnitude at least of order V/M_{Pl}^2, which would spoil this condition.

Very few proposals have been put forward to solve this problem which would not be relying on some sort of fine tuning [1]. Our aim is to investigate in detail the proposal of Stewart [5,6] in which loop corrections can flatten the inflaton potential without any significant fine-tuning. We will explore in this case the region of parameter space which is allowed by the observed magnitude and spectral index of the curvature perturbation.

THE RUNNING MASS MODELS

In the model proposed by Stewart [5], slow-roll inflation occurs, with the following Renormalization Group (RG) improved potential for the canonically normalized

---

1) For a general discussion and references on inflation see [1]

CP478 , *COSMO—98*, edited by David O. Caldwell

inflaton field ϕ;

$$V = V_0 + \frac{1}{2}m_\phi^2(\phi)\phi^2 + \frac{1}{2}m_\psi^2(\phi)\psi^2 + \frac{1}{4}\lambda(\phi)\phi^2\psi^2 \cdots . \tag{2}$$

The constant term V_0 is supposed to come from the supersymmetry breaking and to dominate at all relevant field values. Non–renormalizable terms, represented by the dots, give the potential a minimum at large ϕ, but they are supposed to be negligible during inflation. The last two terms also vanish during inflation, since $\psi = 0$, but are responsible for the hybrid exit from the inflationary period.

The inflaton mass-squared and all the other parameters depend on the renormalization scale Q, and following [5,6] we have taken $Q = \phi$, where now ϕ denotes the classical v.e.v of the inflaton field during inflation. Such choice for the renormalization scale minimizes the one loop correction to the potential since the main contribution goes like $\ln(\phi/Q)$ for ϕ larger than any other scale. If this is not the case some other Q will be appropriate and the simplification we have made is no more viable. We will assume that the inflaton v.e.v. is the dominant scale up to the end of inflation.

At the Planck scale, $m_\phi^2(M_{Pl})$ is supposed to be negative, with the generic magnitude

$$|m_0^2| = |m_\phi^2(M_{Pl})| \sim V_0 \tag{3}$$

coming from supergravity corrections [4,7].

If there were no running, this would give $|\eta| \sim 1$, preventing slow–roll inflation. But at field values below the Planck scale, the RG drives $m_\phi^2(\phi)$ to small values, corresponding to $|\eta(\phi)| \ll 1$, and slow–roll inflation can take place there.

Since in this model the η parameter changes considerably as ϕ decreases, slow-roll inflation is assumed to continue until some epoch ϕ_{end}, when $\eta(\phi)$ becomes of order 1. Then ϕ start a brief phase of fast–roll up to the critical value ϕ_c, where the mass term of the other field ψ becomes negative and inflation finally ends when the fields settle down in the true vacuum.

THE SCALE DEPENDENCE OF THE SPECTRAL INDEX

The scale dependence of the spectral index in this kind of models is strictly related to the RG equation of the inflaton mass. The case of a gauge coupling dominated running has been studied in [6–8] while the Yukawa dominated running has been considered in [9]. To illustrate the general behavior, let us consider the linear approximation for the running inflaton mass in the slow roll regime and use

$$m_\phi^2(\phi) \simeq m_\star^2 + c\ln(\phi/\phi_\star) \tag{4}$$

where the $\star$ denotes values of the variables where V' vanishes and the constant c is small and proportional to the relevant coupling. We have then that the relevant observational quantities can all be written as function of the three parameters

$$c = -\frac{dm_\phi^2}{d\ln(\phi)}|_{\phi=\phi_\star} = 2m_\phi^2(\phi_\star) \quad (5)$$

$$\tau = -|c|\ln(\phi_\star) \quad (6)$$

$$\sigma = \lim_{\phi\to\phi_\star} ce^{cN(\phi)}\ln(\phi_\star/\phi), \quad (7)$$

where $N(\phi)$ is the number of e-folds from ϕ to ϕ_{end} in the slow roll approximation (we are assuming the fast roll regime does not change this number substantially):

$$N(\phi) = \int_{\phi_{end}}^{\phi} \frac{V}{V'}d\phi. \quad (8)$$

We obtain then for the spectral index in the linear approximation:

$$n(N) - 1 = 2\sigma e^{-cN} - 2c; \quad (9)$$

while the COBE normalization imposes a constraint on V_0:

$$\frac{V_0^{1/2}}{M_{Pl}^2} = 5.3\times 10^{-4}|\sigma|\exp\left[-\frac{\tau}{c} - cN_{COBE} - \frac{\sigma}{c}e^{-cN_{COBE}}\right]. \quad (10)$$

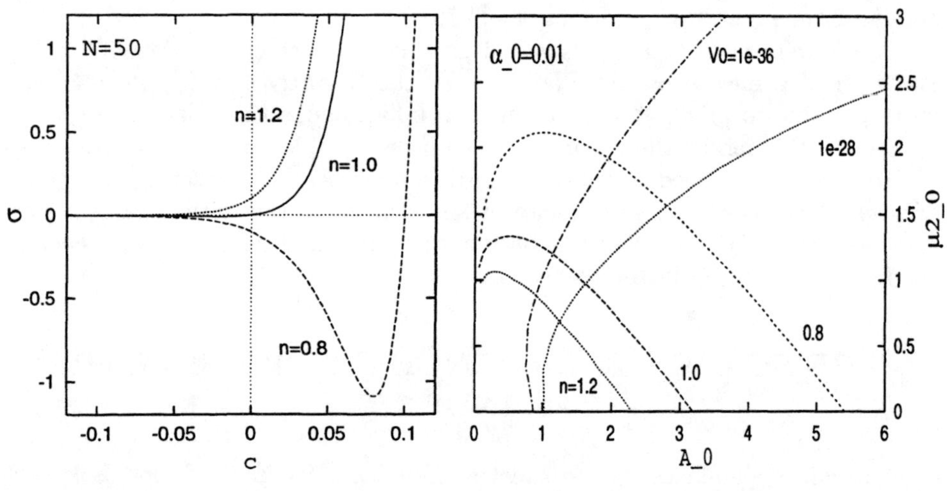

FIGURE 1. (a) Lines of constant spectral index in the $\sigma - c$ plane for N = 50; (b) Similar lines in the case of gauge dominated running of the inflaton mass in the plane $\mu_0^2 = |m_\phi^2(M_{Pl})|M_{Pl}^2/V_0$ vs $A_0 = 2(a/|b|)\,\tilde{m}^2(M_{Pl})M_{Pl}^2/V_0$ ($\tilde{m}$ is the gaugino mass and a, b are group factors of order 1) for a gauge coupling $\tilde{\alpha}_0 = |b|/(2\pi)\,\alpha(M_{Pl}) = 0.01$. Also the lines of constant V_0 are displayed in units of M_{Pl}^4 for $N_{COBE} = 45$. This region corresponds to positive σ and c [8].

So for every particular model, the experimental constraints limit the range of the parameters allowed. We see in Fig. 1a the region in the $\sigma - c$ plane compatible with a spectral index $|n-1| \leq 0.2$ at $N = 50$. In the case where the running is dominated by an $SU(N)$ gauge group we can easily translate this region into physical parameters, as shown in Fig. 1b. Notice that in this particular case, an inflaton mass squared of order V_0/M_{Pl}^2 at the Planck scale is acceptable.

CONCLUSIONS

Quantum corrections can be strong enough to cancel the supergravity contribution of order V_0/M_{Pl}^2 to the inflaton mass and allow slow roll inflation to take place. The requirement to have the spectral index in the experimental range constraints tightly the parameter space of the specific models.

Since the running of the inflaton mass has to be substantial to give the cancellation, in this class of models the spectral index has a significant variation on cosmological scales, usually within the reach of the Planck satellite [10]. Such scale dependence can be parametrized by eq. (9).

ACKNOWLEDGMENTS

I am very grateful and indebted to David H. Lyth and Leszek Roszkowski with whom this work has been done. I would like to thank David Caldwell and the organizers of COSMO 98 for the exciting workshop and for financial support.

This work was supported by PPARC grant GR/L40649.

REFERENCES

1. Lyth D. H., and Riotto A., hep-ph/9807278.
2. Dine M., Fischler W., and Srednicki M., *Nucl. Phys.* **B189**, 575 (1981).
3. Coughlan G. D., Holman R., Ramond P., and Ross G. G., *Phys. Lett.***140B**, 44 (1984).
4. Copeland E. J., Liddle A. R., Lyth D. H., Stewart E. D., and Wands D., *Phys. Rev.* **D49**, 6410 (1994).
5. Stewart E. D., *Phys. Lett.* **B391**, 34 (1997).
6. Stewart E. D., *Phys. Rev.* **D56**, 2019 (1997).
7. Covi L., Lyth D. H., and Roszkowski L., hep-ph/9809310.
8. Covi L., and Lyth D. H., hep-ph/9809562.
9. Covi L., hep-ph/9812232
10. home page at `http://astro.estec.esa.nl/Planck`

A Quantum Field Theory Warm Inflation Model

Arjun Berera

Department of Physics and Astronomy, Vanderbilt University, Nashville, TN 37235, USA

Abstract. A quantum field theory warm inflation model is presented that solves the cosmological horizon problem. An interpretation of the model is given from supersymmetry and superstring theory.

It has been known for a long time to us [1] (for a review of earlier work see [2]) and perhaps a longer time to Nature that inflation is a very attractive solution to the cosmological puzzles. Yet despite its simple picture, a dynamical realization of inflation has proven to be an arduous task. For a long time cosmologists adhered to the notion that a de Sitter expansion regime would necessitate a rapid depletion of radiation energy density ρ_r thus creating a supercooled environment during inflation. Warm inflation cosmology [3] has clarified this misconception by demonstrating the viability of concurrent radiation production during an inflationary regime. Furthermore, the warm inflation picture has a couple of immediate conceptional advantages. Firstly the dynamics is completely free of questions about the quantum-to-classical transition. The scalar inflaton field is in a well defined classical state, thus immediately justifying the application of a classical evolution equation. Also, the fluctuations of the inflaton, which are the metric perturbations [4], are classical. Secondly the dynamics underlying warm inflation is based on the best understood nonequilibrium regime, the state perturbed from thermal equilibrium. In this regime, a self-consistent prescription for dynamics is well defined. These two points imply the dynamics is free of conceptual ambiguity, thus permitting a clear road towards a theory. Notwithstanding, the challenge is to find models that satisfy the requirements of this prescription.

In this talk, a quantum field theory warm inflation model is presented, based on the analysis in [5,6], that solves the horizon/flatness problems. The model obtains, from the elementary dynamics of particle physics, cosmological scale factor trajectories that begin in a radiation dominated regime, enter an inflationary regime and then smoothly exit back into a radiation dominated regime, with nonnegligible radiation throughout the evolution.

CP478 , *COSMO—98*, edited by David O. Caldwell

The basic idea of our implementation of warm inflation is quite simple; a scalar field, which we call the inflaton, interacts with several other fields through shifted couplings $g^2(\varphi - M_i)^2\chi_i^2$ and $g(\varphi - M_i)\bar{\psi}\psi$ to bosons and fermions repectively. The mass sites M_i are distributed over some range. As the inflaton relaxes toward its minimum energy configuration, it will decay into all fields that are light and coupled to it. In turn this generates an effective viscosity. That this indeed happens has been demonstrated in detail in Refs. [5,7]. In order to satisfy one of the requirements of a successful inflation (60 or so e-folds), overdamping must be very efficient. The purpose of distributing the masses M_i is to increase the interval for φ in which light particles emerge through the shifted couplings.

The basic Lagrangian that we will consider is of a scalar field ϕ interacting with $N_M \times N_\chi$ scalar fields χ_{jk} and $N_M \times N_\psi$ fermion fields ψ_{jk},

$$\begin{aligned}\mathcal{L}[\phi, \chi_{jk}, \bar{\psi}_{jk}, \psi_{jk}] &= \frac{1}{2}(\partial_\mu\phi)^2 - \frac{m^2}{2}\phi^2 - \frac{\lambda}{4!}\phi^4 \\ &+ \sum_{j=1}^{N_M}\sum_{k=1}^{N_\chi}\left\{\frac{1}{2}(\partial_\mu\chi_{jk})^2 - \frac{f_{jk}}{4!}\chi_{jk}^4 - \frac{g_{jk}^2}{2}\left(\phi - M_j\right)^2\chi_{jk}^2\right\} \\ &+ \sum_{j=1}^{N_M}\sum_{k=1}^{N_\psi}\left\{i\bar{\psi}_{jk}\,\partial\!\!\!/\psi_{jk} - h_{jk}(\phi - M_j)\bar{\psi}_{jk}\psi_{jk}\right\}, \end{aligned} \tag{1}$$

where all coupling constants are positive: λ, $f_{jk}, g_{jk}^2, h_{jk} > 0$. For simplicity, we consider in the following $f_{jk} = f$, $g_{jk} = h_{jk} = g$. Also, we will set $N_\psi = N_\chi/4$, which along with our choice of coupling implies a cancelation of radiatively generated vacuum energy corrections in the effective potential [8]. We call this kind of model a distributed mass model (DMM), where the interaction between ϕ with the χ_{jk} and ψ_{jk} fields establishes a mass scale distribution for the χ_{jk} and ψ_{jk} fields, which is determined by the mass parameters $\{M_i\}$. Thus the χ_{jk} and ψ_{jk} effective field-dependent masses, $m_{\chi_{jk}}(\phi, T, \{M\})$ and $m_{\psi_{jk}}(\phi, T, \{M\})$, respectively, can be constrained even when $\langle\phi\rangle = \varphi$ is large. The mass sites are chosen to be $M_i = iT_M/g$, where T_M is a constant that is of order the temperature T during warm inflation.

The above Lagrangian has been realized from an effective N=1 global SUSY theory with superpotential

$$W(\Phi, \{X_i\}) = 4m\Phi^2 + \lambda\Phi^3 + \sum_{i=1}^{N_M}\left[4\mu_i X_i^2 + f_i X_i^3 + \lambda_i'\Phi^2 X_i + \lambda_i''\Phi X_i^2\right], \tag{2}$$

which represents an inflaton interacting with the modes of a string. Here Φ is a single chiral superfield which represents the inflaton and X_i $i = 1, \ldots, N_M$ are a set of chiral superfields that interact with the inflaton. All the superfields will have their antichiral superfields $\bar{\Phi}$, $\{\bar{X}_i\}$ appearing in kinetic and Hermetian conjugate (h.c.) terms. In the chiral representation the expansion of the superfields in terms of the Grassmann variable θ is $\Phi = \phi + \theta\psi + \theta^2 F$ and $X_i = \chi_i + \theta\psi_i + \theta^2 F_i$, $i = 1, \ldots, N_M$. Here $\phi = (\phi_1 + i\phi_2)/\sqrt{2}$ and $\chi_i = (\chi_1 + i\chi_2)/\sqrt{2}$ are complex scalar fields as well as F and $\{F_i\}$, and ψ and $\{\psi_i\}$ are Weyl spinors. By definition, the inflaton Φ characterizes the state of the vacuum energy through a nonzero amplitude in the bosonic sector $\langle\phi\rangle \equiv \varphi \neq 0$.

The DM-model is realized for the case $\mu_i = gM_i/2$, $\lambda_i^{'} = 0$, and $\lambda_i^{''} = -2g$, for which the masses of the χ_i, ψ_i fields are respectively $m_{\chi_i}^2 = g^2(\varphi - M_i)^2 - 2gm\varphi - (3g\lambda/4)\varphi^2$ and $m_{\psi_i}^2 = g^2(\varphi - M_i)^2$. At $\varphi = 0$, the masses of the χ_i, ψ_i pair are equal, which is required by supersymmetry. On the other hand, a nonzero inflaton field amplitude, $\varphi \neq 0$, implies a soft breaking of supersymmetry, which in turn permits mass differences. It has been checked that the soft breaking terms do not cause any problems for the results in [6].

The hierarchy of mass levels in the above model Eqs. (1) and (2) has a reminiscient similarity to the mass levels of a string. Clearly the above superpotential, thus the DM-model, captures this basic feature of strings. Also, since the DM-model can be derived from F-term SUSY, i.e. the superpotential, it is a natural model in the technical sense of renormalizability. Further details on matters related to the SUSY origin of the DM-model and its string interpretation can be found in [9].

The next task is to derive the effective equation of motion for φ that descibes dissipative dynamics. The basic idea underlying dissipative dynamics is very simple. The decay products of ϕ and fields to which it couples create, in a sense, a viscous fluid which in turn acts to slow the motion of φ. The 1-loop effective equation of motion for the scalar field ϕ is obtained by setting $\phi = \varphi + \eta$ in Eq. (1) and imposing $\langle\eta\rangle = 0$. Then from Weinberg's tadpole method the 1-loop evolution equation for φ (for homogeneous field) is

$$\ddot{\varphi} + 3H\dot{\varphi} + m^2\varphi + \frac{\lambda}{6}\varphi^3 + \frac{\lambda}{2}\varphi\langle\eta^2\rangle$$
$$+ g^2 \sum_i^{N_M} \sum_j^{N_\chi} (\varphi - M_i)\langle\chi_{ij}^2\rangle + g \sum_i^{N_M} \sum_j^{N_\chi/4} \langle\psi_{ij}\bar{\psi}_{ij}\rangle = 0 \,. \qquad (3)$$

In the above, the term $3H\dot{\varphi}$ describes the energy red-shift of φ due to the expansion of the Universe. This term comes naturally once we start with a background expanding metric for Eq. (1). In the warm-inflation regime of interest here, the thermalization condition must hold, which requires the characteristic time scales (given by the inverse of the decay width) for the fields in Eq. (1) to be faster than the expansion time scale, $H \gg \Gamma$, where Γ are decay widths given in [5,6]. In this case, the calculation of the (renormalized) thermal averages in Eq. (3) can be approximated just as in the Minkowski space-time case. A systematic perturbative evaluation of the averages in the adiabatic, strong dissipative regime was presented in [5] and re-derived in [10] with extension to fermions. Based on the systematic perturbative approach, the effective equation of motion for φ is

$$\ddot{\varphi} + V'_{\text{eff}}(\varphi, T) + \eta(\varphi)\dot{\varphi} = 0 \ , \tag{4}$$

where $V'_{\text{eff}}(\varphi, T) = \partial V_{\text{eff}}(\varphi, T)/\partial\varphi$ is the field derivative of the 1-loop finite temperature effective potential, which can be computed by the standard methods. $\eta(\varphi) \equiv \eta^{\text{B}}(\varphi) + \eta^{\text{F}}(\varphi)$ is a field dependent dissipation; their explicit expressions are given in [5,10]. The above equation of motion is subject to the thermalization condition and the adiabatic condition, that the dynamic time-scale for φ must be much larger than the typical collision time-scale ($\sim \Gamma^{-1}$), $|\varphi/\dot{\varphi}| \gg \Gamma^{-1}$.

The model outlined above has been analyzed in the regime where $V'(\varphi, T) \approx \lambda\varphi^3/6$ dominates. To enforce this condition, further constraints are imposed on the parameter space. There is insufficient space to present the complete solution here, but it can be found in [5,6]. To summaries the results, we find observationally large e-folds $N_e > 60$ in the regime $g < 1$, $N \sim 10$, $\varphi/T \sim 10^3$, and $\lambda \sim 10^{-9}$. In addition, a large number of mass sites M_i are necessary $N_M \sim 10^3$. The total number of particle fields necessary for the dynamics in this regime, $NN_M \sim 10^4$, is not inconsistent with the particle content of excited states in string theory [9].

In summary, the model described above has two appealing features. Firstly, since the dynamics is derived from a first principles treatment of thermal field theory, which drives inflation from the natural dynamics of a scalar field, slow roll is a consequence of the dynamics and not an input. Secondly, the model offers an interesting connection to high energy unification, through its relation to superstrings. Finally an aside, it is interesting to examine whether small warm inflations, $N_e \sim 1$, can be implemented as reheating phases after supercooled inflation.

REFERENCES

1. A. H. Guth, Phys. Rev **D23**, 347 (1981); A. Albrecht and P. J. Steinhardt, Phys. Rev. Lett. **48**, 1220 (1982); A. Linde, Phys. Lett. **108B**, 389 (1982).
2. K. A. Olive, Phys. Rep. **190**, 307 (1990).
3. A. Berera, Phys. Rev. Lett. **75**, 3218 (1995); A. Berera, Phys. Rev. **D54**, 2519 (1996).
4. A. Berera and L. Z. Fang, Phys. Rev. Lett. **74**, 1912 (1995).
5. A. Berera, M. Gleiser and R. O. Ramos, Phys. Rev. **D58**, (1998) 123508.
6. A. Berera, M. Gleiser and R. O. Ramos, (**hep-ph/9809583**).
7. A. Hosoya and M. Sakagami, Phys. Rev. **D29**, 2228 (1984); M. Morikawa, Phys. Rev. **D33**, 3607 (1986); M. Gleiser and R. O. Ramos, Phys. Rev. **D50**, 2441 (1994).
8. L. Dolan and R. Jackiw, Phys. Rev. **D9**, 3320 (1974).
9. A. Berera and T. W. Kephart, VAND-TH-98-19, (**hep-ph-9811295**).
10. J. Yokoyama and A. Linde, (**hep-ph/9809409**).

Remarks on Inflation

Tanmay Vachaspati

Department of Physics, Case Western Reserve University, 10900 Euclid Avenue, Cleveland, OH 44106-7079, USA.

Abstract. It has been shown that sub-Planckian models of inflation require initial homogeneity on super-Hubble scales under certain commonly held assumptions. Here I remark on the possible implications of this result for inflationary cosmology.

The observed homogeneity of the universe on superhorizon scales can be explained if there was a period of accelerated expansion (inflation) in the early universe that took a relatively small homogeneous patch of space and blew it up to encompass a volume larger than what we observe today. In a recent paper [1], Mark Trodden and I addressed the issue of how large the initial homogeneous patch has to be. The result we obtained is that there is a lower bound on the inflationary horizon, H_{inf}^{-1}, which depends on the pre-inflationary cosmology. In particular, if the pre-inflationary cosmology is a Friedman-Robertson-Walker (FRW) universe, we must have $H_{inf}^{-1} > H_{FRW}^{-1}$. If we further assume that, to get inflation with Hubble constant H_{inf}^{-1}, the appropriate inflationary conditions (homogeneity, vacuum domination etc.) must be satisfied over a region of physical size L which is larger than H_{inf}^{-1}, then we obtain $L > H_{FRW}^{-1}$. Hence the conditions for inflation need to be satisfied on cosmological scales specified by the pre-inflationary epoch. In the particular case of a radiation dominated FRW, the problem seems to be even more severe since the causal horizon coincides with H_{FRW}^{-1}. Hence, it is clear that to solve the homogeneity problem, inflationary models in which inflation emerges from a non-inflationary, classical epoch, must assume large-scale homogeneity as an initial condition.

Perhaps more important than the result itself is the fact that we have identified the conditions under which such a derivation is possible. The key assumptions are that Einstein's equations and the weak energy conditions are valid, and that spacetime topology is trivial. (We also assume that singularities apart from the big bang are absent.) If these conditions hold, one would conclude that sub-Planckian inflation alleviates the large-scale homogeneity problem but does not *solve* it[1].

---

[1] The word "solve" may hold different meanings for different individuals. My view is that a solution should not assume the result it purports to obtain.

CP478 , *COSMO—98*, edited by David O. Caldwell

Note that there are two related issues that are brought to the forefront in the above discussion. The first is - do inflationary models (within the stipulated assumptions) solve the homogeneity problem? - and my answer to this question is in the negative. The second is - can inflation occur, given that it seems to require unlikely initial conditions? Here the answer is in the positive - inflation can indeed occur as long as we have a suitable physical theory. However, one may wish to further consider the likelihood of having no inflation or the probabilities with which various kinds of inflation can occur, and this is where the questions become difficult. Hopefully some of the issues involved will become clearer by the end of this article.

A way around the result in [1] is to consider inflationary models in which inflation starts at the Planck epoch (for example, chaotic inflation [2]) since these do not contain a classical pre-inflationary epoch. Hence, at least as far as classical physics goes, inflation is imposed as an initial condition in these models. These initial conditions are sometimes justified by using quantum cosmology – if the wavefunction of the universe is "highly peaked" around the inflationary initial conditions, one might say that these conditions are favored. However, what if the wavefunction only has a small tail around the inflationary initial conditions? I do not think that that would exclude an inflating epoch of our universe since, it could be argued, that most of the other non-inflating universes (where the peak of the wave-function is) would not be able to harbor observers such as us. Hence the argument appears to be inconclusive at present. On the other hand, anthropic arguments are essential to any theory in which the creation of universes is probabilistic. As our understanding of astrophysics and gravity improves, the arguments are likely to get sharpened. As a very simple example of possible forthcoming refinements, if we assume that life can only exist on planets, then an understanding of the cosmological conditions leading to maximal planetary formation will help in narrowing down a measure for calculating probabilities.

Guth [3] has given a persuasive argument for believing that inflation took place in the early universe in spite of any required unlikely initial conditions. He likens inflationary cosmology to the evolution of life. Today we observe a rich variety of life forms, and also a large homogeneous universe. It is hard to imagine how all the miraculous forms of life could have been created directly. Similarly, it is hard to explain the direct creation of our universe. These wonders are easier to comprehend in terms of evolutionary theory - life started out in the shape of some very simple molecules, which then inevitably evolved into the present forms of life. Similarly, a small patch of the universe that satisfied some suitable properties underwent inflation and inevitably evolved into our present universe. So Guth makes the correspondence shown in Fig. 1.

I find this to be a very compelling argument for the existence of an inflationary phase of the universe. The discussion in Ref. [1] impacts on this correspondence in the first stage - what is the chance of getting a small patch with the correct conditions? This is the same question that biologists may ask - what is the chance of getting the first few molecules from which life can follow?

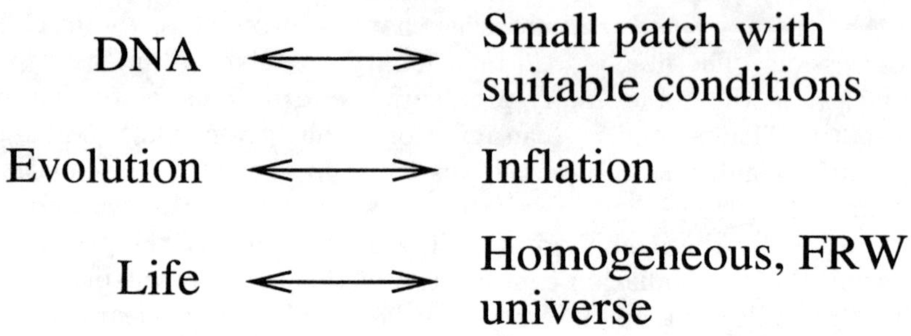

FIGURE 1. The correspondence proposed by Alan Guth.

In the theory of evolution, the formation of the right kind of molecules would depend on the geological and climatic conditions at the time. For example, the occurrence of lightening storms could facilitate chemical reactions that could enhance the probability of forming the molecules. Similarly we have found that the chance of getting a small patch in the universe with all the right conditions for inflation is very low, but that there are conditions under which this probability can be enhanced. So I would like to propose a further extension to the correspondence as shown in Fig. 2.

Our philosophy in adopting inflation as a paradigm is that it greatly enhances the probability for the creation of the universe that we see, even though it is at the expense of invoking a fundamental scalar field - the inflaton. Then, since the introduction of the other factors in the extended correspondence can further enhance the chances of creating an inflating universe, this same philosophy guides us to include them as part of the paradigm.

I should add that the preceding argument is not completely obvious, as Guth and Linde explain, since inflation washes out the initial conditions and a scheme to obtain more likely initial conditions may only provide an infinitesimal increase in the final probability of observing a homogeneous universe. However, if we consider a physical theory in which direct creation of the universe, sub-horizon inflation, as well as super-horizon inflation are all possible, and in which the inflationary phases occur with the same expansion rates, it seems clear that the most spatial volume will be produced by the sub-horizon inflation. This will be true even in the case when the inflation is "eternal".

At this conference Koffman has raised the practical question of how these developments should affect the way inflationary cosmology is studied. I would first of all suggest recognition of the fact that current inflationary models do not "solve" the homogeneity problem but only alleviate it. To solve the problem, one must resort to relaxing at least one of the common assumptions. Linde has been advocating the path of Planckian inflation for a variety of reasons. The path of quantum cosmology has also been taken by a number of researchers but, based on what I discussed above, these approaches appear to leave open a number of difficult questions that

Lightening storms, volcanic eruptions, aliens, etc.	⟷	Non-minimal scalars, quantum effects, spatial topology, etc.

FIGURE 2. A proposed extension to Guth's correspondence.

need to be answered before we can claim to have understood the homogeneity of the universe. To me two other approaches seem more promising. The first is to study the conditions under which quantum effects can give rise to violations of the weak energy condition (such as pursued by Ford and collaborators [4]), and subsequently to creation of baby universes (for an early related paper see [5]). The second approach is to explore modifications of the classical Einstein equations. This also ties in with the fact that essentially any high energy theory that is being considered these days does include such modifications.

If quantum effects or extensions to Einstein's equations lead to the conclusion that inflation can begin from a microphysical patch, profound consequences will follow since it would open the possibility of creating universes in the laboratory. One could further imagine that stellar collapse, for example, may lead to baby universe creation, and that there may be other universes lurking in the centers of galaxies.

It is a pleasure to thank Alan Guth, Andrei Linde and Alex Vilenkin for describing their views on inflation and also for their patience in this matter. I am grateful to Lawrence Krauss and Mark Trodden for discussions. This work was supported by the Department of Energy.

REFERENCES

1. T. Vachaspati and M. Trodden, gr-qc/9811037.
2. A. Linde, *Particle Physics and Cosmology* (Harwood, Chur, Switzerland, 1990).
3. A. Guth, private communication (1998).
4. For a recent paper see, L. H. Ford and T. A. Roman, gr-qc/9901074 (1999).
5. E. Farhi, A. Guth and J. Guven, *Nucl. Phys.* **B339**, 417 (1990).

Inflation after preheating

Gary Felder*

*Department of Physics, Stanford University, Stanford, CA 94305-4060, USA

Abstract. Preheating after inflation may lead to nonthermal phase transitions with symmetry restoration. These phase transitions may occur even if the total energy density of fluctuations produced during reheating is relatively small as compared with the vacuum energy in the state with restored symmetry. As a result, in some inflationary models one may encounter a secondary stage of inflation after preheating.

INTRODUCTION

The theory of cosmological phase transitions is usually associated with the processes of symmetry restoration due to high temperature effects and the subsequent symmetry breaking which occurs as the temperature decreases in an expanding universe [1]. A particularly important version of this theory is the theory of first order cosmological phase transitions developed in [2]. It served as a basis for the first versions of inflationary cosmology [3], as well as for the theory of electroweak baryogenesis [4].

Recently this theory was supplemented by the possibility of *nonthermal* cosmological phase transitions [5,6], i.e. phase transitions driven by fluctuations produced so rapidly that they did not have time to thermalize. In some cases these phase transitions are first order [7]; they occur by formation of bubbles of the phase with spontaneously broken symmetry inside the metastable symmetric phase. But if the lifetime of the metastable state is large enough, one may encounter a short secondary stage of inflation *after* preheating [5].

In this paper I will briefly present the theory of such phase transitions and then give the results of numerical lattice simulations which directly demonstrated the possibility of such a secondary stage of inflation. A more detailed development of the theory can be found in [5]. The work presented here was done in collaboration with Lev Kofman of the Canadian Institute for Theoretical Astrophysics, Andrei Linde of Stanford University, and Igor Tkachev of Purdue University.

THEORY

We have been studying a model with the effective potential

CP478 , *COSMO—98*, edited by David O. Caldwell

$$V(\phi, X) = \frac{\lambda}{4}(\phi^2 - v^2)^2 + \frac{g^2}{2}\phi^2 X^2 . \tag{1}$$

The inflaton scalar field ϕ has a double-well potential and interacts with an N-component scalar field X; $X^2 = \sum_{i=1}^{N} X_i^2$. For simplicity, the field X is taken to be massless and without self-interaction. The fields couple minimally to gravity in a FRW universe with a scale factor $a(t)$. We take $\lambda = 10^{-13}$ [8], and assume that $g^2 \gg \lambda$. We consider $v \sim 10^{-3} M_p \sim 10^{16}$ GeV, which corresponds to the GUT scale.

The potential $V(\phi, X)$ has a minimum at $\phi = v$, $X = 0$ and a maximum at $\phi = X = 0$ with the curvature $V_{\phi\phi} = -m^2 = -\lambda v^2$. This effective potential acquires corrections due to quantum (or thermal) fluctuations of the scalar fields [1,8], $\Delta V = \frac{3}{2}\lambda\langle\phi^2\rangle\phi^2 + \frac{g^2}{2}\langle X^2\rangle\phi^2 + \frac{g^2}{2}\langle\phi^2\rangle\chi^2 + ...$, where I have written only the leading terms depending on ϕ and X. The effective mass squared of the field ϕ is given by

$$m_\phi^2 = -m^2 + 3\lambda\phi^2 + 3\lambda\langle\phi^2\rangle + g^2\langle X^2\rangle \tag{2}$$

Symmetry is restored (i.e. $\phi = 0$ becomes the stable equilibrium point) when the fluctuations $\langle\phi^2\rangle$, $\langle X^2\rangle$ become sufficiently large to make the effective mass squared positive at $\phi = 0$.

The theory of nonthermal phase transitions in this model is developed in [5]. The main idea behind this theory is that symmetry restoration may occur even if the energy density would be insufficient to generate large enough fluctuations of the fields in a thermal distribution. The reason is that preheating preferentially excites low momentum modes of the fields X_i with anomalously high occupation numbers. These excited modes can thus lead to very large fluctuations $\langle X^2\rangle$ relative to what would result if the same energy were distributed among all modes thermally.

It was shown in [5] that the condition for thermal inflation in this model is $g^4 \gg \lambda$ and the degree of this inflation is roughly $10^{-1}\left(\frac{g^4}{\lambda}\right)^{1/4}$. By contrast, the condition for nonthermal inflation is $g^2 \gg \lambda$ and the degree of inflation obtained is $(\frac{g^2}{\lambda})^{1/4}$. These formulas give a good idea of the general behavior of these fields, but in practice there are several important effects which are neglected in these analytic estimates.

First of all, the transfer of energy from the field ϕ to the fields X_i takes a certain amount of time during which the universe expands and the energy density is diluted. The larger the number N of X_i fields the faster this energy will be transferred, so we expect the effects of inflation and phase transitions to be stronger for large N.

These estimates were also made neglecting the effects of backreaction of the particles X_i on the inflaton field and on the expansion of the universe. This backreaction will tend to shut down the resonant production of particles and thus shorten or prevent entirely the occurrence of a secondary stage of inflation. The only way to fully account for all the effects of backreaction and expansion is through numerical lattice simulations.

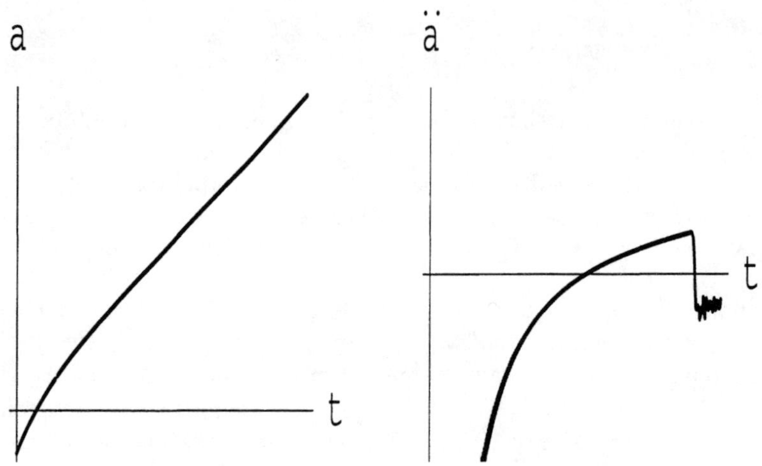

FIGURE 1. Scale factor and its second derivative

SIMULATION RESULTS

A numerical investigation of this model was performed in [7] for $g^2/\lambda \gg 1$. The simulations were done using a staggered leap-frog calculation on a three dimensional lattice with first-order difference methods used for spatial and time derivatives. The details of the lattice simulations used there (and here) can be found in that paper.

The authors of [7] found a strongly first order phase transition. The strength of the phase transition increased with an increase of g^2/λ and of the number N of the fields X_i. However, for the parameters of the model studied in [7] ($g^2/\lambda = 200$) there was no inflation during symmetry restoration. This is not unexpected because the estimates discussed above indicated that the expansion of the universe during the short stage of nonthermal inflation cannot be greater than $(\frac{g^2}{\lambda})^{1/4}$.

We have extended these simulations for $g^2/\lambda = 800$ and $N = 9$. With these parameters the strength of the phase transition became much greater, and there was a short stage of inflation prior to the phase transition. The simulation showed that the oscillations of the inflaton field decreased until the field was trapped near zero, where it remained until the moment of the phase transition when it rapidly jumped to its symmetry breaking value. This transition occurred in a nearly spherical region of the lattice which quickly grew to encompass the entire space. (See the figures in [7].)

Figure 1 shows the scale factor a and its second derivative $\ddot{a}$ as a function of time. While the inflaton is trapped in the false vacuum state $\ddot{a}$ becomes positive, indicating a brief stage of inflation. We can thus conclude that it is possible to achieve inflation for parameters for which this would not have been possible in thermal equilibrium ($g^4 \ll \lambda$). This inflation should be amplified by an increase in

N or $\frac{g^2}{\lambda}$. We are currently doing a simulation with $N = 19$ in which we expect to see a longer stage of inflation.

CONCLUSIONS

We have demonstrated through numerical calculations the occurrence of a stage of inflation driven by symmetry restoration induced by nonthermal fluctuations. This opens up the possibility of secondary stages of inflation occurring in models where it would previously have been considered impossible.

ACKNOWLEDGEMENTS

This work was supported in part by DOE grant DE-FG02-91ER40681 (Task B) (S.K. and I.T.), NSF grants PHY-9219345 (A.L.), PHY-9501458 (S.K. and I.T.), AST95-29-225 (L.K. and A.L.), and by the Alfred P. Sloan Foundation (S.K.).

REFERENCES

1. D.A. Kirzhnits, JETP Lett. **15**, 529 (1972); D.A. Kirzhnits and A.D. Linde, Phys. Lett. **42B**, 471 (1972); Sov. Phys. JETP **40**, 628 (1974); Ann. Phys. **101**, 195 (1976); S. Weinberg, Phys. Rev. **D9**, 3320 (1974); L. Dolan and R. Jackiw, Phys. Rev. **D9**, 3357 (1974).
2. D.A. Kirzhnits and A.D. Linde, Ann. Phys. (N.Y.) **101**, 195 (1976).
3. A.H. Guth, Phys. Rev. **D23**, 347 (1981); A.D. Linde, Phys. Lett. **108B**, 389 (1982); A. Albrecht and P.J. Steinhardt, Phys. Rev. Lett. **48**, 1220 (1982).
4. V.A. Kuzmin, V.A. Rubakov and M.E.Shaposhnikov, Phys. Lett. 155B (1985) 36; M.E. Shaposhnikov, JETP Lett. **44** (1986) 465; Nucl. Phys. **B287** (1987) 757; Nucl. Phys. **B299** (1988) 797.
5. L. Kofman, A. Linde, and A. A. Starobinsky, Phys. Rev. Lett. **76**, 1011 (1996).
6. I. I. Tkachev, Phys. Lett. **B376**, 35 (1996).
7. S. Khlebnikov, L. Kofman, A. Linde, and I. Tkachev, First-order nonthermal phase transition after preheating, Phys. Rev. Lett. **81**, 2012 (1998), hep-ph/9804425.
8. A. D. Linde, *Particle Physics and Inflationary Cosmology* (Harwood, Chur, Switzerland, 1990); E. W. Kolb and M. S. Turner, *The Early Universe* (Addison-Wesley, Redwood City, California, 1990).

Inhomogeneity and the Post-Inflationary Universe

Richard Easther and Matthew Parry

Department of Physics, Brown University, Providence, RI02912

Abstract. We discuss the interaction between perturbations in the inflaton and the background during the preheating phase in simple inflationary models. By numerically solving the Einstein field equations we are able to assess the impact of non-linear gravitational effects on preheating, and to assess the accuracy of perturbative discussions of the preheating epoch.

INTRODUCTION

Inflation drives the primordial universe towards the very special initial state needed for it evolve into the form it is observed to have today. However, to be successful, inflation must terminate gracefully with the energy density of the inflaton field being transferred to radiation, a process known as *reheating*.

Since 1990 [1] it has been realized that reheating can be driven by nonlinear, coherent effects leading to explosive particle production via parametric resonance. The resulting distribution is far from thermal equilibrium, and the process is often called *preheating* (see [2,3], and rcfs. within).

The standard analytic approach to preheating is based on a Fourier expansion of the relevant fields. The equation of motion of the k-th mode has the form of a Mathieu or Lamé equation. Modes corresponding to values of k in resonance bands grow exponentially, and the precise resonance structure depends sensitively on the underlying particle physics. Because it selectively amplifies specific Fourier modes, parametric resonance implies an enhancement in the spatial variation of the fields. While the non-linear backreaction on the field evolution has been examined analytically [2,4] and numerically [5,6] the backreaction of the inhomogeneity on the underlying spacetime metric and its implications for the evolution of the universe has only recently begun to be studied [3,7–13].

We have broken new ground by tackling the back-reaction problem using the full Einstein field equations, and eschewing the use of any perturbative approximations. We show that large metric inhomogeneities can be induced by parametric resonance, so approximations to the full field equations may not adequately describe the evolution of the universe during (and after) preheating. At present, our

CP478 , *COSMO—98*, edited by David O. Caldwell

principal simplifying assumption is that the inhomogeneity lies in a single spatial direction, which allows us to work with a $1+1$ dimensional system of partial differential equations.

Here we briefly survey the growth of metric perturbations during preheating in an inflationary model driven by a $\lambda\phi^4$ potential. When treated analytically, the $m^2\phi^2$ model does not undergo parametric resonance during reheating [11], a result we confirmed numerically [3]. However, $\lambda\phi^4$ possesses a single resonance band, and when we examine the evolution of modes within this band we see observe strong amplification of the corresponding metric perturbations. This work will be more fully described in our forthcoming paper [13].

METRIC AND INITIAL CONDITIONS

We have assumed that the universe has a planar symmetry, so the metric functions depend only on t and z, and are independent of x and y. In [3], we studied reheating after $m^2\phi^2$ inflation using the metric

$$ds^2 = dt^2 - A^2(t,z)\,dz^2 - B^2(t,z)\,(dx^2+dy^2), \tag{1}$$

which describes an inhomogeneous universe in which the dx^2+dy^2 sections have zero spatial curvature. In principle, we could retain this metric for the $\lambda\phi^4$ case, but it is advantageous to work in conformal-like co-ordinates analogous to those which simplify the homogeneous system [4,13]. Specifically, by transforming t and z to η and ζ we write equation (1) as:

$$ds^2 = \alpha^2(\eta,\zeta)\,(d\eta^2 - d\zeta^2) - \beta^2(\eta,\zeta)\,(dx^2+dy^2). \tag{2}$$

In practice we use the co-ordinate transformation to compute A and B from equation (1), as well as α and β, as we evolve the system numerically. The equations of motion will be given in [13], and the numerical techniques used to solve them are similar to those we used in [3]. We focus on initial configurations where a single mode (when viewed in the perturbative context) is excited. We choose $\alpha(0,\zeta)=\beta(0,\zeta)=1$, $\phi(0,\zeta)=\phi_0$ and $\alpha_{,\zeta}(0,\zeta)=\beta_{,\zeta}(0,\zeta)=\phi_{,\zeta}(0,\zeta)=0$. The constraints are solved if

$$\alpha_{,\eta}(0,\zeta) = \frac{\kappa^2}{2C}\left(\frac{\phi_{,\eta}^2}{2}+V(\phi_0)\right)-\frac{C}{2}, \quad \beta_{,\eta}(0,\zeta)=\sqrt{\frac{\kappa^2}{3}\left(\frac{\langle\phi_{,\eta}^2\rangle}{2}+V(\phi_0)\right)}=C. \tag{3}$$

Our choice of C ensures $\langle\alpha_{,\eta}(0,\zeta)\rangle = \langle\beta_{,\eta}(0,\zeta)\rangle$, where $\langle\cdots\rangle$ is a spatial average. The actual inhomogeneity is injected through the inflaton kinetic energy,

$$\phi_{,\eta}(0,\zeta) = \Pi + \epsilon\sin\left(\frac{2\pi k\zeta}{Z}\right), \tag{4}$$

where Z is the length of our "box", and Π is the average initial velocity. Since both t and z are transformed, slices of constant η in the conformal frame are not mapped directly to slices of constant t in the physical frame. The initial density perturbation is on the order of ϵ^2.

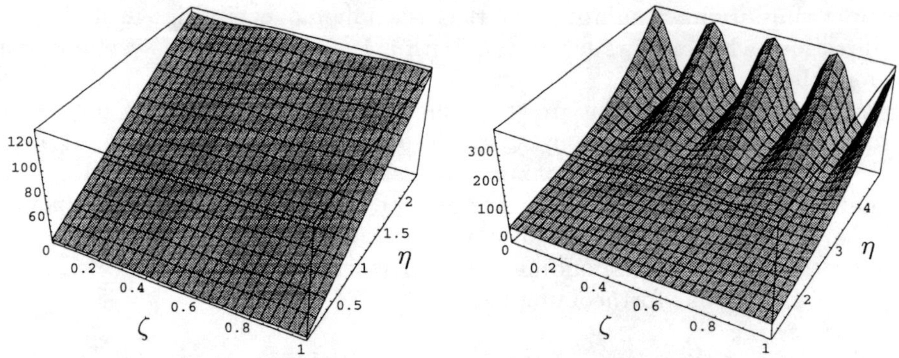

FIGURE 1. The evolution of A (the g_{zz} component of equation (1)) is plotted as a function of η and ζ. The left panel shows the evolution of a mode with k slightly too large to be in resonance, while the right hand case undergoes resonance, and significant inhomogeneity is generated. The units are arbitrary. The initial perturbation is $\epsilon^2 = 10^{-6}$, and the simulations begins at the end of inflation.

RESULTS

When we examined reheating after $m^2\phi^2$ we confirmed that there is no significant resonant amplification, and that the perturbative analysis is valid [3]. However, after $\lambda\phi^4$ inflation there is a single, narrow resonance band, and the scaling properties of the solution ensure that modes which are initially in this band remain there indefinitely.

Fig. 1 shows the evolution of the metric component A for two different perturbations, one inside the resonance band and one outside it. A significant degree of inhomogeneity is generated in the metric by the resonance in the fields. While a simple analytical treatment suggests that resonance lasts indefinitely, the backreac-

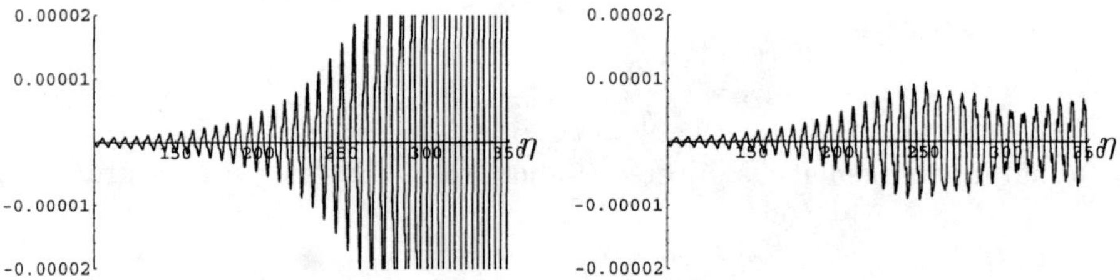

FIGURE 2. The Fourier mode of Φ, the metric perturbation, for a resonant mode is shown: the left panel gives the perturbative result, while the right panel shows the evolution of the mode derived from the full nonlinear analysis.

tion from the nonlinear field evolution eventually halts the resonant amplification. This feature is apparent in our simulation, and is illustrated in Fig. 2, where we plot the metric perturbation, Φ [14]. The termination of resonance coincides roughly with the comparatively sudden growth of the metric perturbation seen qualitatively in the right panel of Fig. 1.

If we solve for an inhomogeneous field, ϕ, while averaging over the spacetime background (corresponding to numerical treatments which include the nonlinear field equations, but assume that the metric is unperturbed) the subsequent evolution of Φ differs from that which we obtain in the presence of the gravitational backreaction. Comparing the two results will allow us isolate the contribution from the inhomogeneous parts of the metric to the overall evolution. We are also extending this work to more general initial perturbations than equation (4), so we can include interactions between different modes and study non-perturbative effects such as the formation of primordial black holes after inflation [13].

ACKNOWLEDGMENTS

We thank Robert Brandenberger and Fabio Finelli for useful discussions. Computational work in support of this research was performed at the Theoretical Physics Computing Facility at Brown University. RE is supported by DOE contract DE-FG0291ER40688, Task A.

REFERENCES

1. J. H. Traschen and R. H. Brandenberger, Phys. Rev. D **42**, 2491 (1990).
2. L. Kofman, A. Linde, and A. Starobinsky, Phys. Rev. D **56**, 3258 (1997).
3. M. Parry and R. Easther, Phys. Rev. D (to appear) (1999).
4. P. B. Greene, L. Kofman, A. Linde, and A. A. Starobinsky, Phys. Rev. D **56**, 6175 (1997).
5. S. Y. Khlebnikov and I. I. Tkachev, Phys. Rev. Lett. **77**, 219 (1996).
6. T. Prokopec and T. G. Roos, Phys. Rev. D **55**, 3768 (1997).
7. H. Kodama and T. Hamazaki, Prog. Theor. Phys. **96**, 949 (1996).
8. T. Hamazaki and H. Kodama, Prog. Theor. Phys. **96**, 1123 (1996).
9. Y. Nambu and A. Taruya, Prog. Theor. Phys **97**, 83 (1996).
10. B. Bassett, D. Kaiser, and R. Maartens, hep-ph/9808404 (1998).
11. F. Finelli and R. Brandenberger, hep-ph/9809490 (1998).
12. B. Bassett, F. Tamburini, D. Kaiser, and R. Maartens, hep-ph/9901319 (1999).
13. R. Easther and M. Parry, BROWN-HET-1152 (in preparation) (1999).
14. V. Mukhanov, H. Feldman, and R. H. Brandenberger, Phys. Rep. **215**, 203 (1992).

Ultra High Energy Cosmic Rays and Inflation.

Igor I. Tkachev

TH Division, CERN, CH-1211 Geneva 23, Switzerland
and
Institute for Nuclear Research, Russian Academy of Sciences, 60th October Anniversary Prosp. 7a, Moscow 117312, Russia

Abstract. Two processes of matter creation after inflation: 1) gravitational creation of superheavy (quasi)stable particles, and 2) non-thermal phase transitions leading to formation of topological defects, may be relevant to the resolution of the puzzle of cosmic rays observed with energies beyond GZK cut-off. Both possibilities are reviewed in this talk.

INTRODUCTION

According to the modern tale, all matter in the Universe was created in reheating after inflation. While this happened really long ago and on very small scales, this process is obviously of such vital importance that one may hope to find some observable consequences, specific for particular models of particle physics. And, indeed, we now believe that there can be some clues left. Among those are: topological defects production in non-thermal phase transitions [1], GUT scale baryogenesis [2], generation of primordial background of stochastic gravitational waves at high frequencies [3], just to mention a few. However, matter appears in many kinds and forms, and it is hard to review all possibilities in one talk. I'll concentrate on a possible relation to a mounting puzzle of the Ultra High Energy Cosmic Rays (UHECR).

When proton (or neutron) propagates in CMB, it gradually looses energy colliding with photons and creating pions [4]. There is a threshold energy for the process, so it is effective for very energetic nucleons only, which leads to the famous Greisen-Zatsepin-Kuzmin (GZK) cutoff of the high energy tail of the spectrum of cosmic rays. All this means that detection of, say, 3×10^{20} eV proton would require its source to be within ~ 50 Mpc. However, many events above the cut-off were observed by Yakutsk, Haverah Park, Fly Eye and AGASA collaborations [5] (for the review see Ref. [6]).

CP478 , *COSMO—98*, edited by David O. Caldwell

Results from the AGASA experiment [7] are shown in Fig. 1. The dashed curve represents the expected spectrum if conventional extragalactic sources of UHECR would be distributed uniformly in the Universe. This curve displays the theoretical GZK cut-off, but we see events which are way above it. (Numbers attached to the data points show the number of events observed in each energy bin.) Note that no candidate astrophysical source, like powerful active galaxy nuclei, were found in the directions of all six events with $E > 10^{20}$ eV [7]

There were no conventional explanation found to these observations, and the question arises, is it indication of the long awayted new physics, at last ?

Many solutions to the puzzle were suggested, which rely on different extensions of the standard model, in one way or the other. Among those are:

- A particle which is immune to CMBR. In this scenario, primary particle is produced in conventional astrophysical accelerators and is able to travel cosmological distances. There are variations to this scheme. This can be a new exotic particle able to produce normal air showers in Earth's atmosphere [8], or this can be an accelerated (anti)neutrino annihilating via Z^0 resonance on the relic neutrinos in a local high density neutrino clump, thus producing energetic gamma or nucleon [9]. Massiveness of neutrino, $m_\nu \sim$ eV, is a necessary requirement in this scheme.

- Another possibility is that UHECR are produced when topological defects destruct near the lab (on the cosmological scale) [10]. Topological defects which were considered in these kinds of scenarios were: strings [11], superconducting strings [10], networks of monopoles connected by strings [12], magnetic monopoles [13].

- Conceptually the simplest possibility is that UHECR are produced (again cosmologically locally) in decays of some new particle [14]. The candidate X-particle must obviously obey constraints on mass, number density and lifetime.

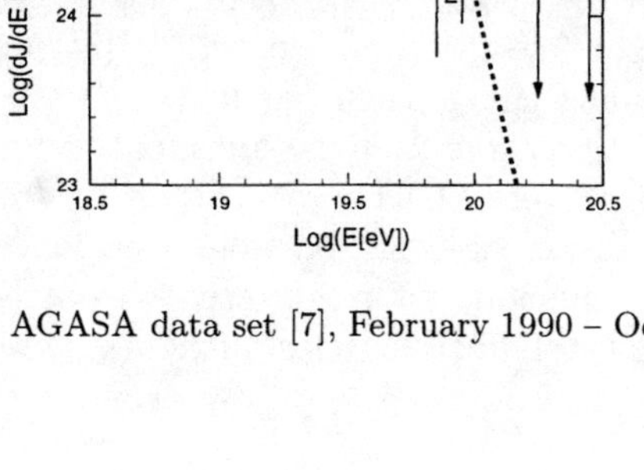

FIGURE 1. AGASA data set [7], February 1990 – October 1997.

UHECR FROM DECAYING PARTICLES

In order to produce cosmic rays in the energy range $E > 10^{11}$ GeV, the decaying primary particle has to be **heavy**, with the mass well above GZK cut-off, $m_X > 10^{12}$ GeV. The lifetime, τ_X, cannot be much smaller than the age of the Universe, $\tau_U \approx 10^{10}$ yr. Given this shortest possible lifetime, the observed flux of UHE cosmic rays will be generated with the rather low density of X-particles, $\Omega_X \sim 10^{-12}$, where $\Omega_X \equiv m_X n_X / \rho_{\rm crit}$, n_X is the number density of X-particles and $\rho_{\rm crit}$ is the critical density. On the other hand, X-particles must not overclose the Universe, $\Omega_X < 1$. With $\Omega_X \sim 1$, the X-particles may play the role of cold dark matter and the observed flux of UHE cosmic rays can be matched if $\tau_X \sim 10^{22}$ yr.

The problem of the particle physics mechanism responsible for a long but finite lifetime of very heavy particles can be solved in several ways. For example, otherwise conserved quantum number carried by X-particles may be broken very weakly due to instanton transitions, or quantum gravity (wormhole) effects [14]. Other interesting models of superheavy long-living particles were found in Refs. [15].

Spectra of UHE cosmic rays arising in decays of relic X-particles were successfully fitted to the data for m_X in the range $10^{12} < m_X/\text{GeV} < 10^{14}$ [16].

Here I address the issue of X-particle abundance. It was noticed [17,18] that such heavy particles are produced in the early Universe from the vacuum fluctuations and their abundance can be correct naturally, if the standard Friedmann epoch in the Universe evolution was preceded by the inflationary stage. This is a fundamental process of particle creation unavoidable in the time varying background and it requires no interactions. Temporal change of the metric is the single cause of particle production. Basically, it is the same process which during inflation had generated primordial large scale density perturbations. No coupling (e.g. to the inflaton or plasma) is needed. All one needs are stable (very long-living) X-particles with the mass of order of the inflaton mass, $m_X \approx 10^{13}$ GeV. Inflationary stage is not required to produce superheavy particles from the vacuum. Rather, the inflation provides a cut off in excessive gravitational production of heavy particles which would happen in the Friedmann Universe if it would start from the initial singularity [18]. Resulting abundance is quite independent of detailed nature of the particle which makes the superheavy (quasi)stable X-particle a very interesting dark matter candidate. New particle needs good name. I like Wimpzilla [19].

Friedmann Cosmology. For particles with conformal coupling to gravity (fermions or scalars with $\xi = 1/6$ in $\xi R\phi^2$ interaction term with the curvature), it is the particle mass which couples the system to the background expansion and serves as the source of particle creation. Therefore, just on dimensional grounds, we expect $n_X \propto m_X^3 a^{-3}$ at late times when particle creation diminishes. In Friedmann cosmology, $a \propto (mt)^\alpha \propto (m/H)^\alpha$ and the anticipated formulae for the X-particles abundance can be parameterised as $n_X = C_\alpha m_X^3 (H/m_X)^{3\alpha}$. It is expansion of the Universe which is responsible for particle creation. Therefore, this equation which describes simple dilution of already created particles is valid when already $H << m_X$. On the other hand particles with $m_X >> H$ cannot be created by

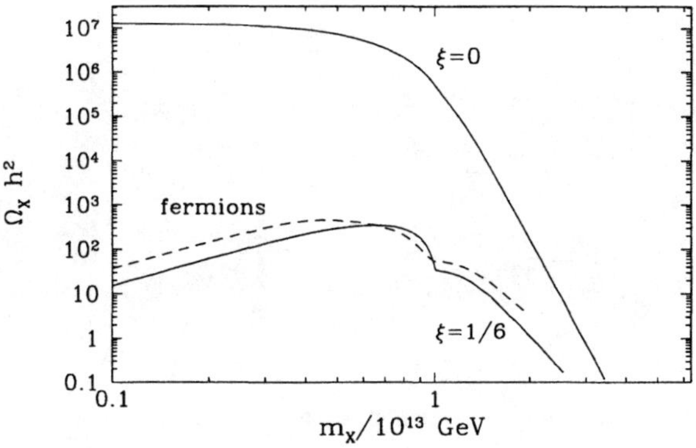

FIGURE 2. Ratio of the energy density in X-particles, gravitationally generated in inflationary cosmology, to the critical energy density is shown as a function of X-particle mass, Ref. [18].

this mechanism. Creation occurs when $H \sim m_X$. Coefficient C_α can be found numerically [18], its typical value is $O(10^{-2})$, and we find that stable particles with $m_X > 10^9$ GeV will overclose the Universe. There is no room for Superheavy particles in our Universe if it started from the initial Friedmann singularity [18], since the value of the Hubble constant is limited from above only by the Planck constant in this case.

Inflationary Cosmology. If there was inflation, the Hubble constant (in effect) did not exceeded the inflaton mass, $H < m_\phi$. The mass of the inflaton field has to be $m_\phi \approx 10^{13}$ GeV as constrained by the amplitude of primordial density fluctuations relevant for the large scale structure formation. Therefore, production of particles with $m_X > H \sim 10^{13}$ GeV has to be suppressed in inflationary cosmology. Results of direct numerical integration of gravitational particle creation in chaotic inflation model with the potential $V(\phi) = m_\phi^2\phi^2/2$ is shown in Fig. 2.

This figure was calculated assuming $T_{\rm R} = 10^9$ GeV for the reheating temperature. (At reheating the entropy of the Universe was created in addition to X-particles. In general, multiply this figure by the ratio $T_{\rm R}/10^9$ GeV and divide it by the fractional entropy increase per comoving volume if it was significant at some late epoch.) Reheating temperature is constrained, $T_{\rm R} < 10^9$ GeV, in supergravity theory [20]. We find that $\Omega_X h^2 < 1$ if $m_X \approx$ (few) $\times\ 10^{13}$ GeV. This value of mass is in the range suitable for the explanation of UHECR events [18]. Gravitationally created superheavy X-particles can even be the dominating form of matter in the Universe today if X-particles are in this mass range [17,18].

TOPOLOGICAL DEFECTS AND INFLATION

Decaying topological defect can naturally produce very energetic particles, and this may be related to UHECR [10] - [13], for recent reviews see [6]. However,

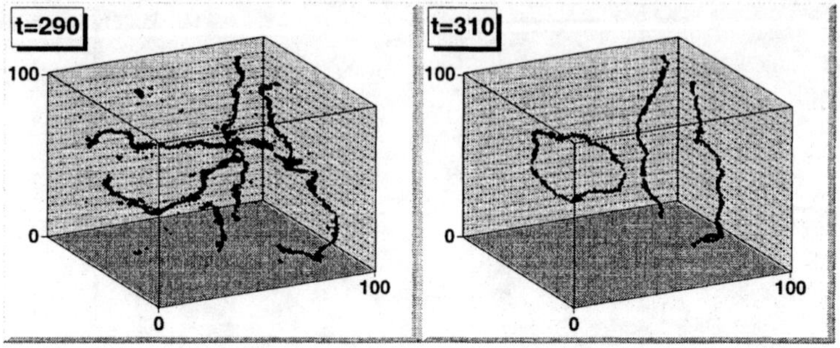

FIGURE 3. String distribution at two successive moments of time.

among motivations for inflation there was the necessity to get rid of unwanted topological defects. And inflation is excellent doing this job. Since temperature after reheating is constrained, especially severely in supergravity models, it might be that the Universe was never reheated up to the point of GUT phase transitions. Topological defects with a sufficiently high scale of symmetry breaking cannot be created. How then topological defects could populate the Universe?

The answer may be provided by non-thermal phase transitions [1] which can occur in preheating [21] after inflation. Explosive particle production caused by stimulated decay of inflaton oscillations lead to anomalously high field variances which restore symmetries of the theory even if actual reheating temperature is small. Defects form when variances are reduced by the continuing expansion of the Universe and phase transition occur. This problem is complicated, and while some features can be anticipated and some quantities roughly estimated, the problem requires numerical study. In recent papers [22] the defect formation and even the possibility of the first order phase transitions during preheating was demonstrated explicitly. Fig. 3 shows string distribution in a simulation with symmetry breaking scale $v = 3 \times 10^{16}$ GeV, when a pair of "infinite" strings and one big loop had formed. Size of the box is comparable to the Hubble length at this time.

CONCLUSIONS

Next generation cosmic ray experiments, which will be soon operational, will tell us which model for UHECR may be correct and which has to be ruled out. One unambiguous signature is related to homogeneity and anisotropy of cosmic rays. If particles immune to CMBR are there, the UHECR events should point towards distant, extraordinary astrophysical sources [23]. If wimpzillas are in the game, the Galaxy halo will be reflected in anisotropy of the UHECR flux [24]. It is remarkable that we might be able to learn about the earliest stages of the Universe's evolution. Discovery of heavy X-particles will mean that the model of inflation is likely correct, or that at least "standard" Friedmann evolution from the singularity is ruled out, since otherwise X-particles would have been inevitably overproduced [18].

REFERENCES

1. L. Kofman, A. Linde, and A. A. Starobinsky, *Phys. Rev. Lett.* **76**, 1011 (1996); I. I. Tkachev, *Phys. Lett.* **B376**, 35 (1996);
2. E. W. Kolb, A. D. Linde and A. Riotto, *Phys. Rev. Lett.* **77**, 4290 (1996); E. W. Kolb, A. Riotto and I. I. Tkachev, *Phys. Lett.* **B423**, 348 (1998).
3. S. Y. Khlebnikov and I. I. Tkachev, *Phys. Rev.* **D56**, 653 (1997).
4. K. Greisen, *Phys. Rev. Lett.* **16**, 748 (1966); G. T. Zatsepin and V. A. Kuzmin, *Pisma Zh. Eksp. Teor. Fiz.* **4**, 114 (1966).
5. N. Hayashida *et. al.*, *Phys. Rev. Lett.* **73**, 3491 (1994); D. J. Bird *et. al.*, *Astroph. J.* **424**, 491 (1994); **441**, 144 (1995); T. A. Egorov *et. al.*, in: *Proc. Tokyo Workshop on Techniques for the Study of Extremely High Energy Cosmic Rays*, ed. M.Nagano (ICRR, U. of Tokyo, 1993).
6. P. Bhattacharjee and G. Sigl, astro-ph/9811011; V. Berezinsky, astro-ph/9811268.
7. Takeda M. et al., *Phys. Rev. Lett.* **81**, 1163 (1998).
8. D. J. Chung, G. R. Farrar, and E. W. Kolb, astro-ph/9707036.
9. T. J. Weiler, hep-ph/9710431; S. Yoshida, G. Sigl, S. Lee, *Phys. Rev. Lett.* **58**, 55 (1998).
10. C. T. Hill, D. N. Schramm and T. P. Walker, *Phys. Rev.* **D36**, 1007 (1987);
11. J. H. MacGibbon and R. H. Brandenberger, *Nucl. Phys.* **B331**, 153 (1990); P. Bhattacharjee and N. C. Rana, *Phys. Lett.* **B246**, 356 (1990).
12. V. Berezinsky, X. Martin and A. Vilenkin, *Phys. Rev.* **D56**, 2024 (1997); V. Berezinsky and A. Vilenkin *Phys. Rev. Lett.* **79**, 5202 (1997).
13. C. T. Hill, *Nucl. Phys.* **B224**, 469 (1983); T. W. Kephart and T. J. Weiler, *Astropart. Phys.* **4**, 271 (1996).
14. V. Berezinsky, M. Kachelriess and A. Vilenkin, *Phys. Rev. Lett.* **79**, 4302 (1997); V. A. Kuzmin and V. A. Rubakov, *Phys. Atom. Nucl.* **61**, 1028 (1998).
15. K. Benakli, J. Ellis, and D. V. Nanopoulos, hep-ph/9803333; T. Han, T. Yanagida and R.-J. Zhang, *Phys. Rev.* **D58**, 095011 (1998); K. Hamaguchi, Y. Nomura, T. Yanagida, *Phys. Rev.* **D58**, 103503 (1998).
16. V. Berezinsky, P. Blasi, A. Vilenkin, *Phys. Rev.* **D58**, 103515 (1998); M. Birkel and S. Sarkar, *Astropart. Phys.* **9**, 297 (1998).
17. D. J. H. Chung, E. W. Kolb and A. Riotto, hep-ph/9802238.
18. V. A. Kuzmin and I. I. Tkachev, *JETP Lett.* **68**, 271 (1998); and hep-ph/9809547.
19. E. W. Kolb, D. J.H. Chung, A. Riotto, hep-ph/9810361.
20. J. Ellis, J. Kim and D. V. Nanopoulos, *Phys. Lett.* **B145**, 181 (1984).
21. L. Kofman, A. Linde, and A. Starobinsky, Phys. Rev. Lett. **73**, 3195 (1994); S. Y. Khlebnikov and I. I. Tkachev, *Phys. Rev. Lett.* **77**, 219 (1996), *Phys. Rev. Lett.* **79**, 1607 (1997), *Phys. Lett.* **B390**, 80 (1997);
22. S. Khlebnikov, L. Kofman, A. Linde, and I. Tkachev, *Phys. Rev. Lett.* **81**, 2012 (1998); I. Tkachev, S. Khlebnikov, L. Kofman, and A. Linde, *Phys. Lett.* **B440**, 262 (1998); S. Kasuya and M. Kawasaki, *Phys. Rev.* **D58**, 083516 (1998).
23. G. R. Farrar and P. L. Biermann, *Phys. Rev. Lett.* **81**, 3579 (1998).
24. S. L. Dubovskii, P. G. Tinyakov *JETP Lett.* **68**, 107 (1998).

Inflationary Reheating and Fermions

Patrick B. Greene

Department of Physics, University of Toronto,
60 St. George Street, Toronto, Ontario M5S 1A7, Canada

Abstract. Coherent oscillations of the inflaton field at the end of inflation can parametrically excite fermions in much the same way that bosons are created in preheating. Although Pauli-blocking prohibits the occupation number of created fermions from growing exponentially, fermion production occurs in a manner significantly different from the expectations of simple perturbation theory. Here, I discuss the nature of fermion production after inflation and possible applications including the efficient transfer of inflaton energy and the production of super-massive fermions during fermionic preheating.

Consider a simple model of chaotic inflation with the potential $\frac{1}{4}\lambda\phi^4$ for an inflaton field ϕ coupled to a massless spin-$\frac{1}{2}$ field ψ by a Yukawa interaction. At the end of inflation, the inflaton field will oscillate coherently about the minimum of its effective potential with an initial amplitude $\phi_o \sim O(0.1M_p)$. In perturbation theory, one treats the homogeneous and quasi-classical inflaton field as a condensate of scalar inflaton particles, each of which can decay into a pair of ψ-particles. Each fermion then carries away half of the energy of a typical inflaton particle, giving a spectrum narrowly peaked around the comoving momentum $k \approx 0.42\sqrt{\lambda}\phi_o$. The inflaton energy is transfered to fermions after $O(\frac{1}{h^2}) \gg 1$ inflaton oscillations.

To investigate fermion production non-perturbatively, we are interested in the equation of motion for the field operator ψ. Following the usual prescription (see [1] and references therein for details), one seeks eigenfunctions of the Dirac equation in the presence of the classical time-dependent source, $\phi(t)$. As the inflaton field is spatially homogeneous, only the temporal part of the eigenmode obeys a non-trivial equation of motion. These modes, $X_k(\tau)$, obey an oscillator-type equation with a *complex* frequency that varies periodically with time:

$$X_k'' + \left(\kappa^2 + qf^2 - i\sqrt{q}f'\right)X_k = 0\ . \qquad (1)$$

Here, the comoving momentum k enters the equation in the combination $\kappa^2 \equiv \frac{k^2}{\lambda\phi_o^2}$ and the character of the solutions is defined by the parameter $q \equiv \frac{h^2}{\lambda}$ (h is the dimensionless Yukawa coupling). The background oscillations enter in the form

CP478 , *COSMO—98*, edited by David O. Caldwell

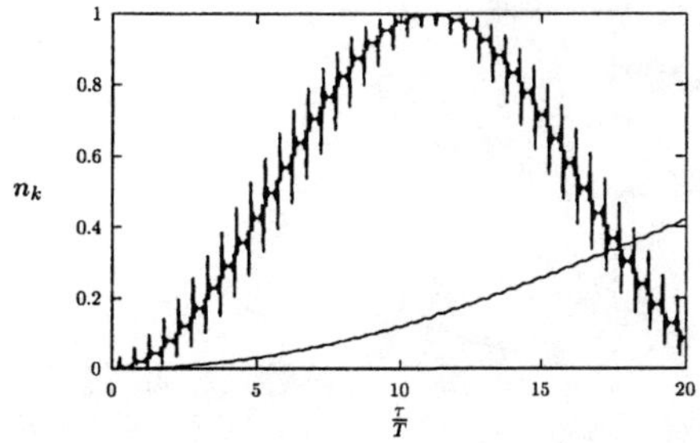

FIGURE 1. *The comoving occupation number n_k of fermions in $\lambda\phi^4$-inflation as a function of time (in units of inflaton oscillations) for $q \equiv \frac{h^2}{\lambda} = 10^{-4}$ and 100 and $\kappa^2 = 0.18$ and 11.9, respectively. The period of the modulation $\frac{\pi}{\nu_k T}$ is about 88 and 22 accordingly.*

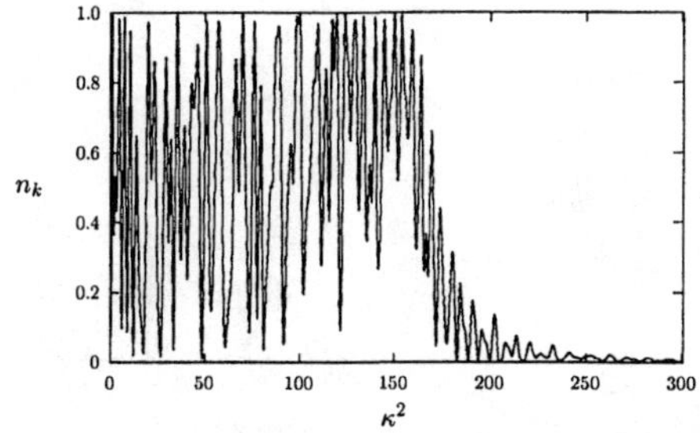

FIGURE 2. *The comoving occupation number of fermions in $m_\phi^2\phi^2$-inflation as a function of (scaled) comoving momentum after 50 inflaton oscillations for initial resonance parameter $q_o = 10^3$. Expansion destroys the details of the resonance band and leads to a fermi-sphere of width $q^{1/4} \sim q_o^{1/4} a^{1/4}$.*

$f(\tau) = cn\left(\tau, \frac{1}{\sqrt{2}}\right)$ having unit amplitude and a period $T = 7.416$. Note that we are working in scaled conformal field and time variables so the effects of expansion do not appear in the equations of motion. We can express the comoving occupation number of particles in a given state through the solutions of eq. (1)

$$n_k(\tau) = \frac{(\Omega_k - \sqrt{q}f)}{2\Omega_k}[\ |X_k'|^2 + \Omega_k^2|X_k|^2 - 2\Omega_k Im(X_k X_k'^*)\ ], \tag{2}$$

where $\Omega_k^2 \equiv \kappa^2 + qf^2$. The energy density of created fermions is $\epsilon_\psi = \frac{1}{2\pi^3}\int d^3k\,\Omega_k\,n_k$.

It turns out that, for all κ^2 and q, the solutions of eq. (1) are periodic in time. This is shown by the numerical solutions for the comoving occupation number in Fig. (1). Furthermore, it is easy to show that the comoving occupation number defined by eq. (2) will obey $n_k \leq 1$ in accordance with the Pauli principle. We see from Fig. (1) that, while the occupation number exhibits some high frequency oscillations (period $< \frac{T}{2}$), the most interesting behavior occurs over longer periods. If we average the occupation number over an inflaton period, $\bar{n}_k(\tau) = \frac{1}{T}\int_\tau^{(\tau+T)} d\tau n_k(\tau)$, the average occupation number is found to obey the simple equation: $\bar{n}_k(\tau) = F_k \sin^2 \nu_k \tau$. For a given theory, i.e. for a given value of the resonance parameter q, $F_k \leq 1$ is a momentum dependent amplitude and ν_k is a momentum dependent frequency. In Fig. (3) the amplitude F_k as a function of κ^2 is plotted for several values of q. We see that the perturbative expectation is only met for $q \leq 10^{-4}$.

In fact, for large q, the fermions are excited up to $\kappa^2 \simeq \sqrt{q}$. This is the same result as for the broad bosonic resonance and can be understood in the same manner. The comoving occupation number, eq. (2), is an adiabatic invariant of the mode equation (1). The condition that the modes X_k evolve non-adiabatically is $\dot{\Omega}_k \geq \Omega_k^2$

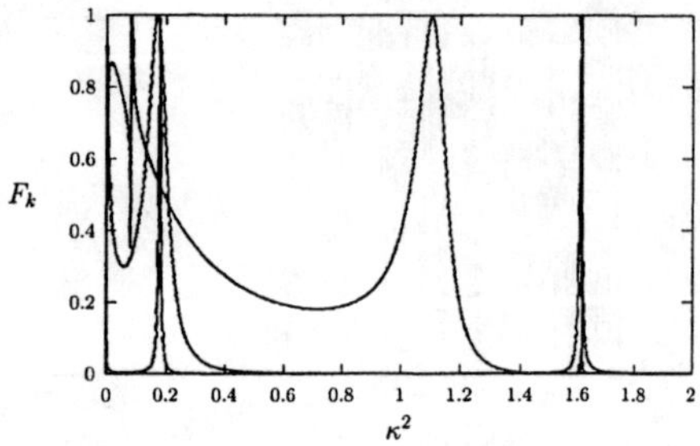

FIGURE 3. *The envelope functions F_k showing the bands of fermion resonance excitation in $\lambda\phi^4$-inflation for $q \equiv \frac{h^2}{\lambda} = 10^{-4}, 10^{-2}$, and 1.0 (the narrowest to broadest band, respectively).*

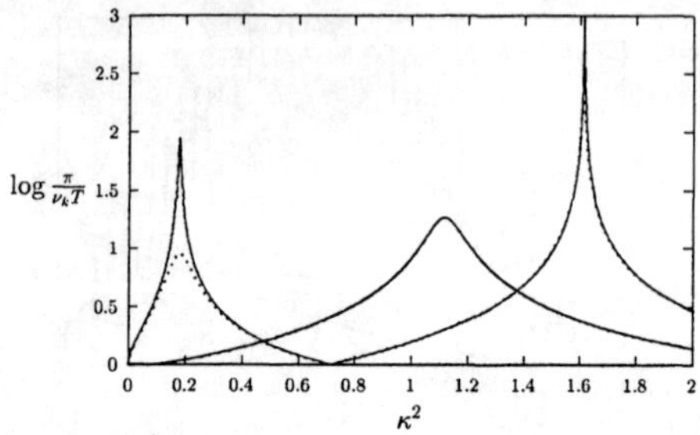

FIGURE 4. *The* log *of the period of modulation $\frac{\pi}{\nu_k T}$ (in units of inflaton oscillations) as a function of κ^2 for $q \equiv \frac{h^2}{\lambda} = 10^{-4}, 10^{-2}$, and* 1.0 *for the light, dotted, and heavy curves respectively.*

which leads to the condition $\kappa^2 \leq \sqrt{q}$ for non-adiabatic evolution, and thus, particle creation. This occurs much more rapidly than one would expect from perturbation theory. In Fig. (4) the period of mode oscillations $\frac{\pi}{\nu_k T}$ is plotted in units of inflaton oscillations. For all q, the bands saturate after only $10 - 100$ inflaton oscillations.

Turning to the energetics for the most interesting case of the broad resonance excitation, $q \gg 1$, we find $\epsilon_\psi \sim 0.1 h^2 q^{1/4} \epsilon_\phi$, where the inflaton energy is $\epsilon_\phi = \frac{1}{4}\lambda\phi_o^4$. In chaotic $\frac{1}{4}\lambda\phi^4$-inflation, $\lambda \simeq 10^{-13}$. If the resonance parameter q is large but the coupling parameter is small, $h \leq 0.1$, only a small fraction of the inflaton's energy will be converted into fermions. Although explosive decay will not occur for this model, more general theories such as hybrid models can have large enough resonance parameters to allow efficient decay to fermions.

If the fermion field has a small bare mass term or if we consider inflation with a $\frac{1}{2}m_\phi^2\phi^2$ potential, the conformal invariance of the theory will be broken. In this case, the occupation number of fermions for $q \gg 1$ no longer evolves periodically but becomes stochastic, rapidly fluctuating between $n_k = 1$ and $n_k = 0$. This destroys the well defined resonance bands depicted in Fig. (3) and can lead to the production of super-massive fermions of mass $m_\psi \leq h\phi_o$. These fermions can come to dominate the energy density of the universe or survive as massive relics.

As a particular example of the changes brought by expansion, consider $m_\phi^2\phi^2$-inflation with a Yukawa coupling to a still massless fermion. In the broad resonance case, one gets a sphere in comoving momentum space with average occupation $\bar{n}_k = \frac{1}{2}$ that *expands* with time. A snapshot of this sphere is shown in Fig. (2).

REFERENCES

1. Greene P. B., and Kofman L., to appear in *Phys. Lett.* **B448**, (1999); Available as eprint hep-ph/9807339.

Formation of Topological Defects during Preheating

Shinta Kasuya and Masahiro Kawasaki

Institute for Cosmic Ray Research, University of Tokyo, Tanashi, Tokyo 188-8502, Japan

Abstract. We consider the formation of topological defects after inflation. In order to take into account the effects of the rescattering of fluctuations, we integrate the classical equation that describes the evolution of a complex scalar field on the two-dimensional lattice with a slab symmetry. The growth of fluctuations during preheating is found not to be enough for defect formation, and rather a long stage of the rescattering of fluctuations after preheating is necessary. We conclude that the topological defects are not formed if the breaking scale η is lager than $\sim (2-3) \times 10^{16}$GeV.

It was recently recognized that topological defects may be formed during preheating [1–5]. Preheating is the very beginning of the reheating process, and occurs because of the parametric resonant effects [6–9]. Its essence is an *induced* effect in the sense that the presence of the produced particles stimulates further decay of the coherently oscillating scalar field into those particles [10]. This phenomenon is thus peculiar to bosonic particles that obey the Bose-Einstein statistics.

Rescatterings [11–15] are quite important to determine the dynamics of the defect formation as suggested by Refs. [16,17], because interactions between particles with different momenta are fully included. To deal with the effect of the rescattering, we must go on to the lattice simulation [11–13]. Therefore, we integrated the equations describing the dynamics in a real space on the lattice. There are some difficulties in the simulations on lattice. One of the difficulties is how we can implement the quantum fluctuations of the fields on the lattice. We rely on the idea of Ref. [11] and do not discuss in detail (actually, it is not sensitive to the initial conditions for the fluctuations if the order of magnitude is properly taken [13]). The other one is the box size and the lattice size. The lattice size must be small enough for identifying the defects. The box size should not be taken too much smaller than the horizon size at the end of the simulation, otherwise we miss the defects whose density is $O(10)$ per horizon. We also take the proper lattice and box sizes so that we can see the resonance effects. From these requirements, we calculate the evolution of a complex scalar field in a two-dimensional lattice abandoning to calculate it in three dimensions because of the lack of machine memories.

CP478 , *COSMO—98*, edited by David O. Caldwell

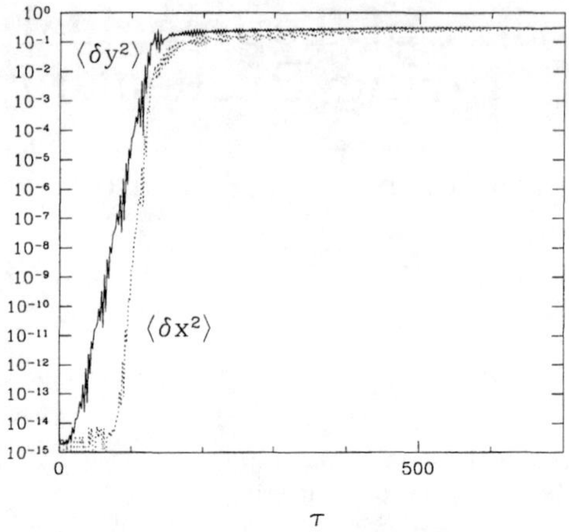

FIGURE 1. Evolution of fluctuations for $\eta = 10^{16}$ GeV.

Let us consider the wine bottle potential

$$V(\Phi) = \frac{\lambda}{2}(|\Phi|^2 - \eta^2)^2, \tag{1}$$

where Φ is the complex scalar field, λ is the very small self coupling constant, and η is the breaking scale. This model has a global U(1) symmetry, and cosmic strings are formed when the symmetry is spontaneously broken.

We plot the evolution of fluctuations for $\eta = 10^{16}$ GeV in Fig. 1. We use the same data as Fig.2 of Ref. [4], but in the logarithmic scale. Then it is obvious that the growth exponent for y-direction is $\mu_y \simeq 0.141$, and for x-direction is twice a larger: $\mu_x \simeq 2\mu_y$. Fluctuations in x-direction grows much faster than those in y-direction because of rescattering effects, but the former does not overtake the latter. This exactly coincides with the results of Ref. [5], and, at least in the preheating stage, there is no difference between the simulations in two and three dimensions (We agree that rescatterings are less effective in two dimensions to some extent after preheating stage).

We have found that fluctuations produced during preheating do not lead to the defect formation even if their amplitudes become the same order of magnitude as that of the homogeneous mode at the end of preheating. Actually, in the case for $\eta = 3 \times 10^{16}$ GeV, the amplitude of fluctuations grows as large as that of the homogeneous mode, but we see no defects. We also have found that the effects of the rescattering of the fluctuations are essential and a rather long duration of the rescattering is necessary for the topological defects to be formed as suggested by Refs. [16,17]. For $\eta = 2 \times 10^{16}$ GeV, we see only a few defects in the horizon size,

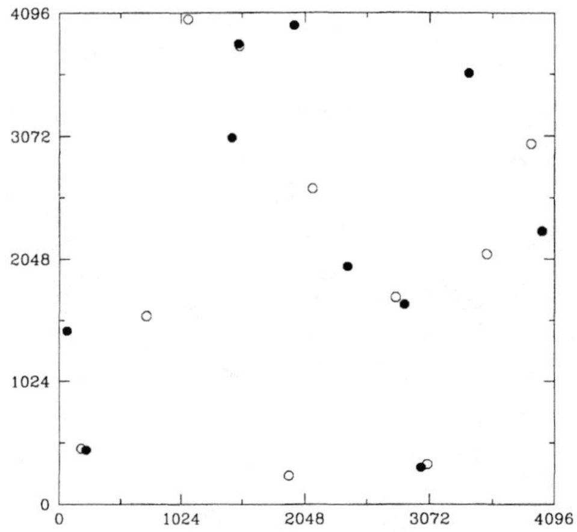

FIGURE 2. Strings (dots) and anti-strings (circles) at $\tau = 2000$ for $\eta = 10^{16}$ GeV.

and they may disappear very soon because of the pairwise appearance of strings and anti-strings. On the other hand, we have found $O(10)$ defects in the horizon for $\eta = 2 \times 10^{16}$ GeV, as is seen in Fig. 2, and also there is very little time dependence of the number per horizon.

Physical reason is as follows. Fluctuations in the resonance mode scatter off each other to make fluctuations with higher and lower momenta (See Fig. 3). However, there are two types of scattering. One receives Bose enhancement effects and the other does not. However, the former processes are the zero-sum game. If high momentum fluctuations are produced, the homogeneous mode is also created and cannot decay further. The latter processes which do not feel the Bose enhancement in the first place make the spectrum broad and smooth. After such processes, the occupation numbers of the mode outside the resonance band grow large enough for the zero mode to decay into all the modes. The decay of the homogeneous mode completes in such a complicated way, and the topological defects will be formed through the Kibble-like mechanism later. Therefore, the rescattering timescale for the defect formation should be considerably longer.

As a result, the phases of the field become random in each region of the universe, and the topological defects are formed in a similar way as the Kibble mechanism. These processes take a lot of time so that it is necessary for the field not to fall into the minimum of the potential too soon after preheating. This constrains the value of the breaking scale, since it takes longer time for the field to settles down in the minimum of the potential for smaller breaking scale. We have found that it is necessary to have $\Delta\tau \gtrsim 200$ from our calculations, and we conclude that the topological defects are not formed if $\eta \gtrsim (2-3) \times 10^{16}$GeV.

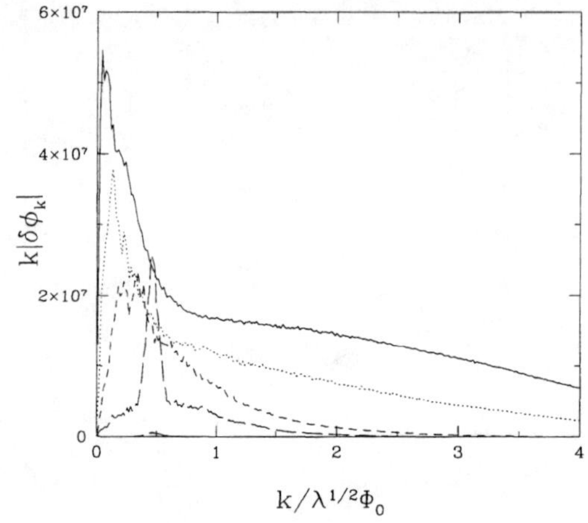

FIGURE 3. Spectrum of the fluctuations for $\eta = 10^{16}$ GeV. These lines shows $\tau = 800, 500, 200, 130, 100$ from top to bottom.

REFERENCES

1. Kofman, L., Linde, A., and Starobinsky, A., *Phys. Rev. Lett.* **76**, 1011 (1996).
2. Tkachev, I.I., *Phys. Lett.* **B376**, 35 (1996).
3. Kasuya, S., and Kawasaki, M., *Phys. Rev.* **D56**, 7597 (1997).
4. Kasuya, S., and Kawasaki, M., *Phys. Rev.* **D58**, 083516 (1998).
5. Tkachev, I., Khlebnikov, S., Kofman, L., and Linde, A., *Phys. Lett.* **B440**, 262 (1998).
6. Kofman, L., Linde, A., and Strarobinsky, A., *Phys. Rev. Lett.* **73**, 3195 (1994).
7. Shtanov, Y., Traschen, J., and Brandenberger, R., *Phys. Rev.* **D51**, 5438 (1995).
8. Boyanovsky, D., de Vega, H.J., Holman, R., Lee, D.-S., and Singh, A., *Phys. Rev.* **D51**,4419 (1995).
9. Yoshimura, M., *Prog. Theor. Phys.* **94**, 873 (1995).
10. Kasuya, S., and Kawasaki, M., *Phys. Lett.* **B388**, 686 (1996).
11. Khlebnikov, S.Yu., and Tkachev, I.I., *Phys. Rev. Lett.* **77**, 219 (1996).
12. Khlebnikov, S.Yu., and Tkachev, I.I., *Phys. Rev. Lett.* **79**, 1607 (1997).
13. Prokopec T., and Roos, T.G., *Phys. Rev.* **D55**, 3768 (1997).
14. Kofman, L., Linde, A., and Strarobinsky, A., *Phys. Rev.* **D56**, 3258 (1997).
15. Greene, P.B., Kofman, L., Linde, A., and Strarobinsky, A., *Phys. Rev.* **D56**, 6175 (1997).
16. Kofman, L., in *Proceedings of the Workshop on Particle Physics and the Early Universe*, edited by L. Roszkowski, (1998), astro-ph/9802221.
17. Kofman, L., in *Proceedings of the 3rd RESCEU International Symposium on Particle Cosmology*, edited by K. Sato *et al*, (1998), hep-ph/9802285.

Global Structure of the Inflationary Universe

Raphael Bousso

Department of Physics, Stanford University, California 94305

Abstract. I describe the quantum creation of black holes in de Sitter space. A single topology change can lead to the formation of multiple black holes and the eventual disconnection of large spacelike sections. Applied to cosmology, this process leads to the production of a large or infinite number of universes in most models of inflation and yields a new picture of global structure.

de Sitter space is the maximally symmetric solution of the vacuum Einstein equations with a cosmological constant Λ. It may be visualized as a (3,1)-hyperboloid embedded in (4,1) Minkowski space. In this spacetime, geodesic observers find themselves immersed in a bath of thermal radiation [1] of temperature $T = \sqrt{\Lambda/3}/(2\pi)$. This raises the question of stability, which was investigated in 1983 by Ginsparg and Perry [2]. They showed that de Sitter space does not possess the classical Jeans instability found in hot flat space. It does, however, possess a semiclassical instability to a spontaneous topology change corresponding to the nucleation of a black hole.

When the black hole appears, it will typically be degenerate; that is, it will have the same size as the cosmological horizon, and will be in thermal equilibrium with it. Ginsparg and Perry argued on an intuitive basis that this equilibrium would be unstable, with quantum fluctuations causing the black hole to be slightly smaller, and hotter, than the cosmological horizon. Then, presumably, the black hole would start to evaporate and eventually disappear. They showed that the time scale between black hole nucleations is vastly larger than the time needed for evaporation (see also Ref. [3]). In this sense, de Sitter space would "perdure".

In a collaboration with S. Hawking, this argument was recently confirmed [4]. We used a model that includes the quantum radiation in the s-wave and large N limit, at one loop. We found, to our surprise, that nearly degenerate Schwarzschild-de Sitter black holes anti-evaporate. But there is a different way of perturbing the degenerate solution which leads to evaporation, and we found that this mode would always be excited when black holes nucleate spontaneously.

The process of black hole creation and subsequent evaporation is shown in Fig. 1

CP478 , *COSMO—98*, edited by David O. Caldwell

(evolution of the spacelike sections) and Fig. 2 (causal structure). Crucially, for

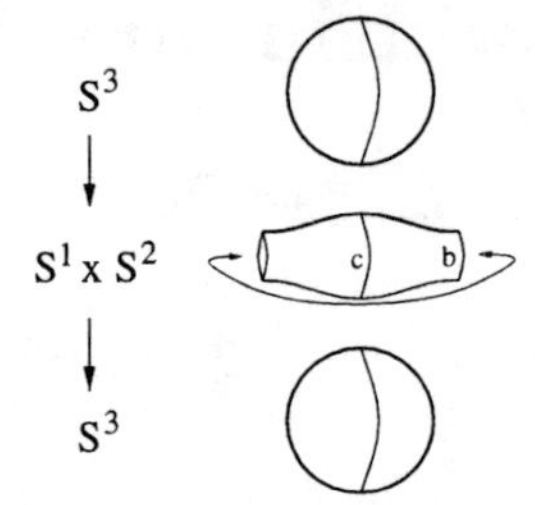

FIGURE 1. Evolution of spacelike hypersurfaces of de Sitter space during the creation and subsequent evaporation of a single black hole. The spontaneous creation of a handle changes the spatial topology from spherical (S^3) to toroidal ($S^1 \times S^2$) with constant two-sphere radius. (The double-headed arrow indicates that opposite ends of the middle picture should be identified, closing the S^1.) If the quantum fluctuations are dominated by the lowest Fourier mode on the S^1, there will be one minimum and one maximum two-sphere radius, corresponding to a black hole (b) and a cosmological horizon (c). This resembles a 'wobbly doughnut' with cross-sections of varying thickness. As the black hole evaporates, the thinnest cross-section decreases in size. Finally the black hole disappears, i.e. the doughnut is pinched at its thinnest place and reverts to the original spherical topology.

the case of a single black hole, the topology reverts to the original de Sitter space after the process is completed.

In Ref. [5], I considered different perturbations of the degenerate black hole solution, which correspond to higher quantum fluctuation modes. Occasionally, one such mode will dominate over lower modes, leading to the presence of *multiple* pairs of apparent black hole horizons in a single nucleation event, and eventually to the formation of several black hole interiors. I showed that they evaporate, and that space will disconnect at the event when a black hole finally disappears. If more than one black hole is present, this leads to the disintegration of the universe.

This process is depicted in Fig. 3, where the evolution of the space-like sections in a process of multiple black hole creation and subsequent evaporation is shown. The corresponding causal diagram is given in Fig. 4. These pictures should be compared to the similar, yet much less dramatic evolution in the case where only one black hole is created, Figs. 1 and 2.

The decay described here can be viewed as a sequence of topology changes. The first topology change is the semiclassical creation of a handle with multiple black hole horizons. It corresponds to a local non-perturbative quantum fluctuation of the metric on the scale of a single horizon volume. It will therefore happen independently in widely separated horizon volumes of de Sitter space. While the black holes evaporate, the asymptotically de Sitter regions between the black holes grow exponentially. Thus, the second topology change, corresponding to the final disappearance of the black holes, yields a number of separate de Sitter spaces which already contain exponentially large spacelike sections. Because it occurs at such

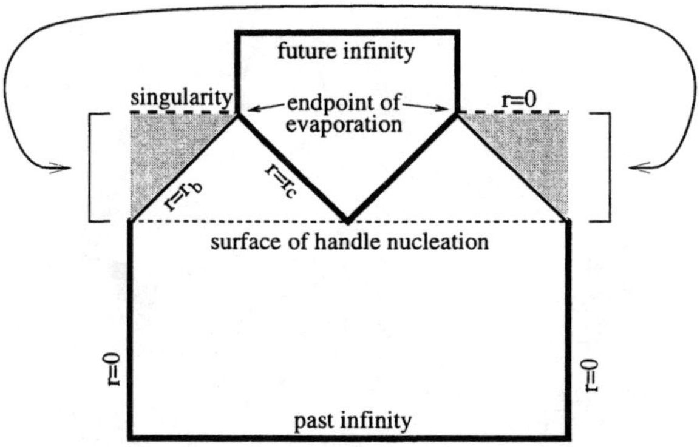

FIGURE 2. Penrose diagram for the process depicted in Fig. 1. The shaded region is the black hole interior. In the region marked by the square brackets the spatial topology is $S^1 \times S^2$, and opposite ends should be identified. The middle picture in Fig. 1 corresponds to the handle nucleation surface shown here. After the black hole evaporates, a single de Sitter universe remains.

large scales, the effect differs from many other considerations of baby universes. Inside the daughter universes, the handle-creation process will again occur locally. Thus a single de Sitter universe decomposes iteratively into an infinite number of disjoint copies of itself, proliferating, as it were, by budding.

According to the inflationary paradigm, the most successful theory of primordial cosmology, our universe went through a period of de Sitter-like expansion before the onset of the radiation and matter dominated eras. Therefore the process described in this paper is relevant to the global structure of the universe: It means that we live in one of many universes that originated in the same primordial inflationary region. There is a large class of models that lead to eternal inflation; in this case, we inhabit one of an infinite number of separate universes produced from a single region.

REFERENCES

1. G. W. Gibbons and S. W. Hawking: *Cosmological event horizons, thermodynamics, and particle creation*. Phys. Rev. D **15**, 2738 (1977).
2. P. Ginsparg and M. J. Perry: *Semiclassical perdurance of de Sitter space*. Nucl. Phys. **B222**, 245 (1983).
3. R. Bousso and S. W. Hawking: *Pair creation of black holes during inflation*. Phys. Rev. D **54**, 6312 (1996), gr-qc/9606052.
4. R. Bousso and S. W. Hawking: *(Anti-)evaporation of Schwarzschild-de Sitter black holes*. Phys. Rev. D **57**, 2436 (1998), hep-th/9709224.
5. R. Bousso: *Proliferation of de Sitter space*. Phys. Rev. D **58**, 083511 (1998), hep-th/9805081.

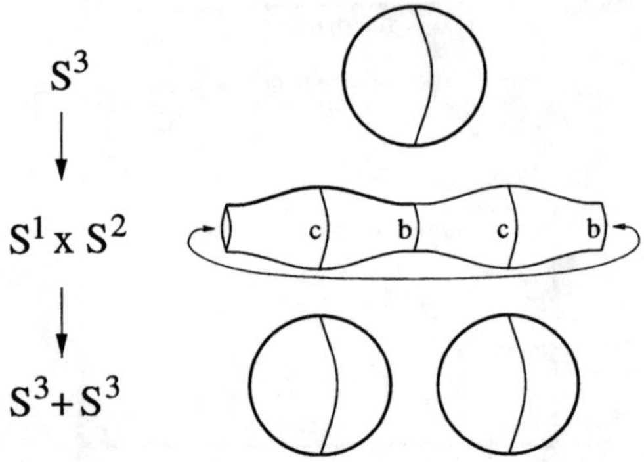

FIGURE 3. Evolution of spacelike hypersurfaces of de Sitter space during the creation of a handle yielding multiple black holes ($n = 2$) and their subsequent evaporation. This should be compared to Fig. 1. If the quantum fluctuations on the $S^1 \times S^2$ handle are dominated by the second Fourier mode on the S^1, there will be two minima and two maxima of the two-sphere radius, seeding to two black hole interiors (b) and two asymptotically de Sitter regions (c). This resembles a 'wobbly doughnut' on which the thickness of the cross-sections oscillates twice. As the black holes evaporate, the minimal cross-sections decrease. When the black holes disappear, the doughnut is pinched at two places, yielding two disjoint spaces of spherical topology, the daughter universes.

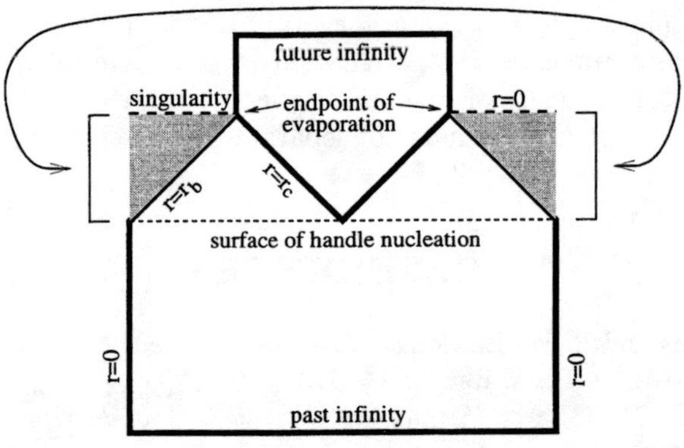

FIGURE 4. Penrose diagram for the process depicted in Fig. 3. The shaded regions are the two black hole interiors. In the region marked by the square brackets the spatial topology is $S^1 \times S^2$, and opposite ends should be identified. The middle picture in Fig. 3 corresponds to the horizon freezing surface shown here. After the black holes evaporate, two separate de Sitter universes remain.

SUPERSTRINGS
AND
COSMOLOGY

From Superstrings to M Theory

John H. Schwarz*

*California Institute of Technology[1]
Pasadena, California 91125

Abstract. In this talk I will survey some of the basic facts about superstring theories in 10 dimensions and the dualities that relate them to M theory in 11 dimensions. Then I will mention some important unresolved issues.

INTRODUCTION

Superstring theory first achieved widespread acceptance during the *first superstring revolution* in 1984-85. There were three main developments at this time. The first was the discovery of an anomaly cancellation mechanism [2], which showed that supersymmetric gauge theories can be consistent in ten dimensions provided they are coupled to supergravity (as in type I superstring theory) and the gauge group is either SO(32) or $E_8 \times E_8$. Any other group necessarily would give uncanceled gauge anomalies and hence inconsistency at the quantum level. The second development was the discovery of two new superstring theories—called *heterotic* string theories—with precisely these gauge groups [3]. The third development was the realization that the $E_8 \times E_8$ heterotic string theory admits solutions in which six of the space dimensions form a Calabi–Yau space, and that this results in a 4d effective theory at low energies with many qualitatively realistic features [4]. Unfortunately, there are very many Calabi–Yau spaces and a whole range of additional choices that can be made (orbifolds, Wilson loops, etc.). Thus there is an enormous variety of possibilities, none of which stands out as particularly special.

In any case, after the first superstring revolution subsided, we had five distinct superstring theories with consistent weak coupling perturbation expansions, each in ten dimensions. Three of them, the *type I* theory and the two heterotic theories, have $\mathcal{N} = 1$ supersymmetry in the ten-dimensional sense. Since the minimal 10d spinor is simultaneously Majorana and Weyl, this corresponds to 16 conserved supercharges. The other two theories, called *type IIA* and *type IIB*, have $\mathcal{N} = 2$ supersymmetry (32 supercharges) [5]. In the IIA case the two spinors have opposite handedness so that the spectrum is left-right symmetric (nonchiral). In the IIB case the two spinors have the same handedness and the spectrum is chiral.

[1] Work supported in part by the U.S. Dept. of Energy under Grant No. DE-FG03-92-ER40701.

CP478 , *COSMO—98*, edited by David O. Caldwell
 1-56396-853-3/99/$15.00

The understanding of these five superstring theories was developed in the ensuing years. In each case it became clear, and was largely proved, that there are consistent perturbation expansions of on-shell scattering amplitudes. In four of the five cases (heterotic and type II) the fundamental strings are oriented and unbreakable. As a result, these theories have particularly simple perturbation expansions. Specifically, there is a unique Feynman diagram at each order of the loop expansion. The Feynman diagrams depict string world sheets, and therefore they are two-dimensional surfaces. For these four theories the unique L-loop diagram is a closed orientable genus-L Riemann surface, which can be visualized as a sphere with L handles. External (incoming or outgoing) particles are represented by N points (or "punctures") on the Riemann surface. A given diagram represents a well-defined integral of dimension $6L + 2N - 6$. This integral has no ultraviolet divergences, even though the spectrum contains states of arbitrarily high spin (including a massless graviton). From the viewpoint of point-particle contributions, string and supersymmetry properties are responsible for incredible cancellations. Type I superstrings are unoriented and breakable. As a result, the perturbation expansion is more complicated for this theory, and the various world-sheet diagrams at a given order (determined by the Euler number) have to be combined properly to cancel divergences and anomalies [6].

M THEORY

In the 1970s and 1980s various supersymmetry and supergravity theories were constructed. (See [7], for example.) In particular, supersymmetry representation theory showed that ten is the largest spacetime dimension in which there can be a supersymmetric Yang–Mills theory, with spins ≤ 1 [8]. This is a pretty (*i.e.*, very symmetrical) classical field theory, but at the quantum level it is both nonrenormalizable and anomalous for any nonabelian gauge group. However, as we indicated earlier, both problems can be overcome for suitable gauge groups (SO(32) or $E_8 \times E_8$) when the Yang–Mills theory is embedded in a type I or heterotic string theory.

The largest possible spacetime dimension for a supergravity theory (with spins ≤ 2), on the other hand, is eleven. Eleven-dimensional supergravity, which has 32 conserved supercharges, was constructed 20 years ago [9]. It has three kinds of fields—the graviton field (with 44 polarizations), the gravitino field (with 128 polarizations), and a three-index antisymmetric tensor gauge field $C_{\mu\nu\rho}$ (with 84 polarizations). These massless particles are referred to collectively as the *supergraviton.* 11d supergravity is also a pretty classical field theory, which has attracted a lot of attention over the years. It is not chiral, and therefore not subject to anomaly problems.[2] It is also nonrenormalizable, and thus it cannot be a fundamental theory. However, we now believe that it is a low-energy effective description of M

[2] Unless the spacetime has boundaries. The anomaly associated to a 10d boundary can be canceled by introducing E_8 supersymmetric gauge theory on the boundary [10].

theory, which is a well-defined quantum theory [11]. This means, in particular, that higher dimension terms in the effective action for the supergravity fields have uniquely determined coefficients within the M theory setting, even though they are formally infinite (and hence undetermined) within the supergravity context.

Intriguing connections between type IIA string theory and 11d supergravity have been known for a long time. If one carries out *dimensional reduction* of 11d supergravity to 10d, one gets type IIA supergravity [12]. In this case dimensional reduction can be viewed as a compactification on a circle in which one drops all the Kaluza–Klein excitations. It is easy to show that this does not break any of the supersymmetries. The field equations of 11d supergravity admit a solution that describes a supermembrane. This solution has the property that the energy density is concentrated on a two-dimensional surface. A 3d world-volume description of the dynamics of this supermembrane, quite analogous to the 2d world volume actions of superstrings, has been constructed [13]. The authors suggested that a consistent 11d quantum theory might be defined in terms of this membrane, in analogy to string theories in ten dimensions.[3] Another striking result was the discovery of double dimensional reduction [14]. This is a dimensional reduction on a circle, in which one wraps one dimension of the membrane around the circle and drops all Kaluza–Klein excitations for both the spacetime theory and the world-volume theory. The remarkable fact is that this gives the (previously known) type IIA superstring world-volume action [15].

For many years these facts remained unexplained curiosities until they were reconsidered by Townsend [16] and by Witten [11]. The conclusion is that type IIA superstring theory really does have a circular 11th dimension in addition to the previously known ten spacetime dimensions. This fact was not recognized earlier because the appearance of the 11th dimension is a nonperturbative phenomenon, not visible in perturbation theory.

To explain the relation between M theory and type IIA string theory, a good approach is to identify the parameters that characterize each of them and to explain how they are related. Eleven-dimensional supergravity (and hence M theory, too) has no dimensionless parameters. As we have seen, there are no massless scalar fields, whose vevs could give parameters. The only parameter is the 11d Newton constant, which raised to a suitable power $(-1/9)$, gives the 11d Planck mass m_p. When M theory is compactified on a circle (so that the spacetime geometry is $R^{10} \times S^1$) another parameter is the radius R of the circle. The parameters of type IIA superstring theory are the string mass scale m_s, introduced earlier, and the dimensionless string coupling constant g_s. An important fact about all five superstring theories is that the coupling constant is not an arbitrary parameter. Rather, it is a dynamically determined vev of a scalar field, the *dilaton,* which is a supersymmetry partner of the graviton. With the usual conventions, one has $g_s = \langle e^{\phi} \rangle$.

---

3) It is now clear that this cannot be done in any straightforward manner, since there is no weak coupling limit in which the supermembrane describes all the finite-mass excitations.

We can identify compactified M theory with type IIA superstring theory by making the following correspondences:

$$m_s^2 = 2\pi R m_p^3 \tag{1}$$

$$g_s = 2\pi R m_s. \tag{2}$$

Conventional string perturbation theory is an expansion in powers of g_s at fixed m_s. Equation (2) shows that this is equivalent to an expansion about $R = 0$. In particular, the strong coupling limit of type IIA superstring theory corresponds to decompactification of the eleventh dimension, so in a sense M theory is type IIA string theory at infinite coupling.[4] This explains why the eleventh dimension was not discovered in studies of string perturbation theory.

These relations encode some interesting facts. The fact relevant to eq. (1) concerns the interpretation of the fundamental type IIA string. Earlier we discussed the old notion of double dimensional reduction, which allowed one to derive the IIA superstring world-sheet action from the 11d supermembrane (or M2-brane) world-volume action. Now we can make a stronger statement: The fundamental IIA string actually *is* an M2-brane of M theory with one of its dimensions wrapped around the circular spatial dimension. No truncation to zero modes is required. Denoting the string and membrane tensions (energy per unit volume) by T_{F1} and T_{M2}, one deduces that

$$T_{F1} = 2\pi R T_{M2}. \tag{3}$$

However, $T_{F1} = 2\pi m_s^2$ and $T_{M2} = 2\pi m_p^3$. Combining these relations gives eq. (1). It should be emphasized that all the formulas in this section are exact, due to the large amount of unbroken supersymmetry.

Type II superstring theories contain a variety of p-brane solutions that preserve half of the 32 supersymmetries. These are solutions in which the energy is concentrated on a p-dimensional spatial hypersurface. (Adding the time dimension, the world volume of a p-brane has $p + 1$ dimensions.) The corresponding solutions of supergravity theories were constructed by Horowitz and Strominger [17]. A large class of these p-brane excitations are called *D-branes* (or Dp-branes when we want to specify the dimension), whose tensions are given by [18]

$$T_{Dp} = 2\pi m_s^{p+1}/g_s. \tag{4}$$

This dependence on the coupling constant is one of the characteristic features of a D-brane. It is to be contrasted with the more familiar g^{-2} dependence of soliton masses (e.g., the 't Hooft–Polyakov monopole). Another characteristic feature of D-branes is that they carry a charge that couples to a gauge field in the Ramond-Ramond (RR) sector of the theory. (Such fields can be described as bispinors.)

---

[4] The $E_8 \times E_8$ heterotic string theory is also eleven-dimensional at strong coupling [10].

The particular RR gauge fields that occur imply that even values of p occur in the IIA theory and odd values in the IIB theory.

D-branes have a number of special properties, which make them especially interesting. By definition, they are branes on which strings can end D stands for *Dirichlet* boundary conditions. The end of a string carries a charge, and the D-brane world-volume theory contains a $U(1)$ gauge field that carries the associated flux. When n Dp-branes are coincident, or parallel and nearly coincident, the associated $(p+1)$-dimensional world-volume theory is a $U(n)$ gauge theory. The n^2 gauge bosons A_μ^{ij} and their supersymmetry partners arise as the ground states of oriented strings running from the ith Dp-brane to the jth Dp-brane. The diagonal elements, belonging to the Cartan subalgebra, are massless. The field A_μ^{ij} with $i \neq j$ has a mass proportional to the separation of the ith and jth branes. This separation is described by the vev of a corresponding scalar field in the world-volume theory.

In particular, the D2-brane of the type IIA theory corresponds to our friend the supermembrane of M theory, but now in a background geometry in which one of the transverse dimensions is a circle. The tensions check, because (using eqs. (1), (2), and (4)) $T_{D2} = 2\pi m_s^3/g_s = 2\pi m_p^3 = T_{M2}$. The mass of the first Kaluza–Klein excitation of the 11d supergraviton is $1/R$. Using eq. (2), we see that this can be identified with the D0-brane. More identifications of this type arise when we consider the magnetic dual of the M theory supermembrane. This turns out to be a five-brane, called the M5-brane.[5] Its tension is $T_{M5} = 2\pi m_p^6$. Wrapping one of its dimensions around the circle gives the D4-brane, with tension $T_{D4} = 2\pi R T_{M5} = 2\pi m_s^5/g_s$. If, on the other hand, the M5-frame is not wrapped around the circle, one obtains the so-called NS5-brane of the IIA theory with tension

$$T_{NS5} = T_{M5} = 2\pi m_s^6/g_s^2.$$

This 5-brane, which is the magnetic dual of the fundamental IIA string, exhibits the conventional g^{-2} solitonic dependence.

To summarize, type IIA superstring theory is M theory compactified on a circle of radius $R = g_s \ell_s$. M theory is believed to be a well-defined quantum theory in 11d, which is approximated at low energy by 11d supergravity. Its supersymmetric excitations (which are the only ones known when there is no compactification) are the massless supergraviton, the M2-brane, and the M5-brane. These account both for the (perturbative) fundamental string of the IIA theory and for many of its nonperturbative excitations. The identities presented here are exact, because they are protected by supersymmetry.

IMPORTANT UNRESOLVED ISSUES

One issue that needs to be settled if superstring theory is to be used for phenomenology is where supersymmetry fits into the story. It is clear that at the string

---

[5] In general, the magnetic dual of a p-brane in d dimensions is a $(d-p-4)$-brane.

scale ($\approx 10^{18}$ GeV) or the Planck scale the underlying theory has maximal supersymmetry (32 conserved supercharges). The question that needs to be answered is at what scales they are broken and by what mechanisms. The traditional picture (which looks the most plausible to me) is that at the compactification/GUT scale ($\approx 10^{16}$ GeV) the symmetry is broken to $\mathcal{N} = 1$ in $d = 4$ (four conserved supercharges), and this persists to the TeV scale, where the final susy breaking occurs. The TeV scale is indicated by three separate arguments: the gauge hierarchy problem, supersymmetric grand unification, and the requirement that the lightest superparticle (LSP) be a cosmologically significant component of dark matter. It would be astonishing if this coincidence turned out to be a fluke. There is other support for this picture such as the mass of the top quark and the ease with which it gives electroweak symmetry breaking. Despite all these indications, we cannot be certain that this picture is correct until it is demonstrated experimentally. As I once told a newspaper reporter: discovery of supersymmetry would be more profound than life on Mars.

Another important issue is the problem of vacuum degeneracy and the stabilization of moduli. Let me explain. The underlying theory is completely unique, with no dimensionless parameters. Nevertheless, typical quantum vacua have continuous parameters, called *moduli*, which arise as the vacuum values of scalar fields. Notable examples are the sizes of extra dimensions and the string coupling constant. For typical string vacua, the effective potential has many flat directions, so there is a continuum, or *moduli space*, of minima. The fields that correspond to the flat directions describe massless spin zero particles. These particles typically interact with roughly gravitational strength, which is a problem because the gravitational force is observed to be pure tensor to better than 1% accuracy. So it seems that we should seek a vacuum without moduli, which is very difficult to do. However, if a realistic vacuum of this type is ever found, it will not have any continuously adjustable parameters, and therefore it will be completely predictive (at least in principle).

Perhaps the most challenging unresolved issue of all, is the *cosmological constant.* This is a term in the effective action that describes the energy density of the vacuum, which is observable in a gravitational theory. Observationally, there are indications that it may be nonzero, but it is extremely small. Taking the 1/4 power, the energy scale is $\leq 10^{-11}$ GeV. In a fundamental theory it receives contributions from many sources such as vacuum condensates and zero point energies. Supersymmetry ensures that boson and fermion zero point energies cancel, so the natural scale would seem to be the TeV susy breaking scale, which is many orders of magnitude too high. This is a fine-tuning problem that is reminiscent of the gauge hierarchy problem. Presumably string theory will provide an elegant solution. Until we know what the relevant mechanism is, it is hard to be confident that there is not an alternative to supersymmetry for solving the gauge hierarchy problem. I believe that when the correct solution to the problem of the cosmological constant is found, it will spark another revolution in our understanding. Recently toy models without supersymmetry that seem to have a vanishing cosmological constant have been constructed by Kachru, Kumar, and Silverstein [19]. These models are far from realistic, and have not led yet to new qualitative understandings. However, they are the best lead we have at the present time.

To conclude, there has been dramatic progress in understanding string theory in the past few years, but there are still crucial issues that remain unresolved. Future experimental discoveries will be essential to help guide our thinking. Sooner (Tevatron) or later (LHC) exciting new phenomena are bound to show up. My bet is on Higgs and superparticles. But if I should turn out to be wrong, that would not mean that string theory is wrong.

REFERENCES

1. Schwarz, J.H., hep-th/9807135.
2. Green, M.B., and Schwarz, J.H., *Phys. Lett.* **149B**, 117 (1984).
3. Gross, D.J., Harvey, J.A., Martinec, E., and Rohm, R., *Phys. Rev. Lett.* **54**, 502 (1985).
4. Candelas, P., Horowitz, G.T., Strominger, A., and Witten, E., *Nucl. Phys.* **B258**, 46 (1985).
5. Green, M.B., and Schwarz, J.H., *Phys. Lett.* **109B**, 444 (1982).
6. Green, M.B., and Schwarz, J.H., *Phys. Lett.* **151B**, 21 (1985); *Nucl. Phys.* **B255**, 93 (1985).
7. Salam, A., and Sezgin, E., eds., *Supergravities in Diverse Dimensions*, reprints in 2 vols., World Scientific (1989).
8. Brink, L., Schwarz, J.H., and Scherk, J., *Nucl. Phys.* **B121**, 77(1977); Gliozzi, F., Scherk, J., and Olive, D., *Nucl. Phys.* **B122**, 253 (1977).
9. Cremmer, E., Julia, B., and Scherk, J., *Phys. Lett.* **76B**, 409 (1978).
10. Hořava, P., and Witten, E., *Nucl. Phys.* **B460**, 506 (1996), hep-th/9510209.
11. Witten, E., *Nucl. Phys.* **B443**, 85 (1995), hep-th/9503124.
12. Campbell. C., and West, P., *Nucl. Phys.* **B243**, 112 (1984); Huq, M., and Namazie, M., *Class. Quantum Grav.* **2**, 293 (1985); Giani, F., and Pernici, M., *Phys. Rev.* **D30**, 325 (1984).
13. Bergshoeff, E., Sezgin, E., and Townsend, P.K., *Phys. Lett.* **B189**, 75 (1987).
14. Duff, M.J., Howe, P.S., Inami, T., and Stelle, K.S., *Phys. Lett.* **B191**, 70 (1987).
15. Green, M.B., and Schwarz, J.H., *Phys. Lett.* **136B**, 367 (1984).
16. Townsend, P.K., *Phys. Lett.* **B350**, 184 (1995), hep-th/9501068.
17. Horowitz, G.T., and Strominger, A., *Nucl. Phys.*, 197 **B360** (1991).
18. Polchinski, J., *Phys. Rev. Lett.* **75**, 4724 (1995), hep-th/9510017; p. 293 in *Fields, Strings, and Duality* (TASI 96), eds. C. Efthimiou and B. Greene, World Scientific 1997, hep-th/9611050.
19. Kachru, S., Kumar, J., and Silverstein, E., hep-th/9807076.

New and Improved Superstring Phenomenology[1]

Joseph D. Lykken

Theoretical Physics Dept.
Fermi National Accelerator Laboratory
P.O. Box 500, Batavia, IL 60510

Abstract. Recent developments in string theory have important implications for cosmology. Topics discussed here are inflation, the cosmological constant, smoothing of cosmological singularities, and dark matter from parallel universes.

LOST IN (MODULI) SPACE

The duality revolution which occurred during the past five years has enormously advanced our knowledge and perspective regarding the theory formerly known as superstrings. There is now overwhelming evidence [1] to the effect that there is a unique string theory (often called M theory), which contains, in addition to strings, a variety of membranes and even particle-like objects. In a kind of Planckian democracy, none of these degrees of freedom can truely be regarded as more fundamental than the others. In fact in one limit of this theory (somewhat confusingly called the M theory limit), the string degrees of freedom are completely absent. This recent explosion of our understanding of nonperturbative string physics nicely complements the impressive edifice of perturbative string knowledge built up during the previous decade. All of which leads to the obvious question:

- If we know so much about string theory, why can't we predict anything?

The main reason for this embarrassing irony is that string theory does not have a unique (consistent, stable,) ground state. This was known to be true at the perturbative level for many years, but only recently have we realized that this disturbing property appears to persist even when we bring to bear our full arsenal of nonperturbative string dynamics. At low energies string theory is described (mostly, at least,) by an effective field theory; without a definite choice for the string vacuum we cannot even specify the degrees of freedom of this low energy theory, let alone the form or parameters of the effective lagrangian.

The problem is that we are lost in moduli space. The effective field theory limit of string theory contains a number of scalar fields, called moduli, with flat potentials. This is not surprising since most known string vacua preserve some spacetime supersymmetry, and moduli are a generic feature of supersymmetric field theories. We can parametrize a moduli space by the vacuum expectation values (vevs) of these scalar fields. As we move around in this moduli space, the effective field theory, defined by shifting by these vevs, can vary enormously. Not only are there variations in couplings, but at special points in moduli space the number and type of light degrees of freedom changes.

We have only just begun to feel our way around the intricate tapestry which represents the full moduli space of consistent string vacua. We have probed around the edges, which represent the various possible perturbative limits of string theory, as well as the 11-dimensional (non-stringy) M theory limit. The interior remains largely *terra incognita*, although string duality relations help us trace the threads connecting these different limits.

[1] Talk presented at the International Workshop on Particle Physics and the Early Universe (COSMO-98), 15-20 Nov, Asilomar, Monterey, CA.

CP478 , *COSMO—98*, edited by David O. Caldwell

We know neither the dimensionality nor the topology of the full string moduli space. Nor do we know the connectivity of this space. The tapestry may be very frayed, with many ragged patches connected to the main body by only a few threads; there may even be "string islands", points or patches of moduli space completely disconnected from the main body.

One (at least) of these points ought to correspond to the Standard Model particle physics and FRW cosmology that we observe at low energies and long distance scales. But where?

MODULI AND COSMOLOGY

It is not suprising that the existence of moduli fields has implications for cosmology. Indeed this is true even for approximate moduli, i.e., scalars whose flat potentials are lifted by nonperturbative effects. Since the number and type of moduli vary as we move around in moduli space, most of what we can say about string cosmology is very dependent on where we think the string ground state is.

Certain moduli, however, have slightly more robust characteristics. Perturbative limits of string theory contain a dilaton, a weakly coupled scalar whose vev determines the string coupling, and thereby the basic relationships between the Planck scale, compactification scales, and gauge couplings. These same perturbative limits also contain a pseudoscalar axion, and indeed axions seem to be generic features of large classes of string vacua. Compactified dimensions in string theory can assume a wide variety of geometries; nevertheless certain features of the modulus describing the "overall" scale of compactification are somewhat generic.

Another interesting class of moduli for cosmology are what I will call "invisible" moduli. The vevs of these scalars describe either dimensionless couplings or new mass and length scales associated with hidden sectors-exotic matter and gauge fields which couple only gravitationally to ordinary matter.

The good news for cosmologists is that string moduli provide natural candidates for the scalar fields that may perform some cosmologically important tasks. These include the inflaton and perhaps quintessence. The bad news for cosmologists is that generic regions of string moduli space won't look good cosmologically. Indeed generically moduli are more likely to be a cosmological headache than a panacea. The devil, furthermore, is in the details, and generically these details are difficult to tackle.

The dilaton, for example, has properties which make it an attractive candidate for the inflaton [2–4]. The dilaton acquires a nonvanishing potential only from nonperturbative effects; the relative gradient in this potential is naturally of order the inverse Planck mass, as desired for slow roll inflation. On the other hand, there are a number of problems with dilaton inflation [5–7]. Specific scenarios require additional assumptions about the nonperturbative contributions to the potential; these assumptions are hard to pin down with our current level of knowledge. In many scenarios there is the additional problem that the dilaton kinetic energy dominates the potential energy. Furthermore, near any of the perturbative limits of string theory, the dilaton is generically unstable; it's vev wants to run off to infinity, producing an infinitely weakly coupled theory. One can postulate nonperturbative fixes for this runaway behavior, but such scenarios are neither rigorous nor robust.

STRING ISLANDS AND THE COSMOLOGICAL CONSTANT

The cosmological constant problem is the most notorious and vexing problem of quantum gravity [8]. Any attempt to unify quantum mechanics with gravity leads (at least naively) to the conclusion that quantum fluctuations in the vacuum (i.e. zero-point energies) must couple to gravity. Since these zero-point energy sums are typically divergent, their natural scale in a quantum field theory is some ultraviolet cutoff U. One thus expects to generate a cosmological constant of order

$$\Lambda \sim U^4 \quad , \tag{1}$$

and that the entropy associated with a system of linear size L scales like

$$S \sim L^3 U^3 \quad . \tag{2}$$

Note that Λ is positive if bosonic modes dominate the sum, and negative if fermionic modes dominate. Supersymmetric vacua have zero cosmological constant, due to bose-fermi cancellations.

The cosmological constant problem arises because any reasonable choice for the ultraviolet cutoff scale U leads to a Λ which exceeds the observational upper bound by a ridiculously large multiple, roughly 10^{-100} (here

I am invoking "roughly" as per standard usage in cosmology, meaning order of magnitude in the exponent). This is because particle physics scales such as the Planck mass (10^{19} GeV), the Standard Model Higgs vev (10^2 GeV), and the apparent supersymmetry-breaking scale ($10^2 - 10^3$ GeV), greatly exceed the energy scale characterizing the current matter density of the universe (10^{-3} eV). This is a disturbing problem, made worse by our desire to allow a rather large effective Λ during an earlier inflationary epoch. It should also be noted that recent ideas about quintessence in no way address the main cosmological constant problem; rather quintessence models evolve one very small effective Λ value to another very small effective Λ value.

String theory, which is (if nothing else) a consistent theory of quantum gravity, ought to give us some profound insight to this problem. Unfortunately even with recent advances of the duality revolution, the cosmological constant problem remains a complete mystery even in string theory.

A possible ray of hope is provided by a suggestion of Witten [9], which ties in nicely [10] to some recent work on the idea of "string islands". Witten observed that in 2+1 spacetime dimensions you can have supersymmetry of the vacuum (and thus Λ=0) without supersymmetry of the spectrum (i.e. no bose-fermi degeneracy for particles). Furthermore, string theory in 2+1 dimensions actually becomes string theory in 3+1 dimensions in the limit where the string coupling (determined by the dilaton vev) goes to infinity. This peculiar phenomenon is similar to that which leads to the 11-dimensional M theory limit. The strongly-coupled 2+1 dimensional string theory has light solitons, which actually behave exactly like a set of light Kaluza-Klein modes. These solitonic degrees of freedom thus represent the degrees of freedom of a third spatial dimension compactified on a circle. In the strong coupling limit the radius of this circle becomes infinite, and a 3+1 dimensional theory results.

This suggests a method for finding non-supersymmetric string vacua with zero cosmological constant, by starting with 2+1 dimensional string vacua which contain a dilaton. Note that it is important for this trick that the 2+1 string vacua do not contain any geometrical moduli associated with compactifications from higher dimensions. Such moduli would invalidate the original argument, leading presumably to a nonzero Λ whose scale is set by the square of the bose-fermi mass splittings divided by the compactification scale; this is too large unless we manage to keep all the mass splittings below about 100 GeV.

Thus the pure version of this trick requires string islands (peninsula?): string vacua which contain the dilaton and its axion partner, but do not contain any geometrical moduli. Surprisingly, such string islands are known to exist even in the weakly-coupled limit of the heterotic string [11]. There is, for example, a 3+1 dimensional string vacuum whose low energy limit is pure 3+1 dimensional $N = 4$ supergravity. Many more examples have been constructed recently [12].

I should emphasize that even if one could exhibit string islands corresponding to non-supersymmetric vacua with zero cosmological constant, it is another matter entirely to show that any such vacuum is consistent with the Standard Model. An important conclusion for cosmologists, however, is that current thinking about making string vacua which are more "realistic" seems to favor reduced sets of moduli. A broader conclusion is that in the long run cosmological considerations are likely to play an important role in resolving mysteries about the vacuum state of string theory.

DELIGHTFUL D BRANES

As mentioned above, string theory abounds with membranes of various types and dimensionalities. Of particular interest are D branes, objects which, considered as backgrounds for string propagation, preserve part of the underlying spacetime supersymmetry. D branes have a number of special properties, and occur with various dimensionalities (thus we have D instantons, D particles, D strings, and D membranes with up to 9 spatial dimensions). From the string point of view a D brane is a soliton whose mass (or tension, or mass per unit volume) is proportional to the inverse of the string coupling g_s. This means, among other things, that D branes become light in those regions of string moduli space where the string coupling is large.

This simple fact has led to a new interpretation for singularities of various fixed spacetime backgrounds in which strings propagate. These singularities are associated with compactifying some of the original 9 or 10 spatial dimensions onto orbifolds, conifolds, or other singular geometries. A D brane can wrap around a D-dimensional closed cycle of this compact space; when this cycle is shrunk to a point a singularity appears, associated with the vanishing mass of the wrapped D brane. This observation provides a generic and physically intuitive mechanism for "smoothing" spacetime singularities.

The obvious question, of course, is whether this D brane smoothing also applies to cosmological singularities. Recent work suggests that the answer is yes [13,14], although perhaps not in all cases [15]. This is an exciting avenue for future research.

DARK MATTER IN PARALLEL UNIVERSES

The mass scale at which string physics becomes stringy is known as the string scale, m_s. For the weakly coupled heterotic string, the string scale can be shown to be about 10^{18} GeV, only about an order of magnitude smaller than the Planck mass, m_p. However in other regions of string moduli space the string scale can be much smaller [16]. Since we don't know where we are in moduli space, we also don't know the value of m_s. The most we can say at present is that m_s is greater than about 1 TeV, due to nonobservation of stringy effects in the Tevatron collider experiments [17].

It is tempting to imagine [17] that perhaps the string scale is not too far above the current lower bound, in the multi-TeV region which will eventually be accessible to colliders. If this bold hypothesis is correct, we are immediately faced with the problem of explaining the small ratio m_s/m_p. In string theory this small ratio is presumably related to certain moduli having very large or very small vevs, as measured in units of m_s. If these moduli are "invisible" moduli of the type discussed earlier, then their existence may have no other direct consequences for observable low energy physics.

On the other hand, this small ratio could be a consequence of large compactified dimensions [18,19], through a scaling relation like

$$m_p^2 \sim m_s^{n+2} \, R^n \quad , \tag{3}$$

where R is the size of the large compact dimensions, and n is the number of such dimensions. For m_s of order a TeV and $n \geq 2$, R in the above relation can be as large as 1 mm!

Actually, this form of the large extra dimensions scenario is completely ruled out by particle physics constraints, unless we make an additional bold hypothesis: that the entire Standard Model gauge theory is confined to live on a membrane orthogonal to the large extra dimensions. Since D branes are known to have supersymmetric gauge theories confined to their worldvolumes, this hypothesis fits rather nicely with our current picture of string theory. If correct, the graviton has many massive Kaluza-Klein copies, but the Standard Model particles know of the existence of large extra dimensions only through coupling to gravity. There are, not surprisingly, many interesting cosmological implications of this scenario [19–21].

One intriguing observation is that, if the Standard Model gauge theory is confined to some configuration of branes, then there may be other gauge theories confined to other brane configurations, separated from us in one or more of the large extra dimensions. Such hidden sectors are very much like parallel universes, except that they are gravitationally coupled to the visible universe. If these other "brane-worlds" contain stable matter, planets, stars, galaxies, etc., these will all appear to us as dark matter. Since the laws of (non-gravitational) physics could be quite different in these parallel worlds, qualitatively new forms of macroscopic matter may also be produced. It would be interesting to determine the current observational bounds on (i) dark "planets" in the vicinity of our solar system, (ii) dark "stars" within our galaxy and the galactic halo, and (iii) the density and distribution of dark "galaxies".

REFERENCES

1. J. Schwarz, these proceedings; *gr-qc/9812029*.
2. P. Binetruy and M.K. Gaillard, *Phys. Rev.* **D34**, 3069 (1986).
3. M. Cvetic and R.L. Davis, *Phys. Lett.* **B296**, 316 (1992).
4. T. Banks, M. Berkooz, G. Moore, S. Shenker, P. Steinhardt, *Phys. Rev.* **D52**, 3548 (1995).
5. B. Campbell, A. Linde, and K. Olive, *Nucl. Phys.* **B335**, 146 (1991).
6. R. Brustein and P. Steinhardt, *Phys. Lett.* **B302**, 196 (1993).
7. T. Banks, *hep-th/9601151*.
8. J.D. Cohn, *astro-ph/9807128*.
9. E. Witten, *Int. J. Mod. Phys.* **A10**, 1247 (1995); *Mod. Phys. Lett* **A10** 2153 (1995).
10. S. Kachru and E. Silverstein, *Phys. Rev. Lett.* **80**, 4855 (1998).
11. S. Chaudhuri, G. Hockney, and J. Lykken, *Phys. Rev. Lett.* **75**, 2264 (1995).
12. A. Dabholkar and J.A. Harvey, *hep-th/9809122*.
13. F. Larsen and F. Wilczek, *Phys. Rev.* **D55** 4591 (1997).

14. M. Maggiore and A. Riotto, *hep-th/9811089*.
15. T. Banks, W. Fischler and L. Motl, *hep-th/9811194*.
16. E. Witten, *Nucl. Phys.* **B471** 135 (1996).
17. J. Lykken, *Phys. Rev.* **D54** 3693 (1996).
18. I. Antoniadis, S. Dimopoulos, and G. Dvali, *Nucl. Phys.* **B516** 70 (1998); I. Antoniadis, N. Arkani-Hamed, S. Dimopoulos, and G. Dvali, *Phys. Lett.* **B436** 257 (1998).
19. N. Arkani-Hamed, S. Dimopoulos, and G. Dvali, *hep-ph/9807344*.
20. N. Arkani-Hamed, S. Dimopoulos, and J. March-Russell, *hep-ph/9809124*.
21. A. Linde and N. Kaloper, *hep-th/9811141*.

Testing string theory by probing the pre-bangian Universe

Gabriele Veneziano*

* *Theoretical Physics Division, CERN*
CH - 1211 Geneva 23

Abstract. After recalling why superstring theory suggests a new cosmological principle of "asymptotic past triviality", I will argue that classical (quantum) gravitational instabilities can inflate (warm up) an asymptotic-past-trivial Universe. I will then discuss how near-future observations could provide a window through which we can probe the pre-bangian Universe and thus test string theory both at short and at large distances.

INTRODUCTION

Earlier at this Conference, the pre-big bang (PBB) scenario was quickly dismissed as being: i) misguided; ii) fine-tuned; iii) non-predictive. It so happens that this talk was prepared precisely in such a way as to answer those three "allegations" in the order: i) by arguing that superstring theory's short and large-distance properties strongly motivate the PBB scenario; ii) by showing that inflation follows, without fine-tuning, from generic initial conditions of the PBB type; iii) by claiming that the PBB model is so predictive that, at least in its minimal form, it may soon be ruled out experimentally. These three topics will be dealt with in the following three sections. The last one will present some conclusions.

SUPERSTRING-BASED MOTIVATIONS

Short-distance inspiration

Since the classical (Nambu–Goto) action of a string is proportional to the area

CP478, *COSMO—98*, edited by David O. Caldwell

A of the surface it sweeps, its quantization must introduce a quantum of length λ_s through:

$$S/\hbar = A/\lambda_s^2 \, . \tag{1}$$

By replacing Planck's constant in quantum string theory [1], λ_s plays the role of a minimal observable length, an ultraviolet cut-off. Physical quantities are expected to be bound by appropriate powers of it, e.g.

$$H^2 \sim R \sim G\rho < \lambda_s^{-2} \; , \; k_B T/\hbar < c\lambda_s^{-1} \; , \; R_{comp} > \lambda_s \, . \tag{2}$$

In other words, in quantum string theory (QST), relativistic quantum mechanics should solve the singularity problems in much the same way as non-relativistic quantum mechanics solved the singularity problem of the hydrogen atom by putting the electron and the proton a finite distance apart. Also, QST gives us a framework where we can ask daring questions such as: What was there before the big bang?

Large-distance inspiration

Even at large distance (low-energy, small curvatures), superstring theory does not automatically give Einstein's general relativity. Rather, it leads to a scalar–tensor theory. The new scalar particle/field ϕ, the so-called dilaton, is unavoidable in string theory; it gets reinterpreted as the radius of a new dimension of space in so-called M-theory [2]. The dilaton is massless to all orders in perturbation theory, i.e. as long as supersymmetry remains unbroken. Is the dilaton a problem or an opportunity? Possibly both: while we can try to avoid its potential dangers, we may use some of its properties to our advantage ... Let me discuss how.

In string theory, ϕ controls the strength of all forces, gravitational and gauge alike. One finds, typically:

$$l_P^2/\lambda_s^2 \sim \alpha_{gauge} \sim e^{\phi} \, , \tag{3}$$

expressing the basic unification of all forces in string theory. Note that, in our conventions, the weak-coupling region corresponds to $\phi \ll -1$. In order not to contradict precision tests of the equivalence principle and of the constancy of the gauge and gravitational couplings in the recent past, we require [3] the dilaton to have a mass and to be frozen at the bottom of its own potential *today*. This does not exclude, however, the possibility of the dilaton having evolved cosmologically (after all, the metric did!) within the weak-coupling region, where it was practically massless. The amazing (yet simple) observation [4] is that, by so doing, the dilaton may have inflated the Universe!

A simplified argument, which, although not completely accurate, captures the essential physical point, consists in writing the ($k = 0$) Friedmann equation, $3H^2 =$

$8\pi G\rho$, and in noticing that a growing dilaton (meaning through (3) a growing G) can drive the growth of H even if the energy density of standard matter decreases in an expanding Universe. This new kind of inflation (characterized by growing H and ϕ) has been termed dilaton-driven inflation (DDI). The basic idea of pre-big bang cosmology [4–6] is thus clear: the dilaton started at very large negative values (where it is massless), grew according to its field equations while inflating the Universe, and finally settled, sometime in our recent past, at its present value ($\phi = \phi_0$) corresponding to the minimum of its potential.

DDI is not just possible. It exists as a class of cosmological solutions thanks to the duality symmetries of string cosmology. Under a prototype example of these symmetries, the so-called scale-factor duality [4], an FRW cosmology evolving (at lowest order in derivatives) from a singularity in the past, is mapped into a DDI cosmology going towards a singularity in the future. Of course, the lowest-order approximation breaks down *before* either singularity is reached. A (stringy) moment away from their respective singularities, these two branches should join smoothly to give a single non-singular cosmology (the so-called graceful exit). Since the DDI phase is characterized by growing coupling and curvature, it must itself have originated from a regime in which both quantities were very small. We take this as the main lesson/hint to be learned from low-energy string theory and raise it to the level of a new cosmological principle [7].

Asymptotic Past Triviality

The concept of asymptotic past triviality (APT) is quite similar to that of "asymptotic flatness", familiar from general relativity. The main differences consist in making only assumptions concerning the asymptotic past (rather than future or space-like infinity) and in the additional presence of the dilaton. It seems physically (and philosophically) satisfactory to identify the beginning with simplicity (see e.g. entropy-related arguments concerning the arrow of time). What could be simpler than a trivial, empty and flat Universe? Nothing of course! The problem is that such a Universe, besides being uninteresting, is also non-generic. By contrast, asymptotically flat/trivial Universes are initially simple, yet generic (i.e. not fine-tuned) in the precise mathematical sense that they involve the right number of "integration constants" to describe a general solution. We will ask whether these APT initial data will evolve as to generate a physically interesting big bang-like state at some later time, and argue that it is precisely what should be expected, owing to well-known classical and quantum gravitational instabilities.

INFLATION AS GRAVITATIONAL COLLAPSE

The assumption of APT entitles us to treat the early history of the Universe through the classical field equations of the low-energy (because of the small curvature), tree-level (because of the weak coupling), effective action of string theory. Even then, the problem of determining the properties of a generic solution to the field equations is a formidable one. Very luckily, however, through a field redefinition, we are able to map our problem into one that has been much investigated, both analytically and numerically, in the literature: that of the gravitational collapse of a massless scalar field. Such a system has been considered by many authors, in particular by Christodoulou [8], precisely in the regime of interest to us. In line with the APT postulate, one assumes, in the analogue gravitational collapse problem, very "weak" initial data with the aim of finding under which conditions gravitational collapse later occurs. Gravitational collapse means that the metric (and the volume of 3-space) shrinks to zero at a space-like singularity. However, typically, the dilaton blows up at that same singularity. By undoing the field redefinition one can show that gravitational collapse becomes DDI in the original string-cosmology problem.

How generically does gravitational collapse take place? The singularity theorems by Hawking and Penrose [9] state that, under some general assumptions, singularities are inescapable in GR. All but one of those assumptions are automatically satisfied in the case at hand. Only the existence of a closed trapped surface needs to be imposed. Rigorous results [8] show that this condition cannot be waived: sufficiently weak initial data do not lead to closed trapped surfaces, to collapse, or to singularities. Sufficiently strong initial data do. But where is the borderline? This is not known in general, but precise criteria do exist for particularly symmetric space-times, e.g. for those endowed with spherical symmetry. However, no matter what the general collapse/singularity criterion will eventually turn out to be, we do know that: i) it cannot depend on an over-all additive constant in ϕ; ii) it cannot depend on an over-all multiplicative factor in $g_{\mu\nu}$. This is a simple consequence of the invariance of the effective equations describing the system in this regime under shifts of the dilaton and rescaling of the metric.

We conclude that, generically, some regions of space will undergo gravitational collapse, will form horizons and singularities therein, but nothing, at the level of our approximations, will be able to fix either the size of the horizon or the value of ϕ at the onset of collapse. When this is translated into the original cosmological problem, one is describing, in the region of space-time within the horizon, a period of DDI in which both the initial value of the Hubble parameter and that of ϕ are left arbitrary. These two initial parameters determine the range of validity of our description, since both curvature and coupling increase during DDI and, therefore, the low-energy and/or tree-level description is bound to break down at some point. The smaller the initial Hubble parameter (i.e. the larger the initial horizon size) and the initial coupling, the longer we can follow DDI through the effective equations and the larger the number of reliable e-folds that we gain.

This does answer, in my opinion, the objections raised recently [10] to the PBB scenario according to which it is fine-tuned. The situation here actually resembles that of chaotic inflation [11]. Given some generic (though APT) initial data, we should ask which is the distribution of sizes of the collapsing regions and of couplings therein. Then, only the "tails" of these distributions, i.e. those corresponding to sufficiently large, and sufficiently weakly coupled regions, will produce Universes like ours, the rest will not. A basic difference between the large numbers needed in (non-inflationary) FRW cosmology and the large numbers needed in PBB cosmology should be stressed. In the former, the ratio of two classical scales, i.e. of total curvature to its spatial component, which is expected to be $O(1)$, has to be taken as large as 10^{60}. In the latter, the above ratio is initially $O(1)$ in the collapsing/inflating region, and ends up being very large in that region thanks to DDI. However, the common order of magnitude of these two classical quantities is a free parameter, and is taken to be much larger than a (classically irrelevant) quantum scale.

In conclusion, we may summarize recent progress on the problem of initial conditions by saying that [7]: **Dilaton-driven inflation in string cosmology is as generic as gravitational collapse in general relativity.**

OBSERVABLE RELICS AND HEATING THE PRE-BANG UNIVERSE

PBB relics

Since there are already several review papers on this subject (e.g. [12]), I will limit myself to a short summary:

- For gravitational waves and dilatons, one obtains quite steep spectra [13] and thus small contributions at large scales. The reverse is also true: at short scales, the expected yield in a stochastic background of gravitational waves is much larger than in standard inflationary cosmology, where one expects $\Omega_{GW} < 10^{-14}$. Values of Ω_{GW} in the range of 10^{-6}–10^{-7} are possible in some regions of parameter space, which, according to some estimates of sensitivities [14], could be inside detection capabilities in the near future.
- Vacuum fluctuations of gauge bosons are not amplification in standard cosmology, since a conformally flat metric (of the type forced upon by inflation) decouples from the electromagnetic (EM) field precisely in $D = 3 + 1$ dimensions. As a very general remark, the only background field that can, through its cosmological variation, amplify EM (more generally gauge-field) quantum fluctuations is the effective gauge coupling itself. By its very nature, in the pre-big bang scenario the effective gauge coupling inflates together with space during the PBB phase. It is thus automatic that any efficient PBB inflation

brings together a huge variation of the effective gauge coupling and thus a very large amplification of the primordial EM fluctuations [15–17]. This can possibly provide the long-sought origin for the primordial seeds of the observed galactic magnetic fields. Notice, however, that, unlike GW, EM perturbations interact considerably with the hot plasma of the early (post-big bang) Universe. Thus, converting the primordial seeds into those that may have existed at the proto-galaxy formation epoch is by no means a trivial exercise. Work is in progress to try to adapt existing codes [18] to the evolution of our primordial seeds.

- Fluctuations of the so-called universal axion of string theory, i.e. of the supersymmetric partner of the dilaton, can be large even at large scales [19]. Also, unlike the GW spectrum, that of axions is very sensitive to the cosmological behaviour of internal dimensions during the DDI epoch. On one side, this makes the model less predictive. On the other, it tells us that axions represent a window over the multidimensional cosmology expected generically from string theories, which must live in more that four dimensions. Curiously enough, the axion spectrum becomes exactly HZ (i.e. scale-invariant) when all the nine spatial dimensions of superstring theory evolve in a rather symmetric way [16]. In situations near this particularly symmetric one, axions are able to provide a new mechanism [20] for generating large-scale CMB anisotropy and LSS. This model, being of the isocurvature type, bears some resemblance to the one recently advocated by Peebles [21] and, like his, is expected to contain some calculable amount of non-Gaussianity, as well as some characteristic acoustic peak structure.

- Many other perturbations, which arise in generic compactifications of superstrings, have also been studied, and lead to interesting spectra. For lack of time, I will refer to the existing literature [16,17].

Heat and entropy as a quantum gravitational instability

Before closing this section, I wish to recall how one sees the very origin of the hot big bang from a cold start in the PBB scenario. One can easily estimate the total energy stored in the quantum fluctuations, which were amplified by the pre-big bang backgrounds. The result is, roughly, $\rho_{quantum} \sim N_{eff}\ H^4_{max}$, where N_{eff} is the effective number of species that are amplified and H_{max} is the maximal curvature scale reached. We expect $H_{max} \sim M_s = \lambda_s^{-1}$, and we know that, in heterotic string theory, N_{eff} is in the hundreds. It is tempting to assume [16] that, precisely when the dilaton reaches a value such that $\rho_{quantum}$ is critical, the Universe will enter the radiation-dominated phase. This constraint gives, typically, $e^{\phi_{exit}} \sim 1/N_{eff}$, i.e. a value for the dilaton close to its present value [16]. The entropy density σ in these quantum fluctuations can also be estimated by using some general results [22]. One finds:

$$\sigma \sim N_{eff} H_{max}^3 \,. \tag{4}$$

It is easy to check that, at the assumed time of exit, this entropy saturates a recently proposed bound [23], which also turns out to be a physically acceptable value for the entropy of the Universe just after the big bang: a large entropy ($\sim 10^{90}$) on the one hand, a small entropy for the total mass and size of the observable Universe on the other. Thus, PBB cosmology neatly explains why the Universe, at the big bang, *looks* so fine-tuned, and provides for it a natural arrow of time [23].

CONCLUSIONS

- Pre-big bang cosmology is a "top–down" rather than a "bottom–up" approach to cosmology. This should not be forgotten when testing its predictions.
- It does not need to invent an inflaton, or to fine-tune its potential: inflation is "natural" thanks to the duality symmetries of string cosmology and to its relation to gravitational collapse.
- The problem of initial conditions "decouples" from the singularity problem; it is classical, scale-free, and unambiguously defined. Issues of fine-tuning can be addressed and, I believe, answered.
- PBB cosmology is a tight, highly constrained scenario. As such, it makes a number of predictions:
  - the tensor contribution to $\Delta T/T$ should be very small;
  - some non-Gaussianity in $\Delta T/T$ correlations is expected, and calculable.
  - the axion-seed mechanism should lead to a characteristic acoustic-peak structure, which is being calculated;
  - it should be possible to convert the predicted seed magnetic fields into observables by using some reliable code for their late evolution;
  - a characteristic spectrum of stochastic gravitational waves is expected to surround us, and could be large enough to be measurable within a decade or so.

REFERENCES

1. G. Veneziano, *The Challenging Questions*, Erice, 1989, ed. A. Zichichi (Plenum Press, New York, 1990), p. 199.
2. See, for instance, E. Witten, Nucl. Phys. **B443** (1995) 85;
   P. Horava and E. Witten, Nucl. Phys. **B460** (1996) 506.
3. T.R. Taylor and G. Veneziano, Phys. Lett. **B213** (1988) 459.
4. G. Veneziano, Phys. Lett. **B265** (1991) 287.
5. M. Gasperini and G. Veneziano, Astropart. Phys. **1** (1993) 317, Mod. Phys. Lett. **A8** (1993) 3701, Phys. Rev. **D50** (1994) 2519.
6. An updated collection of papers on the PBB scenario is available at http://www.to.infn.it/~gasperin/.
7. A. Buonanno, T. Damour and G. Veneziano, *Pre-big bang bubbles from the gravitational instability of generic string vacua*, hep-th/9806230, and references therein.
8. D. Christodoulou, Comm. P. Appl. Math. **56** (1993) 1131, and references therein.
9. R. Penrose, Phys. Rev. Lett. **14** (1965) 57; S. W. Hawking and R. Penrose, Proc. Roy. Soc. Lond. **A314** (1970) 529.
10. M. Turner and E. Weinberg, Phys. Rev. **D56** (1997) 4604; N. Kaloper, A. Linde and R. Bousso, Phys. Rev. **D59** (1999) 043508.
11. A. Linde, Phys. Lett. **129B** (1983) 177.
12. G. Veneziano, in *String Gravity and Physics at the Planck Energy Scale*, Erice, 1995, eds. N. Sanchez and A. Zichichi (Kluver Academic Publishers, Boston, 1996), p. 285; M. Gasperini, ibid., p. 305.
13. R. Brustein, M. Gasperini, M. Giovannini and G. Veneziano, Phys. Lett. **B361** (1995) 45; R. Brustein et al., Phys. Rev. **D51** (1995) 6744.
14. P. Astone et al., Phys. Lett. **B385** (1996) 421; B. Allen and R. Brustein, Phys. Rev. **D55** (1997) 970.
15. M. Gasperini, M. Giovannini and G. Veneziano, Phys. Rev. Lett. **75** (1995) 3796; D. Lemoine and M. Lemoine, Phys. Rev. **D52** (1995) 1955.
16. A. Buonanno, K. A. Meissner, C. Ungarelli and G. Veneziano, JHEP **1** (1998) 4;
17. R. Brustein and M. Hadad, Phys. Rev. **D57** (1998) 725.
18. R. M. Kulsrud, R. Cen, J. P. Ostriker and D. Ryu, Ap. J. **480** (1997) 481.
19. E.J. Copeland, R. Easther and D. Wands, Phys. Rev. **D56** (1997) 874;
    E.J. Copeland, J.E. Lidsey and D. Wands, Nucl. Phys. **B506** (1997) 407.
20. R. Durrer, M. Gasperini, M. Sakellariadou and G. Veneziano, Phys. Lett. **B436** (1998) 66; Phys. Rev. **D59** (1999) 043511; M. Gasperini and G. Veneziano, Phys. Rev. **D59** (1999) 043503.
21. P. J. E. Peebles, *An isocurvature CDM cosmogony, I and II*, astro-ph/9805194, and astro-ph/9805212.
22. M. Gasperini and M. Giovannini, Phys. Lett. **B301** (1993) 334; Class. Quant. Grav. **10** (1993) L133; R. Brandenberger, V. Mukhanov and T. Prokopec, Phys. Rev. Lett. **69** (1992) 3606; Phys. Rev. **D48** (1993) 2443.
23. G. Veneziano, *Pre-bangian origin of our entropy and time arrow*, hep-th/9902126, and references therein.

Initial Conditions in Brany and Stringy Cosmology

Nemanja Kaloper[1]

Department of Physics, Stanford University, Stanford, CA 94305-4060

Abstract. We discuss the issue of initial conditions for inflation in models with sub-millimeter internal spaces and in Pre-Big Bang. The models based on sub-millimeter internal spaces are consistent with the inflationary paradigm with their initial conditions if inflation takes place before the stabilization of the internal space. In contrast, the Pre-Big Bang model requires special initial conditions to make our universe.

In this note, we will review the issue of initial conditions in cosmological models which may emerge from theories of branes and strings. We will consider the models of the world on a wall with sub-millimeter internal spaces and the Pre-Big Bang (PBB) scenario. Our conclusions for these two classes of models are rather different. In models with sub-millimeter internal dimensions, the initial universe needs to be smooth over distances which are within the order of magnitude of the string length, provided that inflation takes place before the stabilization of the internal space, which is consistent with the inflationary paradigm. In contrast, in PBB, to produce our universe, the initial universe must be already very large, and smooth at scales which are huge compared to the string length.

I MODELS WITH SUB-MILLIMETER INTERNAL SPACES

There has been considerable interest recently in higher dimensional theories with internal dimensions which are large compared to the fundamental Planck length of the theory [1]- [3]. If the fundamental theory is higher-dimensional, with the unification scale essentially set by gauge theory, the higher-dimensional fundamental Planck scale may be many orders of magnitude smaller than the Planck scale in a world of only four dimensions ($4D$). This is because after the internal space is stabilized, the effective $4D$ Planck mass is given by $M_4^2 \sim \left(\frac{r_0}{l_{D+4}}\right)^D M_{D+4}^2$ [1]. If

[1]) Plenary talk presented at COSMO-98 Worskshop at Asilomar, Monterey, CA, USA, November 15-20, 1998.

CP478 , *COSMO—98*, edited by David O. Caldwell

$\frac{r_0}{l_{D+4}} >> 1$, the reduced Planck mass may appear to be many orders of magnitude larger than the fundamental Planck mass, which is roughly equal to the unification scale of the theory. The hierarchy problem can thus be restated as the question why the internal space is stabilized at some specific radius, which could be large compared to the fundamental scale. The most dramatic proposals [1] postulate the unification scale to be at ~ 1 TeV, thus suggesting that the Great Desert of particle physics is merely a mirage.

In [4,5] it has been argued that having all of inflation occur after compactification may require an extremely light wall inflaton, as compared to the unification scale. Indeed, the effective potential along the wall in this scenario cannot be greater than M^4. Then the Hubble constant during inflation on the wall is given by $H \sim \sqrt{\frac{8\pi V(\phi)}{3M_p^2}} \lesssim \frac{M^2}{M_p}$. Inflation occurs only if the inflaton mass m is smaller than H, which implies the constraint $m \lesssim H \lesssim \frac{M^2}{M_p}$. This bound is completely independent of the number of internal dimensions. For $M \sim 1$ TeV, as proposed in [1], one gets an extremely strong constraint on the inflaton mass, $m \lesssim 10^{-4} eV$ [4,5]. Even if this is allowed, one still one finds very small density perturbations. Alternatively, to fit the COBE data, one needs $m \sim 10^{-10}$ eV using hybrid inflation [7,8].

However, inflation must really occur before the internal dimensions are stabilized. The reason is that the only natural timescale for the beginning of inflation in this model is the fundamental Planck time M^{-1}, when the density of the universe was of the order M^{D+4}. In contrast, at the time of stabilization, the Hubble parameter H at this time is smaller than $M^2/M_p \ll M$, which implies that inflation occurs on a time scale much greater than M^{-1}. Thus the universe must be sufficiently large and homogeneous from the very beginning to survive and not to loose the homogeneity during the long period of time from $t \sim M^{-1}$ to $t \sim H^{-1}$. On the other hand, one cannot simply assume that the universe must be homogeneous at all times, because at the beginning of inflation it must be strongly inhomogeneous: its density at the wall must be many orders of magnitude greater than the density in the bulk. Since M^{-1} is the only natural scale for homogeneity, one can hardly explain from first principles how this specific structure could be formed unless there was a previous stage of inflation, simultaneously in the bulk and on the wall, which could extend the Planck scale M^{-1} to the scale M_p/M^2.

This can be illustrated by a toy model of wall inflation, given by a $3D$ inflating domain wall in a $5D$ space-time. It is supported by the energy-momentum tensor $T^\mu{}_\nu = -\sigma\delta(w)\text{diag}(1,1,1,1,0)$, where σ is the surface tension of the 3D wall in the 5D space-time. We could solve Einstein's equations with this source to determine the 3-brane metric [9,6]:

$$ds_5^2 = (1 - H|w|)^2\Big(-dt^2 + e^{2Ht}(dx^2 + dy^2 + dz^2)\Big) + dw^2 \qquad (1)$$

where H is the Hubble constant along the 3-brane, given by $H = \frac{4\pi\sigma}{3M^3}$. More details can be found in [6].

The solution (1) can be interpreted as an inflating 3-brane, with the event horizon on the brane given by H^{-1}. In the direction transversal to the brane, the metric resembles the Rindler metric, with repulsive gravitational field and Rindler horizon, located at $w = H^{-1}$. Any inflating point on the brane is completely surrounded by an event horizon, at a distance H^{-1} from it. The no-hair theorem for this metric shows that if one perturbs this metric on a scale greater than $O(H^{-1})$, this perturbation is rapidly stretched in all directions by the expansion of the universe, just like in the usual inflationary universe scenario. For example, if instead of the domain wall at the plane $\omega = 0$ one considers a spherical domain wall positioned at $x^2 + y^2 + z^2 + \omega^2 = r^2$ with $r \gg H^{-1}$ inflation will blow up this bubble, stretch its walls, and metric near the wall will be again described by our solution (1). However, if $r \ll H^{-1}$, then the surface tension of the domain wall will shrink its size to zero within the time $t \ll H^{-1}$, and inflation will never happen. This means, as we expected, that in order to find out whether inflation on the wall will happen, we need to properly adjust the initial conditions on a scale H^{-1}. Therefore, unless there is inflation before stabilization of the internal space, it is hard to see how such adjustments can be attained.

In conclusion, we have found evidence that inflation in models $M \sim 1$ TeV must occur when the internal dimensions are small, $\sim M^{-1}$. This is the only possibility to explain the age of the universe at the time of stabilization. When this occurs, the ratio M/M_p may be many orders of magnitude from its poststabilization value, being much closer to unity in the beginning. This can evade the kinematical constraints on the mass of the inflaton [4,5], in effect rendering them completely harmless. The inflationary scenario which accomplishes this in the world-as-a-brane view has been developed in [10,11].

II INITIAL CONDITIONS IN PRE-BIG BANG

The PBB scenario [12] is based on solutions which inflate superexponentially for some time in the string frame. The curvature and the coupling become large, and the evolution proceeds towards the curvature singularity (and not away from it). Assuming that the growth of the curvature and coupling can be stopped, one could match the solution onto a late time Friedmann-Robertson-Walker (FRW) one. The phase of strong curvature would resemble the Big Bang, while the phase before it has been dubbed the PBB [12].

Nevertheless, already in [12] it has been noted that at the onset of the dilaton-dominated inflating phase, the size of the homogeneous domain had to be large in string units, $L_H >> l_s$, and the string coupling had to be small, $g^2 < 1$, but this has not been considered a real problem. More recently, in [13] it was pointed that the other sources of curvature except the dilaton could delay the onset of inflation, and hence require fine tuning. This was disputed in [14–16].

Our analysis [17] has produced precise arguments illustrating the fine tuning problem in PBB. Our results show that if PBB is to solve the horizon and flatness problems, it requires special initial conditions. They are not a solution but just a reformulation of the cosmological problems, and PBB needs to explain them. The details of derivation of these constraints can be found in [17]. The condition for successful inflation can be stated as

$$\left(\frac{\Phi_i}{\Phi_f}\right)^{1/\sqrt{3}} = \left(\frac{|t_i|}{l_s}\right)^{1+1/\sqrt{3}} \geq 10^{30} \tag{2}$$

This constraint can be translated into the following requirements for the horizon, mass, entropy of the universe, and the coupling constant g at the onset of inflation:

$$L_i \geq 2 \times 10^{19} l_s \qquad M_i \geq 10^{72} M_s \qquad S \geq 2 \times 10^{38} \qquad g_0^2 \leq 10^{-53} \tag{3}$$

If the universe at the onset of PBB inflation were to appear accidentally, the probability it would satisfy (3) is minute, $P \leq \exp(-10^{38})$ at best. Further, it would still have to be homogeneous over distances exceeding $10^{19} l_s$, in contrast to the case of standard inflation, which requires homogeneity over length scales of only few l_s. The constraint (2) can be translated in terms of the integral of motion $B = -\dot{\Phi} a^3$:

$$B \geq 4 \times 10^{38} g_0^{-2} \geq 4 \times 10^{91} \tag{4}$$

If PBB is to successfully solve cosmological problems, it must explain the origin of these numbers. Merely saying that these numbers are required for inflation only amounts to restating the cosmological problems in a different fashion, and no more.

After these concerns have been raised, several papers have appeared with more details about these issues. In [18,19], a newest version of the PBB scenario was proposed, which considers the collapse of classical dilaton waves into black holes. However the black holes so formed must be extremely huge, $M \sim 10^{72} M_S$, to fit our universe inside of them. In [20], numerical studies have been conducted on the viability of PBB, and in [21] more general initial conditions have been studied. Other studies of the effect of anisotropies have been done in [22]. The current status is that PBB still needs to input the constraints $g^{-2} \geq 10^{53}$ and $B \geq 10^{38} g^{-2} \geq 10^{91}$. We believe that the need to introduce such extraordinarily large parameters represents a major challenge to the PBB cosmology.

Acknowledgements

The author wishes to thank N. Arkani-Hamed, R. Bousso, S. Dimopoulos, T. Damour, I.I. Kogan, A. Linde, J. March-Russell and G. Veneziano for useful conversations. This work has been supported in part by NSF Grant PHY-9219345.

REFERENCES

1. N. Arkani-Hamed, S. Dimopoulos and G. Dvali, *Phys. Lett.* **B429**, 263 (1998); eprint hep-ph/9807344; I. Antoniadis, N. Arkani-Hamed, S. Dimopoulos and G. Dvali, *Phys. Lett.* **B436**, 257 (1998); N. Arkani-Hamed, S. Dimopoulos and J. March-Russell, eprint hep-th/9809124.

2. K. Dienes, E. Dudas and T. Gherghetta, *Phys. Lett.* **B436**, 55 (1998); hep-ph/9806293; hep-ph/9807522; K. Dienes, E. Dudas, T. Gherghetta and A. Riotto, eprint hep-ph/9809406.
3. J.D. Lykken, *Phys. Rev.* D **54**, 3693 (1996); I. Antoniadis and M. Quiros, *Phys. Lett.* **B392**, 61 (1997); R. Sundrum, eprint hep-ph/9805471; hep-ph/9807348; G. Shiu and S.-H. Tye, *Phys. Rev.* D **58**, 106007 (1998); Z. Kakushadze and S.-H. Tye, eprint hep-th/9809147; C. Bachas, hep-ph/9807415, K. Benakli, eprint hep-ph/9809582; G. Dvali and S.-H. Tye, eprint hep-ph/9812483.
4. K. Benakli and S. Davidson, eprint hep-ph/9810280.
5. D. Lyth, eprint hep-ph/9810320.
6. N. Kaloper and A. Linde, eprint hep-th/9811141.
7. A. Linde, *Phys. Lett.* **B129**, 177 (1983).
8. A. Linde, *Phys. Lett.* **B259**, 38 (1991); *Phys. Rev.* D **49**, 748 (1994).
9. A. Vilenkin, *Phys. Lett.* **B133** 177 (1983); J. Ipser and P. Sikivie, *Phys. Rev.*D **30** 712 (1984).
10. N. Arkani-Hamed, S. Dimopoulos, N. Kaloper and J. March-Russell, eprint hep-ph/9903224.
11. N. Arkani-Hamed, S. Dimopoulos, N. Kaloper and J. March-Russell, talk by Savas Dimopoulos at COSMO-98 and contribution to this volume.
12. G. Veneziano, *Phys. Lett.* B **265**, 287 (1991); M. Gasperini and G. Veneziano, *Astropart. Phys.* **1**, 317 (1993).
13. M.S. Turner and E. Weinberg, *Phys. Rev.* D **56**, 4604 (1997).
14. G. Veneziano, *Phys. Lett.* B **406**, 297 (1997).
15. A. Buonanno, K.A. Meissner, C. Ungarelli and G. Veneziano, *Phys. Rev.* D **57**, 2543 (1998).
16. M. Maggiore and R. Sturani, *Phys. Lett.* B **415**, 335 (1997).
17. N. Kaloper, A. Linde and R. Bousso, *Phys. Rev.* D **59**, 043508 (1999).
18. A. Buonanno, T. Damour and G. Veneziano, eprint hep-th/9806230.
19. A. Ghosh, G. Pollifrone and G. Veneziano, *Phys. Lett.* B **440**, 20 (1998).
20. J. Maharana, E. Onofri and G. Veneziano, *JHEP* **9804**, 004 (1998).
21. T. Chiba, eprint gr-qc/9902003.
22. D. Clancy, J.E. Lidsey and R. Tavakol, *Phys. Rev.* D **58**, 044017 (1998); eprint gr-qc/9806065; A. Feinstein and M.A. Vazquez-Mozo, *Phys. Lett.* B **441**, 40 (1998).

OBSERVATIONAL COSMOLOGY

Cosmological Parameters

Michael S. Turner[1]

Departments of Astronomy & Astrophysics and of Physics
Enrico Fermi Institute, The University of Chicago
Chicago, IL 60637-1433

NASA/Fermilab Astrophysics Center
Fermi National Accelerator Laboratory, Box 500
Batavia, IL 60510-0500

Abstract. The discussion of cosmological parameters used to be a source of embarrassment to cosmologists. Today, measurements of the cosmological parameters are leading the way into the era of precision cosmology. The CMB temperature is measured to four significant figures, $T_0 = 2.7277 \pm 0.002\,$K; the Hubble constant is now determined with a reliable error estimate, $H_0 = (65 \pm 5)\,\mathrm{km\,sec^{-1}\,Mpc^{-1}}$; the mass density of baryons is precisely determined by big-bang nucleosynthesis $\Omega_B = (0.019 \pm 0.001)h^{-2}$; and the age of the Universe inferred from the ages of the oldest stars is $14 \pm 1.5\,$Gyr, which is consistent the expansion age. Further, we have the first full accounting of matter and energy in the Universe, complete with a self consistency check. Expressed a fraction of the critical density it goes like this: neutrinos, between 0.3% and 15%; stars, between 0.3% and 0.6%; baryons (total), $5\% \pm 0.5\%$; matter (total), $40\% \pm 10\%$; smooth, dark energy, $80\% \pm 20\%$; totaling to the critical density (within the errors).

INTRODUCTION

Five years ago "precision cosmology" would have been an oxymoron. That is not true today. Our knowledge of the cosmological parameters reflects that change. Once an embarrassment, they are quickly becoming a source of pride.

The COBE FIRAS team led by John Mather determined the temperature of the cosmic microwave background radiation to four significant figures, $T_0 = (2.7277 \pm 0.00001 \pm 0.002)\,$K. Here, the first error is the error relative to the internal reference black body and the second error, which dominates the error budget, is the uncertainty in determining the temperature of the internal reference black body (Fixsen et al, 1996). Had the on-board thermometers performed better, they might have had another couple of significant figures (which boggles the mind!). In any case, this is a truly impressive measurement, cosmological or otherwise.

---

1) My work is supported by the DoE (at Chicago and Fermilab) and by the NASA through grant NAG 5-7092 at Fermilab.

CP478 , *COSMO—98*, edited by David O. Caldwell

For the first time in thirty years, there is agreement (within the uncertainties) on the value of the Hubble constant and a reliable estimate of the uncertainty. *My summary* of the recent measurements is $H_0 = (65 \pm 5)\,\mathrm{km\,s^{-1}\,Mpc^{-1}}$ (see Freedman, 1999 for a more expert assessment of the situation). The improvement is due in large measure to the Hubble Key Project, which has established Cepheid calibrations for the secondary distance indicators and just as importantly stimulated renewed interest in the examining the distance scale. Systematic uncertainties remain; foremost are the distance to the LMC and reddening and metallicity effects on the Cepheid period-luminosity relation.

Similar progress has been made in dating the age of the oldest objects in the Universe: globular clusters, the heavy elements and white dwarfs (see e.g., Chaboyer, 1998). Allowing for a time of a 2 Gyr until the formation of the first generation of stars, the age of the Universe is $t_0 = (14 \pm 1.5)\,\mathrm{Gyr}$. While the hard lower limit to the age of 10 Gyr derives from all three methods, the central value and error is driven by age dating of globular clusters. Here progress has occurred due to better stellar modelling, improvement in the distance scale and better microphysics. Regarding the distance scale; the parallax distance measurements made by Hipparcos played a role, though, probably more important was the fact that it spurred scrutiny of the entire enterprise.

I devote the rest of this article to a summary of our knowledge of the matter and energy content of the Universe.

AN INVENTORY OF MATTER AND ENERGY

Preliminaries

The quantity and composition of matter and energy in the Universe is of fundamental importance in cosmology. The fraction of the critical energy density contributed by all forms of matter and energy,

$$\Omega_0 \equiv \frac{\rho_{\rm tot}}{\rho_{\rm crit}} = \sum_i \Omega_i\,, \tag{1}$$

determines the geometry of the Universe:

$$R^2_{\rm curv} = \frac{H_0^{-2}}{\Omega_0 - 1}\,. \tag{2}$$

Here, subscript '0' denotes the value at the present epoch, $\rho_{\rm crit} = 3H_0^2/8\pi G \simeq 1.88h^2 \times 10^{-29}\,\mathrm{g\ cm^{-3}}$, Ω_i is the fraction of critical density contributed by component i (e.g., baryons, photons, stars, etc) and $H_0 = 100h\,\mathrm{km\,s^{-1}\ Mpc^{-1}}$. The sign of $R^2_{\rm curv}$ specifies the spatial geometry: positive for a 3-sphere, negative for a 3-saddle and 0 for the flat space.

The present rate of deceleration of the expansion depends upon Ω_0 as well as the composition of matter and energy,

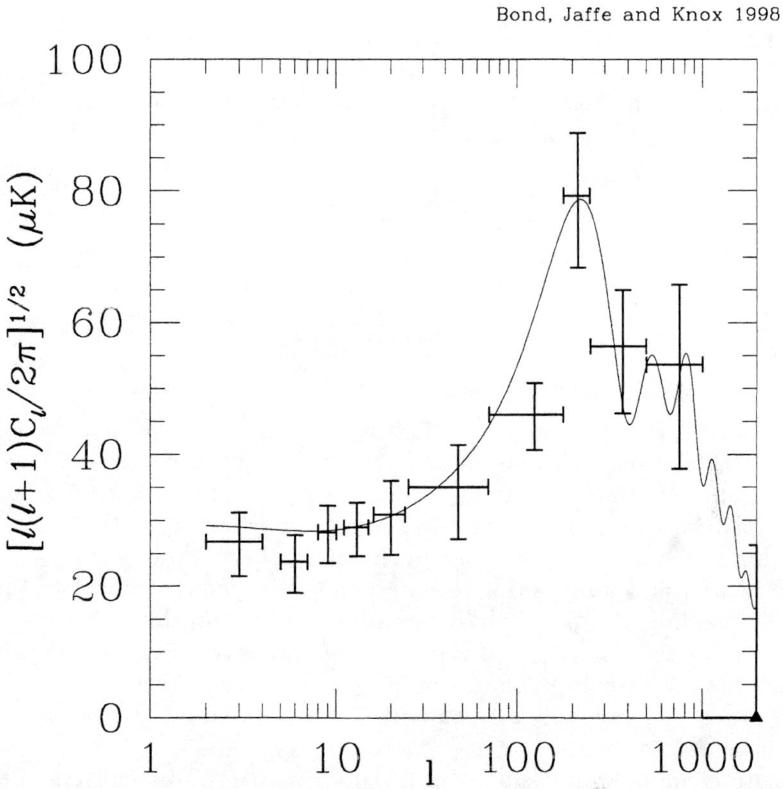

FIGURE 1. Summary of CBR anisotropy measurements, binned to reduce error bars and visual confusion. The temperature variations across the sky have been expanded in spherical harmonics, $\delta T(\theta, \phi) = \sum_i a_{lm} Y_{lm}(\theta, \phi)$ and $C_l \equiv \langle |a_{lm}|^2 \rangle$. In plain language, this plot shows the size of the temperature variation between two points on the sky separated by angle θ (ordinate) vs. multipole number $l = 200°/\theta$ ($l = 2$ corresponds to 100°, $l = 200$ corresponds to $\theta = 1°$, and so on). The theoretical curve is for the ΛCDM model with $H_0 = 65\,\mathrm{km\,s^{-1}\,Mpc^{-1}}$ and $\Omega_M = 0.4$, which is a good fit to the data. Note also the evidence for the first of a series of "acoustic peaks." The presence of these acoustic peaks is a key signature of the density perturbations of quantum origin predicted by inflation (Figure courtesy of L. Knox).

$$q_0 \equiv \frac{(\ddot{R}/R)_0}{H_0^2} = \frac{1}{2}\Omega_0 + \frac{3}{2}\sum_i \Omega_i w_i \,. \tag{3}$$

The pressure of component i, $p_i \equiv w_i \rho_i$; e.g., for baryons $w_i = 0$, for radiation $w_i = 1/3$, and for vacuum energy $w_i = -1$.

The fate of the Universe – expansion forever or recollapse – is not directly determined by Ω_0 and/or q_0. It depends upon precise knowledge of the composition of *all* components of matter and energy, for all times in the future. Recollapse occurs only if there is a future turning point, that is an epoch when the expansion rate,

$$H^2 = \frac{8\pi G}{3}\sum_i \rho_i - \frac{1}{R_{\rm curv}^2}\,, \tag{4}$$

vanishes and

$$\frac{\ddot{R}}{R} = -\frac{4\pi G}{3}\sum_i \rho_i[1 + w_i] \tag{5}$$

is less than zero. In a universe comprised of matter alone, a positively curved universe ($\Omega_0 > 1$) eventually recollapses and a negatively curved universe ($\Omega_0 < 1$) expands forever. However, exotic components complicate matters: a positively curved universe with positive vacuum energy can expand forever, and a negatively curved universe with negative vacuum energy can recollapse.

The quantity and composition of matter and energy in the Universe is also crucial for understanding the past. It determines the relationship between age of the Universe and redshift, when the Universe ended its early radiation dominated era, and the growth of small inhomogeneities in the matter. Ultimately, the formation and evolution of large-scale structure and even individual galaxies depends upon the composition of the dark matter and energy.

Measuring the quantity and composition of matter and energy in the Universe is a challenging task. Not just because the scale of inhomogeneity is so large, around 10 Mpc; but also, because there may be components that remain exactly or relatively smooth (e.g., vacuum energy or relativistic particles) and only reveal their presence by their influence on the evolution of the Universe itself.

Radiation

Because the cosmic background radiation (CBR) is known to be black-body radiation to very high precision (better than 0.005%) and its temperature is known to four significant figures, $T_0 = 2.7277 \pm 0.002\,$K, its contribution is very precisely known, $\Omega_\gamma h^2 = 2.480 \times 10^{-5}$. If neutrinos are massless or very light, $m_\nu \ll 10^{-4}\,$eV, their energy density is equally well known because it is directly related to that of the photons, $\Omega_\nu = \frac{7}{8}(4/11)^{4/3}\Omega_\gamma$ (per species) (there is a small 1% positive correction to this number; see Dodelson & Turner, 1992).

It is possible that additional relativistic species contribute significantly to the energy density, though big-bang nucleosynthesis (BBN) severely constrains the amount (the equivalent of less than 0.4 of a neutrino species; see e.g., Burles et al, 1999) unless they were produced by the decay of a massive particle after the epoch of BBN.

In any case, we can be confident that the radiation component of the energy density today is small. The matter contribution (denoted by Ω_M), consisting of particles that have negligible pressure, is the easiest to determine because matter clumps and its gravitational effects are thereby enhanced (e.g., in rich clusters the matter density typically exceeds the mean density by a factor of 1000 or more). With this in mind, I will decompose the present matter/energy density into two components, matter and vacuum energy,

$$\Omega_0 = \Omega_M + \Omega_\Lambda \, , \tag{6}$$

ignoring the contribution of the CBR and ultrarelativistic neutrinos. I will use vacuum energy as a stand in for a smooth component (more later). Vacuum energy and a cosmological constant are indistinguishable: a cosmological constant corresponds to a uniform energy density of magnitude $\rho_{\rm vac} = \Lambda/8\pi G$.

Total Matter/Energy Density: $\Omega_0 = 1 \pm 0.2$

There is a growing consensus that the anisotropy of the CBR offers the best means of determining the curvature of the Universe and thereby Ω_0, cf., Eq. (2). This is because the method is intrinsically geometric – a standard ruler on the last-scattering surface – and involves straightforward physics at a simpler time (see e.g., Kamionkowski et al, 1994).

At last scattering baryonic matter (ions and electrons) was still tightly coupled to photons; as the baryons fell into the dark-matter potential wells the pressure of photons acted as a restoring force, and gravity-driven acoustic oscillations resulted. These oscillations can be decomposed into their Fourier modes; Fourier modes with $k \sim lH_0/2$ determine the multipole amplitudes a_{lm} of CBR anisotropy. Last scattering occurs over a short time, making the CBR is a snapshot of the Universe at $t_{\rm ls} \sim 300,000\,$yrs. Each mode is "seen" in a well defined phase of its oscillation. (For the density perturbations predicted by inflation, all modes the have same initial phase because all are growing-mode perturbations.) Modes caught at maximum compression or rarefaction lead to the largest temperature anisotropy; this results in a series of acoustic peaks beginning at $l \sim 200$ (see Fig. 1). The wavelength of the lowest frequency acoustic mode that has reached maximum compression, $\lambda_{\rm max} \sim v_s t_{\rm ls}$, is the standard ruler on the last-scattering surface. Both $\lambda_{\rm max}$ and the distance to the last-scattering surface depend upon Ω_0, and the position of the first peak $l \simeq 200/\sqrt{\Omega_0}$. This relationship is insensitive to the composition of matter and energy in the Universe.

CBR anisotropy measurements, shown in Fig. 1, now cover three orders of magnitude in multipole number and involve more than twenty experiments. COBE is the most precise and covers multipoles $l = 2 - 20$; the other measurements come from balloon-borne, Antarctica-based and ground-based experiments using both low-frequency ($f < 100\,\mathrm{GHz}$) HEMT receivers and high-frequency ($f > 100\,\mathrm{GHz}$) bolometers. Taken together, all the measurements are beginning to define the position of the first acoustic peak, at a value that is consistent with a flat Universe. Various analyses of the extant data have been carried out, indicating $\Omega_0 \sim 1 \pm 0.2$ (see e.g., Lineweaver, 1998). It is certainly too early to draw definite conclusions or put too much weigh in the error estimate. However, a strong case is developing for a flat Universe and more data is on the way (Python V, Viper, MAT, Maxima, Boomerang, DASI, and others). Ultimately, the issue will be settled by NASA's MAP (launch late 2000) and ESA's Planck (launch 2007) satellites which will map the entire CBR sky with 30 times the resolution of COBE (around 0.1°) (see Page & Wilkinson, 1999).

Matter: $\Omega_M = 0.4 \pm 0.1$

Baryons

For more than twenty years big-bang nucleosynthesis has provided a key test of the hot big-bang cosmology as well as the most precise determination of the baryon density. Careful comparison of the primeval abundances of D, ^{3}He, ^{4}He and ^{7}Li with their big-bang predictions defined a concordance interval, $\Omega_B h^2 = 0.007 - 0.024$ (see e.g., Copi et al, 1995; for another view, see Hata et al, 1995).

Of the four light elements produced in the big bang, deuterium is the most powerful "baryometer" – its primeval abundance depends strongly on the baryon density ($\propto 1/\rho_B^{1.7}$) – and the evolution of its abundance since the big bang is simple – astrophysical processes only destroy deuterium. Until recently deuterium could not be exploited as a baryometer because its abundance was only known locally, where roughly half of the material has been through stars with a similar amount of the primordial deuterium destroyed. In 1998, the situation changed dramatically, launching BBN into the precision era of cosmology.

Over the past four years there have been claims of upper limits, lower limits, and determinations of the primeval deuterium abundance, ranging from $(\mathrm{D/H}) = 10^{-5}$ to $(\mathrm{D/H}) = 3 \times 10^{-4}$. In short, the situation was confusing. In 1998 Burles and Tytler clarified matters and established a strong case for $(\mathrm{D/H})_P = (3.4 \pm 0.3) \times 10^{-5}$ (Burles & Tytler, 1998a,b). That case is based upon the deuterium abundance measured in four high-redshift hydrogen clouds seen in absorption against distant QSOs, and the remeasurement and reanalysis of other putative deuterium systems. In this enterprise, the Keck I 10-meter telescope and its HiRes Echelle Spectrograph have played the crucial role.

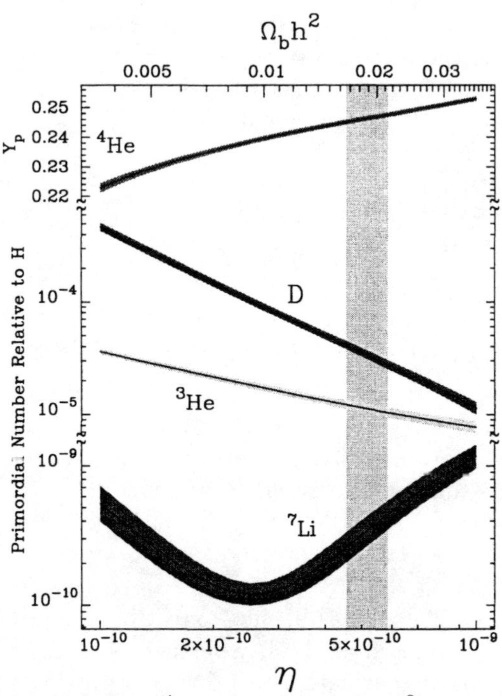

FIGURE 2. Predicted abundances of ^{4}He (mass fraction), D, ^{3}He, and ^{7}Li (number relative to hydrogen) as a function of the baryon density; widths of the curves indicate "2σ" theoretical uncertainty. The dark band highlights the determination of the baryon density based upon the recent measurement of the primordial abundance of deuterium (Burles & Tytler, 1998a,b), $\Omega_B h^2 = 0.019 \pm 0.0024$ (95% cl); the baryon density is related to the baryon-to-photon ratio by $\rho_B = 6.88\eta \times 10^{-22}\,\mathrm{g\,cm^{-3}}$ (from Burles et al, 1999).

The Burles – Tytler measurement turns the previous factor of three concordance range for the baryon density into a 10% determination of the baryon density, $\rho_B = (3.8 \pm 0.4) \times 10^{-31}\,\mathrm{g\,cm^{-3}}$ or $\Omega_B h^2 = 0.02 \pm 0.002$ (see Fig. 2), with about half the error in ρ_B coming from the theoretical error in predicting the BBN yield of deuterium. [A very recent analysis has reduced the theoretical error significantly, and improved the accuracy of the determination of the baryon density, $\Omega_B h^2 = 0.019 \pm 0.0012$ (Burles et al, 1999).]

This precise determination of the baryon density, based upon the early Universe physics of BBN, is consistent with two other measures of the baryon density, based upon entirely different physics. By comparing measurements of the opacity of the Lyman-α forest toward high-redshift quasars with high-resolution, hydrodynamical simulations of structure formation, several groups (Meiksin & Madau, 1993; Rauch et al, 1997; Weinberg et al, 1997) have inferred a lower limit to the baryon density, $\Omega_B h^2 > 0.015$ (it is a lower limit because it depends upon the baryon density squared divided by the intensity of the ionizing radiation field). The second test involves the height of the first acoustic peak: it rises with the baryon density (the higher the baryon density, the stronger the gravitational force driving the acoustic oscillations). Current CBR measurements are consistent with the Burles – Tytler baryon density; the MAP and Planck satellites should ultimately provide a 5% or better determination of the baryon density, based upon the physics of gravity-driven acoustic oscillations when the Universe was 300,000 yrs old. This will be an important cross check of the BBN determination.

Weighing the dark matter

Since the pioneering work of Fritz Zwicky and Vera Rubin, it has been known that there is far too little material in the form of stars (and related material) to hold galaxies and clusters together, and thus, that most of the matter in the Universe is dark (see e.g. Trimble, 1987). Weighing the dark matter has been the challenge. At present, I believe that clusters provide the most reliable means of estimating the total matter density.

Rich clusters are relatively rare objects – only about 1 in 10 galaxies is found in a rich cluster – which formed from density perturbations of (comoving) size around 10 Mpc. However, because they gather together material from such a large region of space, they can provide a "fair sample" of matter in the Universe. Using clusters as such, the precise BBN baryon density can be used to infer the total matter density (White et al, 1993). (Baryons and dark matter need not be well mixed for this method to work provided that the baryonic and total mass are determined over a large enough portion of the cluster.)

Most of the baryons in clusters reside in the hot, x-ray emitting intracluster gas and not in the galaxies themselves, and so the problem essentially reduces to determining the gas-to-total mass ratio. The gas mass can be determined by two methods: 1) measuring the x-ray flux from the intracluster gas and 2) mapping the Sunyaev - Zel'dovich CBR distortion caused by CBR photons scattering off hot electrons in the intracluster gas. The total cluster mass can be determined three independent ways: 1) using the motions of clusters galaxies and the virial theorem; 2) assuming that the gas is in hydrostatic equilibrium and using it to infer the underlying mass distribution; and 3) mapping the cluster mass directly by gravitational lensing (Tyson, 1999). Within their uncertainties, and where comparisons can be made, the three methods for determining the total mass agree (see e.g., Tyson, 1999); likewise, the two methods for determining the gas mass are consistent.

Mohr et al (1998) have compiled the gas to total mass ratios determined from x-ray measurements for a sample of 45 clusters; they find $f_{\rm gas} = (0.075 \pm 0.002)h^{-3/2}$. Carlstrom (1999), using his S-Z gas measurements and x-ray measurements for the total mass for 27 clusters, finds $f_{\rm gas} = (0.06 \pm 0.006)h^{-1}$. (The agreement of these two numbers means that clumping of the gas, which could lead to an overestimate of the gas fraction based upon the x-ray flux, is not a problem.) Invoking the "fair-sample assumption," the mean matter density in the Universe can be inferred:

$$\begin{aligned}\Omega_M = \Omega_B/f_{\rm gas} &= (0.3 \pm 0.05)h^{-1/2}\ ({\rm Xray}) \\ &= (0.25 \pm 0.04)h^{-1}\ ({\rm S-Z}) \\ &= 0.4 \pm 0.1\ (\text{my summary})\,. \end{aligned} \tag{7}$$

I believe this to be the most reliable and precise determination of the matter density. It involves few assumptions, and most of them have now been tested (clumping, hydrostatic equilibrium, variation of gas fraction with cluster mass).

Supporting evidence for $\Omega_M = 0.4 \pm 0.1$

This result is consistent with a variety of other methods that involve different physics. For example, based upon the evolution of the abundance of rich clusters with redshift, Henry (1999) finds $\Omega_M = 0.45 \pm 0.1$ (also see, Bahcall & Fan, 1998 and N. Bahcall, 1999). Dekel and Rees (1994) place a low limit $\Omega_M > 0.3$ (95% cl) derived from the outflow of material from voids (a void effectively acts as a negative mass proportional to the mean matter density).

The analysis of the peculiar velocities of galaxies provides an important probe of the mass density averaged over very large scales (of order several hundred Mpc). By comparing measured peculiar velocities with those predicted from the distribution of matter revealed by redshift surveys such as the IRAS survey of infrared galaxies, one can infer the quantity $\beta = \Omega_M^{0.6}/b_I$ where b_I is the linear bias factor that relates the inhomogeneity in the distribution of IRAS galaxies to that in the distribution of matter (in general, the bias factor is expected to be in the range 0.7 to 1.5; IRAS galaxies are expected to be less biased, $b_I \approx 1.$). Recent work by Willick & Strauss (1998) finds $\beta = 0.5 \pm 0.05$, while Sigad et al (1998) find $\beta = 0.9 \pm 0.1$. The apparent inconsistency of these two results and the ambiguity introduced by bias preclude a definitive determination of Ω_M by this method. However, Dekel (1994) quotes a 95% confidence lower bound, $\Omega_M > 0.3$, and the work of Willick & Strauss seems to strongly indicate that Ω_M is much less than 1; both are consistent with $\Omega_M \sim 1$.

Finally, there is strong, but circumstantial, evidence from structure formation that Ω_M is around 0.4 and significantly greater than Ω_B. It is based upon three different lines of reasoning. First, there is no viable model for structure formation without a significant amount of nonbaryonic dark matter. The reason is simple: in a baryons-only model, density perturbations grow only from the time of decoupling, $z \sim 1000$, until the Universe becomes curvature dominated, $z \sim \Omega_B^{-1} \sim 20$; this is not enough growth to produce all the structure seen today with the size of density perturbations inferred from CBR anisotropy. With nonbaryonic dark matter and $\Omega_M \gg \Omega_B$, dark-matter perturbations begin growing at matter – radiation equality and continue to grow until the present epoch, or nearly so, leading to significantly more growth and making the observed large-scale structure consistent with the size of the density perturbations inferred from CBR anisotropy.

Second, the transition from radiation domination at early times to matter domination at around 10,000 yrs leaves its mark on the shape of the present power spectrum of density perturbations, and the redshift of matter – radiation equality depends upon Ω_M. Measurements of the shape of the present power spectrum based upon redshift surveys indicate that the shape parameter $\Gamma = \Omega_M h \simeq 0.25 \pm 0.05$ (see e.g., Peacock & Dodds, 1994). For $h \sim 2/3$, this implies $\Omega_M \sim 0.4$. (If there are relativistic particles beyond the CBR photons and relic neutrinos, the formula for the shape parameter changes and $\Omega_M \gg 0.4$ can be accommodated; see Dodelson et al, 1996). Third, the Ly-α forest, seen in the absorption spectra of high-redshift quasars, gives us a unique view of the Universe at a time when structure was just becoming nonlinear. Based upon a comparison of high-resolution N-body simulations with 28 Keck HIRES spectra, Weinberg et al (1998) infer $\Omega_M = 0.34 \pm 0.1$.

Mass-to-light ratios and Ω_M

The most mature approach to estimating the matter density involves the use of mass-to-light ratios, the measured luminosity density, and the deceptively simple equation,

$$\langle \rho_M \rangle = \langle M/L \rangle \mathcal{L} \,, \tag{8}$$

where $\mathcal{L} = 2.4h \times 10^8 \, L_{B\odot} \, \mathrm{Mpc}^{-3}$ is the measured (B-band) luminosity density of the Universe. Once the average mass-to-light ratio for the Universe is determined, Ω_M follows by dividing it by the critical mass-to-light ratio, $(M/L)_{\rm crit} = 1200h$ (in solar units). Though it is tantalizingly simple – and it is far too easy to take any measured mass-to-light ratio and divide it by $1200h$ – this method does not provide an easy and reliable method of determining Ω_M.

Based upon the mass-to-light ratios of the bright, inner regions of galaxies, $(M/L)_* \sim$ few, the fraction of critical density in stars (and closely related material) has been determined, $\Omega_* \simeq (0.003 \pm 0.001)h^{-1}$ (see e.g., Faber & Gallagher, 1979). Persic & Salucci (1992) derive a similar value based upon the observed stellar-mass function. Luminous matter accounts for only a tiny fraction of the total mass density and only about a tenth of the baryons.

CNOC (Carlberg et al, 1996, 1997) have done a very careful job of determining a mean cluster mass-to-light ratio, $(M/L)_{\rm cluster} = 240 \pm 50$, which translates to an estimate of the mean matter density, $\Omega_{\rm cluster} = 0.20 \pm 0.04$. Because clusters contain thousands of galaxies and cluster galaxies do not seem *radically* different from field galaxies, one is tempted to take this estimate of the mean matter density seriously. However, it is significantly smaller than the value I advocated earlier, $\Omega_M = 0.4 \pm 0.1$. Which estimate is right?

I believe the higher number, based upon the cluster baryon fraction, is more reliable and that we should be surprised that the CNOC number is so close, closer than we had any right to expect! After all, only a small fraction of galaxies are found in clusters and the luminosity density $\mathcal{L}$ itself evolves strongly with redshift and corrections for this effect are large and uncertain. (We are on the tail end of star formation in the Universe: 80% of star formation took place at a redshift greater than unity.) While the value for Ω_M derived from the cluster baryon fraction also relies upon clusters, the underlying assumption is far weaker and much more justified, namely that clusters provide a fair sample of matter in the Universe.

Even if mass-to-light ratios were measured in the red (they typically are not), where the light is dominated by low-mass stars and reflects the integrated history of star formation rather than the current star-formation rate as it does in the blue, one would still require the fraction of baryons converted into stars in clusters to be identical to that in the field to have agreement between the CNOC estimate and that based upon the cluster baryon fraction. Apparently, the fraction of baryons converted into stars in the field and in clusters is similar, but not identical.

To put this in perspective and to emphasize the shortcomings of the mass-to-light technique, had one used the cluster mass-to-x-ray ratio and the x-ray luminosity density, one would have inferred $\Omega_M \sim 0.05$. A factor of two discrepancy based upon optical mass-to-light ratios does not seem so bad. Enough said.

Missing energy found!

The results $\Omega_0 = 1 \pm 0.2$ and $\Omega_M = 0.4 \pm 0.1$ are in apparent contradiction, suggesting that one or both are wrong. However, prompted by a strong belief in a flat Universe, theorists have explored the remaining logical possibility: a dark, exotic form of energy that is smoothly distributed and contributes 60% of the critical density (Turner et al, 1984; Peebles, 1984). Being smoothly distributed its presence would not have been detected in measurements of the matter density. The properties of this missing energy are severely constrained by other cosmological facts, including structure formation, the age of the Universe, and CBR anisotropy. So much so, that a smoking-gun signature for the missing energy was predicted (see e.g., Turner, 1991): accelerated expansion.

In 1998 evidence for accelerated expansion was presented in the form of the magnitude – redshift (Hubble) diagram for fifty-some type Ia supernovae (SNe Ia) out to redshifts of nearly 1. Two groups, the Supernova Cosmology Project (Perlmutter et al, 1998) and the High-z Supernova Search Team (Riess et al, 1998; Schmidt et al, 1998), working independently and using different methods of analysis, each found evidence for accelerated expansion. Perlmutter et al (1998) summarize their results as a constraint to a cosmological constant (see Fig. 4),

$$\Omega_\Lambda = \frac{4}{3}\Omega_M + \frac{1}{3} \pm \frac{1}{6}. \tag{9}$$

For $\Omega_M \sim 0.4 \pm 0.1$, this implies $\Omega_\Lambda = 0.85 \pm 0.2$, or just what is needed to account for the missing energy!

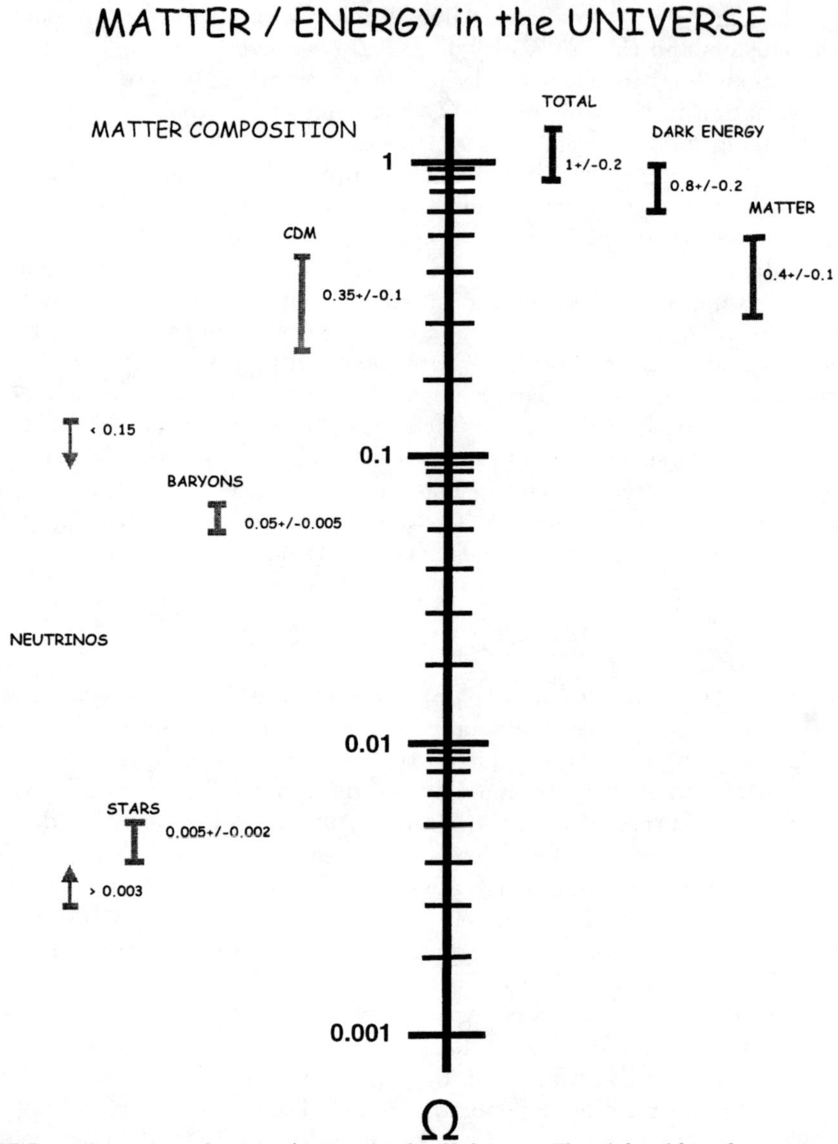

FIGURE 3. Summary of matter/energy in the Universe. The right side refers to an overall accounting of matter and energy; the left refers to the composition of the matter component. The contribution of relativistic particles, CBR photons and neutrinos, $\Omega_{\rm rel}h^2 = 4.170 \times 10^{-5}$, is not shown. The upper limit to mass density contributed by neutrinos is based upon the failure of the hot dark matter model of structure formation (White, Frenk & Davis, 1983; and Dodelson et al, 1996) and the lower limit follows from the evidence for neutrino oscillations (Fukuda et al, 1998). Here H_0 is taken to be $65\,{\rm km\,s^{-1}\,Mpc^{-1}}$.

Recently, two other studies, one based upon the x-ray properties of rich clusters of galaxies (Mohr et al, 1999) and the other based upon the properties of double-lobe radio galaxies (Guerra et al, 1998), have reported evidence for a cosmological constant (or similar dark-energy component) that is consistent with the SN Ia results (i.e., $\Omega_\Lambda \sim 0.7$).

There is another powerful test of an accelerating Universe whose results are more ambiguous. It is based upon the fact that the frequency of multiply lensed QSOs is expected to be significantly higher in an accelerating universe (Turner, 1990). Kochanek (1996) has used gravitational lensing of QSOs to place a 95% cl upper limit, $\Omega_\Lambda < 0.66$; and Waga and Miceli (1998) have generalized it to a dark-energy component with negative pressure: $\Omega_X < 1.3 + 0.55w$ (95% cl), both results for a flat Universe. On the other hand, Chiba and Yoshii (1998) claim evidence for a cosmological constant, $\Omega_\Lambda = 0.7^{+0.1}_{-0.2}$, based upon the same data. From this I conclude: 1) lensing excludes Ω_Λ larger than 0.8, and 2) when larger objective surveys of gravitational-lensed QSOs are carried out (e.g., the Sloan Digital Sky Survey), there is the possibility of uncovering another smoking-gun for accelerated expansion.

By far, the strongest evidence for dark energy is the SN Ia data. The statistical errors reported by the two groups are smaller than possible systematic errors. Thus, the believability of the results turn on the reliability of SNe Ia as one-parameter standard candles. SNe Ia are thought to be associated with the nuclear detonation of Chandrasekhar-mass white dwarfs. The one parameter is the rate of decline of the light curve: The brighter ones decline more slowly (the so-called Phillips relation; see Phillips, 1993). The lack of a good theoretical understanding of this (e.g., what is the physical parameter?) is offset by strong empirical evidence for the relationship between peak brightness and rate of decline, based upon a sample of nearby SNe Ia. It is reassuring that in all respects studied, the distant sample of SNe Ia appear to be similar to the nearby sample. For example, distribution of decline rates and dispersion about the Phillips relationship. The local sample spans a range of metallicity, likely spanning that of the distant sample, and further, suggesting that metallicity is not an important second parameter.

At this point, it is fair to say that if there is a problem with SNe Ia as standard candles, it must be subtle. Cosmologists are even more inclined to believe the SN Ia results because of the preexisting evidence for a "missing-energy component" that led to the prediction of accelerated expansion.

Cosmic concordance

With the SN Ia results we have for the first time a complete and self-consistent accounting of mass and energy in the Universe (see Fig. 3). The consistency of the matter/energy accounting is illustrated in Fig. 4. Let me explain this exciting figure. The SN Ia results are sensitive to the acceleration (or deceleration) of the expansion and constrain the combination $\frac{4}{3}\Omega_M - \Omega_\Lambda$. (Note, $q_0 = \frac{1}{2}\Omega_M - \Omega_\Lambda$;

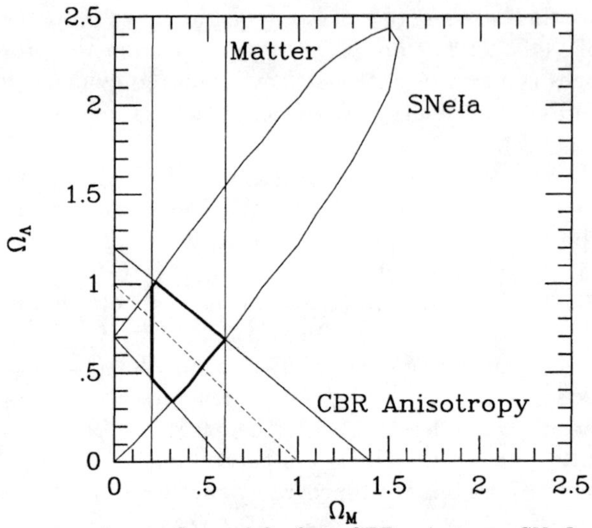

FIGURE 4. Two-σ constraints to Ω_M and Ω_Λ from CBR anisotropy, SNe Ia, and measurements of clustered matter. Lines of constant Ω_0 are diagonal, with a flat Universe shown by the broken line. The concordance region is shown in bold: $\Omega_M \sim 1/3$, $\Omega_\Lambda \sim 2/3$, and $\Omega_0 \sim 1$. (Particle physicists who rotate the figure by 90° will recognize the similarity to the convergence of the gauge coupling constants.)

$\frac{4}{3}\Omega_M - \Omega_\Lambda$ corresponds to the deceleration parameter at redshift $z \sim 0.4$, the median redshift of these samples). The (approximately) orthogonal combination, $\Omega_0 = \Omega_M + \Omega_\Lambda$ is constrained by CBR anisotropy. Together, they define a concordance region around $\Omega_0 \sim 1$, $\Omega_M \sim 1/3$, and $\Omega_\Lambda \sim 2/3$. The constraint to the matter density alone, $\Omega_M = 0.4 \pm 0.1$, provides a cross check, and it is consistent with these numbers. Cosmic concordance!

But there is more. We also have a consistent and well motivated picture for the formation of structure in the Universe, ΛCDM. The ΛCDM model, which is the cold dark matter model with $\Omega_B \sim 0.05$, $\Omega_{\rm CDM} \sim 0.35$ and $\Omega_\Lambda \sim 0.6$, is a very good fit to all cosmological constraints: large-scale structure, CBR anisotropy, age of the Universe, Hubble constant and the constraints to the matter density and cosmological constant (Krauss & Turner, 1995; Ostriker & Steinhardt, 1995; Turner, 1997).

CONCLUDING REMARKS

The cosmological parameters T_0, H_0, t_0, and Ω_i are now all measured with reliable and respectable error estimates. This is a remarkable achievement for a science about which it used to be said, "the errors are always in the exponents." Already, the consistency between H_0, t_0 and Ω_i provides an important crosscheck on the consistency of the hot big-bang framework, which it passes with flying colors.

Over the next two decades we can expect to see even more progress. After the MAP and Planck missions and the completion of the Sloan Digital Sky Survey, it is not unreasonable to expect a 1% determination of H_0 and Ω_0, and a 5% determination of Ω_M and Ω_B, with significant crosschecks. In addition, we should have a much better understanding of the mysterious dark energy. The COBE FIRAS measurement of T_0 has set a standard that is unlikely to be exceeded for a long time. These precision measurements of the cosmological parameters will test the big-bang framework and general relativity as well as testing the paradigm that aspires to extend it, Inflation + Cold Dark Matter.

These are exciting times in cosmology!

REFERENCES

1. Bahcall, N. 1999, astro-ph/9901076.
2. Bahcall, N. & Fan, X. 1998, Astrophys. J., 504, 1.
3. Burles, S. & Tytler, D. 1998a, Astrophys. J., 499, 699.
4. Burles, S. & Tytler, D. 1998b, Astrophys. J., 507, 732.
5. Burles, S., Nollett, K., Truan, J. & Turner, M.S. 1999, Phys. Rev. Lett., in press.
6. Carlberg, R. et al 1996, Astrophys. J., 436, 32.
7. Carlberg, R. et al 1997, Astrophys. J., 478, 462.
8. Carlstrom, J. 1999, Physica Scripta, in press.
9. Chaboyer, B. 1998, Phys. Rep., in press (astro-ph/9808200).
10. Chiba, M. & Yoshii, Y. 1998, Astrophys. J., in press (astro-ph/9808321).
11. Copi, C.J., Schramm, D.N. & Turner, M.S. 1995, Science, 267, 192.
12. Dekel, A. 1994, Ann. Rev. Astron. Astrophys., 32, 371.
13. Dekel, A. & Rees, M. 1994, Astrophys. J. Lett., 422, L1.
14. Faber, S. & Gallagher, J. 1979, Ann. Rev. Astron. Astrophys., 17, 135.
15. Fixsen, D.J. et al 1996, Astrophys. J., 473, 576.
16. Freedman, W. 1999, Physica Scripta, in press.
17. Fukuda, Y. et al (SuperKamiokande Collaboration) 1998, Phys. Rev. Lett., 81, 1562.
18. Guerra, E.J., Daly, R.A. & Wan, L. 1998, Astrophys. J., submitted (astro-ph/9807249)
19. Hata, N. et al 1995, Phys. Rev. Lett., 75, 3977.
20. Henry, P. 1999, in this volume.
21. Kamionkowski, M., Spergel, D.N. & Sugiyama, N. 1994, Astrophys. J. Lett., 426, L57.
22. Kochanek, C. 1996, Astrophys. J., 466, 638.
23. Krauss, L. & Turner, M.S., 1995, Gen. Rel. Grav., 27, 1137.
24. Lineweaver, C. 1998, Astrophys. J. Lett., 505, 69.
25. Mohr, J., Mathiesen, B. & Evrard, A.E. 1998, Astrophys. J., submitted.
26. Mohr, J. et al 1999, in preparation.
27. Ostriker, J.P. & Steinhardt, P.J. 1995, Nature
28. Page, L. & Wilkinson, D.T. 1999, Rev. Mod. Phys. 71, S173.

29. Peebles, P.J.E. 1984, Astrophys. J., 284, 439.
30. Perlmutter, S. et al 1998, Astrophys. J., in press (astro-ph/9812133).
31. Persic, M. & Salucci, P. 1992, Mon. Not. R. astron. Soc., 258, 14p.
32. Phillips, M. M. 1993, Astrophys. J. Lett., 413, L105.
33. Riess, A. et al 1998, Astron. J., 116, 1009.
34. Schmidt, B. et al 1998, Astrophys. J., 507, 46.
35. Sigad, et al 1998, Astrophys. J., 495, 516.
36. Trimble, V. 1987, Ann. Rev. Astron. Astrophys., 25, 425.
37. Turner, E.L. 1990, Astrophys. J. Lett., 365, L43.
38. Turner, M.S. 1983, Phys. Rev. D, 28, 1243.
39. Turner, M.S. 1991, Physica Scripta, T36, 167.
40. Turner, M.S. 1997, in Critical Dialogues in Cosmology, ed. N. Turok (World Scientific, Singapore), p. 555.
41. Turner, M.S. 1998, Pub. Astron. Soc. Pac., in press.
42. Turner, M.S., Steigman, G. & Krauss, L. 1984, Phys. Rev. Lett., 52, 2090.
43. Tyson, J. A. 1999, Physica Scripta, in press.
44. Weinberg, D. et al 1997, Astrophys. J., 490, 546.
45. Waga, I. & Miceli, A.P.M.R. 1998, astro-ph/9811460.
46. White, S.D.M., Frenk, C. & Davis, M. 1983, Astrophys. J. Lett., 274, L1.
47. White, S.D.M. et al, 1993, Nature 366, 429.
48. Willick, J. & Strauss, M. 1998, Astrophys. J., 507, 64.

Cosmological Parameters from Supernovae: Two Groups' Results Agree

Saul Perlmutter* [1] and Adam Riess† [2]

*Lawrence Berkeley National Laboratory and the Center for Particle Astrophysics, U.C. Berkeley
†Department of Astronomy, U.C. Berkeley
Berkeley, California 94720

Abstract. Measurements of type Ia supernova apparent brightnesses over a range of distances allows a measurement of changes in the universe's expansion rate. This measurement yields information about the mass density and vacuum energy density ("cosmological constant") of the universe. We present results from two research groups that are in excellent agreement. Both groups' results indicated with high statistical confidence that the Universe is currently accelerating ($q_0 < 0$) and that the cosmological constant (Ω_Λ) makes an important contribution to the total energy density. We discuss the techniques involved in these measurements, and the systematic uncertainties due to supernova evolution, grey extinction, weak lensing, selection bias, sample contamination, a local void, and methodology. To date, none of the sources of systematic uncertainty that have been identified and estimated can reconcile the data with $\Omega_\Lambda = 0$ and $q_0 > 0$.

Introduction: Summary of the Technique

In this presentation we will speak for two different supernova groups, both of which are measuring the cosmological parameters using supernovae. This way we can use our time most effectively to present the whole story, particularly because both groups are reaching the same conclusion. You can find more details about each group's work in the

---

[1] for the Supernova Cosmology Project: G. Aldering, B.J. Boyle, P.G. Castro, W.J. Couch, S. Deustua, S. Fabbro, R.S. Ellis, A.V. Filippenko, A. Fruchter, G. Goldhaber, A. Goobar, D.E. Groom, I.M. Hook, M. Irwin, A.G. Kim, M.Y. Kim, R.A. Knop, J.C. Lee, T. Matheson, R.G. McMahon, H.J.M. Newberg, C. Lidman, P. Nugent, N.J. Nunes,R. Pain, N. Panagia, C.R. Pennypacker, S. Perlmutter, R. Quimby, P. Ruiz-Lapuente, B. Schaefer, and N. Walton.
[2] for the High-Z Supernova Search: P. Challis, A. Clocchiatti, A. Dierks, A.V. Filippenko, P. Garnavich, R. Gilliand, C. Hogan, S. Jha, R. Kirshner, B. Leibundgut, M. Phillips, D. Reiss, A. Riess, B. Schmidt, R. Schommer, C. Smith, J. Spyromilio, C. Stubbs, N. Suntzeff, and J. Tonry.

CP478, *COSMO—98*, edited by David O. Caldwell
1999 American Institute of Physics 1-56396-853-3

publications listed on their Web pages: `http://supernova.lbl.gov` and `http://cfa-www.harvard.edu/cfa/oir/Research/supernova/HighZ.html`.

Supernovae make a particularly elegant cosmological tool. By measuring their brightnesses and redshifts it is possible to determine the cosmological parameters in a rather straightforward way. One way to describe the measurement is that the brightness can be taken as an indicator of how far away the supernova is—and hence how far back in time we are looking—while the redshift can tell you how much the universe has expanded since that time. (While the photons of light are traveling to us in an expanding universe they stretch to longer wavelength by just the same proportion as the universe has stretched.) Thus in a "Hubble diagram" plot of redshift against magnitude (brightness) of supernovae we can look for different curvatures depending on whether the universe was expanding faster or slower in the past. If it was expanding faster in the past (i.e. the redshifts are larger than expected for a given supernova brightness), then the expansion has been decelerating, presumably due to the gravitational effect of more mass density in the universe. Conversely, if there was less redshift in the past than expected, this would be due either to less mass density (if the deceleration is less than expected) or perhaps to a cosmological constant (if there is in fact no deceleration, but an acceleration).

The brightness of a supernova is of course a changing quantity. The supernova explosion appears as a brightening point of light that rises to its peak brightness in a few weeks and fades away in a few months. We will be discussing the brightest of the supernova types, the type Ia supernova, and generally using its peak brightness whenever we refer to the brightness of a supernova.

The idea of using supernovae for cosmological measurements goes back over 60 years. Baade [1], in 1938, first pointed out that supernovae are very similar in brightness to each other and hence could provide a tool for cosmological measurements. In the late '70s, Tammann [2] and Colgate [3] both suggested that type I supernovae might be used to measure the deceleration parameter, once the Space Telescope became available. In the 1980s, after the sub-classification of type Ia supernova was recognized, an intensive search for high redshift supernovae was conducted by a Danish team in Chile. However, after two years of intensive searching it was only possible to find and follow a single type Ia supernova, and even that one was discovered weeks after its maximum light [4].

It was thus obvious that type Ia supernovae presented a real problem as an experimental tool: they are rare – only two explode in a typical galaxy per millennium; they are random – so it is particularly difficult to schedule telescope time on the largest telescopes (which must be scheduled 6 months in advance); and they vary rapidly – they are difficult to catch before they reach their maximum light. The Supernova Cosmology Project first developed a technique to make these supernovae useful cosmological tools [5]. By observing many wide fields in a couple of nights, it is possible to image tens of thousands of galaxies. If one returns three weeks later and observes the exact same galaxies, it is possible to guarantee the discovery of a dozen supernovae in these new images. Extensive image analysis software makes it possible to hunt through the tens of thousands of galaxies and find the spots of light that were bright in the second set of images that were not seen three weeks earlier. With only a three-week gap between the two sets of images almost all of the supernovae are found on their way up towards their maximum brightness. By scheduling the first set of images just after new moon and the second set of images just before the next new moon, it is possible then to guarantee a "batch" of supernovae all ready to be observed while they are still brightening just before new moon so they can be followed with both spectroscopy and photometry during the dark time of the month, the best time for observations on the large telescopes.

This strategy worked well. Each semester for the past four years a new batch of supernova could be discovered and followed. Both groups have been using this strategy to generate data sets of high redshift supernovae, with lightcurves and spectroscopy [6–8]. In fact, the technique has been so successful in guaranteeing supernova discoveries that it has even been possible to begin scheduling the Hubble space telescope weeks in advance. (For the Hubble space telescope, it is necessary to predict in what square degree of sky the supernovae will be discovered—an apparently impressive feat of prognostication.)

Width-Luminosity Relation and K Corrections

To put these high redshift supernovae on the Hubble plot of redshift vs. apparent magnitude, it is necessary to understand two complications. First, the supernovae are not all identical standard candles, that is, they do not all reach exactly the same brightness at maximum light. A few are a little bit brighter and a few are a little bit fainter than the typical supernovae. In the early 1990s, Mark Phillips [9] pointed out that this range of brightness could be recognized by the timescale of the supernova's lightcurve. Phillips characterized this timescale by the rate of decline – the magnitude drop in the first 15 days – which he called Δm_{15}. He found that this decline rate accurately predicted the brightness of the supernovae. Since then Riess, Press, and Kirshner [10,11] demonstrated an elegant statistical technique to characterize this timescale, called the "lightcurve shape method." This technique adds or subtracts different amounts of a "correction template," thus adding to or taking away from the shoulders of a standard lightcurve template, and broadening the timescale or shrinking the timescale. With this approach, the amount of correction template added or subtracted predicts the brightness of the supernova. Perlmutter et al. [5,6] has used a third approach to characterizing this lightcurve timescale, called the "stretch factor." It appears that the supernova lightcurves can all be fit almost perfectly by linearly stretching or compressing the time axis of a standard lightcurve template [12]. With this approach, the linear stretch factor on the timescale, s, predicts the brightness of the supernova.

Not only does the timescale of the supernova work in a single filter color, but it is also possible to characterize the lightcurve shape in multiple colors. This means that if one knows the timescale (e.g. the stretch factor) of a given supernova one can predict the color on a given day in its lightcurve. By comparing this expected color to the observed color, it is possible to measure how much the supernova has been reddened by dust, thus indicating how much it has been dimmed by dust [11]. After correcting a heterogeneous group of low redshift supernovae (whose relative distances can be measured from their redshift) using this lightcurve timescale correction we find a remarkably homogeneous corrected sample. All three approaches to the lightcurve timescale can give dispersions as low as approximately 10 percent in brightness, after correcting for this width luminosity relationship and the reddening/extinction of the supernovae by dust. This remarkably good calibration has been shown using low-redshift supernovae discovered by the Calan/Tololo supernova survey [13], as well as with an extensive new set of low-redshift supernova lightcurves [14] observed and studied at the Mt. Hopkins telescopes.

The other complication in plotting the high redshift and low redshift supernovae on a single graph comes about because of the different portion of the spectrum of a supernova that can be seen at low redshift through a blue filter and at high redshift through a red filter. Although we can observe very similar portions of the redshifted spectrum in this way, there are small differences that must be taken into account. We calculate these differences, called the "cross-filter K correction" [15] (cf. [16] and [17] for discussions of single-filter K corrections) by multiplying the filter transmission function with the spectrum of the supernova chosen for the appropriate day in its lightcurve and for the appropriate stretch factor of its lightcurve. We find that it is possible to keep the uncertainties in this calculation below two percent [18].

Separating Ω_M and Ω_Λ

We are now ready to look for the cosmological parameters on the plot of redshift vs. "corrected" apparent magnitude—corrected for lightcurve width and extinction. However, we do not know if any change in curvature on this plot is due to the mass energy density (Ω_M) in the universe or the vacuum energy density (Ω_Λ)—the cosmological constant—working against the mass density. Goobar and Perlmutter [19] proposed an approach to separate these two effects. The idea is that at a given redshift the apparent magnitude of a supernova picks out a diagonal strip in the plane of cosmological constant vs. mass density. However, since the mass density and the cosmological constant enter into the equation for apparent magnitude with different powers of redshift, we find that this diagonal strip rotates as supernovae are measured at higher redshifts. Thus it is possible to look for an overlap between measurements at different redshifts and begin to narrow in on a unique value for the cosmological constant and for the mass density – not simply a linear combination. (See [20] for a first demonstration of this approach with supernovae at $z \sim 0.4$ and $z \sim 0.8$.)

Results

We can see the results of this measurement as we put more and more supernovae onto the plot of magnitude vs. redshift as shown in Figure 1. The Ω_M-versus-Ω_Λ results from supernovae measured by the Supernova Cosmology Project [21] are shown on the left panel of Figure 2, while the results from the High-Z Supernovae Search group [22] are shown on the right panel.

It is important to notice that on the plane of mass density vs. cosmological constant we see that whether the universe expands forever or recollapses eventually is mostly determined by the sign of the cosmological constant. (There is one small corner of this plane in which the mass density is high enough to cause the universe to recollapse before the universe expands enough to the dilute the mass density and make it insignificant compared to the vacuum energy density.) Most of us who have taken a course in cosmology in the past 25 years has been taught that a spatially closed universe is a universe that will recollapse eventually, while a spatially open universe will expand forever. However it is clear on this plot that once the cosmological constant has been introduced into the problem all four possibilities can be found, i.e. the universe can be closed and yet expand forever and vice versa.

The most important conclusions from these graphs are as follows: first, if the universe is flat then there is very strong and clear evidence for a positive cosmological constant. The high redshift supernovae are on average approximately 40 percent fainter than they would appear if the universe were flat with a zero cosmological constant. Second, these measurements provide a direct determination of the age of the universe in units of the inverse Hubble constant. As shown in Figure 3, the confidence regions lie almost parallel to the isochrons of constant age. They pick out a value of about 14.5 billion years for a Hubble constant of 63, or approximately 15 billion years in the case of a flat universe. You will also notice that the confidence region is almost entirely on the accelerating side of the $q_0 = 0$ line. (The exact linear combination of mass density and cosmological constant that is most accurately measured is not the same linear combination as the deceleration parameter, $q_0 \equiv \Omega_M/2 - \Omega_\Lambda$. For the redshift range of this sample, the measurement is most sensitive to the linear combination $0.8\Omega_M - 0.6\Omega_\Lambda$.)

Finally, there is still strong statistical evidence (99 percent confidence) for a cosmological constant in the case of an open universe, that is, with no constraint for the universe to be flat. However, in this case the high-redshift supernovae are only 20 percent or so fainter than for a low-mass $\Lambda = 0$ universe. Since this difference is still statistically significant, the main concern is the systematic error in the measurement, which we discuss next.

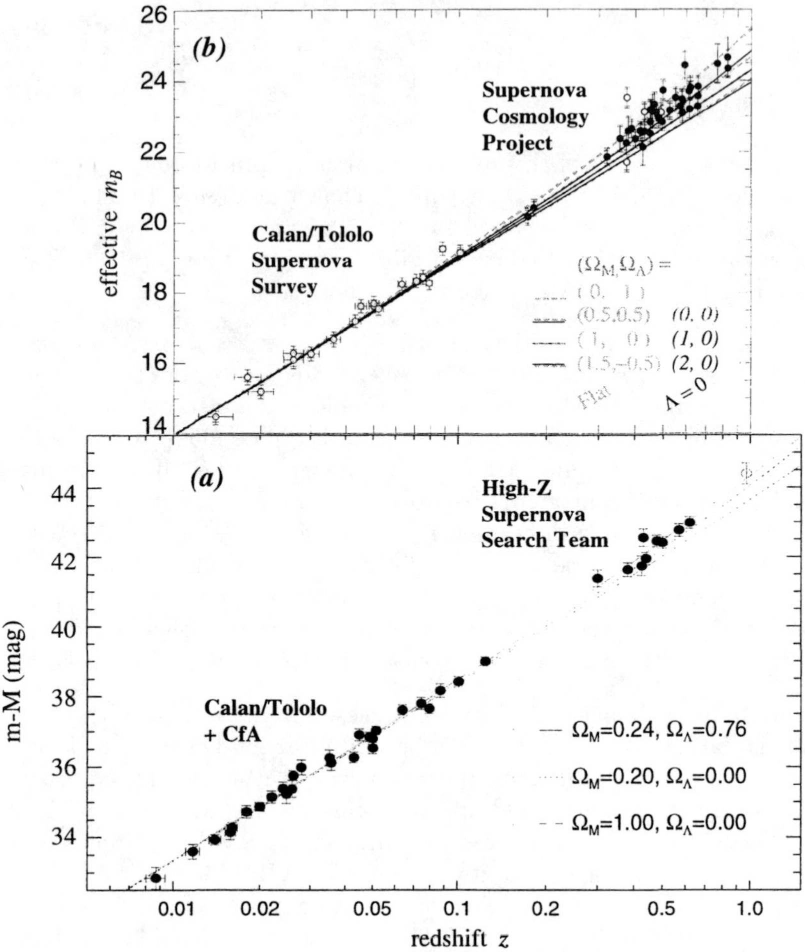

FIGURE 1. Hubble diagram (a) for 10 high-redshift Type Ia supernovae from the High-Z Supernova Search, and 27 low-redshift Type Ia supernovae from both the Calán/Tololo Supernova Survey and the CfA supernova survey; and (b) for 42 high-redshift Type Ia supernovae from the Supernova Cosmology Project, and 18 low-redshift Type Ia supernovae from the Calán/Tololo Supernova Survey. All magnitudes are corrected for the SN Ia lightcurve width-luminosity relation. The curves in panel (a) are for $(\Omega_M, \Omega_\Lambda) = (0.24, 0.76)$ on top, $(0.2, 0)$ in middle, and (1,0) on bottom. The solid curves in panel (b) are the theoretical $m_B^{\text{effective}}(z)$ for a range of cosmological models with zero cosmological constant: $(\Omega_M, \Omega_\Lambda) = (0, 0)$ on top, $(1, 0)$ in middle and (2,0) on bottom. The dashed curves are for a range of flat cosmological models: $(\Omega_M, \Omega_\Lambda) = (0, 1)$ on top, $(0.5, 0.5)$ second from top, $(1, 0)$ third from top, and (1.5,-0.5) on bottom.

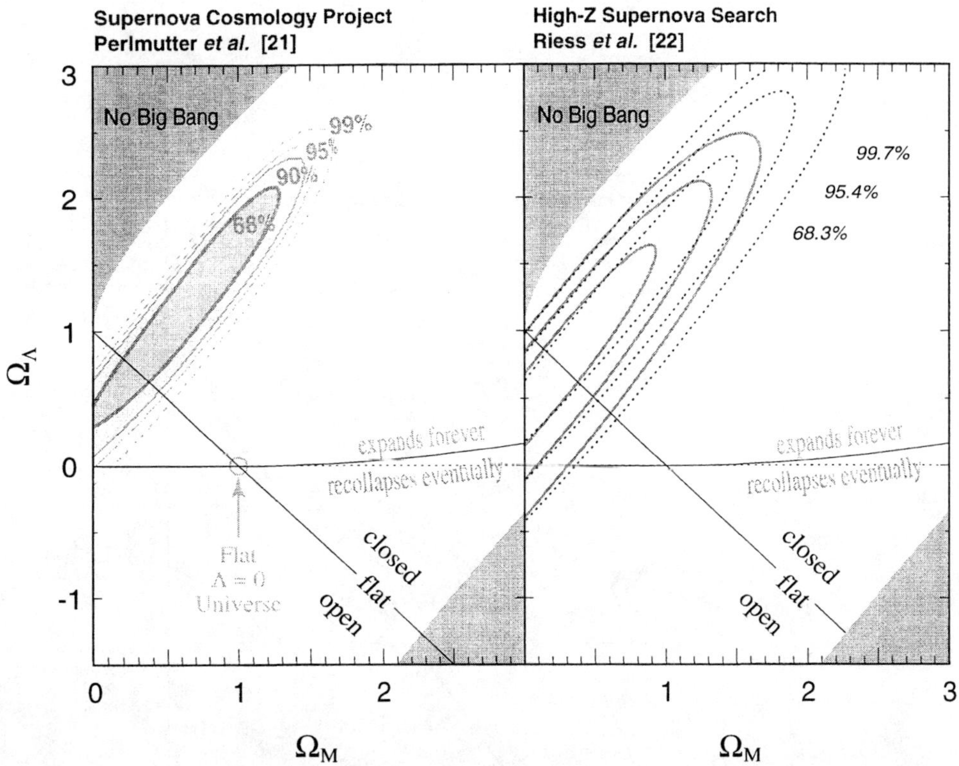

FIGURE 2. Best-fit confidence regions in the Ω_M–Ω_Λ plane for (left panel) the Supernova Cosmology Project results and for (right panel) the High-Z Supernova Search results. The dotted contours in the right panel represent the results excluding one supernova at z=0.97 for which there is not a spectrum to confirm the Type Ia supernova identification. These plots from the two groups are in excellent agreement. (A minor caveat here is that these results are not *completely* independent, since the two groups' analyses share many low-redshift Calan/Tololo supernovae with relatively high statistical weight and the result shown in the right panel includes two high-redshift supernovae from the Supernova Cosmology Project, one low-weighted and one with typical statistical weight.) Note that the spatial curvature of the universe—open, flat, or closed—is not determinative of the future of the universe's expansion, indicated by the near-horizontal solid line. In cosmologies above this near-horizontal line the universe will expand forever, while below this line the expansion of the universe will eventually come to a halt and recollapse. This line is not quite horizontal because at very high mass density there is a region where the mass density can bring the expansion to a halt before the scale of the universe is big enough that the mass density is dilute with respect to the cosmological constant energy density. The upper-left shaded region, labeled "no big bang," represents "bouncing universe" cosmologies with no big bang in the past. The lower right shaded region corresponds to a universe that is younger than the oldest heavy elements, for any value of $H_0 \geq 50$ km s^{-1} Mpc^{-1}.

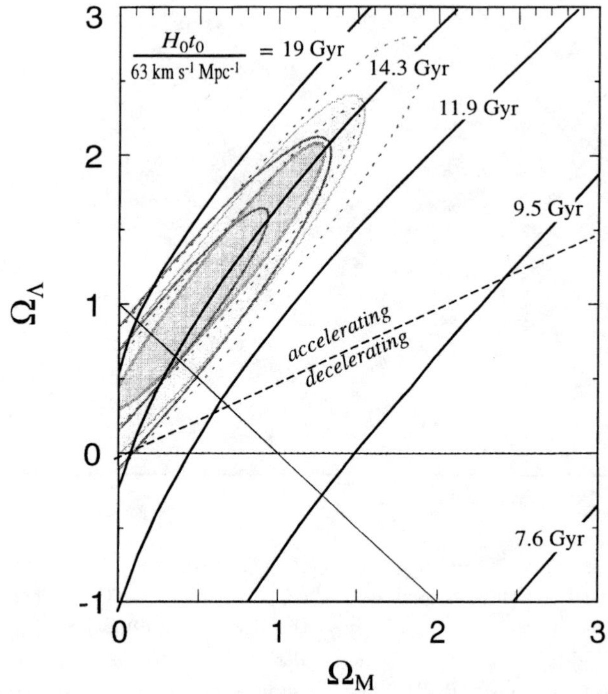

FIGURE 3. Isochrones of constant $H_0 t_0$, the age of the universe relative to the Hubble time, H_0^{-1}, with the best-fit confidence regions in the Ω_M–Ω_Λ plane for both the Supernova Cosmology Project (filled contours) and the High-Z Supernova Search (open contours and dashed contours) results. The isochrones are labeled for the case of $H_0 = 63$ km s^{-1} Mpc^{-1}, representing a typical value found from studies of SNe Ia. If H_0 were taken to be 10% larger, the age labels would be 10% smaller. The diagonal line labeled accelerating/decelerating is drawn for $q_0 \equiv \Omega_M/2 - \Omega_\Lambda = 0$, and divides the cosmological models with an accelerating or decelerating expansion at the present time.

Systematic Uncertainties or What could be wrong with the supernova measurements?

Given the startling implications of the results for the energy content of the Universe, both teams have been very wary of possible systematic errors which could otherwise explain the results we are seeing. In other words, what could be wrong with the supernova measurements?

The candidates we consider most likely to be causing systematic errors (and which keep us up at night the longest) are: evolution, extinction (both interstellar and intergalactic), weak lensing, selection bias, a local void, sample contamination, and the analysis approach. We will here discuss each of these. In considering the nature and size of possible systematic errors it is important to keep in mind the size of the effect or margin both teams see. The high-redshift type Ia supernovae are observed to be 20 to 30 percent dimmer than expected in a low mass Universe ($\Omega_M = 0.2$) *without a positive cosmological constant*, i.e. one which is not accelerating [21,22]. The size of the effect is much larger in a flat cosmology.

Evolution

Evolution in astronomy occurs when an astrophysical object changes its characteristics over the course of the lifetime of the Universe. Woe to the astronomer who has relied on the universality of such characteristics as a measuring tool only to find that they evolve. Such evolution plagued previous workers who tried to infer the global deceleration rate from brightest cluster galaxies in the 1970's [23].

Fortunately, SNe Ia are events which occur on stellar scales, not galaxian scales and are therefore less likely to be subject to the known evolution of stellar populations. In the nearby sample, SNe Ia are observed to occur in both early and late type galaxies. These galaxies span a larger range of metallicity, stellar age, and interstellar environments than is expected to occur for galaxies back to a redshift of a half. Yet after correction for the light curve shape/luminosity relationship and extinction, no significant Hubble diagram residuals are seen which correlate with host galaxy morphology. This evidence suggests that our distance estimates are insensitive to variations in the supernova progenitor environment [8].

Both teams have been impressed by the tremendous similarity of the observed characteristics of low-redshift and high-redshift SNe Ia. The most striking similarities are revealed in the spectra. Figure 4 (left panel) shows the impressive homogeneity of 5 spectra SNe Ia at approximately 5 days before maximum. One of these objects is a high-redshift SN Ia, but if the labels were covered it would not be possible to distinguish this object from any irregularity of its features [22].

More rigorous tests show that statistically speaking, this high-redshift SN Ia is indistinguishable from its low-redshift brethren. At even higher redshifts (Figure 4, right panel) the same result seems to hold [20]. A supernova spectrum is a complex result of the abundance, velocity, and temperature of the ashes of the burnt material

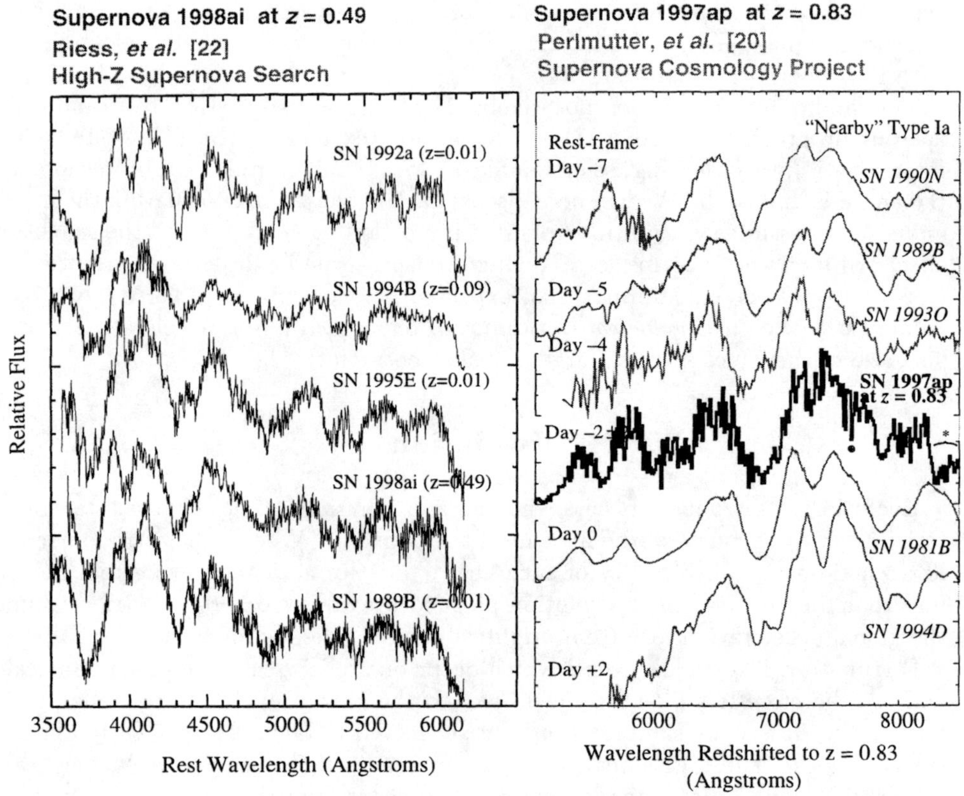

FIGURE 4. Comparisons of low-redshift and high-redshift SN Ia spectra. The left panel compares a variety of low-redshift supernovae spectra, all observed approximately 5 days before maximum light, to a high-redshift $z = 0.49$ supernovae (second from bottom) also observed approximately 5 days (in the supernova's restframe) before maximum. The right panel shows a high-redshift $z = 0.83$ supernova spectrum, observed approximately 2 days (in the supernova's restframe) before maximum, set into a time series of low-redshift supernova spectra. (Note that the left panel plots the spectra as a function of the restframe wavelength, while the right panel redshifts all the spectra as they would appear at redshift $z = 0.83$.) **Right panel: reprinted with permission from Nature 391, 51, Copyright 1998 Macmillan Magazines Limited.**

at the photosphere. It is difficult to imagine (or calculate) a way to retain the primary features of the SN Ia spectrum while altering the explosion energy or peak luminosity by about 20% to 30%. A joint endeavor by both groups has further demonstrated that the time evolution of SN Ia spectra is also indistinguishable at high and low redshifts. Successive spectra provide information about increasingly deeper layers of the supernova (like a CAT scan) and so this test shows high and low-redshift SNe Ia to be structurally similar as well. In addition, both groups have found that the distribution of high-redshift light curves and colors could have been drawn from the low redshift population [21,22]. One of us (AGR) is also working on a sensitive test of evolution by comparing the time interval between explosion and maximum light for high-redshift SNe Ia from the Supernova Cosmology Project [12] and low-redshift SNe Ia from amateur astronomers' images.

The best theoretical models also suggest that little evolution is expected among SNe Ia over the look back time to a redshift of a half. Hoflich, Wheeler, & Thielemann [24] have calculated synthetic spectra of SNe Ia with solar and one-third solar metalicities and have found little difference between these models over optical wavelengths where both teams have observed the SNe Ia. Unfortunately, current lack of understanding of the progenitors of SNe Ia as well as uncertainties about the explosion characteristics and synthesized composition severely detracts from any theoretical indications that SNe Ia do not evolve.

Lensing

Weak lensing can cause high-redshift SNe Ia to appear dimmer than expected, but the impact of this effect on the data is likely to be negligible. Most lines of sight through the Universe are relatively empty compared to the mean and typically result in a relative deamplification of light. The typical deamplification of SN Ia light is only significant when material in galaxies is contained in MACHOs and when the overall mass density of the Universe is high [25]. Yet, as seen in Figures 1 and 2, the SNe Ia data constrain differences between Ω_M and Ω_Λ so that the high values of Ω_M that are required to make lensing important also demand even higher values of Ω_Λ. For the case of $\Omega_M = 0.2$ at a redshift of a half, if all the matter was contained in MACHO's, the typical deamplification would only be about 0.02 mag or 1% in distance. However, weak lensing by compact objects does skew the observed distribution of SN Ia luminosities and suggests a way to directly measure the mean ratio of compact-to-diffuse material in intervening galaxies. Both groups are interested in pursuing this "Cosmological MACHO Experiment" in the future.

Extinction with Reddening and Gray Extinction

When we try to think of a non-cosmological explanation for the dimming of supernova light we invariably turn to a consideration of a famous pitfall of optical astronomy: extinction. Both teams currently measure SN Ia colors to correct for mean SN Ia luminosity. To test this possibility we discarded all SNe with an ambiguous classification and found no change in our conclusions [21,22].

Finally, we note that two independent teams with their own supernovae and analysis methods and have arrived at a nearly identical result (see Figure 3) from this experiment [21,22]. This suggests to us that the mechanics of the measurement have been performed correctly and that the results reflect the outcome that anyone can expect from this experiment.

Our account ledger of known systematic effects results in a total budget of less than 3% distance uncertainty. This is far smaller than the 10%-15% shift in distance (i.e. 20%-30% in brightness) needed to disregard the cosmological constant and an accelerating Universe.

An Unforgeable Signature of Cosmology

But what about systematics which we are either unaware of or are difficult to quantify like evolution? There is a unique signature of the cosmology predicted by our data which is difficult to forge by systematic errors. If the Universe is currently accelerating under the influence of a cosmological constant, this is a relatively recent phenomenon. At greater redshifts when the Universe was smaller, gravitating mass dominated the dynamics of the Universe. Thus, measurements of SNe Ia at redshifts greater than unity will reveal the era of deceleration when the cosmological constant was relatively unimportant. If, however, we have been led astray by a systematic effect, it is likely that this effect will continue to grow with increasing redshift and the data will show a large divergence from the reported cosmological scenario. The size of this difference at a redshift of one is about as large as the effect of a cosmological constant appears to be at a redshift of a half. We therefore eagerly await future observations such as the lightcurve measurements–currently in progress–for the supernova recently discovered by the Supernova Cosmology Project at $z = 1.2$ [28].

Other Sources of Evidence for a Flat Universe

Finally, it is important to remember that while we focused most of our attentions on these small potential systematic errors, the strongest conclusion—in the flat universe case—is a much larger effect. There are other sources of cosmological evidence which suggest that the flat universe case is in fact the more interesting one. In particular, the evidence from the cosmic microwave background is steadily becoming more relevant to this problem. By identifying the position of the first "Doppler" peak in the CBR power spectrum, it is beginning to be possible to rule out the low-mass-density zero-cosmological-constant solution. (See [29], [30], [31], and [32]).

Problems with a Cosmological Constant and Alternatives

There are two aspects of a positive cosmological constant that are disturbing from a theoretical point of view. First, why should the cosmological constant be non-zero and yet so small? If one makes a rough estimate of the vacuum energy density expected from quantum mechanical zero-point energies at the scale of supersymmetry breaking, one would find a cosmological constant approximately 51 orders of magnitude too large. The longtime assumption has been that some cancellation must occur to bring this number down to zero. Now, presumably, this cancellation would have to be precise to a factor of 10^{51}, but then not complete to bring the cosmological constant to zero. The other surprise is that the effect of the current mass energy-density of the universe on the expansion rate is now comparable (within a factor of $\sim$3) to the effect of the vacuum energy-density. Over the lifetime of the universe, the mass density has been dropping by orders of magnitude—as the cube of the scale of the universe—while the vacuum energy density has stayed constant. Why are they now so close?

For both of these reasons there has been an attempt made to find alternative sources of energy density besides the vacuum energy density. Dynamical scaler fields have been proposed which might be recognized by a different equation of state, that is, a different ratio of pressure to density (see, e.g., [33]). We can set limits on this equation of state in the case of a flat universe using the supernovae. The results from both teams suggest an equation of state between -0.6 and -1, with the somewhat more negative equations of state corresponding to mass densities somewhat larger than 0.2 [21,31].

ACKNOWLEDGMENTS

The observations described in this paper were obtained as visiting/guest astronomers at the Cerro Tololo Inter-American Observatory 4-meter telescope, operated by the National Optical Astronomy Observatory under contract to the National Science Foundation; the Keck I and II 10-m telescopes of the California Association for Research in Astronomy; the Wisconsin-Indiana-Yale-NOAO (WIYN) telescope; the European Southern Observatory 3.6-meter telescope; the Isaac Newton and William Herschel Telescopes, operated by the Royal Greenwich Observatory at the Spanish Observatorio del Roque de los Muchachos of the Instituto de Astrofisica de Canarias; the Hubble Space Telescope, and the Nordic Optical 2.5-meter telescope. We thank the dedicated staff of these observatories for their excellent assistance in pursuit of this project. We thank Gary Bernstein and Tony Tyson for developing and supporting the Big Throughput Camera at the CTIO 4-meter; this wide-field camera was important in the discovery of many of the high-redshift supernovae by both groups. This work was supported in part by the Physics Division, E. O. Lawrence Berkeley National Laboratory of the U. S. Department of Energy under Contract No. DE-AC03-76SF000098, by the National

Science Foundation's Center for Particle Astrophysics, University of California, Berkeley under grant No. ADT-88909616. A.G.R. acknowledges the support of the Miller Institute for Basic Research in Science.

REFERENCES

1. Baade 1938, ApJ, 88, 285
2. Tammann, G. A. 1979. In ESA/ESO Workshop on Astronomical Uses of the Space Telescope, F. Macchetto, F. Pacini, & M. Tarenghi, ed., Geneva: ESO, p. 329
3. Colgate, S. 1979, ApJ, 232, 404
4. Nørgaard-Nielsen, et al. 1989, Nature, 339, 523
5. Perlmutter, S., et al. 1995. In Presentations at the NATO ASI in Aiguablava, Spain, LBL-38400, page I.1; also published in Thermonuclear Supernova, P. Ruiz-Lapuente, R. Canal, and J.Isern, editors, Dordrecht: Kluwer, page 749 (1997)
6. Perlmutter, S., et al. 1997, Ap.J., 483, 565
7. Perlmutter, S., et al. 1998. LBL-42230, astro-ph/9812473
8. Schmidt, B. P., et al. 1998, Ap.J., in press
9. Phillips, M. M. 1993, Ap.J.Lett., 413, L105
10. Riess, A. G., Press, W. H., & Kirshner, R. P. 1995, ApJ, 438, L17
11. Riess, A. G., Press, W. H., & Kirshner, R. P. 1996, ApJ, 473, 88
12. Goldhaber. G. 1999, in preparation
13. Hamuy, M., Phillips, M. M., Maza, J., Suntzeff, N. B., Schommer, R. A., & Aviles, R. 1996, AJ, 112, 2391
14. Riess, A. G., et al. 1999, AJ, (in press)
15. Kim, A., Goobar, A., & Perlmutter, S. 1996, PASP, 108, 190
16. Oke, J. B. & Sandage, A. 1968, ApJ, 154, 21
17. Hamuy, M., Phillips, M.M., Wells, L.A., & Maza, J. 1993, PASP, 105, 787
18. Nugent, P., et al. 1999, in preparation
19. Goobar, A., & Perlmutter, S. 1995, ApJ, 450, 14
20. Perlmutter, S., et al. 1998, Nature, 391, 51 and erratum (on author list), 392, 311
21. Perlmutter, S., et al., ApJ,(in press), astro-ph/9812133
22. Riess, A. G., et al. 1998, AJ, 116, 1009
23. Sandage, A. & Hardy, E. 1973, ApJ, 183, 743
24. Höflich, P., Wheeler, J. C., & Thielemann, F. K. 1998, ApJ, 495, 617
25. Holz, D. E., & Wald, R. M. 1998, Phys. Rev. D, 58, 063501
26. Hatano, K., Branch, D., & Deaton, J. 1997, astro-ph/9711311
27. Zehavi, I., Riess, A. G., Kirshner, R. P., & Dekel, A. 1998, ApJ, 503, 483
28. Aldering, G., et al. 1998, IAU Circular 7046
29. Tegmark, et al. 1998, astro-ph/9805117.
30. Lineweaver, C. H. 1998, ApJ, Lett., in press
31. Garnavich, P., et al. 1998, Ap.J., 509, in press
32. Efstathiou, G., et al. 1998, MNRAS, submitted.
33. Caldwell, R. R., Dave, R., & Steinhardt, P. J. 1998, Phys. Rev. Lett., 80, 1582

Type Ia Supernovae for Cosmology A Theorist's View

S. E. Woosley*

*Astronomy Department, UCSC[1]

Abstract.

Although one may use Type Ia supernovae strictly as empirical standard candles, our theoretical understanding of these events is very primitive and this should be a cause for concern. Perhaps the most surprising result to the model builder is that, observationally, the light curves from these events are characterized to such high accuracy by a single parameter. Most models nowadays have at least two parameters, e.g., the mass of ^{56}Ni synthesized *and* its spatial distribution in the explosion. If there are multiple independent parameters, and especially if these parameters have different evolutions with redshift, the use of Type Ia supernovae to determine cosmological distances would be in jeopardy.

STANDARD CANDLES?

To zeroth order we all understand that most Type Ia supernovae (SN Iae) are standard candles. Observationally, for those that have independent distance determinations, the peak luminosity typically varies from event to event by only a factor of a few, perhaps an order of magnitude in the extreme. These observations are consistent with a simple standard model in which a white dwarf accretes to a critical mass, 1.38 $M_\odot$, or more precisely a critical central density near 2×10^9 g cm^{-3}, experiences a degenerate thermonuclear runaway in its middle, and explodes. A certain fraction of the white dwarf must burn to heavy elements in order to unbind the star (0.35 $M_\odot$) and make it expand with the observed velocities, (at least 0.6 $M_\odot$; e.g., Woosley 1990). The highly degenerate nature of the matter tells us that burning will go mostly to nuclear statistical equilibrium. The fuel, carbon and oxygen, has equal numbers of neutrons, so a lot of ^{56}Ni (a doubly magic, Z = N nucleus, in the iron group) will be produced. It is known that to good accuracy, the luminosity of the light curve at peak is proportional to the mass of ^{56}Ni synthesized (Arnett, Branch, & Wheeler 1985). So one might expect SN Ia, all starting from

[1] This research was supported by the National Science Foundation, (AST-97-31569) and by the NASA Theory Program (NAG5-8128.)

CP478 , *COSMO—98*, edited by David O. Caldwell

a similar initial condition, to all produce a common outcome, i.e., make the same mass of ^{56}Ni and be standard candles.

However, we also understand that, beyond zeroth order, SN Iae are *not* standard candles. The picture of the previous paragraph is too simple and the mass of ^{56}Ni made in the explosion is not constant. Even if we ignore the possibility of other classes of models, especially the so called "sub-Chandrasekhar models", the carbon-oxygen ratio in the white dwarf at the time it explodes will vary, as will the white dwarf's accretion rate, metallicity, and rotation rate. These physical circumstances translate into different model characteristics – ignition density, transition density to detonation, if there is one, and time to reach that density. These variable characteristics, in turn, lead to a range of ^{56}Ni masses and explosion energies (e.g., Höflich & Khokhlov 1996).

Because the explosion involves the propagation of a turbulent Rayleigh-Taylor unstable burning from, there will also be some element of chaos in the outcome from very similar models. For example, a small perturbation in the location where the flame begins may make a big difference in the final ^{56}Ni mass and explosion energy (Niemeyer, Hillebrandt, & Woosley 1997).

In summary, the first order variations in the SN Ia brightness distribution are also understood in terms of a variable mass of ^{56}Ni made in the explosion. But are there higher order terms that might be important?

WHY IS THERE A PHILLIP'S RELATION?

People attempt to account for the diversity in SN Ia light curves using an observed relationship between peak luminosity and light curve shape called the "Phillip's Relation". This may be expressed in several forms – template fitting, decline rate - peak luminosity relations, light curve "time stretching" – but all have in common the observed correlation between light curve and decline rate. Bright light curves decline slower than faint ones.

A little thought will show that this relation is actually the opposite of what one expects from the simplest model the reader may have in mind from the previous section. In the simplest one parameter model, the initial binding energies of the white dwarfs are all the same. Each white dwarf burns carbon and oxygen to ^{56}Ni (and only ^{56}Ni), but produces a variable amount of ^{56}Ni. The brightness of the supernova is then proportional to this ^{56}Ni mass. The kinetic energy of the explosion is also greater the more ^{56}Ni one makes. Burning more carbon and oxygen to nickel releases more energy. The diffusion time is shorter for more rapid expansion, and so one expects the brightest supernovae to be the most rapidly expanding and therefore those with the *fastest* decline rate – *exactly the opposite of what is observed.*

Part of the solution to this paradox is that the Phillip's Relation does not have its origin solely in hydrodynamics and nuclear burning, but also in atomic physics and radiation transport (Pinto & Eastman 1999; Pinto 1999; Höflich & Khokhlov 1996).

Iron group elements have enormous ultraviolet opacities, but smaller opacities in the optical. The more ^{56}Ni one makes, the greater the heating rate of the gas and the hotter the trapped radiation. Hot radiation diffuses slower, so for two identical masses of iron group elements expanding at the same rate, the one with the larger ^{56}Ni mass fraction will have a light curve that decays slower.

A second important deviation from the overly simple model described above is that the products of the combustion need not be just ^{56}Ni. At high density, electron capture leads to the production of other iron group isotopes, especially ^{54}Fe and ^{58}Ni. At lower density, the heat capacity of the plasma rises and the nuclear burning fails to go to completion. Silicon and sulfur, in particular, are produced in great abundance (and are seen in the spectrum). The energy yield per gram for these products is almost as much as when ^{56}Ni is produced. Thus the kinetic energy of the explosion is not simply related to the mass of ^{56}Ni that is made. Lower ^{56}Ni masses can, in some models, be accompanied by larger quantities of these other inert species. Hence fainter supernovae can, in practice, be faster.

But what are the implications of this more complex picture of SN Iae as standard candles? Now the explosion energy is decoupled, at least partly, from the ^{56}Ni mass. In addition, the distribution of opacity in the supernova can be quite complex. And everything – the ^{56}Ni mass, the ^{54}Fe and ^{28}Si masses, the explosion energy, and the opacity distribution can change from event to event (and from place to place?).

The amount of ^{56}Ni relative to ^{54}Fe, for example, is particularly sensitive to the central ignition density. Higher densities lead to more electron capture and the production of more ^{54}Fe and ^{58}Ni. Higher ignition densities are expected for white dwarfs with slower accretion rates. Increased metallicity may also favor a larger ignition density because of a neutrino cooling process called the "URCA" process, or simply because nucleosynthesis in metal-rich stars produces more ^{54}Fe. The accretion rate may also be sensitive to the metallicity (of the companion star). We conclude that a variety of environmental circumstances could lead to a difficult-to-calculate variation in both ^{54}Fe and the ratio ^{54}Fe/^{56}Ni.

The same is also true for ^{28}Si (and ^{32}S), though not for the same reasons. The production of ^{28}Si is sensitive to the speed at which the nuclear flame sweeps through the white dwarf. High ^{28}Si production is favored by an initially slow combustion that burns only a small fraction of the white dwarf mass, causes a large expansion, and then makes a transition to a detonation that burns a lot of mass at a density near 10^7 g cm^{-3}. This transition to detonation is very poorly understood and neither the initial slow propagation of the flame, nor the exact density where the transition to detonation occurs (if such a transition *ever* occurs) has ever been calculated from first principles. However, it is to be expected that the flame speed might be higher if the carbon to oxygen ratio is bigger. There are also some indications that the transition to detonation may be easier if the carbon abundance is high (this makes the critical mass for the "detonator" smaller; e.g., Niemeyer & Woosley 1997).

A number of parameterized models that assume different values for the ignition density and detonation transition density have been calculated by Höflich &

Khokhlov (1996). For certain choices, it is indeed possible to recreate the Phillip's Relation. However, one tentative conclusion of Höflich et al. (1996) was that "an intrinsic spread in the supernova properties exists that cannot be accounted for by any single relation between visual brightness and post-maximum decline." It seems that SN Iae *models* at least, are *not* a single parameter family.

But is the uncorrelated diversity in parameters employed in these model studies physical? Can one independently vary, for example, the ignition density and the density for transition to detonation, or does one depend on the other? If the first order correction to the simplest model is a variable ^{56}Ni mass, then are there other *independent* variables – the masses of ^{54}Fe and ^{28}Si and the distributions of all three – ^{54}Fe, ^{56}Ni, and ^{28}Si – with radius, that will affect the light curve? Naively one thinks so, but lacking a physical model of the explosion, it is hard to say.

WHY NOT JUST BUILD A MODEL?

So why not just calculate a first principles model? Even if we don't know the exact characteristics of the progenitor, why not just take a white dwarf accrete matter at a given rate onto its surface, map the problem into a computer grid and run on, say, one of the big ASCI machines at LANL or LLNL? Then one could at least determine the co-dependencies of some of the parameters we have been discussing.

Leaving aside the important issue of how the ignition occurs in the first place (at the center?, off-center?, at one point? at 10 points? Niemeyer, Hillebrandt, & Woosley 1997), it is currently impossible to follow the propagation of a turbulent, Rayleigh-Taylor unstable flame through a star. Of course computers get faster every day, but for some time to come, the models will lack the necessary dimensionality (3D) and resolution to follow the flame through the star and see its possible transition to detonation. The flame thickness ranges from $\sim 10^{-5}$cm at the beginning to cm at late times. Thus the flame may become highly wrinkled and folded. The turbulent length scales range from the size of the star ($\sim 10^8$ cm) to the Kolmogorov scale, a fraction of a cm. Rayleigh Taylor instabilities span a similar large range of scales. To try to calculate turbulence in a 2D code badly misrepresents the cascade to smaller length scales (this has been dramatically shown for the case of classical novae by Kercek, Hillebrandt, & Truran 1998). A 3D calculation is necessary, but the simplest brute force method would require a prohibitive number of zones. The radiation transport calculation is also extremely difficult (Pinto & Eastman 1999).

This does not mean that the problem cannot be done though, only that theorists must be much more clever than they have been. Some of the same novel numerical techniques being used by the chemical combustion community are now being applied to supernovae. We are learning, perhaps not too surprisingly, that their problems resemble ours in many ways. If we really understood the combustion of gasoline in your car engine, we might be well on the way towards understanding supernovae as well. But for now that understanding is lacking.

CAN WE TRUST THE EMPIRICAL RESULTS?

In the meantime, theorists cannot dispute the findings of observers here; they can only advise caution. We are surprised to find that SN Iae are so regular after a simple single parameter adjustment.

The finite Λ results depend upon the assumption that the *diversity* observed in a nearby sample of supernovae fully encompasses the *evolution* in supernovae spanned by observations at large distances, that is the relationship between peak luminosity and decline rate does not *itself* evolve. We have made the case in §2 that there might be multiple parameters. If there are, those parameters might easily have different evolutions, for example, with metallicity, initial mass function (IMF), and star formation rate. In some cases the effects could be subtle – low metallicity would change the radii of companion stars in binary systems and possibly the accretion rate. A shift in IMF would change the carbon mass fraction, etc.

The observers should, and apparently will continue to search for greater diversity in their nearby sample, both by looking at a greater number of supernovae and by observing these supernovae in other wavebands. It is already known that SN Iae are not so regular in the infrared as in the optical. Understanding the systematics of this variation may help us to better quantify the higher order variations of the optical maximum.

Meanwhile theorists will continue to seek a better physical understanding of supernovae, both because it is needed, and because the careful observations reported at this meeting and elsewhere, and their cosmic implications warrant it.

REFERENCES

1. Arnett, W. D., Branch, D., & Wheeler, J. C., *Nature*, **314**, 337, (1985).
2. Höflich, P., & Khokhlov, A., *ApJ*, **457**, 500, (1996).
3. Höflich, P., et al., *ApJL*, **472**, L81, (1996).
4. Kercek, A., Hillebrandt, W., & Truran, J. W., *Astron. and Ap.*, in press, see astro-ph/9811259 (1998).
5. Niemeyer, J. C., & Woosley, S. E., *ApJ*, **475**, 740, (1997).
6. Niemeyer, J. C., Hillebrandt, W., & Woosley, S. E., *ApJ*, **471**, 903, (1997).
7. Pinto, P. A., private communication.
8. Pinto, P. A., & Eastman, R. G., *Astrophysical Journal*, in press, (1999).
9. Woosley, S. E., in *Supernova*, ed. A. G. Petschek, (Springer Verlag: New York), 182, (1990).

From the Cosmological Microwave Background to Large-Scale Structure

Joseph Silk and Eric Gawiser

Departments of Astronomy and Physics, and Center for Particle Astrophysics, University of California, Berkeley, CA 94720

Abstract.
The shape of the primordial fluctuation spectrum is probed by cosmic microwave background fluctuations which measure density fluctuations at $z \sim 1000$ on scales of hundreds of Mpc and from galaxy redshift surveys, which measure structure at low redshift out to several hundred Mpc. The currently acceptable library of cosmological models is inadequate to account for the current data, and more exotic models must be sought. New data sets such as SDSS and 2DF are urgently needed to verify whether the shape discrepancies in $P(k)$ will persist.

INTRODUCTION

Our understanding of primordial fluctuations in the early universe was revolutionized first with inflation and then by the actual detection of temperature fluctuations in the cosmic microwave background (CMB). Inflation gave the spectrum of fluctuations, but not the normalization. The COBE-DMR experiment measured the amplitude of temperature fluctuations on an angular scale that was acausal at last scattering, and hence directly probed inflationary fluctuations, and in particular the fluctuation strength. More than twenty subsequent experiments have plugged the causal gap, measuring fluctuations over angular scales less than or of order the acoustic peak at $\lambda = 220\,\Omega^{1/2}$ that corresponds to the maximum sound horizon in the early universe.

The shape of the fluctuation spectrum is now being probed. The CMB fluctuations measure density fluctuations at $z \sim 1000$ on scales of hundreds of Mpc. At large redshifts one degree subtends a comoving scale of 100 Mpc. A complementary measure arises from galaxy redshift surveys. These measure variations in the luminous matter density out to ~ 300 Mpc at the present epoch. One can combine, after choosing a model, the CMB and large-scale structure (LSS) measures of the fluctuation spectrum. Here we will describe the current status of our understanding of the shape of the primordial fluctuation spectrum.

CP478 , *COSMO—98*, edited by David O. Caldwell

It is customary to use a two-parameter fit to the LSS power: σ_8, the normalization at $8\,h^{-1}$ Mpc, and Γ, measure of shape relative to CDM and nearly equal to $\Omega\, h$ for CDM. Given the several data sets, each with a number of independent data points, this may be an unnecessarily restrictive approach. Of course any data set is imperfect, with possible systematic errors, and the data sets have different selection biases. However, provided the bias is scale-independent, one can renormalize the data sets and examine detailed shape constraints. One has to decide whether to compare a nonlinear power spectrum with the data or whether to correct the data for nonlinearity and compare the data with linear theory. We will employ the latter approach here.

CMB: STATUS OF THE THEORY

Inflation-generated curvature fluctuations provide the paradigm for interpreting the cosmic microwave background temperature anisotropies. There are three components to $\delta T/T$, schematically summarized as

$$\frac{\delta T}{T} = |\delta\phi + \delta\rho_\gamma + \delta v|\,.$$

These are the gravitational potential, intrinsic and Doppler contributions from the last scattering surface. The combined effect of the first two terms results in the Sachs-Wolfe effect $\delta T/T = \frac{1}{3}|\delta\phi|$ which represents the only superhorizon contributions to $\delta T/T$. Inflationary initial conditions then require that the Fourier component $|\delta T/T|_k \propto |\cos(kv_s t_{ls})|$ at the last scattering epoch t_{ls}, where k is the wavenumber and v_s is the sound speed, must be constant as $k \to 0$. Inflation of course specifies the phases of density fluctuations that decompose to sound waves of wavelength less than that of the maximum sound horizon, $v_s t_{ls}$. Longer wavelengths correspond to power-law growing modes at horizon crossing. The wave just entering the horizon at last scattering has a peak at wavenumber $n\pi/v_s t_{lss}$, $n = 1$, and a succession of waves crest at $n = 2,\ 3, \ldots$ before damping sets in as the photon mean free path increases relative to the wavelength. The first acoustic peak projects to $\delta T/T$ on angular scales $\sim\ \Omega^{1/2}$ degree, and is a robust measure of the curvature of the universe. Doppler peaks are 90° out of phase and of lower amplitude, so they fill in the troughs of the acoustic oscillations as measured by the radiation power spectrum. Peak heights are determined in large part by choice of Ω_B and Ω_Λ. An increase in Ω_B enhances the wave compression and reduces the rarefaction phases. An increase in Ω_Λ enhances the ratio of radiation to matter in a flat model, and thereby boosts the peak potential decay and the low ℓ power via the early integrated Sachs-Wolfe effect. Of course increasing the spectral index n also raises the peak height. Peak heights are lowered by reionization and secondary scattering. Not all of these degeneracies are removed by examining the higher peaks. For example, combination of Ω_B and Ω_Λ at specified Ω is nearly degenerate in peak height and peak location since the angular size-redshift relation depends only on Ω.

Lensing by nonlinear and quasilinear foreground structure redistributes the peak power towards very high ℓ in a way that breaks the degeneracies in the CMB. One can search for this effect with interferometer experiments [1] at $\ell \gtrsim 1000$ or else by correlating the CMB fluctuations with large-scale power from either large redshift surveys such as the SDSS [2] or via weak lensing distortions of the CMB [3] or LSS [4].

RECONSTRUCTION OF THE PRIMORDIAL POWER SPECTRUM FROM THE CMB

The inflationary, approximately scale-free, power spectrum $P(k) = Ak^n$, $n \approx 1$, is modified by the transition from radiation to matter domination, since in matter-dominated epochs different growth occurs on subhorizon scales: in the radiation era, there is no subhorizon growth of fluctuations. This modifies $P(k)$ to $P(k) \propto k^{-3}$ on larger scales. The transition occurs at the horizon during matter-radiation equality, namely $12(\Omega h^2)^{-1}$ Mpc. The Boltzmann equation can be solved for temperature fluctuations, mode by mode, and the solutions for $\delta T/T$ are scaled to agree with the quoted errors for each CMB experiment. For each specified cosmological model, we then infer the power spectrum amplitude over scales that correspond to the deprojection on the sky of the experimental window function [6]. We confirm that standard CDM ($\Omega_{CDM} = 1$, $h = 0.5$) fits the CMB data rather poorly, with best fit renormalization that corresponds to a high value of σ_8. The ΛCDM model ($\Omega = 1$, $\Omega_m = 0.4$, $h = 0.6$) gives a reasonably good fit with an acceptable value of σ_8. Much stronger constraints however come when these fits are combined with LSS data.

RECONSTRUCTION OF $P(K)$ FROM LARGE-SCALE STRUCTURE DATA

There are several large-scale structure data sets that one may use to reconstruct $P(k)$. Galaxy redshift surveys include the Las Campanas Survey of 25,000 galaxies [7], the PSCz survey of 1,500 galaxies [8], and the SSRS2/CfA2 survey of 7,000 galaxies [9]. There is also the APM cluster survey which probes to $\sim 300\,h^{-1}$ Mpc [10] and the real space inversion of the 2D APM galaxy survey [11]. The local mass function of clusters [12] has been used to measure $\sigma_8\Omega^{0.6}$, and the high redshift cluster abundance [13] has been used to break the degeneracy between σ_8 and Ω. Peculiar velocities and large scale bulk flows also yield $\sigma_8\Omega^{0.6}$ in a completely bias-independent approach, although systematic uncertainties remain large [14].

The redshift space surveys can be corrected in a straightforward way, for peculiar velocities on small scales and bulk flows on large scales, to derive the real space $P(k)$ by assuming a cosmological model [15]. Correction of data in the non-linear regime is best done by numerical simulation, but can be performed using an empirical formulation calibrated to numerical simulations based on a smooth interpolation from spherical collapse by a factor of 2 in radius on cluster scales [16]. Renormalization of the various measures of $P(k)$ is effected by assuming that all measurements are subject to a scale-independent bias, allowed to be independent for each probe of $P(k)$.

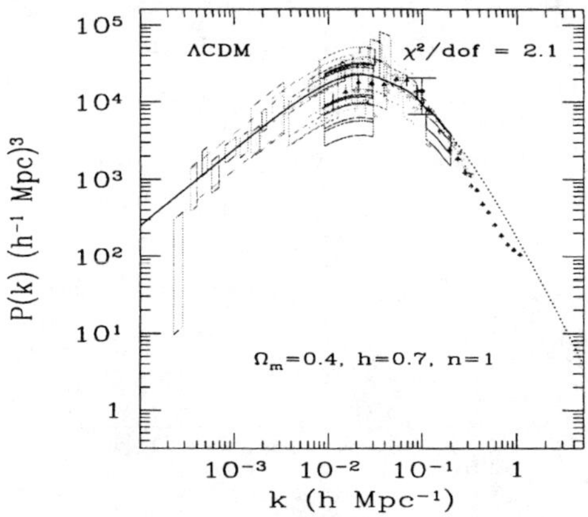

FIGURE 1. Constraints from LSS and CMB on ΛCDM model.

CONFRONTATION OF $P(K)$ WITH CMB AND LSS

Model fitting to LSS alone results in the following conclusions. Of course the standard COBE-normalized CDM model fails completely. Without a large scale-dependent bias factor on 10 – 100 Mpc scales, peculiar velocities and the galaxy cluster abundance are greatly overpredicted. Low density models circumvent these problems. The cluster abundance, evolution and baryon fraction are all in satisfactory agreement with observations [13].

However the combined CMB/LSS fits to $P(k)$ lead [6] to a surprising conclusion. The surprise is that almost all models, while occasionally faring better than sCDM, still provide unacceptable fits to all of the data. Consider for example ΛCDM, currently favored by the SN Ia Hubble diagram. The reduced χ^2 is 2.1 for 70 degrees of freedom. Data set by data set, one still has a problem. For example the values of χ^2(d.o.f.) are APM clusters: 25(8); LCRS 17(5); APM 44(9); IRAS 16(9). The acceptable data sets are CfA for 2 d.o.f., cluster abundances/peculiar velocities for 3 d.o.f. and CMB for 34 d.o.f.

The only mildly acceptable model (reduced $\chi^2 = 1.2$) is CHDM, hot and cold dark matter with $\Omega = 1$. This model overpredicts current cluster abundances and underpredicts the small number of high redshift, luminous x-ray clusters (2 at $z = 0.5$, 1 at $z = 0.8$). However the cluster evolution constraint is disputed [17], and the local normalization is not necessarily robust. The cluster baryon fraction provides an independent and powerful constraint that favors $\Omega_m \approx 0.3$. Of course this rests on the reasonably plausible assumption that clusters provide a fair sample of the baryon fraction of the universe. This need not necessarily be true if gas has had a complex history prior to cluster formation: *e.g.* the gas may have been preheated as is suggested by recent considerations of the entropy of intracluster gas [18]. This would reduce the baryon fraction, but one can equally well imagine scenarios for cluster formation in dense sheets or filaments where the baryon fraction was already enhanced.

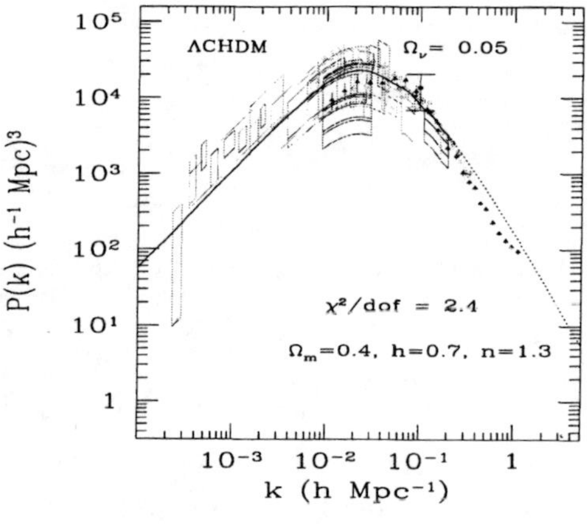

FIGURE 2. ΛCDM model with $\Omega_\nu = 0.05$. A blue tilt of the primordial power spectrum ($n = 1.3$) is necessary to counteract the damping of small-scale perturbations by free-streaming of the massive neutrinos, which makes the peak of the model fall even farther below that of the data unless $n > 1$. Even with this best-fit value of n, the fit to the data is worse than with no HDM, because CMB observations disfavor such a high value of n.

Consider the model preferred by the combination of SNIa and cluster constraints, namely ΛCDM. Figure 1 shows the ΛCDM power spectrum compared with observations of Large-Scale Structure and CMB anisotropy. One can pose the following question: does adding a hot component improve the marginally acceptable LSS fit? We find that as an admixture of HDM is added to the dominant CDM, the combined fit to CMB and LSS deteriorates (Figures 2,3). The reason is that low density CDM models have a $P(k)$ peak that is longward of the apparent peak in the APM data. Adding HDM only exacerbates the mismatch.

Neutrino masses imprint a distinct signature on $P(k)$. This will eventually be a measurable probe of the neutrino mass, from LSS as well as from CMB. Indeed the LSS probe may potentially be more powerful [19]. There is more dynamical range available in probing $P(k)$ with LSS on the neutrino free streaming scale, where the primary signature should be present. Even present data is sensitive to a neutrino mass of around an eV: for example we find that the fit changes significantly between 0.1 and 1 eV. If ΛCDM is in fact the right model, our analysis yields an indirect upper limit on the mass of the most massive neutrino species of $m_\nu \leq 2$eV. While there are considerable systematic uncertainties in this approach, it is promising as a complement to the direct evidence for mass difference between neutrino species from SuperKamiokande [20] and the solar neutrino problem [21], and is already beginning to conflict with results from LSND [22] that require a large mass difference.

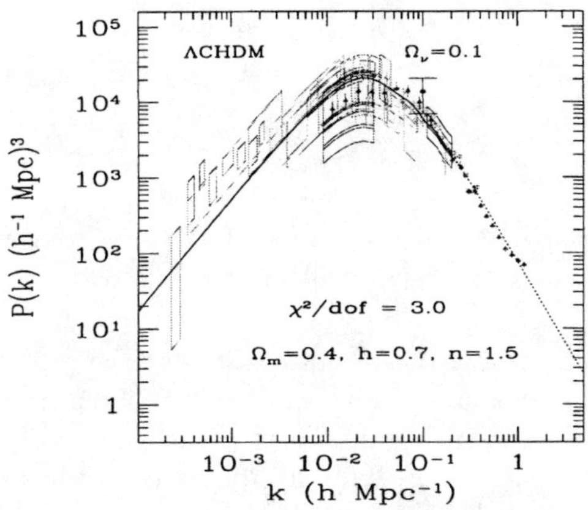

FIGURE 3. ΛCDM model with $\Omega_\nu = 0.10$. The best-fit value of n is now 1.5. The fit to the data worsens as more HDM is added.

One class of exotic models is the following. Take a model that fits all constraints except for the shape. The best contender for such a model is ΛCDM. Inspection of the LSS constraints reveals that there is a deficiency of large-scale power near 100 Mpc. One can add an ad hoc feature on this scale from considering inflationary models with multiple scalar fields (see Figure 4). This could be generated for example by [5] incomplete coagulation of bubbles of new phase in a universe that already has been homogenized by a previous episode of inflation. One can tune the bubble size distribution to be sharply peaked at any preferred scale. This results in nongaussian features and excess power where needed. The non-gaussianity provides a distinguishing characteristic.

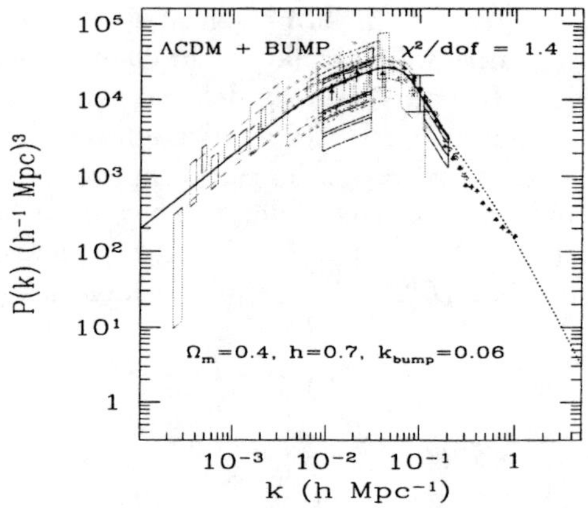

FIGURE 4. Constraints from LSS and CMB on ΛCDM model with a broad enhancement centered at $k = 0.06h^{-1}$Mpc added to the primordial power spectrum.

Other suggestions that fit both CMB and LSS data appeal to an inflationary relic of excess power from broken scale invariance, arising from double inflation in a ΛCDM model, which results in a gaussian feature that is essentially a step in $P(k)$ at the desired wavenumber [23]. This improves the fit in much the same way as adding a hot component to CDM improves the empirical fit. While such ad hoc fits may seem unattractive, one could argue that other aspects of cosmological model building are equally ad hoc, such as postulating a universe in which Λ is only becoming dynamically important at the present epoch. Clearly one has to accommodate such arguments in order to fit the data, if the data is indeed accepted at face value. Moreover there are positive side effects that arise from the tuned void approach. The bubble-driven shells provide a source of overdensities on large scales. Rare shell interactions could produce nongaussian massive galaxies or clusters at low or even high redshift: above a critical surface density threshold gas cooling would help concentrate gas and aid collapse. If massive galaxies were discovered at say $z > 5$ or a massive galaxy cluster at $z > 2$ this would be another indication that the current library of cosmological models is inadequate. New data sets such as SDSS and 2DF are urgently needed to verify whether the shape discrepancies in $P(k)$ will persist.

SUMMARY

If the data are accepted as mostly being free of systematics and ad hoc additions to the primordial power spectrum are avoided, there is no acceptable model for large-scale structure. No one LSS data set can be blamed. Perhaps it is best to wait for improved data. The Sloan and 2DF surveys are already acquiring galaxy redshifts. However another philosophy is to search for more exotic models. Consider for example the primordial isocurvature mode. This has the advantage of forming primordial black holes of stellar mass, since normalization to large-scale structure and the present spectrum over $10-50$ Mpc requires a spectral index that generates nonlinear fluctuations at roughly the epoch of the quark-hadron phase transition, when the horizon contained approximately one solar mass [24] (hence the primordial black holes may be the possibly observed MACHOs). The goodness of fit of this model to the combined CMB/LSS data is similar to that of the ΛCDM model. One cannot distinguish with current data between an exotic isocurvature model and ΛCDM, although neither model is satisfactory. To improve on this, clearly something even more exotic is required. It may be that independent observations will force us in this direction.

ACKNOWLEDGEMENTS

We gratefully acknowledge support by NSF and NASA. This discussion is identical to a paper presented at the 1998 Nobel Symposium on Particle Physics and the Universe, to be published in Physica Scripta.

REFERENCES

1. B. Metcalf and J. Silk, ApJ 492L1 (1998)
2. Y. Wang, D. N. Spergel and M. A. Strauss, preprint astro-ph/9802231, ApJ, in press (1999)
3. M. Zaldarriaga and U. Seljak, preprint astro-ph/9810257 (1998)
4. W. Hu and M. Tegmark, preprint astro-ph/9811168 (1998)
5. L. Amendola, C. Baccigalupi, R. Konoplich, F. Occhionero and S. Rubin, Phys. Rev. D 54 7199 (1996)
6. E. Gawiser and J. Silk, Science 280 1405 (1998)
7. H. Lin et al., ApJ 471 617 (1996)
8. W. Saunders et al., Proc. of *XVIIth Rencontres de Moriond: Extragalactic Astronomy in the Infrared*, ed. G. A. Mamon, T. X. Thuan, J. T. T. Van, (1997), p.431

9. L. N. Da Costa et al., ApJ 437 L1 (1994)
10. H. Tadros, G. Efstathiou, G. Dalton, MNRAS 296 995 (1998)
11. E. Gaztañaga and C. M. Baugh, MNRAS 294 229 (1998)
12. P. T. P. Viana and A. R. Liddle, MNRAS 281 323 (1996)
13. N. A. Bahcall, X. H. Fan and R. Y. Cen, ApJ 485L 53 (1997)
14. T. Kolatt and A. Dekel, ApJ 479 592 (1997)
15. J. A. Peacock and S. J. Dodds, MNRAS 267 1020 (1994).
16. J. A. Peacock and S. J. Dodds, MNRAS 280 L19 (1996).
17. A. Blanchard, J. Bartlett and R. Sadat preprint astro-ph/9809182 (1998)
18. M. Balogh, A. Babul and D. R. Patton, preprint astro-ph/9809159 (1998)
19. W. Hu, D. Eisenstein, and M. Tegmark, Phys. Rev. Lett., 80 5255 (1998)
20. Y. Fukuda et al., preprint hep-ex/9805021, Phys. Rev. Lett. in press (1998)
21. J. N. Bahcall, P. I. Krastev, and A. Yu. Smirnov, preprint hep-ph/9807216, (1998)
22. C. Athanassopoulos et al., Phys. Rev. Lett. 77 3082 (1996)
23. J. Lesgourgues, D. Polarski and A. A. Starobinsky, MNRAS 297 769 (1998)
24. N. Sugiyama and J. Silk, preprint (1998)

Anisotropy in the Microwave Background

Martin White

Departments of Physics and Astronomy, University of Illinois at Urbana-Champaign
1110 West Green St, Urbana, IL 61801-3080

Abstract. The ten's of micro-Kelvin variations in the temperature of the cosmic microwave background (CMB) radiation across the sky encode a wealth of information about the Universe. The full-sky, high-resolution maps of the CMB that will be made in the next decade should determine cosmological parameters to unprecedented precision and sharply test inflation and other theories of the early Universe.

Over the last few years it has become common place to speak of the Cosmic Microwave Background (CMB) anisotropy as the premier laboratory for cosmology and early universe physics. In the titles to most talks the word "anisotropy" is often absent, it being understood that the talk will be about the anisotropy. This is an interesting phenomenon. When we speak of the CMB we could speak about 3 major properties. Firstly, the *existence* of the CMB is one of the pillars of the hot big bang model of cosmology. Secondly, the *black body* spectrum of the CMB, the most perfect black body ever measured in nature, confirms the cosmological origin of the CMB and puts extraordinarily strong constraints on early energy injection in the universe (e.g. through decaying particles, see [1]). These first two properties show that the CMB has already delivered important cosmological information. The fact that our current focus is the third area, the anisotropy, indicates the high level of promise that this area holds.

Like much of cosmology, the CMB is a data driven subject. However, in this proceedings I focus on the theory behind the CMB anisotropies[1], and the current status of theoretical efforts, rather than on the CMB data. It is a blessing of this field that numerous experimental efforts underway will make any statements about the experimental situation obsolete before they reach print. As an overly brief summary of the current status: the current data are in good agreement with our general paradigm and support a spatially flat universe with an almost scale-invariant spectrum of adiabatic fluctuations. Departures from that statement in

1) Because of space, I have referenced primarily work that I have been involved in. Much more representative referencing can be found in those sources.

CP478, *COSMO—98*, edited by David O. Caldwell

any direction fit the data less well, though at present large error bars and theorist's ingenuity limit the strength of statements that can be made.

To understand CMB anisotropies it is helpful to note several general points

- The universe was once hot and dense. At these early times ($\sim 10^5$yr after the bang) the plasma was highly ionized. Thomson ($\gamma - e$) scattering was rapid and tightly coupled the CMB photons to the "baryons" ($p + e$). In the limit that this scattering was rapid, the mean free path was small, a fluid approximation is valid. Thus we speak of the photon-baryon fluid.

- The observed large-scale structure grew through gravitational instability from small perturbations at early times. These density perturbations imply, through Poisson's equation, small perturbations in the gravitational field.

- Combining the above two observations, we infer that the fundamental modes of the system would be gravity sourced sound waves in the fluid. The equation of motion for the sound waves can be derived by taking the tight-coupling limit of the equations of radiative transfer. In this case (see below)

$$[m_{\rm eff}\, \Delta T_k']' + \frac{k^2}{3}\Delta T_k = -F_k \tag{1}$$

  where F_k is the gravitational forcing term, $m_{\rm eff}$ describes the inertia of the fluid, and primes denote derivatives with respect to (conformal) time.

- Finally, recombination (when ps captured es to form hydrogen and the universe became neutral) occurred suddenly, but not instantaneously. With the decrease in the free e^- density the mfp for γs rises from essentially zero to the size of the observable universe. The γs travel freely to us, giving us a snapshot of the fluid at a fixed instant in time. The energy density, or temperature, fluctuations in the fluid are seen as CMB temperature differences (anisotropy) across the sky.

While this way of looking at the anisotropy is physically clear, it is not how the calculations are actually done. Remember that the fluctuations are observed to be small ($\delta \sim 10^{-5}$ c.f. $\alpha_{QED} \sim 10^{-2}$). Thus one writes down the Einstein, fluid and radiative transfer equations, expands about an exact solution and truncates the expansion at linear order. This procedure gives a set of coupled ODEs which describe the evolution of each (independent) Fourier[2] mode. While in some cases the equations can be solved analytically, usually a numerical solution is performed.

Since the Fourier modes decouple in linear theory it is advantageous to work in the Fourier basis in the observations also. Unfortunately the sky is curved, so plane waves are not the natural basis. But a "curved sky Fourier expansion" can still be performed: $\Delta T/T = \sum_\ell a_{\ell m} Y_{\ell m}$. We then discuss the properties of the $a_{\ell m}$,

---

[2]) In hyperbolic geometries the Fourier decomposition needs to be generalized, but this is a technical point.

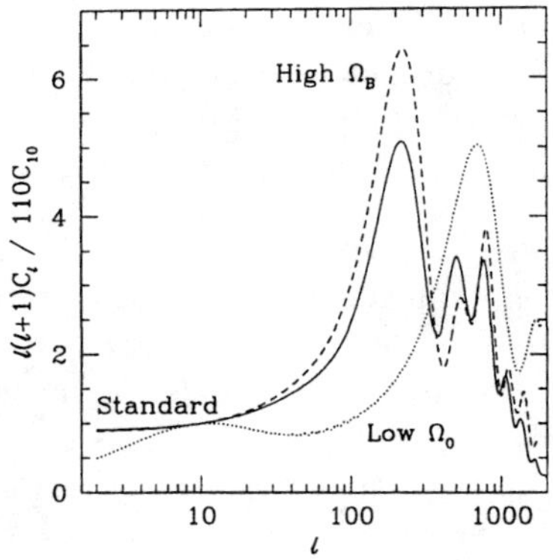

FIGURE 1. The dependence of the angular power spectrum upon some of the cosmological parameters. The solid line shows a "standard" CDM model. The dashed line shows the effect of doubling the baryon density while the dotted line shows the effect of introducing spatial curvature.

known as the multipole moments. By definition $\langle a_{\ell m}\rangle = 0$, so the first non-vanishing correlator is the two-point function. Since the $Y_{\ell m}$ are a complete orthonormal basis, $\langle a_{\ell' m'} a^*_{\ell m}\rangle \propto \delta_{\ell'\ell}\delta_{m'm}$ and by rotational symmetry the proportionality constants can only depend on ℓ. Thus we write $\langle |a_{\ell m}|^2\rangle = C_\ell$. If the fluctuations are Gaussian, having specified the mean and variance we have completely specified the model. For more general distributions the higher moments also need to be specified.

We show in Fig. 1 a typical C_ℓ curve for a standard cold dark matter (CDM) model. A readable introduction to the physics can be found in Refs. [2–6] among others. The precise shape of the power spectrum depends upon cosmological parameters as well as the underlying density perturbations and thereby encodes a wealth of information; see Fig. 1.

The current theoretical situation can be summarized as follows:

- The formalism for computing C_ℓ (and the higher moments) for any FRW spacetime and any model of structure formation exists [7]. Since this is essentially a statement in General Relativity, the proof can be made quite rigorous.

- Not every model has been calculated, but for those where independent calculations have been done (mostly CDM models) independent codes agree to $\mathcal{O}(1\%)$.

- The spectrum encodes information on the cosmology and the model of structure formation and can be measured with exquisite precision.

- Model dependent parameter extraction can simultaneously fit a dozen parameters to an accuracy of $\mathcal{O}(10\%)$, e.g. [8].

- This stunning promise has overshadowed an important additional fact however. Even if our models do not fit every nuance of the observed data, model independent constraints on the parameters exist [9] as do cosmology independent tests of the model of structure formation [9–11].

For some time the promise of the CMB to strongly constrain numerous cosmological parameters has been evident. Both the measurements and the calculations can be done with high precision, and the theories predict a rich structure to the spatial power spectrum. A multi-parameter fit of theory to data, assuming that the fit is good, then allows simultaneous constraints on the model parameters. It is important to understand however that the predictions for cosmological parameter estimation depend both on the assumed theory and on the parameter space which one searches. As an example, for the *MAP* satellite scheduled to launch next year, a fit to a 7 parameter family ΛCDM model gives errors on $\Omega_{\rm B}h^2$ of 4%, $\Omega_{\rm mat}h^2$ of 7%, Ω_Λ of 14% and the optical depth to reionization of 14%. The tensor-to-scalar ratio is essentially unconstrained, as is the tensor spectral index. In combination these constraints, and the assumed spatial flatness, allow a constraint on the Hubble constant of 14%. If we allow *both* curvature and a cosmological constant then the error on Ω_Λ goes up by a factor of 2 and on the Hubble constant by 4!

Much of the effort in parameter estimation of late has focussed on numerical issues (where much earlier work was deficient), on combining CMB observations with other measurements and on extending the parameter space. The latter allows you to explore in detail the relationships that exist in "favored" models that can be relaxed. As an example of the power of the CMB, combined with other measurements, and of extending the parameter space, let us consider a possible measurement of the fluctuations in the $2K$ neutrino background. The hot big bang model predicts that this background must be present, and it too will have a fluctuation spectrum. In the left panel of Fig. 2 we show a calculation of this spectrum, from Ref. [12]. The y-axis scale is $\Delta T/T \sim 10^{-5}$ as for the photons.

Detecting this neutrino signal directly is almost impossible, and detecting the *fluctuations* in the neutrino signal even more so. However, I show in the right panel of Fig. 2 [13] a possible future 1σ "detection" of the fluctuations in the neutrino background inferred from a combination of large-scale structure and CMB data. Clearly this particular example is somewhat fanciful. The "detection" is extremely marginal and the assumptions going into the calculation quite optimistic. However considering how hard it is to do this detection any other way, it serves to illustrate the power of combinations of astrophysical measurements.

As another (topical) example, I show in Fig. 3 (also taken from Ref. [13]) the 1σ limits on the equation of state and fraction of critical density in "dark energy" such as a cosmological constant or dynamical scalar field (sometimes known as xCDM or quintessence).

These somewhat random examples should illustrate the power of the CMB to constrain cosmological parameters, under the assumption that our current models provide a good fit to the data. Of course it may always turn out that while the

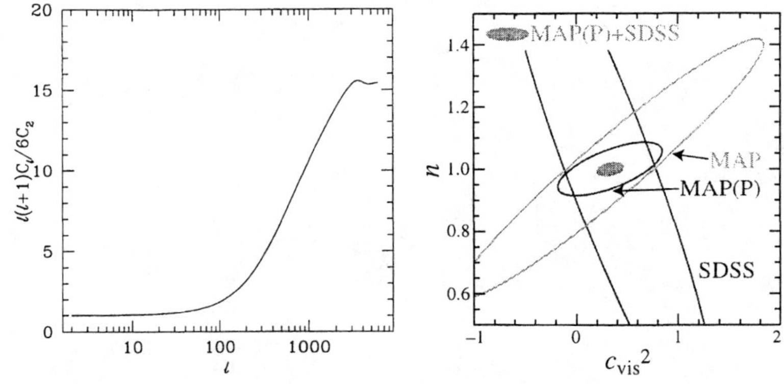

FIGURE 2. (left) The anisotropy spectrum of the $2K$ cosmic neutrino background. The small quadrupole, $\ell = 2$, anisotropy in the neutrino background acts as a source of viscosity in the fluid. At early times when the neutrino energy density was higher than today, this viscosity altered the evolution of the gravitational potentials and hence the CMB anisotropy. (right) A possible future 1σ "detection" of the neutrino signal. The ellipses show expected 1σ error contours from the experiments listed. The x-axis is the viscosity speed of sound. The difference between 0 and 1/3 in $c_{\rm vis}^2$ is the signal of the $\ell = 2$ anisotropy in the neutrino spectrum of the left panel.

paradigm within which we are working is correct, our models are deficient in some detail which prevents a good fit to the data. The strategy in this case is to relax our assumptions and try to reconstruct the model from the observed spectrum. Perhaps eventually the missing ingredient can be found and utopia regained. In the meantime all is not lost.

There exist several model-independent measurements of the cosmological parameters [9]. As an example I show in Fig. 4 two different models of structure formation, both in a critical density universe. The models are chosen not to be good fits to the data, but to be very different from each other. In relativistic perturbation theory there are two kind of perturbations: adiabatic and isocurvature. Any fluctuation can be decomposed into these two basis modes. The solid line in the left panel of Fig. 4 is an example of a pure adiabatic model while the dashed line is an example of a pure isocurvature model. Note that while many things are different in these two models, there are two things that remain fixed. First the damping tail is at the same angular scale in both models ($\ell \sim 1500$). Secondly the separation between e.g. the 2nd and 3rd peaks is the same in both models. The first statement is easy to understand. The damping comes from photon diffusion during the time it takes the universe to recombine. Perturbations on scales smaller than the photon diffusion scale are erased, leading to the damping of power at high-ℓ. Clearly this process is independent of the source of the fluctuations. The second statement is also easy to understand. The photon-baryon fluid behaves like an oscillator with a natural

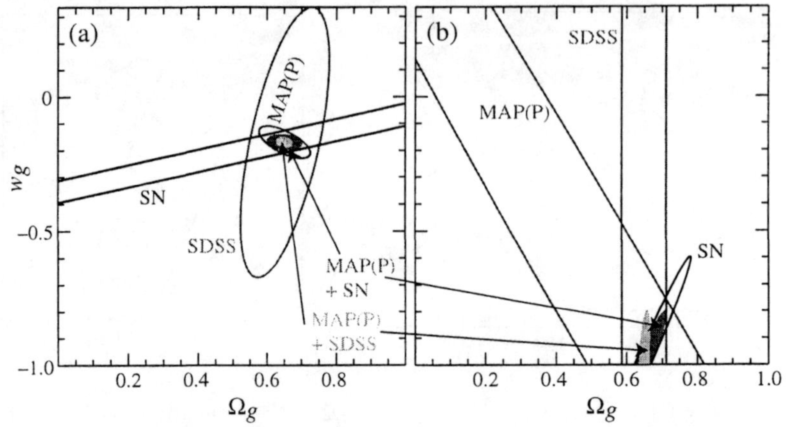

FIGURE 3. The expected 1σ contours on the equation of state $w = p/\rho$ and fraction of critical density Ω_g of a generalized dark energy candidate. Note that the errors do not depend strongly on the assumed equation of state. A model with $w_g = -1$ (right panel) is indistinguishable from a cosmological constant.

frequency. Once the "bell" is struck it wants to ring at that natural frequency. Thus even though the driving forces in the adiabatic and isocurvature models are different, the "ringing" of the higher peaks proceeds at the same frequency. So the peak spacing is fixed.

While the peak spacing ℓ_A and the damping scale ℓ_D are (nearly) independent of the model of structure formation, they do depend on the cosmology. Specifically on the mapping between physical scales at the surface of last scattering ($z \sim 10^3$) and angles on the sky. They are thus probes of the angular diameter distance to last scattering, as shown in the right panel of Fig. 4 for the case of an open universe. More general constraints in the $\Omega_m - \Omega_\Lambda$ plane or the $\Omega_g - w$ plane can be found in [14,13] respectively.

To turn the problem around one can look for test of the model of structure formation independent of the cosmology. In this vein I should mention that CMB based tests of inflation exist which do not require a measurement of the tensor signal [10,11]. In general sharp tests of the model can be performed if information about the polarization of the CMB is obtained (as both MAP and Planck intend). More information on this can be found in Ref. [15].

If our models are close to correct, high precision studies of the CMB anisotropy will revolutionize cosmology. If our models are wrong, one could not hope for a better data set with which to find the right path. We are all eagerly awaiting imminent experimental advances in this field.

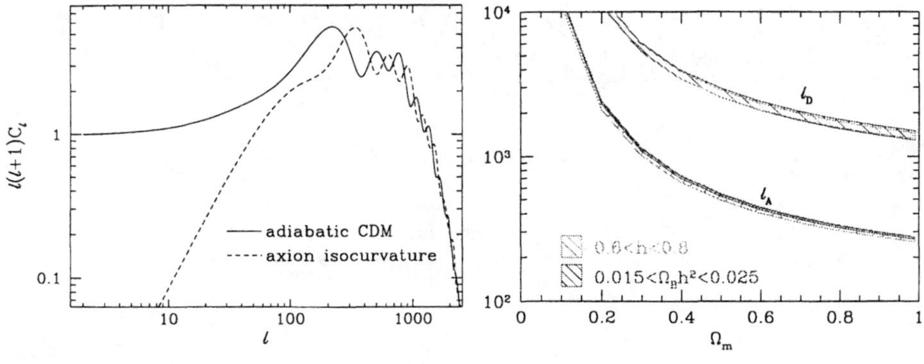

FIGURE 4. (left) The anisotropy spectrum for two different models of structure formation: an adiabatic model (solid line) and an isocurvature model (dashed line). Notice that the damping scale and the distance between the 2nd and 3rd peaks is model independent. (right) the dependence of the damping scale ℓ_D and the acoustic peak scale ℓ_A on the angular diameter distance back to $z \sim 10^3$. In this example, as the universe is made more open by lowering Ω_m without introducing a cosmological constant, the structure shifts to smaller scales or higher ℓ.

REFERENCES

1. Smoot G., and Scott D., in "Review of Particle Properties", p. 127, Caso C., et al., *The European Physical Journal* **C3**, 1 (1998).
2. Lawrence C.R., Scott D., and White M., [astro-ph/9810446]
3. Bennett C.L., Turner M.S., and White M., *Phys. Today* **50**, 32 (1997).
4. Hu W., Sugiyama N., and Silk J., *Nature* **386**, 37 (1997) [astro-ph/9604166]
5. Scott D., and White M., *Gen. Rel. and Gravitation* **27**, 1023 (1995) [astro-ph/9505102]
6. White M., Scott D., and Silk J., *Ann. Rev. Astron. Astrophys.* **32**, 319 (1994).
7. Hu W., Seljak U., White M., and Zaldarriaga M., *Phys. Rev.* **D57**, 3290 (1998) [astro-ph/9709066]
8. Eisenstein D.J., Hu W., and Tegmark M., [astro-ph/9807130]
9. Hu W., and White M., *ApJ* **471**, 30 (1996) [astro-ph/9602019]
10. Hu W., and White M., *Phys. Rev. Lett.* **77**, 1687 (1996) [astro-ph/9602020]
11. Hu W., Spergel D.N., and White M., *Phys. Rev.* **D55**, 3288 (1997) [astro-ph/9605193]
12. Hu W., Scott D., Sugiyama N., and White M., *Phys. Rev.* **D52**, 5498 (1995) [astro-ph/9505043]
13. Hu W., Eisenstein D.J., Tegmark M., and White M., *Phys. Rev.* **D59**, 023512 (1999) [astro-ph/9806362]
14. White M., *ApJ* **506**, 495 (1998) [astro-ph/9802295]
15. Hu W., and White M., *New Astronomy* **2**, 323 (1997) [astro-ph/9706147]

Model-Independent Measurement of the Primordial Power Spectrum

Yun Wang, David N. Spergel, & Michael A. Strauss

Department of Astrophysical Sciences
Princeton University, Princeton, NJ 08544

Abstract.
In inflationary models with minimal amount of gravity waves, the primordial power spectrum of density fluctuations, $A_S^2(k)$, together with the basic cosmological parameters, completely specify the predictions for the cosmic microwave background (CMB) anisotropy and large scale structure. Here we show how we can strongly constrain both $A_S^2(k)$ and the cosmological parameters by combining the data from the Microwave Anisotropy Probe (MAP) and the galaxy redshift survey from the Sloan Digital Sky Survey (SDSS). We allow $A_S^2(k)$ to be a free function, and thus probe features in the primordial power spectrum on all scales. MAP and SDSS have scale-dependent measurement errors that decrease in opposite directions on astrophysically interesting scales; they complement each other and allow the measurement of the primordial power spectrum independent of inflationary models, giving us valuable information on physics in the early Universe, and providing clues to the correct inflationary model.

At present, inflation provides the simplest and most compelling solution to the problems of the standard cosmology: the horizon problem, the flatness problem, and the structure formation problem. A generic prediction of inflationary models is the existence of primordial adiabatic Gaussian random-phase density fluctuations. The properties of these fluctuations are completely determined by their power spectrum $A_S^2(k)$. There currently exists a broad range of inflationary models; they make different predictions for the primordial power spectrum $A_S^2(k)$.

It is customary to parametrize $A_S^2(k)$ as a featureless power law, $A_S^2(k) \propto k^{n_S-1}$. However, there are models for which the power-law index would change with scale [1]. There are other models which predict primordial power spectra with broken scale invariance, which can not be parametrized by a featureless power law [2]. There are physical reasons as well as tentative observational evidence [3] which point to the possibility that the primordial power spectrum has breaks in its power-law form. Given our ignorance of the true form of $A_S^2(k)$, we feel that one should allow $A_S^2(k)$ to be a free function. In this paper we show how we can measure $A_S^2(k)$ without assuming that it is given by any specific model [4].

We parametrize $A_S^2(k)$ as a step function, with amplitudes a_i ($i = 1, 2, ..., 20$) in

CP478 , *COSMO—98*, edited by David O. Caldwell

$\log k$ bins ($k_1 = 0.001\,h\mathrm{Mpc}^{-1}$, and k_2-k_{20} are equally spaced in $\log k$ from k_1 to $k = 0.5h\,\mathrm{Mpc}^{-1}$). The a_i ($i = 1, 2, ..., 20$) are taken to be independent variables. For definiteness, we adopt a model in which the true primordial power spectrum is $A_S^2(k) = 1$ (the scale-invariant Harrison-Zel'dovich spectrum), and the cosmological parameters are $h = 0.5$ (dimensionless Hubble constant), $\Omega_m = 1$ (matter density in units of the critical density), $\Lambda = 0$ (cosmological constant), $\Omega_b = 0.05$ (baryon density in units of the critical density), $\tau_{ri} = 0.05$ (reionization optical depth), and $b = 0.83$ (galaxy bias parameter).

The upcoming data from the Microwave Anisotropy Probe (MAP) [5] and the Sloan Digital Sky Survey (SDSS) [6] provide a unique opportunity for constraining inflationary models. It is useful to expand the temperature fluctuations in the CMB into spherical harmonics: $\delta T/T(\hat{\mathbf{r}}) = \sum_{l,m} a_{T,lm} Y_{lm}(\hat{\mathbf{r}})$, where $\hat{\mathbf{r}}$ is the unit direction vector in the sky, and

$$C_{Tl} \equiv \langle |a_{T,lm}|^2 \rangle = (4\pi)^2 \int \frac{dk}{k} A_S^2(k) \, |\Delta_{Tl}(k, \tau = \tau_0)|^2 . \tag{1}$$

$\Delta_{Tl}(k, \tau = \tau_0)$ is an integral over conformal time τ of the sources that generate the CMB fluctuations, and τ_0 is the conformal time today.

The power spectrum of mass fluctuations in the linear regime today is

$$P(k) = P_0 \, k A_S^2(k) \, T^2(k), \tag{2}$$

where P_0 is a normalization constant, and $T(k)$ is the transfer function, which depends on physics at matter-radiation equality and decoupling. The galaxy redshift survey from the SDSS will allow a determination of $P_G(k)$, the galaxy power spectrum in redshift space; it differs from $P(k)$ due to the bias between mass and galaxy density fields and redshift distortion effects induced by peculiar velocities. We write $P_G(k) = b^2 \left(1 + \frac{2}{3}\beta + \frac{1}{5}\beta^2\right) P(k) \equiv b_{\mathit{eff}}^2 P(k)$, where $\beta \equiv \Omega_m^{0.6}/b$. We take nonlinear effects into consideration by using analytical formulae [4]. We assume the bias b to be constant, which is a good approximation on large scales [7]. On small scales, bias must be scale-dependent and nonlinear. Different galaxy samples (spirals, ellipticals, etc) can be used to test for more complicated bias laws.

Both C_{Tl} and $P_G(k)$ are proportional to the amplitude of $A_S^2(k)$. Fig. 1 shows the accuracy in the determination of the bin amplitudes of $k\,A_S^2(k)$; the 1-σ error bars are shown for using (a) MAP temperature data only, and (b) SDSS data only. The errors have been estimated using the Fisher matrix [4]. Because of the finite resolution of the MAP satellite, the errors on the C_{Tl} increase rapidly with wavenumber for $k \gtrsim 0.2\,h\,\mathrm{Mpc}^{-1}$ (see Fig.1a), while the errors on $P_G(k)$ from the SDSS decrease with wavenumber (see Fig.1b), because on small scales one can average over many k-modes. MAP and SDSS thus quite naturally complement each other in the determination of $A_S^2(k)$. By combining MAP temperature and SDSS data, the primordial power spectrum can be determined to around 16% accuracy for $k \sim 0.01\,h\,\mathrm{Mpc}^{-1}$, and to around 1% accuracy for $k \sim 0.1\,h\,\mathrm{Mpc}^{-1}$, assuming that the cosmological parameters are known [4].

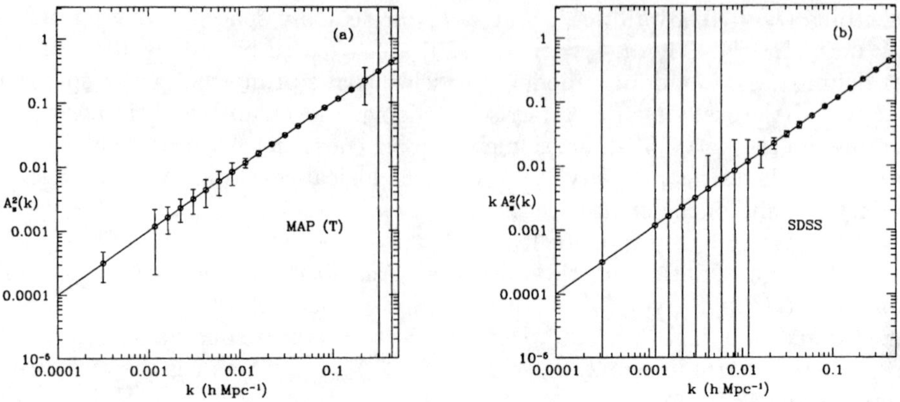

FIGURE 1. The primordial power spectrum with 1-σ error bars (assuming that the cosmological parameters are known) for (a) MAP temperature data only; (b) SDSS data only.

The determination of the primordial power spectrum bin amplitudes a_i does depend on the a priori knowledge of the cosmological parameters. We now estimate four parameters h, Ω_Λ, Ω_b, and τ_{ri}, in addition to the a_i. The a_i and the cosmological parameters cannot be simultaneously determined using MAP temperature data only, because τ_{ri} is degenerate with the a_i for $k \gtrsim 0.01\,h\,\mathrm{Mpc}^{-1}$. However, polarization data removes this degeneracy. Figure 2(a) shows the resulting error bars on the primordial power spectrum if both the temperature and polarization data from MAP are used; they are significantly larger than in Figure 1(a) where only temperature data is used.

Our ability to use the MAP polarization data may be limited by our ability to remove foregrounds. It is therefore important to have an alternative to the MAP polarization data in parameter estimation. Figure 2(b) shows the effect of combining the MAP temperature data with the SDSS data in the analysis, adding the galaxy bias b_{eff} as a fifth parameter. The uncertainty in the determination of the a_i increases by a factor up to 3 on small scales relative to the case in which the cosmological parameters are known *a priori*. This difference only enters through error bars that are only of the order of a few percent. The errors in the a_i are not significantly increased by the inclusion of cosmological parameters in the parameter estimation, because most of the information on the cosmological parameters comes from small angular scales, where the power spectrum can be determined to great accuracy from the SDSS data. Comparison of Figure 2(a) and Figure 2(b) illustrates the importance of combining the MAP data with the SDSS data in the determination of the primordial power spectrum bin amplitudes a_i [4].

Assuming that inflation occurred, all the obtainable information about the inflationary model is contained in the primordial power spectrum $A_S^2(k)$. From the combination of the MAP and SDSS data, we can obtain a measurement of $A_S^2(k)$

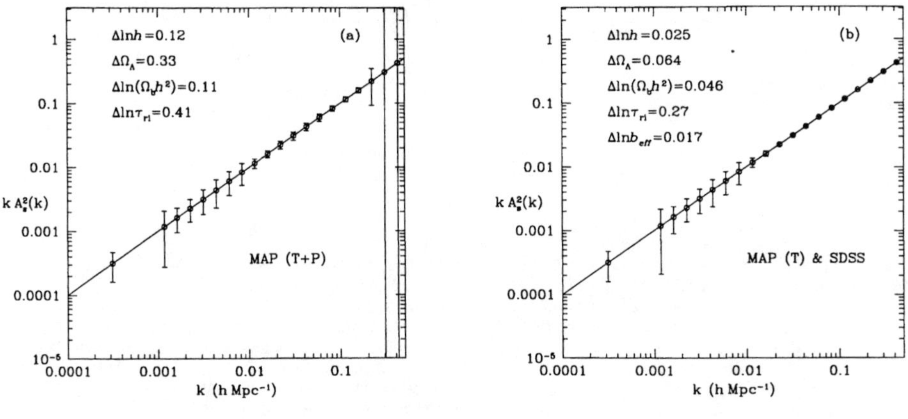

FIGURE 2. The primordial power spectrum with 1-σ error bars. (a) Combined MAP temperature and polarization data, with four cosmological parameters (h, Ω_Λ, Ω_b, τ_{ri}) included in the parameter estimation; (b) Combined MAP temperature and SDSS data, with five cosmological parameters (h, Ω_Λ, Ω_b, τ_{ri}, b) included in the parameter estimation.

without reference to any specific inflationary model, assuming that the primordial fluctuations are adiabatic. If the values of the cosmological parameters that best fit the data are close to what we expect, but the primordial power spectrum differs significantly from the predictions of the current inflationary models, then it will be an indication of new physics in the early universe, and it will provide a solid starting point for building new inflationary models.

This work is supported in part by NSF grant AST98-02802, the MAP/MIDEX project, and NSF grant AST96-16901.

REFERENCES

1. Wang, Y., *Phys. Rev.,* **D50**, 6135 (1994)
2. Holman, R., Kolb, E.W., Vadas, S.L., & Wang, Y., *Phys. Rev.* **D43**, 3833 (1991) and *Phys. Lett.* **B269**, 252 (1991); Adams, J.A., Ross, G.G., & Sarkar, S., *Nucl. Phys. B* **503**, 405 (1997); Lesgourgues, J., Polarski, D., & Starobinsky, A.A., *Nucl. Phys. B* **497**, 479 (1997)
3. Peacock, J.A., *MNRAS* **284**, 885 (1997).
4. Wang, Y., Spergel, D.N., & Strauss, M.A., astro-ph/9802231, *ApJ*, in press (1999).
5. Bennett, C.L., et al., *BAAS* **29**, 1353 (1997).
6. Gunn, J. E., & Weinberg, D. H. 1995, in *Wide-Field Spectroscopy and the Distant Universe*, ed. S. J. Maddox & A. Aragón-Salamanca (Singapore: World Scientific), 3
7. Scherrer, R.J., & Weinberg, D.H., *ApJ* **504**, 607 (1998).

CMB Polarization Constraints on Inflation

William H. Kinney[1]

University of Florida Dept. of Physics
Gainesville, FL 32611

Abstract. In this talk I consider cosmic microwave background (CMB) polarization as a tool for constraining models of cosmological inflation [1]. I use NASA's MAP satellite and the ESA's Planck Surveyor as examples of the observational accuracy that will be achievable in the next few years. I come to two main conclusions. First, cosmic variance in temperature-only measurements severely limits the ability to detect a small tensor/scalar ratio, and measurement of polarization allows significant improvement in the ability to study models that predict very small tensor contributions to the CMB. Second, reionization of the universe can significantly degrade the sensitivity of parameter estimation if temperature anisotropy alone is measured, but measurement of polarization effectively removes the parameter degeneracy between tensor amplitude and reionization optical depth.

I STATISTICS OF CMB MEASUREMENTS: TEMPERATURE AND POLARIZATION

What observations of the cosmic microwave background measure is anisotropy in the temperature of the radiation as a function of direction. It is natural to expand the anisotropy on the sky in spherical harmonics:

$$\frac{\delta T\left(\theta,\phi\right)}{T_0} = \sum_{l=0}^{\infty}\sum_{m=-l}^{l} a_{lm}^{T} Y_{lm}\left(\theta,\phi\right), \tag{1}$$

where $T_0 = 2.728°K$ is the mean temperature of the CMB. Inflation predicts that each a_{lm}^T will be Gaussian distributed with mean $\left\langle a_{lm}^T \right\rangle = 0$ and variance

$$\left\langle a_{l'm'}^{T*} a_{lm}^{T} \right\rangle = C_{Tl}\delta_{ll'}\delta_{mm'}, \tag{2}$$

where angle brackets indicate an average over realizations. For Gaussian fluctuations, the set of C_{Tl}'s completely characterizes the temperature anisotropy. The spectrum of the C_{Tl}'s is in turn dependent on cosmological parameters such as Ω_0, H_0, Ω_B and so forth, so that observation of CMB temperature anisotropy can serve as an exquisitely precise probe of cosmological models. Observational uncertainties in the spectrum of C_l's translate in a straightforward way to observational uncertainties in cosmological parameters. The parameters of interest for testing inflation models are the tensor/scalar ratio r and the spectral index n.

Temperature anisotropy, however, is not the whole story. The cosmic microwave background is also expected to be ***polarized*** due to the presence of fluctuations. The symmetric, trace-free polarization tensor $\mathcal{P}_{ab}$ can be expanded as [2]

$$\frac{\mathcal{P}_{ab}}{T_0} = \sum_{l=0}^{\infty}\sum_{m=-l}^{l}\left[a_{lm}^{E} Y_{(lm)ab}^{E}\left(\theta,\phi\right) + a_{lm}^{B} Y_{(lm)ab}^{B}\left(\theta,\phi\right)\right], \tag{3}$$

---

[1] Electronic address: `kinney@phys.ufl.edu`

CP478 , *COSMO—98*, edited by David O. Caldwell

where the $Y^{E,B}_{(lm)ab}$ are electric- and magnetic-type tensor spherical harmonics, with parity $(-1)^l$ and $(-1)^{l+1}$, respectively. Unlike a temperature-only map, which is described by the single multipole spectrum of C^T_l's, a temperature/polarization map is described by four nonvanishing spectra

$$\left\langle \left|a^T_{lm}\right|^2 \right\rangle \equiv C_{Tl}, \ \left\langle \left|a^E_{lm}\right|^2 \right\rangle \equiv C_{El}, \ \left\langle \left|a^B_{lm}\right|^2 \right\rangle \equiv C_{Bl}, \ \left\langle a^{T*}_{lm} a^E_{lm} \right\rangle \equiv C_{Cl}. \tag{4}$$

Since scalar density perturbations have no "handedness," it is impossible for scalar modes to produce B-mode (pseudoscalar) polarization. Only tensor fluctuations (or foregrounds [3]) can produce a B-mode.

I take as a "fiducial" model COBE normalization [4] with $\Omega_B = 0.05$ and $h = 0.5$, and a reionization optical depth of $\tau_{ri} = 0.05$, corresponding to reionization at a redshift of about $z \sim 13$. Fixed parameters are $\Omega_0 = 1$ and $\Omega_\Lambda = 0$, consistent with inflation. For the purpose of constraining inflation, I will be interested in error ellipses projected onto the $r - n$ plane.

It is particularly significant that even with a *perfect* measurement, the observational uncertainties in the C_l's do not vanish, an effect known as *cosmic variance.* Cosmic variance is simply a finite sample size effect coming from the fact that we have only a single sky to measure, and is more severe at small l. It is in overcoming cosmic variance that precision measurement of CMB polarization holds the most dramatic promise.

II INFLATION MODELS AND OBSERVATIONAL CONSTRAINTS

CMB polarization can be used directly as a probe of the causal structure of the universe [5]. Correlations in the polarization on scales larger than the horizon at last scattering can *only* be produced by a period of inflationary expansion. If observed, such correlations would provide a "smoking gun" for inflation, regardless of any model-dependent assumptions. In this talk, however, I wish to move from inflation in general to inflation in the particular and examine the distinct predictions of different models. It is convenient to define a general classification scheme, or "zoology" for models of inflation. A generic single-field potential can be characterized by two independent mass scales: a "height" Λ^4, corresponding to the vacuum energy density during inflation, and a "width" μ, corresponding to the change in the field value $\Delta\phi$ during inflation. Most (but not all) models can be divided into three general types: *large-field,* with potentials $V(\phi) = \Lambda^4 \exp(\phi/\mu)$ and $V(\phi) = \Lambda^4 (\phi/\mu)^p$, *small-field* with potentials $V(\phi) = \Lambda^4 [1 - (\phi/\mu)^p]$, and *hybrid* with potentials $V(\phi) = [1 + (\phi/\mu)^p]$. Each category of potential occupies a different region in the $r - n$ plane.

Figure 1 and Figure 2 show 2σ error ellipses in the $r - n$ plane for MAP and Planck respectively, superimposed on the predictions for the various inflation scenarios. Errors for temperature only and temperature + polarization are shown. Observation of polarization significantly improves the accuracy of the parameter determination in both cases. This is largely due to removal of the parameter degeneracy between tensor/scalar ratio and reionization optical depth τ_{ri}, as shown in Figure 3 for Planck. Figure 4 shows the advantage of measuring polarization with respect to overcoming cosmic variance limits to detecting a very small tensor contribution to the CMB. A roughly three-fold improvement in sensitivity over that currently planned for Planck will make it possible to distinguish between $p = 2$ and $p > 2$ small-field models.

ACKNOWLEDGMENTS

This work has benefited greatly from conversations with Edward Kolb, Marc Kamionkowski, Scott Dodelson, Albert Stebbins, Gary Hinshaw, Mathias Zaldarriaga, and Lloyd Knox.

This work was supported in part by DOE and NASA grant NAG5-7092 at Fermilab and DOE grant DE-FG05-86ER-40272 at University of Florida.

REFERENCES

1. For a more complete discussion and a list of references, see W. H. Kinney, Phys. Rev. D **58**: 123506 (1998) [astro-ph/9806259].
2. M. Kamionkowski, A. Kosowsky, and A. Stebbins, Phys. R ev. D **55**, 7368 (1997).
3. M. Zaldarriaga and U. Seljak, Report No. astro-ph/9803150.
4. E. F. Bunn and M. White, Astrophys. J. **480**, 6 (1997).
5. D. N. Spergel and M. Zaldarriaga, Phys. Rev. Lett. **79** , 2180 (1997).

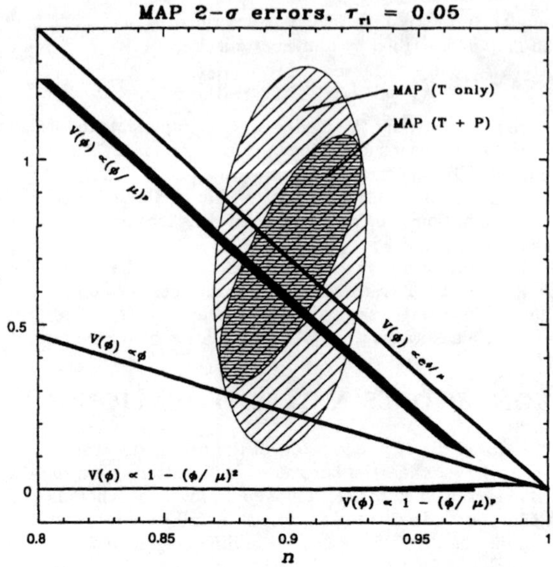

FIGURE 1. 2σ error ellipses in the $r - n$ plane for MAP ($\tau_{\rm ri} = 0.05$), superimposed on the predictions of various generic inflation models.

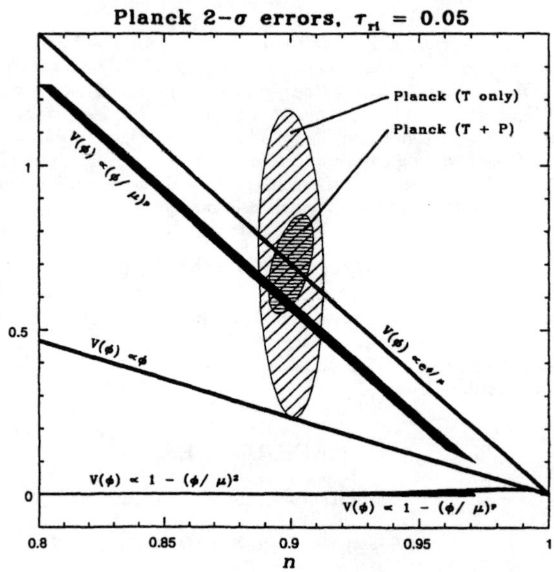

FIGURE 2. Error ellipses for Planck ($\tau_{\rm ri} = 0.05$). As with MAP, measurement of polarization significantly improves parameter estimation.

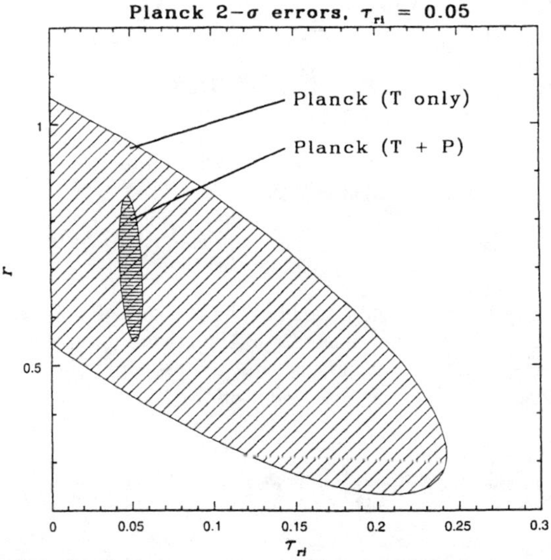

FIGURE 3. Error ellipses for Planck in the $\tau_{ri} - r$ plane. The parameter degeneracy between reionization optical depth and tensor/scalar ratio is removed completely by including polarization information, a significant advantage.

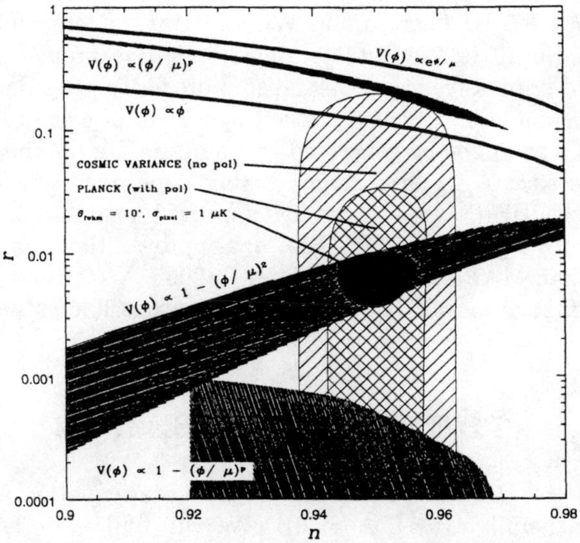

FIGURE 4. The $r - n$ plane on a logarithmic scale, highlighting the predictions of small-field models. Error ellipses are for the cosmic variance limit of a T-only measurement, Planck (with polarization), and a hypothetical experiment with the same $10'$ angular resolution as Planck but a factor of three higher sensitivity. This is sufficient to detect $r \sim 0.01$, which makes is possible to distinguish between $p = 2$ and $p > 2$ small-field models.

Position of First Doppler Peak

Paul H. Frampton

University of North Carolina,
Chapel Hill, NC 27599-3255.

Abstract. A analytic derivation is made of the position of the first Doppler peak for the cosmic background radiation anisotropy. The resulting depndences on Ω_m and Ω_Λ agree well with numerical results showing that the principal physics is the geometry of the geodesics since last scatter at recombination time.

CBR TEMPERATURE ANISOTROPY

One key issue in both cosmology and in fundamental theory of quantum gravity is the cosmological constant. As is well explained in standard reviews of the cosmological constant [1,2] the theoretical expectation for Λ exceeds its observational value by 120 orders of magnitude. The Cosmic Background Radiation (CBR) was discovered [3] in 1965 by Penzias and Wilson. But detection of its temperature anisotropy waited until 1992 when [4,5] the Cosmic Background Explorer (COBE) satellite provided impressive experimental support for the Big Bang model. COBE results are consistent with a scale-invariant spectrum of primordial scalar density fluctuations, such as might be generated by quantum fluctuations during inflation [6–12]. COBE's success inspired many further experiments with higher angular sensitivity than COBE ($\sim 1^o$). NASA has approved a satellite mission MAP (Microwave Anisotropy Probe) for 2000. ESA has approved the Planck surveyor - even more accurate than MAP - a few years later in 2005. With these experiments, the location of the first accoustic (Doppler) peak and possible subsequent peaks will be resolved.

THE SPECIAL CASE $\Lambda = 0$

When $\Lambda = 0$, the Einstein-Friedmann cosmological equation can be solved analytically (not generally true if $\Lambda \neq 0$). We will find $l_1 \sim 1/\sqrt{\Omega_m}$ as follows. Take:

$$ds^2 = dt^2 - R^2 \left[d\Psi^2 + sinh^2 \ \Psi \ d\Theta^2 + sinh^2 \ \Psi \ sin^2 \ \Theta \ d\Phi^2 \right] \tag{1}$$

CP478 , *COSMO—98*, edited by David O. Caldwell

For a geodesic $ds^2 = 0$ and so:

$$\frac{d\Psi}{dR} = \frac{1}{R} \tag{2}$$

The Einstein equation is

$$\left(\frac{\dot{R}}{R}\right)^2 = \frac{8\pi G\rho}{3} + \frac{1}{R^2} \tag{3}$$

so that

$$\dot{R}^2 R^2 = R^2 + aR \tag{4}$$

with $a = \Omega_0 H_0^2 R_0^3$ and hence

$$\frac{d\Psi}{dR} = \frac{1}{\sqrt{R^2 + aR}} \tag{5}$$

This can be integrated to find

$$\Psi_t = \int_{R_t}^{R_0} \frac{dR}{\sqrt{(R + a/2)^2 - (a/2)^2}} \tag{6}$$

The substitution $R = \frac{1}{2}a(coshV - 1)$ leads to

$$\Psi_t = cosh^{-1}(\frac{2R_0}{a} - 1) - cosh^{-1}(\frac{2R_t}{a} - 1) \tag{7}$$

Using $sinh(cosh^{-1}x) = \sqrt{x^2 - 1}$ gives

$$sinh\Psi_t = \sqrt{\left(\frac{2(1-\Omega_0)}{\Omega_0} + 1\right)^2 - 1} - \sqrt{\left(\frac{2(1-\Omega_0)R_t}{\Omega_0 R_0} + 1\right)^2 - 1} \tag{8}$$

The second term of Eq.(8) is negligible as $R_t/R_0 \to 0$ With the metric of Eq.(1) the angle subtended now by the horizon then is

$$\Delta\Theta = \frac{1}{H_t R_t sinh\Psi_t} \tag{9}$$

For $Z_t = 1,100$, the red-shift of recombination one thus finds

$$l_1(\Lambda = 0) \simeq \frac{2\pi Z_t^1/2}{\sqrt{\Omega_m}} \simeq \frac{208.4}{\sqrt{\Omega_m}} \tag{10}$$

This is plotted in Fig. 1 of our work [13].

THE GENERAL CASE $\Lambda \neq 0$

When $\Lambda \neq 0$

$$\dot{R}^2 R^2 = -kR^2 + aR + \Lambda R^4/3 \tag{11}$$

It is useful to define the contributions to the energy density $\Omega_m = 8\pi G\rho/3H_0^2$, $\Omega_\Lambda = \Lambda/3H_0^2$, and $\Lambda_C = -k/H_0^2 R_0^2$. These satisfy

$$\Omega_m + \Omega_\Lambda + \Omega_C = 1 \tag{12}$$

Then

$$l_1 = \pi H_t R_t sinh \Psi_t \tag{13}$$

where

$$\Psi_t = \sqrt{\Omega_C} \int_1^\infty \frac{dw}{\sqrt{\Omega_\Lambda + \Omega_C w^2 + \Omega_m w^3}} \tag{14}$$

After changes of variable one arrives at

$$l_1 = \pi \sqrt{\frac{\Omega_0}{\Omega_C}} \sqrt{\frac{R_0}{R_t}} sinh \left(\sqrt{\Omega_C} \int_1^\infty \frac{dw}{\sqrt{\Omega_\Lambda + \Omega_C w^2 + \Omega_m w^3}} \right) \tag{15}$$

(For positive curvature ($k = +1$) replace *sinh* by *sin*). For the case $k = 0$, the flat universe predicted by inflation, with $\Omega_C = 0$ Eq.(15) reduces to

$$l_1 = \pi \sqrt{\Omega_m} \sqrt{\frac{R_0}{R_t}} \int_1^\infty \frac{dw}{\sqrt{\Omega_\Lambda + +\Omega_m w^3}} \tag{16}$$

These are elliptic integrals, easily do-able by Mathematica. They resemble the formula for the age of the universe:

$$A = \frac{1}{H_0} \int_1^\infty \frac{dw}{w\sqrt{\Omega_\Lambda + \Omega_C w^2 + \Omega_m w^3}} \tag{17}$$

In Fig. 2 of [13] there is a plot of l_1 versus Ω_m for $\Omega_C = 0$. In Fig. 3 are the main result of the iso-l_1 lines on a $\Omega_m - \Omega_\Lambda$ plot for general Ω_C with values of l_1 between 150 and 270 in increments $\Delta l_1 = 10$. The final Fig. 4 of [13] gives a three-dimensional plot of $\Omega_m - \Omega_\Lambda - l_1$.

We can look at the cumulative world data on C_l versus l. Actually even the existence of the first Doppler peak is not certain but one can see evidence for the rise and the fall of C_l. In Fig. 2 of [14] we see such 1998 data and with some licence say that $150 \leq l_1 \leq 270$. The exciting point is that the data are expected to improve markedly in the next decade. In Fig. 3 of [14] there is an artist's impression of both MAP data (expected 2000) and Planck data(2006); the former should pin down l_1 with a small error and the latter is expected to give accurate values of C_l out to $l = 1000$. But even the spectacular accuracy of MAP and Planck will specify only one iso-l_1 line in the $\Omega_m - \Omega_\Lambda$ plot and not allow unambiguous determination of Ω_Λ. Fortunately this ambiguity can be removed by a completely independent set of observations - of high-Z supernovae. The present data suggest [15,16] $\Omega_m \simeq 0.3$ and $\Omega_\Lambda \simeq 0.7$. The non-zero Λ implies that we live in a special cosmic era: Λ was negligible in the past but will dominate the future giving exponential growth $R \sim e^{\Lambda t}, t \to \infty$. This cosmic coincidence is addressed by quintessence [17–19]. The principal point of our own work in [13] is that the value of l_1 depends almost completely only on the geometry of geodesics since recombination, and little on the details of the accoustic waves, since our iso-l_1 plot agrees well with the numerical results of White *et al.* [20].

ACKNOWLEDGEMENT

This work, carried out with Y.J. Ng and R. Rohm, was supported in part by the US Department of Energy under Grant No. DE-FG05-85ER41036.

REFERENCES

1. S. Weinberg, Rev. Mod. Phys. **61,** 1 (1989).
2. Y.J. Ng, Int. J. Mod. Phys. **D1,** 145 (1992).
3. A.A. Penzias and R.W. Wilson, Ap. J. **142,** 419 (1965).
4. G.F. Smoot *et al.*, Ap. J. Lett. **396,** L1 (1992).
5. K. Sanga *et al.*, Ap. J. **410,** L57 (1993).
6. J.M. Bardeen, P.J. Steinhardt and M.S. Turner, Phys. Rev. **D28,** 679 (1983).
7. A.A. Storobinsky, Phys. Lett. **B117,** 175 (1982).
8. A.H. Guth and S.-Y. Pi, Phys. Rev. Lett. **49,** 1110 (1982).
9. S.W. Hawking, Phys. Lett. **B115,** 295 (1982)
10. A.H. Guth, Phys. Rev. D28, 347 (1981).
11. A.D. Linde, Phys. Lett. **B108,** 389 (1982).
12. A. Albrecht and P.J. Steinhardt, Phys. Rev. Lett. **48,** 1220 (1982).
13. P.H. Frampton, Y.J. Ng and R. Rohm. Mod. Phys. Lett. **A13,** 2541 (1998).
14. M. Kamionkowski. *astro-ph/9803168.*
15. S.J. Perlmutter *et al.*, (The Supernova Cosmology Project). *astro-ph/9608192.*
16. S.J. Perlmutter *et al.*, (The Supernova Cosmology Project). *astro-ph/9712212.*
17. P.J.E. Peebles and B. Ratra, Ap. J. Lett. **325,** L17 (1988).
18. I. Zlatev, L. Wang and P.J. Steinhardt. *astro-ph/9807002.*
19. C. Kolda and D. Lyth. *hep-ph/9811375.*
20. M. White and D. Scott, Ap.J. **459,** 415 (1996), *astro-ph/9508157.* W. Hu and M. White, Ap. J. **471,** 30 (1996), *astro-ph/9602019*; Phys. Rev. Lett. **77,** 1687 (1996), *astro-ph/9602020.* M. White, Ap. J. **506,** 495 (1998), *astro-ph/9802295.*

Where is COBE maps' non-Gaussianity?

João Magueijo[1], Pedro G. Ferreira[2,3]1, and Krzysztof M. Górski[4,5]

[1] *Theoretical Physics, Imperial College, Prince Consort Road, London SW7 2BZ, UK*
[2] *Theory Group, CERN, CH-1211, Geneve 23 ,Switzerland*
[3] *CENTRA, Instituto Superior Tecnico, Lisbon, Portugal*
[4] *Theoretical Astrophysics Center, Juliane Maries Vej 30, DK-2100, Copenhagen Ø, Denmark*
[5] *Warsaw University Observatory, Warsaw, Poland*

Abstract. We review our recent claim that there is evidence of non-Gaussianity in the 4 Year COBE DMR data. We present some new results concerning the effect of the galactic cut upon the non-Gaussian signal. These findings imply a localization of the non-Gaussian signal on the Northern galactic hemisphere.

I EVIDENCE FOR CMB NON-GAUSSIANITY

In a recent paper [1] we showed that the 4 Year COBE DMR data exhibits evidence of non Gaussianity at a high confidence level. We made use of statistical tools described in more detail in [2,3]. Since then our result has been corroborated by two other groups [4,5]. In this review we revisit our analysis, and the tests to which we have subjected it, including some new results.

In our analysis we propose, and work with, an estimator for the *normalized bispectrum*, denoted by I^3_ℓ. We refer the reader to [1] for its definition. We then applied this estimator to the inverse noise variance weighted, average maps of the 53A, 53B, 90A and 90B *COBE*-DMR channels, with monopole and dipole removed, at resolution 6, in ecliptic pixelization. We use the extended galactic cut of [6], and [7] to remove most of the emission from the plane of the Galaxy. We apply our statistics to the DMR maps before and after correction for the plausible diffuse foreground emission outside the galactic plane as described in [8], and [9].

By means of Monte Carlo simulations we also found the distributions $P_\ell(I^3_\ell)$ for what we should have seen assuming a Gaussian signal, which is then processed by the experimental set up associated with DMR. These $P(I^3_\ell)$ were inferred from 25000 realizations (see Fig. 1). The observed I^3_ℓ and the distributions $P_\ell(I^3_\ell)$ are plotted in Fig. 1. One immediately notices the presence of a significant deviant.

In order to quantify this deviant we define the goodness of fit statistic

$$X^2 = \frac{1}{N}\sum_\ell X^2_\ell = \frac{1}{N}\sum_\ell(-2\log P_\ell(I^3_\ell) + \beta_\ell), \tag{1}$$

CP478 , *COSMO—98*, edited by David O. Caldwell

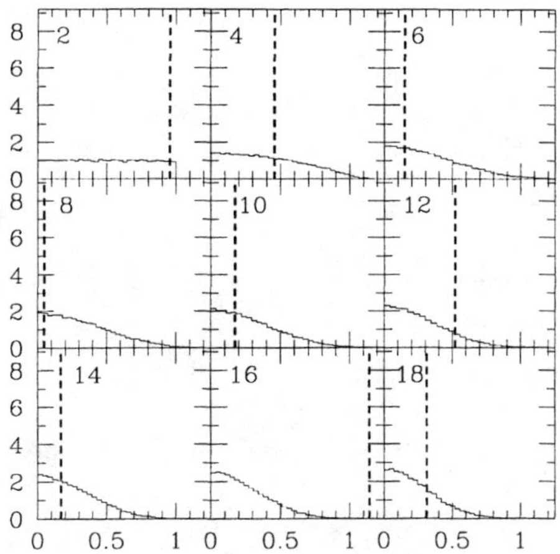

FIGURE 1. The vertical thick dashed line represents the value of the observed I_ℓ^3. The solid line is the probability distribution function of I_ℓ^3 for a Gaussian sky with extended galactic cut and DMR noise.

where the constants β_ℓ are defined so that for each term of the sum $\langle X_\ell^2 \rangle = 1$. The definition reduces to the usual chi squared for Gaussian P_ℓ. We build a X^2 for the *COBE*-DMR data from the $P_\ell(I_\ell^3)$ inferred from Monte Carlo simulations, taking special care with the numerical evaluation of the constants β_ℓ. We call this function X^2_{COBE}. We then find its distribution $F(X^2_{COBE})$ from 10000 random realizations. This is very well approximated by a χ^2 distribution with 12 degrees of freedom. We then compute X^2_{COBE} with the actual observations and find $X^2_{COBE} = 1.81$. One can compute $P(X^2_{COBE} < 1.81) = 0.98$. Hence, it would appear that we can reject Gaussianity at the 98% confidence level.

II IS IT A SYSTEMATIC EFFECT?

We checked that this result could not be due to the following systematics:

1. Foregrounds contamination:
   - Dust (using the DIRBE sky maps and also the Schlegel *et al* dust model)
   - Synchrotron (with the Haslam template)
   - Foreground corrected maps (effect persists on corrected maps)
2. Noise model:
   - Anisotropic sky coverage
   - Noise correlations between different pixels
   - Analysis of noise templates
3. Galactic cut:

- Dependence on shape ("custom" versus constant elevation)
- Dependence on elevation
- Dependence on monopole and dipole subtraction, before or after the cut, with or with out galaxy.

4. Possible small residual errors in corrections for
   - Spurious offsets induced by the cut.
   - Instrument susceptibility to the Earth magnetic field.
   - Callibration errors .
   - Errors due to incorrect removal of the COBE Doppler and Earth Doppler signals.
   - Errors in correcting for emissions from the Earth, and eclipse effects.
   - Artifacts due to uncertainty in the correction for the correlation created by the low-pass filter on the lock-in amplifiers (LIA) on each radiometer
   - Errors due to emissions from the moon, and the planets.
5. Assumptions in Monte Carlos:
   - Dependence on power spectrum tilt
   - Dependence on smooth versus discontinuous power spectrum
   - Dependence on beam shape
   - Dependence on pixelization.

In fact, the confidence level quoted above reflects the worse line up of systematics. If we try to correct for systematics, in general the confidence level for rejecting Gaussianity is enlarged to beyond 99%, as we describe in more detail in [3]

III WHERE IS THE NON-GAUSSIANITY?

We now concentrate on a subset of tests involving the galactic cut which we applied to our result. Changing the galactic cut affects sample variance besides eliminating possible contaminations from the map. We considered extensive variations of the cut, including additions of polar cap cuts to the extended cut.

We found that cuts from the pole affect the result more than cuts from the equator. This suggests that the effect may be localized near the Poles. We therefore decided to compare the effect of applying cuts only in the North or South galactic poles. We considered cuts down to 60° (2668 pixels excluded). We find the curious result that cutting Northern caps is more damaging for the nonGaussian spike than cutting Southern caps (fig. 2). Indeed the first few Southern cap cuts appear to increase the spike.

Non-Gaussianity could in principle be localized in Fourier space without being localized in real space[1]. Examples of such behaviour are given in [10]. We believe

---

[1] See New Scientist, Ed comment to letter, 12/12/98, for a layman's opinion.

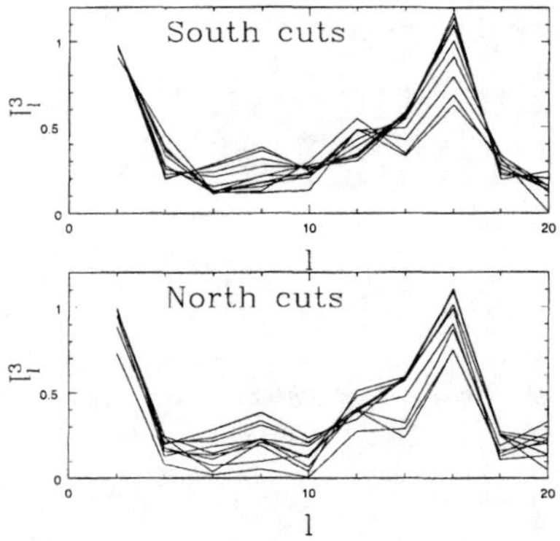

FIGURE 2. The effect of combining the extended cut with polar cap cuts down to 60° in the North or South hemispheres.

that the signal we have found is essentially localized in Fourier space. However the results we have just presented suggest that our signal may indeed be also localised in real space, around the Northern galactic cap.

Acknowledgements JM would like to thank the organizers for an excellent meeting. We thank JNICT, NASA-ADP, NASA-COMBAT, NSF, RS, Starlink, TAC for support.

REFERENCES

1. Ferreira, P. G., Górski, K. M., Magueijo, J. (1998) *Ap. J. Lett.* **503** L1-L4.
2. J. Magueijo, P. Ferreira, K. Gorski, in "The CMB and the Planck Mission", proceedings of the UIMP98 meeting held in Santander, Spain, June 22-26, 1998. astro-ph/9810414.
3. Magueijo, J., Ferreira, P.G., Górski, K.M., in preparation (1999).
4. D. Novikov, H. Feldman, S. Shandarin, astro-ph/9809238.
5. J. Pando, D. Valls-Gabaud, L.-. Fang, astro-ph/9810165.
6. Banday, A.J., Górski, K.M., Bennett, C.L., Hinshaw, G., Kogut, A., Lineweaver, C., Smoot, G.F., and Tenorio, L. (1997) *Ap.J.*, **475**, 393
7. Bennet, C.L., Banday, A.J., Górski, K.M., Hinshaw, G., Jackson, P.D., Keegstra, P., Kogut, A., Smoot, G.F., Wilkinson, D.T., Wright, E. L. (1996) *Ap. J.* **464**, 1
8. Kogut, A., Banday, A. J., Bennett, C. L., Górski, K. M., Hinshaw, G., Smoot, G. F., Wright, E. L. (1996) *Ap. J.* **464** L5
9. Górski, K. M., Hinshaw, G., Banday, A. J., Bennett, C. L., Wright, E. L., Kogut, A., Smoot, G. F., Lubin, P. (1996) *Ap. J. Lett.* **464**, L11
10. P.Ferreira and J.Magueijo, *Phys.Rev.* **D 55** 3358 (1997).

Generating Non-Gaussian Adiabatic Fluctuations from Inflation

D.S. Salopek[1 2]

Department of Physics & Astronomy
6224 Agricultural Road, University of British Columbia
Vancouver, Canada V6T 1Z1

Abstract. As the quality of cosmological data continue to improve, it is natural to test the statistics of primordial fluctuations: are they Gaussian or non-Gaussian? I review a model which generates non-Gaussian adiabatic fluctuations from inflation. Current investigations suggest that there may possibly be a non-Gaussian signal in large angle cosmic microwave background anisotropy data. Statistics of microwave anisotropies could thus serve as a powerful probe of the very early Universe.

INTRODUCTION

The inflationary expansion of the very early universe acts like a microscope that magnifies short scale quantum fluctuations to encompass our entire observable Universe. In a sense, microwave background anisotropy maps such as the COBE DMR maps provide an image of the vacuum state that arose from inflation. If the field interactions during the inflationary epoch are linear, then one expects that the statistics of vacuum fluctuations will be Gaussian — this is the situation that most researchers are hoping for. However, there are indeed models where significant non-linearities arise which would produce a non-Gaussian signal. Here, I will review such an inflation model which produces non-Gaussian adiabatic fluctuations [1].

By considering 3-point correlations of spherical harmonic amplitudes in COBE DMR maps, Ferreira *et al* [2] have found evidence of a non-Gaussian signal (see also Magueijo [3]). They point out a non-Gaussian signal associated with the $\ell = 16$ multi-pole of a spherical harmonic expansion. They are not able to associate the signal with any known foreground emission, although they are concerned that the power takes a mysterious dip at this same value of ℓ. Their work requires further investigation.

1) E-mail: dsscosmos@yahoo.com
2) Current mailing address: Space-Time Institute, 14915-105 Avenue, Edmonton, Alberta, Canada T5P 4M2

CP478 , *COSMO—98*, edited by David O. Caldwell

The computation of nonlinear effects during inflation is a challenging problem. However, nonlinearities are essential to the formulation of the inflationary scenario. For example, at the end of inflation, the inflaton must transfer its energy to a bath of radiation and matter; nonlinearities must thus be significant. The 'heating' of the Universe is an interesting problem that is currently under investigation (see, *e.g.*, Kofman [9]). The computation of nonlinear effects during the slow-roll period of inflation (*e.g.*, before heating of the Universe) faces two severe theoretical obstacles:

(1) The complete field theory that encompasses inflation is not known. String theory may provide the required framework although this remains to be seen.

(2) Quantum gravity effects are important during inflation. For example, it is generally agreed that the quantum generation of tensor modes (gravitational waves) during inflation is a quantum gravity effect although it is only a 'linear' effect.

Soon after the COBE detection in 1992, it was pointed out by Salopek [6] and Davis *et al* [7] and subsequently verified by many others that tensor modes from inflation could indeed contribute significantly to COBE's signal. In particular, Salopek [6] pointed out that at most 50% of COBE's signal could be attributed to tensor modes arising from power-law inflation otherwise there would not be enough fluctuations to account for the observed clustering of galaxies. (For a careful analysis of how current cosmological data restricts tensor modes, consult Zibin *et al* [8]; they confirm the above 50% bound that was deduced over five years earlier. In this conference, Kinney [9] explains how the measurement of polarization by future microwave background anisotropy experiments will help to determine the contribution of tensor modes more accurately.)

In order to bypass severe problems associated with nonlinear quantum effects of gravity, Stewart and Parry and I advocate solving the Hamilton-Jacobi equation for general relativity [10] which provides a nonlinear semiclassical description of gravity interacting with matter. We attempt to imitate the historical development of the theory of atomic spectra. Before the development of the quantum theory in 1926, the semiclassical theory of Bohr and Sommerfeld provided a useful although imperfect description of various atoms.

The non-Gaussian model described below employs several unique features of the Hamilton-Jacobi equation. It has a distinct observational signature: it produces a non-Gaussian distribution for cosmic microwave temperature anisotropies. This model is constrained by the analysis of DMR maps by Kogut *et al* [4] and Ferreira *et al* [2]. Typically each non-Gaussian model produces a specific signature which allows one to discriminate among the different models. This property provides a powerful probe of the very early Universe.

HAMILTON-JACOBI EQUATION FOR COSMOLOGY

In a Hamilton-Jacobi (HJ) formulation, there are just two laws that govern the inflationary epoch:

$$\mathcal{H}(x) = 0 \,, \quad \mathcal{H}_i(x) = 0 \,. \tag{1}$$

The first legislates that the **total energy density vanishes** at every point in the Universe, whereas the second legislates that the **total momentum density vanishes** at every point. In fact, it is a great accomplishment that one may write two mathematical laws that govern the very early Universe. Explicitly, the energy and momentum densities may be written in terms of the scalar fields, ϕ_a, the 3-metric, γ_{ij}, and their respective momenta, π^{ϕ_a}, π^{ij}:

$$\mathcal{H} = \kappa\,\gamma^{-1/2}\,[2\gamma_{ik}\,\gamma_{jl} - \gamma_{ij}\gamma_{kl}]\pi^{ij}\,\pi^{kl} + \frac{\kappa}{2}\sum_a \gamma^{-1/2}\left(\pi^{\phi_a}\right)^2 + \kappa\gamma^{1/2}V(\phi_a) +$$

$$\left\{-\frac{1}{2\kappa}\gamma^{1/2}\,R + \frac{1}{2\kappa}\gamma^{1/2}\sum_a \phi_{a|i}\phi_a{}^{|i}\right\}, \tag{2a}$$

$$\mathcal{H}_i = -2\left(\gamma_{ij}\,\pi^{jl}\right)_{,l} + \pi^{kl}\gamma_{kl,i} + \sum_a \pi^{\phi_a}\phi_{a,i}\,, \tag{2b}$$

where $\kappa = 8\pi G$ and R denotes the Ricci curvature of the 3-metric. The object of principal interest is the generating functional $\mathcal{S} \equiv \mathcal{S}[\gamma_{ij}(x), \phi_a(x)]$ which assigns a number (in general complex) to each field configuration $\phi_a(x)$ on a space-like hypersurface described by the 3-metric $\gamma_{ij}(x)$. The momenta are given by functional derivatives of the generating functional:

$$\pi^{\phi_a} = \frac{\delta\mathcal{S}}{\delta\phi_a}\,, \qquad \pi^{ij} = \frac{\delta\mathcal{S}}{\delta\gamma_{ij}}\,. \tag{2c}$$

The constraints equations (1) are self-contained equations for the generating functional $\mathcal{S}$, which is just the 'phase' of the wavefunctional in the semiclassical limit.

Strong-Coupling Limit: $G \to \infty$

To solve for the generating functional, $\mathcal{S}$, one must resort to approximation methods. In the limit that Newton's constant $G = \kappa/(8\pi)$ is large, one may safely neglect the terms appearing within the braces {} of eq.(2a). The resulting system, *Strongly-Coupled Gravity and Matter*, is exactly solvable [11]. Its simplest solution is of the form

$$\mathcal{S} = -2\int d^3x\,\gamma^{1/2}\,H(\phi_a) \tag{3a}$$

where the function $H \equiv H(\phi_a)$ is a function of the scalar fields and it satisfies the separated Hamilton-Jacobi equation (SHJE):

$$H^2 = \frac{2}{3}\sum_a\left(\frac{\partial H}{\partial\phi_a}\right)^2 + \frac{1}{3}V(\phi_a)\,. \tag{3b}$$

Explicit general solutions of the SHJE exist for potentials of the exponential type

$$V(\phi) = V_0 \exp\left(\sum_a B_a\,\phi_a\right)\,; \tag{4}$$

that is when $\ln V(\phi)$ is linear, H is exactly solvable. More complicated potentials are obtained by gluing together different exponential potentials together [1].

NON-GAUSSIAN ADIABATIC FLUCTUATIONS

For inflation with two scalar fields, non-Gaussian fluctuations arise in a model where one glues continuously three exponential potentials of the type given by eq.(4). Such a model is illustrated in Fig.1. The resulting distribution in large angle temperature anisotropies is shown in Fig. 2. It is strongly non-Gaussian. The impact of similar models on the distribution of galaxies has been investigated by Fan and Bardeen [12] as well as Moscardini *et al* [13].

The challenge in the future is to see if the parameters of the model may be adjusted to fit the non-Gaussian signal of Ferreira *et al* [2] and yet be consistent with the limits of Kogut *et al* [4].

DISCUSSION

Recent analysis suggests that the COBE DMR map may have a non-Gaussian signal [2]. It is definitely possible to construct inflation models which produce non-Gaussian fluctuations [1]. By studying the statistics of microwave background

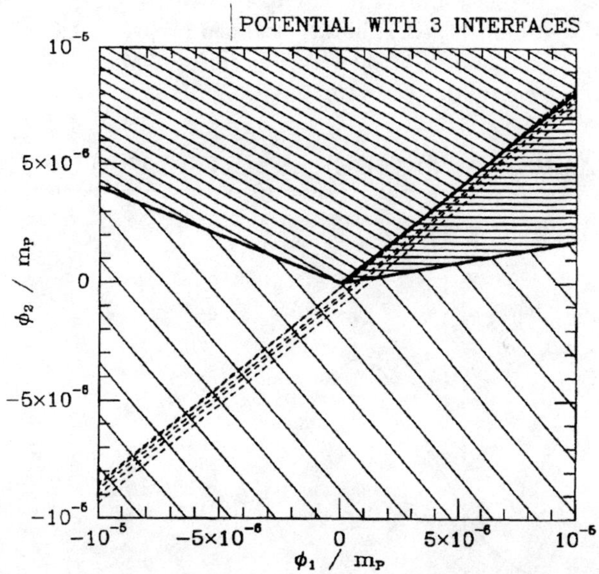

FIGURE 1. Non-Gaussian fluctuations may be generated during inflation when two scalar fields pass over a sharp but continuous feature in their potential. The light solid curves are lines of uniform scalar field potential; the heavy lines are the boundaries between the 3 regions where $\ln V$ is linear. By choosing the initial conditions, the scalar field trajectories (broken lines) which start in the lower left hand corner may pass sufficiently near the origin, and nonlinear effects become important.

fluctuations, one hopes to develop a powerful probe of the very early Universe. The situation should improve dramatically when the next generation of microwave background anisotropy satellites such as MAP and PLANCK are launched.

REFERENCES

1. Salopek, D.S., *Phys. Rev.* **D45**, 1139 (1992).
2. Ferreira, P.G., Magueijo, J., and Gorski, K.M., astro-ph/9803256 (1998).
3. Magueijo, J., these proceedings (1999).
4. Kogut, A., Banday, A.J., Bennett, C.L., Gorski, K.M., Hinshaw, G., Smoot, G.F., and Wright, E.L., *Astrophy. J.* **464**, L5 (1996); *ibid* L29 (1996).
5. Kofman, L., these proceedings (1999).
6. Salopek, D.S. *Phys. Rev. Lett.*, **69**, 3602 (1992).
7. Davis, R.L. *et al*, *Phys. Rev. Lett.* **69**, 1856 (1992).
8. Zibin, J.P., Scott, D., and White, M., University of British Columbia preprint (1999).
9. Kinney, W. these proceedings (1999).
10. Parry, J., Salopek, D.S. and Stewart, J.M., Phys. Rev. **D49**, 2872 (1994).
11. Salopek, D.S., *Class. Quant. Grav.* **15**, 1185 (1998); *ibid* **16**, 299 (1999).
12. Fan, Z.H., and Bardeen J.M., Predictions of a Non-Gaussian Model for Large Scale Structure, UW-PT-92-11, KEK library scanned preprint: 9207442 (1992).
13. Moscardini, L., Borgani, S., Coles, P., Lucchin F., Matarrese S., Messina, A., and Plionis, M., *Astrophys. J. Lett.* **413**, 55 (1993).

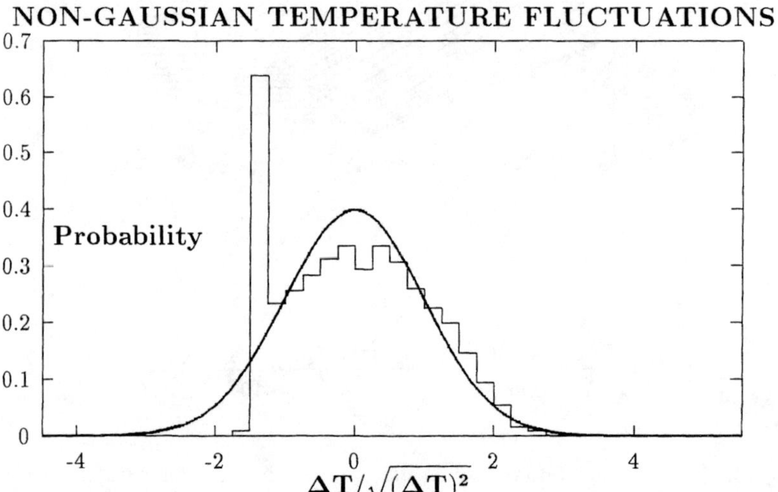

FIGURE 2. For the model described in Fig. 1, a very significant non-Gaussian histogram in large angle cosmic microwave maps is produced. This extreme distribution is shown for illustrative purposes. By adjusting the input parameters, one may arbitrarily adjust the height and placement of the left peak. For comparison, the dark curve is the best fit Gaussian.

Infrared Galaxies

Andrew W. Blain

Cavendish Laboratory, Madingley Road, Cambridge, CB3 0HE, UK

Abstract. To study large-scale structure in the Universe a full census of the contents are required. This is even more important when the processes of galaxy formation are being investigated. In the last year the population of distant galaxies that emit most of their energy in the infrared waveband have been studied in unprecedented detail. The intensity of background radiation at wavelengths between 3 mm and 10 μm has been determined to within a factor of about 2, and much of this background radiation has been resolved into individual galaxies. These galaxies are largely hidden from view in the optical waveband by absorption due to interstellar dust. By combining this new knowledge with the ever growing body of information gathered using optical telescopes, the process of galaxy formation is slowly being revealed. Great opportunities for studying the distant Universe are promised by advances in instrumentation in the millimeter (mm), submm and far-infrared wavebands over the next decade.

INTRODUCTION

Progress in observational cosmology has been unprecedentedly rapid over the last five years, with the advent of the *Hubble Deep Fields*, Lyman-break galaxies and the Keck telescopes. However, there is still no detailed understanding of how and when galaxies formed. In the optical waveband progress should continue, due to the large number of 8-m-class telescopes being commissioned, the increasing sophistication of multi-object spectrographs and new wide-field digital surveys.

A large amount of the energy output of galaxies is emitted in the far-infrared waveband, as revealed by the *IRAS* satellite [1]. This is thermal emission from interstellar dust grains, which efficiently absorb blue and ultraviolet photons, whether from young high-mass stars or active galactic nuclei (AGN). Heated to temperatures of a few tens of Kelvin, the rest-frame emission spectrum of the dust peaks at wavelengths close to 100 μm. In the local Universe, a comparable amount of energy is emitted in the optical and far-infrared wavebands. However, because the *IRAS* survey extended only to redshifts $z \simeq 0.2$, until recently it was impossible to investigate the reprocessed dust radiation from distant galaxies directly, either individually or as an integrated background [2].

CP478 , *COSMO—98*, edited by David O. Caldwell

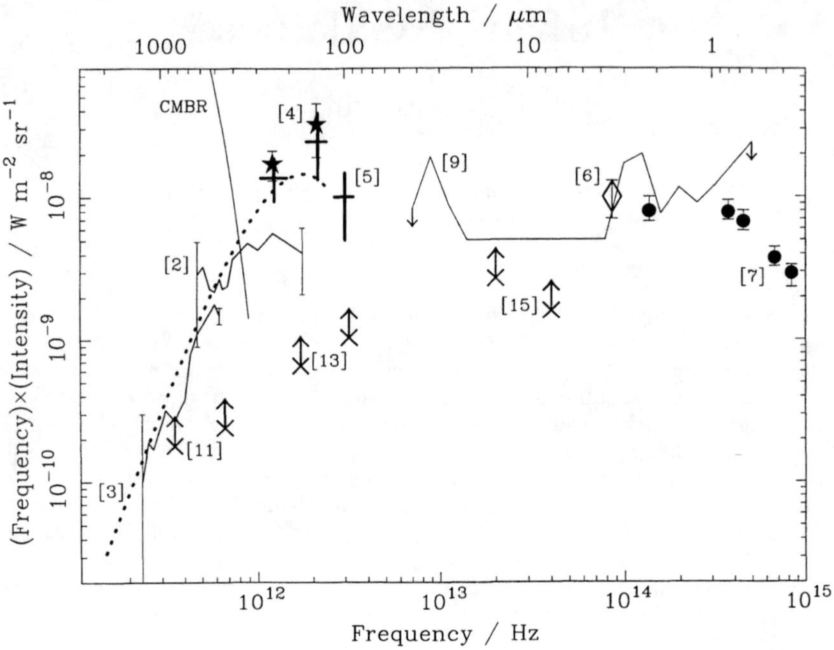

FIGURE 1. The background radiation intensity from the mm to optical waveband. References are numbered. Only [2] was obtained before 1998. Note that at wavelengths of 15 and 850 μm [11,15] most of the background has probably been resolved into its constituent galaxies.

BACKGROUNDS AND DISCRETE SOURCES

Within the last year our knowledge of the quantity of energy re-radiated by dust in distant galaxies has increased dramatically (see Figure 2). Using the FIRAS and DIRBE instruments on the *COBE* satellite, and the *HST (Hubble Space Telescope)*, the intensity of extragalactic background radiation has been determined to within a factor ~ 2 between wavelengths of 1 μm and 2 mm [3–8]. Upper limits to the background intensity from γ–γ/e$^+$e$^-$ pair production along the line of sight to nearby AGN have also been tightened [9]. Two years ago, there was only a single background estimate [2]. The new observations represent a tremendous advance.

The first results of direct surveys to detect the faint galaxies that contribute to the infrared background radiation intensity have also been published this year. Varying fractions of the background have been resolved into individual galaxies using three sets of instruments. At long wavelengths of 850 and 450 μm, the SCUBA bolometer array camera [10] at the JCMT (James Clerk Maxwell Telescope) has been used to image the sky behind distant clusters of galaxies. By exploiting their gravitational lensing effect, most of the 850-μm background has been resolved into individual galaxies [11,12]. At shorter wavelengths of 175 and 95 μm, that are inaccessible from the ground, the *ISO* satellite has resolved about 10% of the background [13,14]. Extremely faint galaxies at mid-infrared wavelengths of 15 and 7 μm have also been determined using *ISO* by exploiting gravitational lensing [15]. The lower limit to the background derived from the 15-μm observations is close to the γ–γ upper limit [9], suggesting that a large fraction of the 15-μm background has been resolved. Several other submm-wave SCUBA surveys have been carried out [16–18], and an interesting upper limit to the surface density of galaxies has been imposed at a wavelength of 2.8 mm [19].

The prime motivation for submm-wave galaxy surveys is their ability to select high-redshift galaxies preferentially. The form of the dust emission spectrum of galaxies is broadly similar to the envelope of the background limits shown in Figure 1. If a high-redshift galaxy is observed at a wavelength of several hundred microns, then its rest-frame spectrum is sampled at a wavelength considerably closer to the peak of the emission spectrum. As a result, the detected flux density is increased; this effect is described as a negative k-correction. In some cases this k-correction is sufficiently large to overcome the inverse square law dimming of the galaxy with increasing redshift, and so the flux density is almost constant over reasonable redshifts $0.5 \lesssim z \lesssim 10$. This effect is illustrated in Figure 2, for a galaxy with an intrinsic luminosity of $5 \times 10^{12}\,L_\odot$ and a dust spectrum that provides a good fit to the properties of the background and surface density of dusty galaxies [20]. Note that the k-correction is important only at observing wavelengths longer than about 200 μm.

The spectrum of a dusty galaxy is a power-law at wavelengths longer than a few hundred microns with a temperature dependent turn-over at shorter wavelengths. Measurements of flux density both above and below the wavelength of the turn-over are thus required to determine the dust temperature, which in turn fixes the total luminosity. Hence, there is a very strong case for carrying out galaxy surveys at long submm wavelengths, to exploit the k-correction effect, and then conducting follow-up observations of the detected galaxies at shorter wavelengths. For a galaxy discovered in a survey at a wavelength of 500 μm, a sensitive follow-up observation at 100 μm is much more useful than one at 1000 μm.

The recent detection of well defined populations of distant dusty galaxies, especially at wavelengths of 850 and 15 μm [11,15], confirms that a population of dusty galaxies exists at large redshifts. The background radiation intensity is dominated by galaxies with redshifts $z \simeq 1$, and so the measured background spectrum is not very useful for probing the detailed properties of more distant galaxies. In contrast, the properties of the galaxies detected by SCUBA, especially their redshift distribution [20], will be extremely helpful in this respect.

The best studied galaxy discovered by SCUBA, SMM J02399−0136, is at $z = 2.8$ [21,22]. Most of the other galaxies detected by SCUBA have plausible identifications on deep *HST* images [23], and so are likely to lie at $z \lesssim 5$. Thanks to the power of the 10-m Keck telescopes, attempts to determine redshifts for 16 SCUBA-selected galaxies [23] are making considerable progress [24], despite the uncertain fraction of their luminosity that is absorbed by interstellar dust.

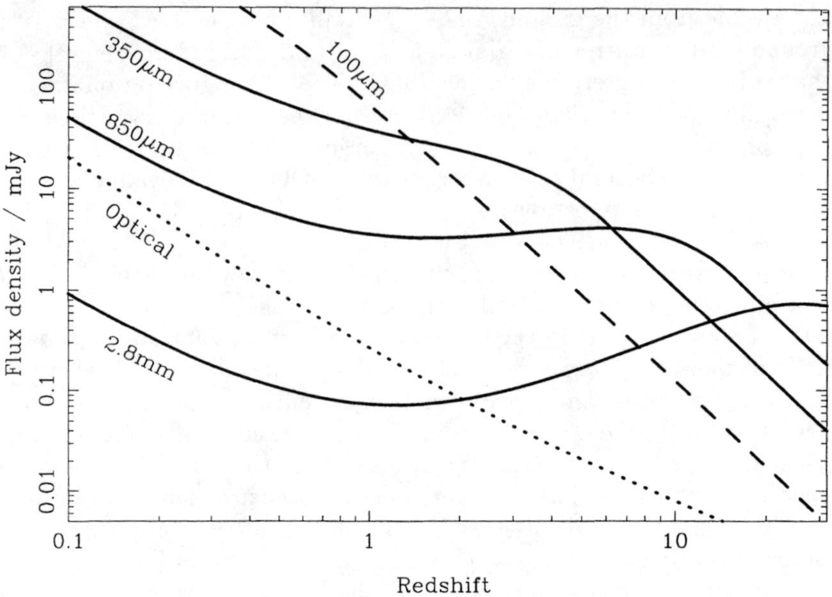

FIGURE 2. The flux density expected from a galaxy with a fixed rest-frame spectrum as a function of redshift. In the optical (or radio) and mid-infrared wavebands, the received energy falls steadily with redshift. However, at wavelengths longer than 200 μm, the effect of redshifting the steep far-infrared dust spectrum counteracts the dimming effect of the inverse square law, and can lead to a brightening of the template galaxy as the redshift increases. This effect is responsible for the unique sensitivity of mm/submm-wave surveys to high-redshift galaxies/AGN.

FUTURE INSTRUMENTATION

COBE has done an excellent job of mapping the extragalactic infrared background. However, there is still a great deal of progress to be made in the detection and study of a large sample of galaxies at wavelengths between several mm and 5 μm. The value of the recent contribution of SCUBA [10] cannot be overstated; however, SCUBA is just the first submm-wave instrument with the sensitivity required to map the distant Universe. SCUBA currently detects a galaxy every few hours, and at 850 μm is limited by source confusion at its 14-arcsec resolution in a few tens of hours of integration in a single field [25]. Great progress is being made in mm/submm-wave instrumentation, improving both mapping speed and angular resolution. For example, the BOLOCAM instrument [26], destined for the 10-m Caltech Submillimeter Observatory (CSO) and the 50-m Large Millimeter Telescope (LMT), should increase the detection rate by a factor of 100–1000 (see Figure 3). Large ground-based interferometer arrays, such as the MMA, will obtain sub-0.1-arcsec resolution submm-wave images of galaxies, and will be able to detect unknown galaxies at a rate similar to that of the LMT. This resolution will be cru-

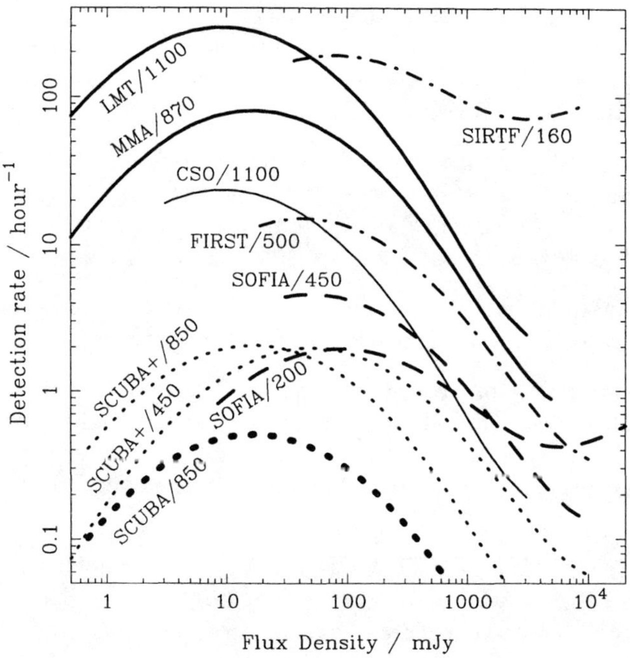

FIGURE 3. The detection rate expected in surveys using future mm/submm/far-infrared instruments. The lines are labeled with the instrument name and the survey wavelength in microns. The left and right end of the lines correspond to the flux density of the confusion limit [25] and the brightest source expected over the whole sky respectively. Note that different points on the same line correspond to independent surveys with different flux density limits and scanning rates. The detection rate expected for *SIRTF* at a wavelength of 70 μm is even greater than that at 160 μm, although the typical redshift of the detected galaxies will be less. A 5×10^{12}-$L_\odot$ galaxy at $z = 3$ is expected to produce flux densities of 0.7, 1.5, 6.2, 8.9 and 6.1 mJy at wavelengths of 1100, 850, 450, 200 and 160 μm respectively.

cial for following up all observations made using other telescopes at coarser angular resolution. To observe at wavelengths shorter than several hundred microns, air- or space-borne instruments are required, and are being planned and constructed.

The best strategy for exploiting the great improvements in survey speed and angular resolution of the full range of future mm/submm/far-infrared instruments cannot yet be finalized, and will remain subject to modification until they all are in service with verified sensitivities. Several points about the future of submm-wave galaxy surveys are clear, however:

1. The space-borne 0.3-m *WIRE* and 0.85-m *SIRTF* will provide extremely large samples of galaxies/AGN at far-/mid-infrared wavelengths. The areas of their surveys would be ideal targets for future surveys at longer wavelengths.

2. The 2.5-m SOFIA airborne observatory will be very timely, and sufficiently flexible and adaptable in operation, to provide the initial far-/mid-infrared follow-up observations of galaxies detected in the mm/submm waveband using SCUBA and BOLOCAM.

3. Within 10 years, submm-wave surveys using the LMT, MMA, *Planck Surveyor* and 3.5-m *FIRST* will provide huge samples of submm selected high-redshift galaxies/AGN. The *Planck* all-sky map will contain at least as rich a variety of new objects as that obtained by *IRAS* 15 years ago.

4. All detailed follow-up observations, and any morphological studies, will require the sub-arcsecond angular resolution of a large interferometer array.

Once a large sample of 10^5 or more galaxies/AGN are obtained using these instruments, the well developed techniques of optical observational cosmology, such as the analysis of distribution functions, can be applied to investigate the evolution of galaxies and large-scale structure, independent of the effects of dust obscuration.

GRAVITATIONAL LENSES

There is another positive feature of galaxy surveys in the mm/submm waveband. Because the chance of a foreground galaxy being directly aligned on the line of sight to a galaxy – and thus a strong magnification being produced – increases with redshift, and there is a unique high-redshift selection bias in submm-wave surveys, the probability of submm detected galaxies being gravitationally lensed is greater as compared with other wavebands [27]. The most relevant instrument for a lens survey is *Planck Surveyor*, which will detect the brightest lenses over the whole sky [28]. A large sample of gravitational lenses offers the opportunity both to study galaxies that would otherwise be too faint to detect, and to investigate the geometry of the Universe. In order to select the subsample of the detected galaxies that display the distorted and multiply imaged structures typical of a lensed galaxy on arcsecond angular scales, MMA imaging will again be essential.

CONCLUSIONS

Within the last year, the intensity of extragalactic background radiation from the mm to the near-infrared waveband has been determined for the first time. This energy was released at shorter wavelengths by star-forming galaxies and AGN during the process of galaxy formation, and has been reprocessed by interstellar dust. The coincident detection of populations of distant dusty galaxies/AGN at six wavelengths from 850 to 7 μm has provided much more information about the properties of distant infrared galaxies. The population of galaxies discovered using SCUBA at 850 μm is particularly interesting. Most of the 850-μm background radiation intensity detected by *COBE* can be accounted for by the existing SCUBA galaxies, indicating that dust-enshrouded high-redshift galaxies/AGN were rare, but very luminous. The next generation of mm/submm-wave instruments – ground-based, air- and space-borne – will provide much larger samples of distant infrared galaxies. The array of data they provide will be comparable in both volume and quality with that obtained in the optical/near-infrared waveband.

ACKNOWLEDGMENTS

This work has benefited greatly from observations using SCUBA at the JCMT in collaboration with Ian Smail, Rob Ivison and Jean-Paul Kneib. I thank Jackie Davidson, Jason Glenn, Malcolm Longair, Priya Natarajan et les garcons SCUBA.

REFERENCES

1. Sanders, D. B., and Mirabel, I. F., *Annual Reviews of Astronomy and Astrophysics*, **34**, 749-792 (1996).
2. Puget, J.-L., et al., *Astronomy and Astrophysics*, **308**, L5-L8, (1996).
3. Fixsen, D. J., et al., *Astrophysical Journal*, **508**, 123-128 (1998).
4. Schlegel, D. J., et al. *Astrophysical Journal*, **500**, 525-553 (1998).
5. Hauser, M. G., et al., *Astrophysical Journal*, **508**, 25-43 (1998).
6. Dwek, E., and Arendt, R., *Astrophysical Journal*, in press (1998; astro-ph/9809239).
7. Pozzetti, L., et al., *MNRAS*, **298**, 1133-1144 (1998).
8. Bernstein, R., Freedman, W., and Madore, B., *Astrophysical Journal*, submitted.
9. Stanev, T., and Franceschini, A., *Astrophysical Journal*, **494**, L159-L162 (1998).
10. Holland, W. S., et al., *MNRAS*, in press (1999; astro-ph/9809122).
11. Smail, I., Ivison, R. J., and Blain, A. W., *Astrophysical Journal*, **490**, L5-L9 (1997).
12. Blain, A. W., et al., *Astrophysical Journal*, submitted.
13. Kawara, K., et al., *Astronomy and Astrophysics*, **336**, L9-L12.
14. Puget, J.-L., et al., *Astronomy and Astrophysics*, in press (1999; astro-ph/9812039).
15. Altieri, B., et al., *Astronomy and Astrophysics*, in press (1999; astro-ph/9810480).
16. Barger, A. J., et al., *Nature*, **394**, 248-251 (1998).
17. Hughes, D., et al., *Nature*, **394**, 241-247 (1998).
18. Eales, S. A., et al., *Astrophysical Journal*, in press (1999; astro-ph/9808040).
19. Wilner, D. J., and Wright, M. C. H., *Astrophysical Journal*, **488**, L67-L70 (1997).
20. Blain, A. W., et al., *MNRAS*, in press (1999; astro-ph/9806062).
21. Ivison, R. J., et al., *MNRAS*, **298**, 583-593 (1998).
22. Frayer, D. T., et al., *Astrophysical Journal*, **506**, L7-L10 (1998).
23. Smail, I., et al., *Astrophysical Journal*, **507**, L21-L24 (1998).
24. Barger, A. J., et al., in preparation.
25. Blain, A. W., Ivison, R. J., and Smail, I., *MNRAS*, **296**, L29-L34 (1998).
26. Glenn, J., et al., in Proc. SPIE, **3357**, 326-334 (1998).
27. Blain, A. W., *MNRAS*, **283**, 1340-1348 (1996).
28. Blain, A. W., *MNRAS*, **297**, 511-516 (1998).

Is the History of Star Formation Compatible with CDM?

Joel R. Primack*, Rachel S. Somerville†

* *Physics Department, University of California, Santa Cruz, CA 95060 USA*
† *Racah Institute of Physics, The Hebrew University, Jerusalem 91904, Israel*

Abstract. Observations now probe the star formation history of the Universe back to a redshift of $z \sim 5$. We investigate whether the predictions of semi-analytic models of galaxy formation based on hierarchical Cold Dark Matter (CDM) type models are in agreement with these direct observations and also with the "fossil" evidence contained in constraints on the ages of present day early-type galaxies. Previous models [1] predicted that the star formation rate density falls off rather steeply at $z \gtrsim 2$, and correspondingly that the majority of the stars in the Universe formed at relatively low redshift. We investigate the effect of including a bursting mode of star formation, assuming that galaxy-galaxy mergers trigger starbursts and using the merger rate that arises naturally in the CDM merging hierarchy. The resulting starbursts substantially increase the global star formation rate at high redshift, leading to predictions that are in good agreement with the star formation rate density at $z \sim 3$ obtained from sub-millimeter observations (SCUBA) and optical/UV estimates after correction for dust extinction. The mass of stars formed at $z \geq 3$ is correspondingly in better agreement with the fossil evidence. We also investigate complementary global quantities such as the mass of cold gas and the average metallicity of cold gas as a function of redshift, and the integrated extra-galactic background light. We argue that these observations, taken together, provide strong constraints on the star formation history of the Universe, and that hierarchical models of the CDM type are in reasonable agreement with these observations when starbursts are included.

I INTRODUCTION

The star formation history of the Universe has been observationally probed using many approaches, both direct and indirect. The star formation rate density has been derived from direct observations using a variety of tracers such as Hα [2–4], [OII] [5], or UV emission [6–8]. Recently, sub-mm observations obtained with the SCUBA instrument were used to estimate the star formation rate at $z \sim 3$ [9]. The compilation of these observations, which extend from $z = 0$ to $z \sim 5.5$, sketches out the history of star formation — see Fig. 1 — and have come to be referred to generically as "the Madau diagram" [7]. However, the interpretation of the Madau diagram is not necessarily straightforward. Because of dust extinction, the optical

CP478 , *COSMO—98*, edited by David O. Caldwell

and UV tracers certainly underestimate the true star formation rate, but by an uncertain factor. The sub-mm flux observed by SCUBA is due to the re-radiation of light that was absorbed by dust at shorter wavelengths, and so provides a useful complement to the optical and UV observations. However, the low resolution of the SCUBA instrument makes it difficult to find unique optical counterparts and measure redshifts to the sources. The measurement of the mass density in the form of cold gas as a function of redshift, derived from observations of damped Lyman-α systems [10,11], provides an indirect and complementary indicator of the star formation history [12], as does the measurement of the average metal content of these systems at various redshifts [13]. The extragalactic background light (EBL), which has been measured from the UV to the sub-mm (reviewed by [14]), is another indirect constraint which may prove to be quite powerful, because it effectively measures the bolometric luminosity density. An alternative method of determining the star formation history of the Universe is to use "fossil evidence" such as the age constraints derived from the small scatter in the observed color-magnitude and fundamental plane relations for early type galaxies and galactic bulges out to $z \simeq 0.5$ (extensive references are given in [15]).

An important question in the study of galaxy formation today is whether any existing theoretical framework can fit all these observations into a self-consistent picture. This is a difficult question, as in order to model these quantities one is forced to consider messy astrophysical processes such as gas dynamics, star formation, supernovae feedback, and absorption and re-radiation by dust. Currently, the only efficient way to do this is using semi-analytic models such as those developed in e.g. [16], [17], and [18]. Previous work (e.g., [17,1]) has left the impression that a "late epoch of star formation" (i.e., the peak in the global star formation history at $z \sim 1.5-2$ and a steep decline at higher redshifts) is a generic feature of CDM-type models. This behaviour is similar to the star formation history sketched out in the original Madau diagram [7], and has been claimed as a success of these models.

However, there is mounting evidence that the true star formation rate history at $z \gtrsim 3$ may be considerably different from the original estimates. This could be due to a number of effects, including dust extinction, a steeper slope on the faint end of the luminosity function (suggested by the results of [19] based on photometric redshifts in the HDF), or unknown selection effects intrinsic to the Lyman-break selection technique. The revised star formation rate density may well be consistent with the SCUBA lower limit at $z = 3$. Moreover, as pointed out by Renzini [15], the integration of the original Madau diagram, like the models of Ref. [1] which have a similar star formation history, would predict that only about 5 percent of the stars in galaxies today formed before a redshift of three, in fairly serious disagreement with the fossil evidence (Renzini has argued that ~ 30 percent of the stars in galaxies today formed before $z = 3$, if cluster galaxies are typical). This might seem to indicate a serious problem with the entire hierarchical, CDM-based picture of structure formation.

However, it turns out that a natural feature of the CDM merging hierarchy, neglected in previous work, can change the situation significantly. This feature is the star formation occurring in starbursts triggered by galaxy-galaxy mergers. We first became interested in this idea as a way to explain the very large luminosities and high star formation rates of the physically small Lyman-break galaxies [21], which it seems to do very successfully. In the remainder of this paper, we will summarize the implications of the starburst model for the properties of Lyman-break galaxies (LBGs), as well as for the global star formation rate density, cold gas density, and metal content of the Universe as a function of redshift. We will conclude that in fact the classic CDM picture is in good agreement with all these observations within the present uncertainties.

II SEMI-ANALYTIC MODELS OF GALAXY FORMATION

Semi-analytic techniques allow one to model the formation and evolution of galaxies in a hierarchical framework, including the effects of gas cooling, star formation, supernova feedback, chemical evolution, galaxy-galaxy merging, and the evolution of stellar populations. The semi-analytic models used here are described in detail in [20], [18], and [21]. These models are in reasonably good agreement with a broad range of local galaxy observations, including the Tully-Fisher relation, the B-band luminosity function, cold gas contents, metallicities, and colors [18]. Our basic approach is similar in spirit to the models originally developed by the Munich [16] and Durham [17] groups, and subsequently elaborated by these groups in numerous other papers (reviewed in [20] and [18]). Significant improvements included in [18] are that we assumed a lower stellar mass-to-light ratio (in better agreement with observed values), included the effects of dust extinction (which are significant in the B-band), and developed an improved "disk-halo" model for supernovae feedback. With these new ingredients, we simultaneously produce a Tully-Fisher relation with the correct zero-point and slope, and a B-band luminosity function with approximately the right normalization and shape on the bright and faint ends.

The framework of the semi-analytic approach is the "merging history" of a dark matter halo of a given mass, identified at $z = 0$ or any other redshift of interest. We construct Monte-Carlo realizations of halo "merger trees" using the method described in [22]. Each branch in the tree represents a halo merging event. When a halo collapses or merges with a larger halo, we assume that the associated gas is shock-heated to the virial temperature of the new halo. This gas then radiates energy and cools. The cooling rate depends on the density, metallicity, and temperature of the gas. Cold gas is turned into stars using a simple recipe, and supernova energy reheats the cold gas according to another recipe. Chemical evolution is traced assuming a constant yield of metals per unit mass of new stars formed. The spectral energy distribution (SED) of each galaxy can then obtained by assuming an IMF and using stellar population models (e.g. [23]; here we use the GISSEL98 models with solar metallicity). It is then straightforward to obtain magnitudes and colours in any desired filter band. Our recipes for star formation, feedback, and chemical evolution contain free parameters, which are set by requiring an average fiducial "reference galaxy" (the central galaxy in a halo with a circular velocity of 220 km s^{-1}) to have an I-band magnitude $M_I - 5\log h = -21.7$ (this requirement fixes the zero-point of the I-band Tully-Fisher relation to agree with observations), a cold gas mass $m_{\rm cold} = 10^{10}h^{-2}M_\odot$, and a stellar metallicity of about solar. The star formation and feedback processes are some of the most uncertain elements of these models, and indeed of any attempt to model galaxy formation. We have investigated several different combinations of recipes for star formation and supernova feedback (sf/fb) and also several cosmologies, discussed in detail in [18] and [21], but here we will discuss results for a single choice of sf/fb recipe (constant efficiency quiescent star formation and the "disk-halo" feedback model) and cosmology ($\sigma_8 = 0.67$ SCDM), and a standard Salpeter IMF.

In modelling starbursts, we assume that they are triggered by galaxy-galaxy mergers. This premise has considerable observational support and is also supported by N-body simulations with gas dynamics and star formation [24]. In our semi-analytic models, galaxies that are within the same large halo may merge according to two different processes. Satellite galaxies lose energy and spiral in to the center of the halo on a *dynamical friction* time-scale. In addition, satellite galaxies orbiting within the same halo may merge with one another according to a modified *mean free path* time-scale. Our modeling of the latter process is based on the scaling formula derived in [25] to describe the results of dissipationless N-body simulations in which galaxy-galaxy encounters and mergers were simulated, covering a large region of parameter space. We [29] are currently investigating the frequency and spatial distribution of halo collisions in very high resolution N-body simulations (based on the method of [30]). We are also extending the studies of gas dynamics and star formation in interacting galaxies using the same approach as [24]. We are covering a larger region of parameter space, and using galaxy properties more appropriate to the high redshift systems we are interested in [31].

The velocity dispersions of observed LBGs can be estimated based on the widths of nebular emission lines such as $H\beta$. Emission lines have been detected for a few of the brightest LBGs from the ground-based sample. The velocity dispersions σ derived from the observed linewidths are $\sigma = 60 - 80$ km s^{-1} for four objects, and ~ 190 km s^{-1} for one object [32]. These values agree well with the burst models, but are in strong disagreement with the no-burst models, which peak at $\sigma \sim 180$ km s^{-1} [21]. These measurements may be affected by several biases, which remain to be unravelled. Linewidths have been obtained only for the brightest LBGs, which may have systematically higher internal velocities. This would increase the discrepancy with no-burst models. The modelling of the velocity dispersion at the small radii probed by the observations (approximately the half-light radius) is also uncertain.

The ages and stellar masses of the burst models [21] also agree well with estimates based on multicolor photometry. LBG colors are well fit by young (< 0.1 Gyr) stellar populations with moderate amounts of dust [28]. These young ages imply that stellar masses are also low ($\sim 10^9 M_\odot$). The LBGs in our no-burst models (and Ref. [1]) are systematically older and more massive than the data indicate.

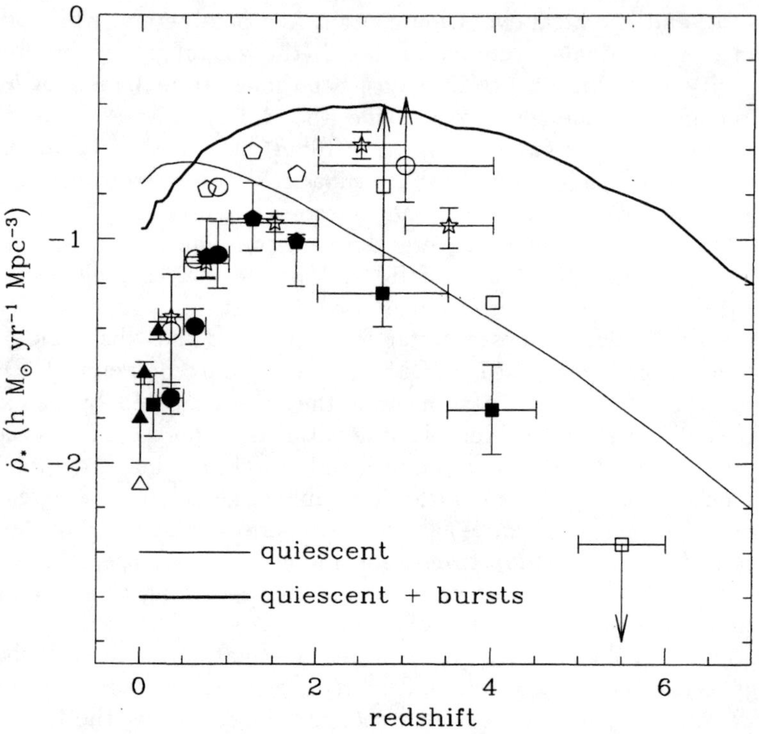

FIGURE 1. The star formation rate density as a function of redshift. The filled triangle at $z = 0$ is from Hα observations of local galaxies [2], with the conversion from Hα flux to star formation rate from Ref. [37]. The open triangle shows the star formation rate from the same observations, with the conversion factor originally used in Ref. [2]. The filled triangle at $z \sim 0.04$ is from new Hα observations by the KISS survey [4], and the filled triangle at $z \sim 0.2$ is from Hα observations of galaxies from the CFRS [3]. The filled square at $z \sim 0.15$ is from 2000 Å fluxes from the UV survey of [8]. Filled circles are determined from the 2800 Å fluxes of CFRS galaxies [6], and filled hexagons from the 2800 Å flux of galaxies in the HDF using photometric redshifts [38]. The star symbols are from HDF photometric redshifts derived by Ref. [19]. Filled squares at $z = 2.75$ and $z = 4$ are from the $\sim$ 1500 Å fluxes of U- and B-dropouts in the HDF [7]. The open symbols show a correction for dust extinction of a factor of two for 2800 Å flux and three for 1500 Å flux, as suggested by [26]. The tip of the vertical arrow at $z \sim 3$ shows a correction of a factor of seven, which may be indicated by recent observational studies of dust extinction in LBGs [27] (also Steidel, this meeting). The large open circle at $z = 3$ shows the lower limit from the SCUBA observations of the HDF region [9]. The fact that the extinction-corrected observations and this lower limit are higher than the prediction of the quiescent model suggests that bursts may be needed to account for the high star formation rate at high redshifts. Note that the discrepancy between the predictions and the data at low redshifts ($z \lesssim 1$) is partly alleviated when the models are corrected downward by about a factor of 2, to correct for the overprediction of the number density of $M \lesssim 0.1 M_*$ halos in the Press-Schechter approximation [39,40].

More photometry and spectra of LBGs will help to clarify whether models including starbursts adequately describe the properties of the LBGs.

Fig. 1 summarizes the direct observational estimates of the star formation rate density of the Universe as a function of redshift, and shows the model predictions for the quiescent and starburst models. The models including only quiescent star formation peak at $z \sim 1.5$ and fall off fairly steeply at higher redshifts, similar to the models of Ref. [1]. The contribution of starburst galaxies to the global star formation rate density is clearly significant at $z \gtrsim 2$. The starburst models exhibit a peak at higher redshift ($z \sim 3$) and are almost flat out to $z \sim 5$. This would appear to be in better agreement with the UV observations derived from Lyman-break galaxies after correction for dust extinction, and with the SCUBA results. The poor behaviour of the models at $z < 1$ is partially due to inaccuracies in the Press-Schechter approximation, and partially a result of our simple assumption of constant quiescent star formation efficiency.

The stars created at high redshift form the population of old stars in present day galaxies. Combining the estimate of the present day stellar mass density from Ref. [41] with Renzini's claim that one third of these stars should be in place by $z = 3$ leads to the estimate that $h\Omega_*(z = 3) = 8 \times 10^{-4}$. If this estimate is correct, the no-burst models, like the models of Ref. [1] or direct integration of the original Madau diagram [7], do not produce enough old stars. The models including starbursts, which are in better agreement with the "new Madau diagram", also produce an old stellar population that is in good agreement with the "fossil evidence". Further evidence supporting this picture is discussed in [21,42].

The semi-analytic models provide all the necessary information to compute the integrated extra-galactic background light (EBL). A first report of our results was given at the VERITAS conference [43]. The EBL from 0.1 to 2 microns is mainly starlight. At $\lambda \lesssim 1\mu$, a significant portion of the light is absorbed by dust and reradiated in the far-IR. The EBL at $\gtrsim 10\mu$ is almost completely due to re-radiation by dust, and is therefore quite sensitive to our assumed dust absorption and emission models. The characteristic shape of the EBL is also sensitive to the assumed cosmology, star formation history, and IMF. With the wealth of multi-band observations that are now becoming available, the EBL will be an important probe of galaxy formation and cosmology.

III SUMMARY

We have shown that our CDM-based hierarchical models are in good agreement with direct observational estimates of the global star formation history, and are compatible with the fossil evidence which indicates that a significant fraction of the stars in galaxies today formed at high redshift. An important new ingredient in our models is the inclusion of starbursts triggered by galaxy-galaxy interactions. It follows naturally in the CDM picture of structure formation that such starburst events are more common at high redshift, and are also brighter because the gas supply has not yet been depleted. The starburst picture accounts naturally for many of the observed features of the high redshift Lyman-break galaxies, including their number density and luminosity distribution, and also their sizes, star formation rates, colors, ages, and line-widths. In addition, the starburst mode results in a significant modification of the star formation history of the universe, moving the peak in the Madau diagram to higher redshift and leading to a much more gradual decline in the star formation rate out to $z \sim 6$. This is in better agreement with the "new Madau diagram" that results when the considerable dust extinction at $z \gtrsim 3$ now favored by observers is taken into account, and with the SCUBA observations.

Many of the galaxy mergers that trigger starbursts will result in the formation of an early-type galaxy or bulge. Small mass starburst remnants may be accreted by larger galaxies, forming a Pop II halo [44] like that of our own Galaxy. At high redshift ($z \gtrsim 2$), merger events occur mainly in and around massive, strongly clustered dark matter halos. The merger-starburst picture therefore predicts simultaneously the strong observed clustering of the Lyman-break galaxies [29], and that the oldest stars in the universe should be found preferentially in dense environments today. The mass of old stars ($z_{\rm form} > 3$) predicted by the starburst models is in good agreement with observational estimates, unlike some previous CDM-based models, which did not produce enough old stars. The gas density and metal content of the universe as a function of redshift, indicated by observations of damped Lyman-α systems, are important complementary constraints on the star formation history and on recipes used to model star formation and supernovae feedback. The integrated extra-galactic background light will be an increasingly important constraint on galaxy formation and the associated processes of dust absorption and re-radiation [43].

Acknowledgments

JRP acknowledges support from NASA and NSF grants at UCSC, and a Forchheimer Visiting Professorship at Hebrew University. RSS acknowledges support from a NASA grant and a GAANN Fellowship at UCSC and a University Fellowship from the Hebrew University, Jerusalem.

REFERENCES

1. C.M. Baugh et al. 1998, ApJ, 498, 504
2. J. Gallego et al. 1995, ApJ, 455, L1
3. L. Tresse, S.J. Maddox 1997, *The Young Universe*, ed. S. D'Odorico, A. Fontana, E. Giallongo, ASP Conf. Series (astro-ph/9712130)
4. C. Gronwall 1998, *Dwarf Galaxies and Cosmology*, eds. Thuan et al., Editions Frontieres, in press (astro-ph/9806240)
5. D. Hogg et al. 1998, ApJ in press, (astro-ph/9804129)
6. S.J. Lilly et al. 1996, ApJ, 460, L1

7. P. Madau et al. 1996, 283, 1388; P. Madau 1996, 7th Annual October Astrophysics Conference in Maryland, "Star Formation Near and Far" (astro-ph/9612157)
8. M.A. Treyer et al. 1997, *The Ultraviolet Universe at Low and High Redshift*, ed. W. Waller, Woodbury: AIP Press (astro-ph/9706223)
9. D. Hughes et al. 1998, Nature, 394, 241
10. K.M. Lanzetta, A.M. Wolfe and D.A. Turnshek 1995, ApJ, 440, 435
11. L.J. Storrie-Lombardi, R.G. McMahon, M.J. Irwin 1996, MNRAS, 283, 79L
12. S.M. Fall, Y.C. Pei, S. Charlot 1996, 464, 43L
13. M. Pettini et al. 1997, ApJ, 486, 665
14. E. Dwek et al. 1998, ApJ, in press (astro-ph/9806129)
15. A. Renzini 1997, ApJ, 488, 35; A. Renzini 1998, in *The Young Universe*, ed. S. D'Odorico, A. Fontana, E. Giallongo, ASP Conf. Series (astro-ph/9801209)
16. G. Kauffmann, S.D.M. White, B. Guiderdoni 1993, MNRAS, 264, 201
17. S. Cole et al. 1994, MNRAS, 271, 781
18. R.S. Somerville, J.R. Primack 1998, MNRAS in press (astro-ph/9802268)
19. M.J. Sawicki, H. Lin, H.K.C. Yee 1997, AJ, 113, 1
20. R.S. Somerville 1997, Ph.D. thesis, UCSC (www.fiz.huji.ac.il/~rachels/thesis.html)
21. R.S. Somerville, J.R. Primack, S.M. Faber 1998, MNRAS, in press (astro-ph/9806228)
22. R.S. Somerville, T.S. Kolatt 1998, MNRAS, in press (astro-ph/9711080)
23. S. Charlot, A. G. Bruzual 1993, ApJ, 405, 538
24. J.C. Mihos, L. Hernquist 1994, ApJ, 425, L13; ibid 1996, ApJ, 464, 641
25. J. Makino, P. Hut 1997, ApJ, 481, 83
26. M. Pettini et al. 1997, ApJ, 478, 536
27. C. Steidel et al. 1998, ApJ, in press (astro-ph/9811399)
28. M. Sawicki, H.K.C. Yee 1998, AJ, 115, 1329
29. T. Kolatt et al. 1998, in prep.
30. A.V. Kravtsov, A.A. Klypin, A.M. Khokhlov 1997, ApJS 111, 73
31. R.S. Somerville, I. Walker, et al. 1998, in prep.
32. M. Pettini et al. 1998, ApJ, in press (astro-ph/9806219)
33. C.C. Steidel et al. 1998, ApJ, 492, 428
34. M. Giavalisco et al. 1998, ApJ, in press (astro-ph/9802318)
35. R.H. Wechsler et al. 1998, ApJ, in press (astro-ph/9712141)
36. Y.P. Jing, Y. Suto 1998, ApJ, 494, L5
37. P. Madau, L. Pozzetti, M. Dickinson 1998, ApJ, 498, 106
38. A.J. Connolly et al. 1997, ApJ, 486, 11L
39. M.A.K. Gross et al. 1998, MNRAS in press (astro-ph/9712142)
40. R.S. Somerville et al. 1998, astro-ph/9807277
41. M. Fukugita et al. 1998, ApJ, 503, 518
42. R.S. Somerville, J.R. Primack, in Proc. Xth Rencontres de Blois 1998, ed. B. Guiderdoni et al. (Editions Frontieres), astro-ph/9811001
43. J.R. Primack, J.S. Bullock, R.S. Somerville, D. MacMinn, in VERITAS Proc., eds. M. Catanese et al., Astroparticle Phys., in press, astro-ph/9812399
44. S.C. Trager et al. 1998, ApJ, 485, 92

Lyman Break Galaxies as Collisional Starbursts

J. Bullock[1], T. Kolatt[1], R. Somerville[2], Y. Sigad[2], P. Jonsson[3], A. Kravtsov[4], A. Klypin[4], J. Primack[1], S. Faber[5] & A. Dekel[2]

[1]*Physics Department, University of California, Santa Cruz, CA 95064*
[2]*Racah Institute of Physics, The Hebrew University, Jerusalem, 91904, Israel*
[3]*Astronomy Department, University of California, Santa Cruz, CA 95064 USA*
[4]*Astronomy Department, New Mexico State University, Box 90001, Dept. 4500, Las Cruces, NM 88003, USA*
[5]*UCO/Lick Observatory, University of California, Santa Cruz, CA 95064 USA*

Abstract. The nature of the galaxies recently discovered at redshift [1,2] $z \gtrsim 3$, when the Universe was only about two billion years old, is an intriguing puzzle. They are forming stars at a high rate, comparable to local "starburst" galaxies [3] , but while the latter are relatively rare, the high-z galaxies are as numerous as local normal galaxies. Since many local starbursts are triggered by galaxy collisions, a much higher collision rate in the past could provide a simple explanation for the high abundance of high-z galaxies [2,4] . Here we use high-resolution cosmological N-body simulations and a hierarchical halo finder to estimate the galaxy collision rate as a function of time in a popular cosmological model (ΛCDM). We find that bright collisional starbursts are indeed frequent enough to account for most of the high-z galaxies. Moreover, many of the objects are of relatively small mass, but they cluster about large-mass halos. They therefore exhibit strong clustering, similar to that observed [5–7] and stronger than that of the relevant massive halos. The collisional starburst scenario [2,4] thus appears consistent with several key properties of high-z galaxies.

Data from the Hubble Deep Field [8] (HDF) and ground-based telescopes [1,2,9,10] have revealed a population of galaxies viewed when the Universe was only $\sim 15\%$ of its current age. The technique used in finding the high-redshift galaxies is based on observing a characteristic spectral attenuation (absorption by neutral hydrogen) shortward of the Lyman limit. These galaxies are therefore referred to as "Lyman-Break Galaxies" (LBGs). The LBGs are observed to be abundant and highly clustered, with comoving number densities and clustering properties at $z \simeq 3$ comparable to those of present-day bright ($\geq L_*$) field galaxies [6,7] . There are competing views regarding the nature of LBGs. In one view [1,7,11,12] most LBGs are large galaxies quiescently forming stars at the bottom of the potential wells of massive dark matter halos. We refer to this idea as the "central quiescent" sce-

CP478 , *COSMO—98*, edited by David O. Caldwell

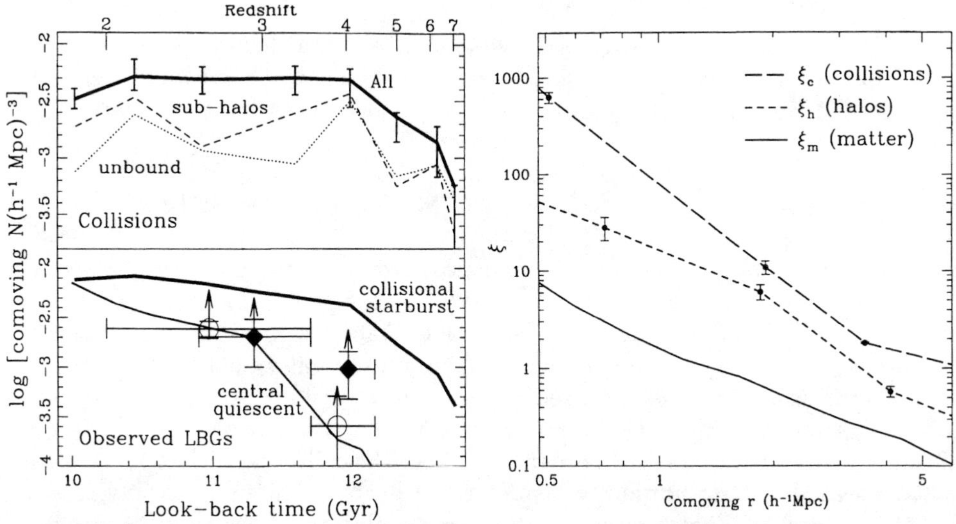

FIGURE 1. (a) Comoving number density of collisions and observed LBGs as a function of look-back time. The solid line in the top panel refers to all halo collisions. The bottom panel presents the number of LBGs actually observed. The thick solid curve represents collisional starbursts brighter than apparent magnitude $\mathcal{M}_{AB} = 25.5$. Observational estimates for LBGs from the HDF and newer ground-based observations with a much larger survey volume are depicted by the open circles and solid diamonds respectively (see text for references). The thin solid line is estimated for the central quiescent scenario (see text). (b) The long-dashed line shows two-point auto-correlation function of DM halos of $M > 7 \times 10^9 h^{-1} M_\odot$ that experienced collisions in the redshift interval $3.9 > z > 2.9$. The lower curve is for the underlying dark matter.

nario. An alternate view [2,4] maintains that LBGs are mainly galactic starbursts triggered by collisions between small, gas-rich galaxies. We refer to this idea as the "collisional starburst" scenario.

Within the central quiescent scenario, there is roughly a one-to-one relationship between the LBGs and massive halos [1,7] . This helps explain the strong clustering of observed LBGs, since in CDM [13] theories of hierarchical structure formation, massive objects are more clustered than low-mass objects. It has been shown that the clustering properties of LBGs at $z \sim 3$ can be reproduced within various CDM models if the LBGs are associated with the most massive collapsed dark-matter (DM) halos at that epoch [14,15,7] . More detailed modeling within the central quiescent framework supports this general concept [11,12] .

Until the modeling reported on here [16], there were no predictions for clustering properties and only rough estimates of number densities of LBGs within the collisional starburst picture. We investigate these questions using collisions between dark matter halos in a ΛCDM N-body simulation. We use semi-analytic estimates of the luminosity of the ensuing bursts in order to directly compare the clustering properties and number densities of these events with the observed properties of

LBGs.

Our simulations make use of the ART code [17], which utilizes an adaptive grid to obtain the unprecedented resolution necessary for identifying collisions between well-resolved galactic halos (or sub-halos which are located within the virial radius of a larger host halo) in a cosmological volume. The cosmological prarmeters are $\Omega_0 = 1 - \Omega_\Lambda = 0.3$, $h = 0.7$, and $\sigma_8 = 1.0$, and the box size is $L_{box} = 30h^{-1}h^{-1}$Mpc. We track the evolution of $N_p = 256^3 \approx 1.7 \times 10^7$ DM particles, and have a mass resolution of $m_p = 1.3 \times 10^8 h^{-1} M_\odot$.

For details of the identification of DM halos and collisions between DM halos see [16]. The top panel in Figure 1a shows the time-evolution of the number density of identified collisions as a function of look-back time, assuming a duration of visibility $\tau_{\rm vis} = 100$ Myr. About one half of the collisions at $z \lesssim 4$ involve sub-halos; such collisions would have been missed without these very high-resolution N-body simulations, which significantly supersede the cruder treatment of collisions in the semi-analytic investigation [4] . In order to connect the collisions with observations, we have assigned a burst luminosity as described in [16]. The upper heavy line in the bottom panel of Figure 1a shows the time-evolution of the number density of observable LBGs with $\mathcal{M}_{\rm AB} < 25.5$, for the collisional starburst model. The *observed* number densities [18,7,10] of LBGs brighter than $\mathcal{M}_{\rm AB} = 25.5$ are shown for comparison, assuming the simulated cosmology. The predicted number densities are somewhat larger than those observed, and thus allow for dust extinction, which we have not included. The arrows on the data points result from assuming a (conservative) factor of three in dust extinction, coupled with the $z \sim 3$ and $z \sim 4$ luminosity functions estimated by recent ground observations [10] .

For comparison, we show the predictions of a central-quiescent type model [16]. This type of model generically predicts a steep fall-off in density towards igher z, in contrast with the collisional scenario, and in apparent disagreement with the newest data [10] (solid diamonds). Redshift evolution is thus a key discriminant between the scenarios. The predictions of the simple model shown here are similar to those of more detailed central-quiescent models [11,4] , but large uncertainties remain.

The two-point auto-correlation function DM halo collisions is shown by the long dashed line in Figure 1b. In the range $1 - 5h^{-1}$Mpc, it can be approximated by a power law $\xi_{\rm c}(r) \simeq (r/r_0)^{-\gamma}$, with $r_0 \simeq 5h^{-1}$Mpc and $\gamma \simeq 2.6$ Shown for comparison is $\xi_{\rm h}(r) \simeq (r/3.5h^{-1}\ {\rm Mpc})^{-2.2}$ for halos more massive than $10^{12}h^{-1}M_\odot$; these are what the massive-halo scenario would identify with LBGs. The lower curve is for the underlying dark matter. The sturbursts are then slightly more clustered than the halos themselves, which, as described in [16], is a result of the collisions being clustered about large mass halos. Given the current uncertainties, both the halo and collision correlation functions are consistent with the parameters derived from observations [7,6] for the simulated cosmology: $r_0 \simeq 6h^{-1}$Mpc and $\gamma \simeq 2$.

The N-body simulations here provide the first quantitative results on the clustering of colliding halos at high redshift and a more accurate measure of their number density than earlier semi-analytic models, which miss roughly half of all collisions. They further show that starbursts associated with collisions of relatively low mass halos are consistent with the observed number density and clustering of bright LBGs at $z \sim 2.5 - 4.5$, at least in the ΛCDM cosmology; other popular cosmologies should be qualitatively similar [4,15] . Finally, they predict that a key test between models is the number of LBGs vs. redshift, which falls off much faster in the central quiescent scenario than for collisional starbursts. This plus further tests, such as the luminosity function and virial mass measurements [4] , should ultimately clearly distinguish between these scenarios.

REFERENCES

1. Steidel, C. , Giavalisco, M. , Pettini, M. , Dickinson, M. & Adelberger, K. 1996, ApJ, 462, L17
2. Lowenthal, J., Koo, D., Guzman, R., Gallego, J., Phillips, A., Faber, S., Vogt, N., & Illingworth, G. 1997, ApJ, 481, 673
3. Heckman, T.M., Robert, C., Leitherer, C., Garnett, D.R. & van der Rydt, F. 1998, ApJ, 503, 646
4. Somerville, R., Primack, J., & Faber, S. 1998, MNRAS, in press (astro-ph/9806228)
5. Steidel, C., Adelberger, K.L., Dickinson, M., Giavalisco, M., Pettini, M. & Kellogg, M. 1998, ApJ 492, 428
6. Giavalisco, M., Steidel, C., Adelberger, K., Dickinson, M., Pettini, M. & Kellogg, M. 1998, ApJ 503, 503
7. Adelberger, K., Steidel, C., Giavalisco, M., Dickinson, M., Pettini, M. & Kellogg, M. 1998, ApJ 505, 18
8. Dickinson, M. 1998, in "The Hubble Deep Field", Ed. M. Livio, S.M. Fall & P Madau
9. Steidel, C., Giavalisco, M., Dickinson, M., & Adelberger, K. 1996, AJ, 112, (352)
10. Steidel, C. C., Adelberger, K. L., Giavalisco, M., Dickinson, M. & Pettini, M. 1998, ApJ submitted (astro-ph/9811399)
11. Baugh, C., Cole, S., Frenk, C. & Lacey, C. 1998, ApJ 498, 501
12. Governato, F., Baugh, C. M., Frenk, C. S., Cole, S., Lacey, C. G., Quinn, T., & Stadel, J. 1998, Nature, 392, 359
13. Blumenthal, G., Faber, S., Primack, J. & Rees, M. 1984, Nature 311, 517
14. Jing, Y. P., Suto, Y. 1998, ApJL, 494, L5
15. Wechsler, R. H., Gross, M. A. K., Primack, J. R., Blumenthal, G. R., & Dekel, A. 1998, ApJ, 506, 19
16. Kolatt, T.S. et al. 1998, Nature, submitted
17. Kravtsov, A. V. , Klypin, A. A. & Khokhlov, A. M. 1997, ApJ Suppl., 111, 73
18. Pozzetti, L. , Madau, P. , Zamorani, G. , Ferguson, H.C. & Bruzual, A. 1998, MNRAS, submitted

Four Testable Predictions of Instanton Cosmology

R.E. Allen

Center for Theoretical Physics, Texas A&M University
College Station, Texas 77843, USA

Abstract. A new cosmological model makes the following predictions: (1) The deceleration parameter q_0 is approximately zero. (2) The mass density parameter Ω_m is less than 1. (3) The universe is spatially closed, but is asymptotically flat as $t \to \infty$, regardless of its matter content. (4) The age of the universe is approximately 15 Gyr if the Hubble parameter h is approximately 0.65.

In a new theory [1], the evolution of the cosmic scale factor $R(t)$ is determined by both the Einstein field equations and an SU(2) cosmological instanton. The instanton dominates in the later universe, making $R(t)$ approximately proportional to t:

$$dR/dt = K \quad , \quad t \to \infty \tag{1}$$

where K is a constant. On the other hand, Einstein gravity dominates in the early universe, giving the usual Friedmann equation

$$(dR/dt)^2 + k = (8\pi G/3)\,\rho R^2 \quad , \quad t \to 0 \tag{2}$$

with k positive (since the universe is spatially a 3-sphere). Suppose that a crossover time t_c is defined by

$$(8\pi G/3)\,\rho R^2 = K^2 \quad , \quad t = t_c. \tag{3}$$

Since ρR^2 decreases with R and t, we then have

$$\rho < \left(\frac{3}{8\pi G}\right)\frac{K^2}{R^2} < \left(\frac{3}{8\pi G}\right)\left(\frac{dR/dt}{R}\right)^2 \quad , \quad t \gg t_c \tag{4}$$

or

$$\Omega_m < 1 \quad , \quad t \gg t_c \tag{5}$$

where

CP478 , *COSMO—98*, edited by David O. Caldwell

$$\Omega_m \equiv \frac{\rho}{\rho_c} \quad , \quad \rho_c \equiv \left(\frac{3}{8\pi G}\right) H^2 \quad , \quad H \equiv \frac{dR/dt}{R}. \tag{6}$$

According to (1), we also have

$$q \equiv -\frac{d^2R/dt^2}{H^2R} \approx 0 \quad , \quad t \gg t_c. \tag{7}$$

With the reasonable assumption that the age of the universe is substantially larger than t_c, the present theory thus makes several testable predictions:

(1) The current deceleration parameter q_0 is approximately zero.

(2) The current mass density parameter Ω_m is less than 1.

Two additional predictions are obtained below:

(3) The universe is spatially closed, but is asymptotically flat as t $\rightarrow \infty$, regardless of its matter content.

(4) The age of the universe is approximately 15 Gyr if the Hubble parameter h is approximately 0.65.

Let us consider the origin of these results in more detail, beginning with the epoch $t \gg t_c$ where the cosmological instanton dominates. In the theory of Ref. 1, this instanton is a topological defect in a GUT Higgs field which condenses in the very early universe. Its order parameter Ψ_s has the symmetry group SU(2)×U(1)× SO(10). The SO(10) gauge symmetry can be ignored in the present context, however, so Ψ_s may be regarded as simply an SU(2)×U(1) order parameter in four-dimensional spacetime. With x^0 chosen to be the radial coordinate, there is a singularity at $x^0 = 0$ which is interpreted as the big bang. This singularity is physically admissible because it is perfectly analogous to the singularity at the center of a superfluid vortex. One can in fact define a "superfluid velocity"

$$v^\mu = -im^{-1}U^{-1}\partial^\mu U \quad \text{where} \quad \Psi_s = n_s^{1/2}U\eta_0 \tag{8}$$

and η_0 is a constant 2-component vector. The field e_α^μ defined by

$$e_\alpha^\mu = v_\alpha^\mu \quad , \quad v^\mu = v_\alpha^\mu \sigma^\alpha \tag{9}$$

is interpreted as the vierbein which determines the geometry of spacetime [1].

If a is the strength of the cosmological instanton centered on $x^0 = 0$, the constraint of fixed SU(2) topological charge requires that [1–3]

$$e_\alpha^\mu = \delta_\alpha^\mu a/mx^0 \quad , \quad e_\mu^\alpha = \delta_\mu^\alpha mx^0/a \tag{10}$$

In addition, larger values of x^0 correspond to larger 3-spheres [1], so $dx^k = x^0 d\overline{x}^k$, where $d\overline{x}^k$ represents a distance within the unit 3-sphere. There are thus two contributions to the expansion of the universe in the present model, with both the 3-space metric tensor g_{kl} and the separation of comoving points increasing as functions of x^0.

The vierbein of (10) leads to a metric with the Robertson-Walker form

$$ds^2 = \eta_{\alpha\beta}\, e^\alpha_\mu e^\beta_\nu \, dx^\mu \, dx^\nu \tag{11}$$

$$= -\left(\frac{mx^0}{a}\right)^2 \left(dx^0\right)^2 + \left[\frac{m\left(x^0\right)^2}{a}\right]^2 d\overline{x}^k d\overline{x}^k \tag{12}$$

$$= -dt^2 + R(t)^2 \left(\frac{dr^2}{1-r^2} + r^2 d\Omega\right) \tag{13}$$

where $\eta_{\alpha\beta} = diag(-1,1,1,1)$ is the Minkowski metric tensor and

$$R(t) = (m/a)\left(x^0\right)^2 \quad , \quad dt = (m/2a)\, d\left(x^0\right)^2 . \tag{14}$$

Then (1) follows immediately, with $K = 2$.

In the later universe, therefore, $R(t)$ is primarily determined by the topology of the cosmological instanton. The mass density ρ has some influence at any finite time t, however, and the *local* metric tensor near a massive object is determined by the Einstein field equations to an extremely good approximation. As usual, these equations follow from the variational principle $\delta S_L/\delta g^{\mu\nu} = 0$ if there are no constraints, where S_L is the full Lorentzian action [1]. In the present theory, there is a constraint imposed by topology on a cosmological scale, but this constraint is of negligible importance on, e.g., a planetary scale. Near a dense concentration of mass in our highly heterogeneous universe, the local deformation of spacetime geometry is still accurately determined by the usual equations of general relativity.

Furthermore, the Einstein field equations are also a very good approximation in the early universe, since the instanton contribution of (1) to $(dR/dt)^2$ is constant, whereas the usual gravitational contribution of (2) grows as $R \to 0$, with $\rho \propto R^{-3}$ (in the matter-dominated era) or R^{-4} (in the radiation-dominated era).

The spatial curvature associated with (13) is proportional to $R(t)^{-2}$ [4]. The present model thus requires a universe that is asymptotically flat, because the topology of the cosmological instanton requires that $R(t) \propto t$ for $t \to \infty$. This is true regardless of the value of Ω_m. Observations of the cosmic microwave background and of large scale structure indicate that the universe is very nearly flat but that $\Omega_m \approx 0.3$ [5].

For $t \gg t_c$, (1) implies that $H^{-1} \approx t$ rather than $\frac{3}{2}t$. The age of the universe is then given by $t_0 \approx H_0^{-1}$ rather than $\frac{2}{3}H_0^{-1}$. With the conventional definition

$H_0 = 100h$ km s^{-1}Mpc^{-1}, a value of $h = 0.65$ [6,7] implies that $t_0 \approx 15$ Gyr rather than 10 Gyr. This larger value is more consistent with recent estimates of $10 - 14$ Gyr for the ages of the oldest globular clusters [8].

The present theory thus makes a number of testable predictions. In particular, the prediction $q_0 \approx 0$ [9] clearly demarcates the present theory from standard cosmology with a flat spacetime, in which $q_0 = 1/2$ [10,11], and from ΛCDM models, in which $q_0 \approx -1/2$ [12]. Quantitative measurements of q_0 are now becoming attainable [13,14], so it should be possible to subject this prediction to convincing tests in the near future.

Acknowledgement

This work was supported by the Robert A. Welch Foundation.

REFERENCES

1. R.E. Allen, *Int. J. Mod. Phys.* A **12**, 2385 (1997), hep-th/9612041.
2. R.E. Allen, proceedings of the CPT-98 Meeting on CPT and Lorentz Symmetry, edited by V.A. Kostelecký (World Scientific, Singapore, 1999); hep-ph/9902228.
3. The e_μ^α are "ordinary" components, after the initial coordinates of Ref. 1 are transformed to four-dimensional spherical coordinates. See S. Weinberg, *Gravitation and Cosmology* (Wiley, New York, 1972), p. 109.
4. C.W. Misner, K.S. Thorne, and J.A. Wheeler, *Gravitation* (W.H. Freeman, San Francisco, 1973)
5. M. Webster, M.P. Hobson, A.N. Lasenby, O. Lahav, and G. Rocha, astro-ph/9802109, and references therein.
6. D. Branch, *Ann. Rev. of Astronomy and Astrophysics* **36**, 17 (1998), astro-ph/9801065, and references therein.
7. W. L. Freedman, J. R. Mould, R. C. Kennicutt, and B. F. Madore, astro-ph/9801080, and references therein.
8. See, for example, B. Chaboyer, P. Demarque, P. J. Kernan, and L. M. Krauss, Astrophys. J. **494**, 96 (1998); astro-ph/9706128.
9. There have been other "coasting universe" proposals, but they involve exotic forms of matter or other *ad hoc* assumptions, whereas here the result $q_0 \to 0$ as $t \to \infty$ emerges automatically within the context of a more general theory. See, e.g., E.W. Kolb, *Astrophys. J.* **344**, 543 (1989); M. Kamionkowski and N. Toumbas, *Phys. Rev. Lett.* **77**, 587 (1996); M. Özer and M.O. Taha, *Nuclear Physics B* **287**, 776 (1987).
10. P.J.E. Peebles, *Principles of Physical Cosmology* (Princeton University Press, Princeton, 1993).
11. E.W. Kolb and M.S. Turner, *The Early Universe* (Addison-Wesley, Redwood City, California, 1990).
12. M.S. Turner, *Nucl. Phys.* **A621**, 522c (1997); astro-ph/9704024.
13. S. Perlmutter et al., Nature **391**, 51 (1998); astro-ph/9712212.
14. P. M. Garnavich et al., Astrophys. J. **493**, L53 (1998); astro-ph/9710123.

Gravitational Waves from Discrete Symmetry Breaking

Marcelo Gleiser* and Ronald Roberts*

*Department of Physics and Astronomy, Dartmouth College, Hanover, NH 03755, USA

Abstract. The breaking of an approximate discrete symmetry, the final stages of a first order phase transition, or a post-inflationary biased probability distribution for scalar fields are possible cosmological scenarios characterized by the presence of unstable domain wall networks. We compute the contribution of these collapsing domains to the stochastic background of gravitational radiation and explore their observability by present and future gravitational wave detectors.

INTRODUCTION

In order to further our understanding of the physical processes that took place early in the history of the Universe we should explore potential events which may have left an imprint detectable by current or future experiments. One exciting possibility is that certain primordial processes generated a stochastic background of gravitational waves. During the past decade or so, several cosmological sources of stochastic gravitational waves have been proposed [1].

All ground-based detectors (interferometric or resonant bar), such as LIGO or VIRGO, probe the frequency interval $10\text{Hz} \lesssim f \lesssim 10^4\text{Hz}$. Lower frequencies are to be probed in space, as with the planned LISA mission, which has a projected sensitivity of $10^{-4}\text{Hz} \lesssim f \lesssim 1\text{Hz}$.

We propose another powerful source of primordial gravitational waves, the non-spherical collapse of bounded domain walls, or bags, which separate different vacuum regions. The results summarized here can be found in more detail in Ref. [2].

We can envisage at least three scenarios where unstable domain wall networks can be generated: i) the breaking of an approximate discrete symmetry in a thermal phase transition; ii) biased fluctuations in a non-thermal post-inflationary scenario; iii) during the final stages of a first order phase transition, as bubbles of the true vacuum percolate, leaving a disconnected network of shrinking false vacuum domains.

CP478 , *COSMO—98*, edited by David O. Caldwell

THE MODEL

We can model all these scenarios with a biased double-well potential for a real scalar field,

$$V(\phi) = \frac{\lambda}{4}\left(\phi^2 - \phi_0^2\right)^2 + h\phi_0\phi\left(\frac{1}{3}\phi^2 - \phi_0^2\right) , \tag{1}$$

where $h > 0$ is a dimensionless constant which biases the potential towards the $+\phi_0$ minimum. The energy density difference between the two minima is $\Lambda = \frac{4}{3}h\phi_0^4$, while the energy density barrier between the maximum at $\phi_0 = -(h/\lambda)\phi_0$ and the global minimum ($+\phi_0$) is $B = \frac{\lambda}{4}\phi_0^4[1 - \mathcal{O}(h/\lambda)]$. Coordinates scale as $\tilde{x} = \sqrt{\lambda}\phi_0\, x$, while energies scale as $\tilde{E} = (\sqrt{\lambda}/\phi_0)\, E$.

We compute the output in gravitational radiation from the collapsing domains within the full, linearized gravity approximation [3]. The reader may also consult Ref. [4], as we follow a similar approach. As toy models to more realistic bag configurations, we considered two types of domains, ellipsoids and deformed Gaussians. We are presently extending the previous analysis (Ref. [2]) to include a wider variety of deformations on the bags. These configurations are used as initial conditions for the Klein-Gordon equation, which dictates their dynamics. We then compute the gravitational radiation emitted as the bags evolve and shrink.

From simple scaling arguments, the integrated energy radiated in gravitational waves can be written as,

$$E_{\rm GW}(R_{\rm av}) = \epsilon(R_{\rm av})G\sigma^2 R_{\rm av}^3 , \tag{2}$$

where σ is the surface energy density for the domains, and $\epsilon(R_{\rm av}) \equiv \epsilon_0\left(\frac{R_{\rm av}}{\xi_0}\right)^\alpha$ is the efficiency parameter, with ϵ_0 a number to be determined. ξ_0 is the $T = 0$ correlation length. We have also shown that the integrated gravitational radiation output scales approximately as [2]

$$E_{\rm GW}(R_{\rm av}) = E_{\rm GW}^0\left(\frac{\phi_0}{M_{Pl}}\right)^2\left(\frac{R_{\rm av}}{\xi_0}\right)^\beta , \tag{3}$$

while the rate of emission is

$$\dot{E}_{\rm GW} \simeq \sqrt{2}\tilde{E}_{\rm GW}^0[2\sqrt{\lambda}]\left(\frac{\phi_0^4}{M_{Pl}^2}\right)\left(\frac{R_{\rm av}}{\xi_0}\right)^{\beta-1} . \tag{4}$$

THE TWO SCENARIOS AND RESULTS

We ivestigated two scenarios: i) only one of the vacua percolates and ii) both vacua percolate. For case i), the spectrum is characterized by very high frequency,

$f^0 \simeq 1.3 \times 10^{10}\sqrt{\lambda}[2\sqrt{\lambda}]\left(\frac{100}{g_*}\right)^{\frac{1}{3}}$ Hz , and the fraction of the energy density in gravitational waes and the detector strain are, respectively,

$$h^2\Omega^0_{\rm GW} \simeq \frac{1.7\times 10^{-3}}{[2\sqrt{\lambda}]^{-3}}\left(\frac{100}{g_*}\right)^{\frac{3}{2}}\left(\frac{\tilde{E}^0_{\rm GW}}{10^4}\right)\lambda\left(\frac{\phi_0}{M_{Pl}}\right)^2 , \tag{5}$$

$$h_c(f) \simeq \frac{4.1\times 10^{-30}}{[2\sqrt{\lambda}]^{-1/2}}\left(\frac{\tilde{E}^0_{\rm GW}}{10^4}\right)^{\frac{1}{2}}\left(\frac{100}{g_*}\right)^{\frac{5}{12}}\left(\frac{\phi_0}{M_{Pl}}\right) , \tag{6}$$

and thus beyond present or projected detector sensitivities.

The situation for case ii), when both vacua percolate, is much more promising. The vacua are initially separated by a convoluted domain wall, with initial average curvature $R_{\rm av} \simeq \xi$. The tension force dominates the wall dynamics for a while, until the vacuum pressure starts accelerating the walls against each other, eventually causing the formation of large, unstable domains. In order for this to happen, the asymmetry must satisfy $h < \frac{3\sqrt{2}}{2}\lambda[2\sqrt{\lambda}]^3\tilde{E}^0_0$, while, for the walls to disappear as their average curvature reaches cosmologically significant scales ($R_{\rm av} \lesssim \lambda_H$), $h \gtrsim (\lambda)^{1/2+(\gamma-2)}(\sqrt{g_*}\phi_0/M_{Pl})^{3-\gamma}\tilde{E}^0_0$. To these bounds we add the percolation constraint, $p_- \geq 0.31$, which becomes, $h \leq 0.15\lambda$ [5].

Writing the average radius of the domains as $R_{\rm av} \equiv a\lambda_H$, the energy density in gravitational waves from collapsing domains is, (a is not the same as the ellipsoidal axis) $\rho_{\rm GW} \simeq \frac{\dot{E}_{GW}}{a^2\lambda_H^2}$, or with $\kappa \equiv 45/32\pi^3$,

$$\rho_{\rm GW} \simeq \frac{2\tilde{E}^0_{\rm GW}\phi_0^4 a^{\beta-3}\lambda^{(\beta-1)/2}}{\kappa^{(3-\beta)/2}g_*^{(\beta-3)/2}}\left(\frac{\phi_0}{M_{Pl}}\right)^{5-\beta} . \tag{7}$$

We have shown that the redshifted frequency is,

$$f^0 \simeq 6.1\times 10^{11}a^{-1}\left(\frac{g_*}{100}\right)^{\frac{1}{6}}\left(\frac{\phi_0}{M_{Pl}}\right) \text{Hz} , \tag{8}$$

while the fraction of the energy density in gravitational waves today is

$$h^2\Omega^0_{\rm GW} \simeq \frac{6.7\times 10^{-\beta+1}\left(\frac{\tilde{E}^0_{\rm GW}}{10^4}\right)\left(\frac{100}{g_*}\right)^{\frac{\beta-\frac{1}{3}}{2}}}{a^{3-\beta}\lambda^{(1-\beta)/2}\kappa^{-\beta/2}}\left(\frac{\phi_0}{M_{Pl}}\right)^{5-\beta} . \tag{9}$$

We have summarized our results in Fig. 1, where the detector strain vs. frequency are plotted for several values of β and compared with the LIGO (initial and advanced) and LISA sensitivities. The solid squares locate an electroweak phase transition at $T = 200$ GeV. The solid triangles, a hypothetical transition at $T = 10^9$ GeV. Continuous curves are for ellipsoids while dashed curves are for Gaussians.

The dot-dashed curve is for a Gaussian with $a = 10^{-2}$. LIGO is mostly sensitive to collapsing walls at $\phi_0 \sim 10^9$ GeV or so, while LISA can probe the electroweak phase transition for a comfortable range of parameters.

It is clear that the stochastic background of gravitational waves left behind from collapsing domain walls can be detectable for several situations of interest, illustrating the potential impact of gravitational wave astronomy on applications of particle physics to the early Universe.

MG was partially supported by the NSF through a Presidential Faculty Fellows Award no. PHY-9453431 and MG and RR by a NASA grant no. NAG5-6613.

REFERENCES

1. B. Allen, in *Les Houches School on Astrophysical Sources of Gravitational Waves*, eds. J. A. Marck and J. P. Lasota (Cambridge University Press, Cambridge, England, 1996) [gr-qc/9604033]. This is a fairly complete, but somewhat out of date, list of references. See also Ref. [2].
2. M. Gleiser and R. Roberts, Phys. Rev. Lett. **81** (1998) 5497.
3. S. Weinberg, *Gravitation and Cosmology* (Wiley, New York, 1972), Ch. 10.
4. A. Kosowsky, M. Turner, and R. Watkins, Phys. Rev. D **45** (1992) 4514; M. Kamionkowski, A. Kosowsky, and M. Turner, Phys. Rev. D **49** (1994) 2837.
5. G. Gelmini, M. Gleiser, and E. W. Kolb, Phys. Rev. D**39** (1989) 1558.

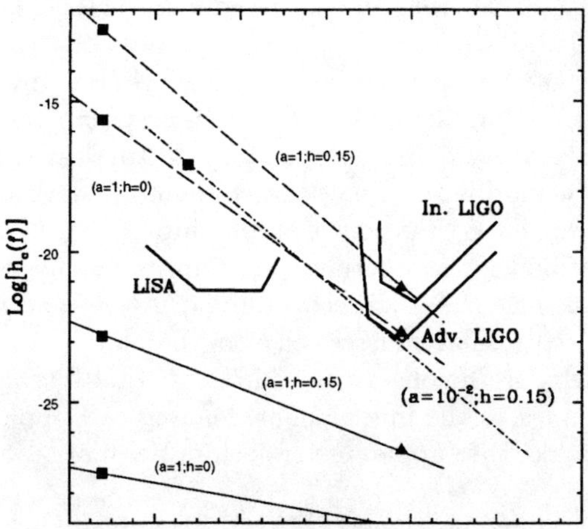

FIGURE 1. Predicted detector strain vs. dominant frequency band from collapsing nonspherical domains. We also display the sensitivities of the LIGO and LISA detectors. The solid cubes are for an electroweak transition at $T = 200$ GeV and the solid triangle for a hypothetical transition at $\phi_0 = 10^9$ GeV.

Observations of Blazars at TeV Energies: Current Status and Future Prospects

Stefan Westerhoff

Santa Cruz Institute for Particle Physics
University of California, Santa Cruz, CA 95064
e-mail: stefan@scipp.ucsc.edu

Abstract. Very High Energy (VHE) γ-ray astronomy [1] is a rather young research field. It represents the relatively unsampled energy region above $\sim$100 GeV. This overview briefly discusses how probing the universe with VHE γ-rays tests astrophysical and cosmological scenarios, and describes some current and future experiments.

INTRODUCTION

The launch of the Compton Gamma-Ray Observatory (CGRO) by NASA in 1991 led to an enormous influx of data from 50 keV to 30 GeV. Attention has focussed on the class of galaxies with strong non-thermal activity in their center caused by Active Galactic Nuclei (AGN). The EGRET detection of γ-emission from about 60 AGN [2,3] has established them as the most luminous class of extragalactic objects at energies above 100 MeV.

Out of the variety of different types of AGN, EGRET only detects *blazars*, a subclass of AGN that comprises radio-loud, flat-spectrum radio quasars and BL Lacertae objects. Blazars are characterized by featureless, polarized continuum emission and a rapid variability on time scales of hours to days at most wavelengths. This variability allows the estimation of upper limits on the size of the emitting regions. The implication that an enormous luminosity is released from a very small volume has motivated the current understanding of AGN as accreting supermassive black holes of $\sim 10^8$-10^{10} solar masses, ejecting bipolar jets with the γ-radiation originating in these jets. Radio observations indicate that blazars are AGN with the radio jets at small angles to the line of sight, leading to Doppler-boosted emission and kinematic effects like the (apparent) superluminal motion of radio knots in the jets.

Although the power-law spectra of EGRET blazars suggest an observable flux even at TeV energies, only three extragalactic objects, Mrk 421 (z=0.031), Mrk 501 (z=0.034), and 1ES 2344+514 (z=0.044), all relatively nearby BL Lacs, have been unambiguously detected at TeV energies [4–6] (see Figure 1 for a TeV skymap).

CP478, *COSMO—98*, edited by David O. Caldwell

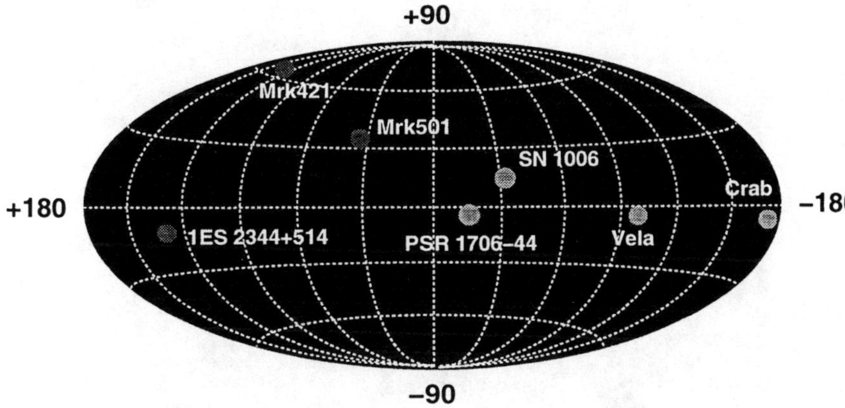

FIGURE 1. Skymap (galactic coordinates) of the TeV sky. Apart from the highly variable blazars [4–6], 4 Galactic objects [7–10], all supernova remnants with a γ-flux constant in intensity and spectrum, have been detected so far.

None of these BL Lacs are strong EGRET sources (1ES 2344+514 has in fact not been detected by EGRET), and strong EGRET sources at high redshifts like 3C 279 at z=0.54 have not been detected at TeV energies.

Whereas the origin of the TeV emission is still a subject of lively debate, the fact that the established TeV blazars are relatively close ($z \leq 0.05$) is generally attributed to the extragalactic background radiation which makes the universe opaque for larger distances and higher energies. Models for TeV emission and the impact of the extragalactic background will be discussed in the next sections.

Another possible reason for the rather small number of established TeV sources is the detection technique. Satellite experiments like EGRET with their small detector area are excellent tools in the MeV to GeV region, but as source spectra tend to be falling power laws, a large effective area is essential for a sufficient rate from point sources in the VHE region. The restriction to earth-bound detectors makes the reconstruction of the primary γ-rays considerably more complicated, as the Earth's atmosphere causes the primary particle to induce a cascade of secondary particles, the so-called air shower.

By far the most successful ground-based technique to date is the imaging atmospheric Cherenkov technique [11], in which the parameters of the primary particle are determined by detecting the Cherenkov radiation produced by the relativistic secondary particles of the air shower cascade. The Whipple telescope [12] on Mt. Hopkins (Arizona) is the pioneer of this technique; other telescopes are located on La Palma, Spain (HEGRA), in Targasonne, France (CAT), and in Woomera, Australia (Cangaroo) (covering the southern sky).

The Whipple telescope consists of a 10 m optical reflector with a camera of 109 photomutipliers (pixels) mounted in the focal plane. With this fine pixellation, it

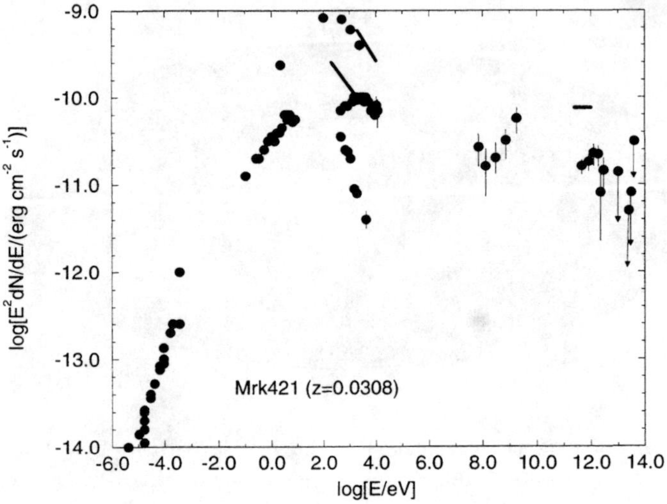

FIGURE 2. Spectral energy distribution (SED) of the blazar Mrk 421 with the two characteristic "bumps" in the X-ray and GeV region. The data sets are not simultaneous, illustrating the intrinsic source variability. The three established TeV sources are high frequency peaked BL Lacs (HBLs), i.e. the low energy peak is in the X-ray region in contrast to low frequency peaked BL Lacs (LBLs).

is possible to *image* the shape of the particle cascade and to efficiently separate the γ-ray signal from the much more abundant hadronic cosmic ray background [13] that cannot be vetoed otherwise.

Air Cherenkov telescopes are *pointed* instruments with a field of view of the order of 3°. As they are restricted to clear, dark nights, their duty cycle is only about 10 %. Optimized for the detailed study of individual sources, they are limited in their ability to continuously monitor large parts of the sky for transient phenomena.

BLAZAR MODELS

With their manifestation as TeV sources, γ-ray emission from Mrk 421, Mrk 501, and 1ES 2344 has now been detected over nearly 18 orders of magnitude in energy, from radio to VHE wavelengths, with a large energy fraction emitted at GeV and TeV energies. Figure 2 shows the spectral energy distribution of Mrk 421 with its two characteristic "bumps". Whereas the lower frequency bump is generally interpreted as the synchrotron radiation of relativistic electrons in the jet, there is no consensus as to the origin of the high frequency bump; both electron- and proton-based models have been suggested. In electron-based models, the relativistic electrons that cause the synchrotron radiation also upscatter the low energy photons in the jet (synchrotron-self-Compton or SSC model [14]) or ambient photons (external Compton or EC models [15]) into high energy photons by inverse Compton scattering. SSC models are generally very successful in reproducing the spectra of high-frequency peaked BL Lacs; the mechanism fails, however, in explaining the spectra of sources with a γ-ray luminosity *exceeding* the optical luminosity, a phenomenon observed in some of the strong EGRET blazars. In addition, recent observations of 18 TeV γ-ray emission from Mrk 501 during the 1997 flaring state [16] pose severe constraints on models based on electron acceleration and inverse-Compton upscattering alone.

In proton-based models [17] (proton-initiated cascade or PIC models), a certain amount of protons enters the jet. These protons are accelerated up to energies of 10^{18} eV [18] (pion production via collisions with matter is negligible because the matter density in radio jets is low) where photo production of pions (via $\Delta(1232)$) is the dominant cooling process. This induces an electromagnetic cascade in the jets. If the energy of the γ-rays is above the critical value E_{crit} with $\tau_{\gamma\gamma}(E_{crit}) = 1$, i.e. the optical depth $\tau_{\gamma\gamma}$ exceeds 1, they produce $e^{\pm}$-pairs which again produce γ-rays. This cascade-cycle is repeated several times until the energy of the γ-rays is below E_{crit}. Then the jet is no longer opaque and the high energy γ-rays escape.

Simultaneous observations of blazars with different instruments over a wide range of energies may provide some answers as to the mechanisms leading to TeV γ-rays. Multifrequency campaigns on Mrk 421 in 1995 and Mrk 501 in 1996 have shown significant correlations between X-ray and γ-ray flaring with time lags of less than a day [19]. This correlation is natural in the context of the SSC model, whereas the proton models are less predictive and further studies are necessary.

The detection of neutrinos from AGN would unambiguously indicate the presence of protons in jets; unfortunately, the predicted flux is lower than the sensitivity of current neutrino detectors [20].

COSMIC INFRARED BACKGROUND

Photons from sources at cosmological distances suffer absorption by e^+e^--pair production with photons of the diffuse extragalactic background light. The cross section for this process peaks near threshold; whereas the well-measured cosmic microwave background makes the universe opaque for γ-rays above ~ 100 TeV, it is the optical to infrared background from ~ 0.1 to 100 μm that affects the propagation of source photons with energies from several hundred GeV to 50 TeV.

Direct measurements of the background at these wavelengths are notoriously difficult because of local effects and dust. It has been proposed [21] that one can infer the diffuse extragalactic background *indirectly* by measuring the expected TeV cutoff in the source spectra of VHE sources. At this early stage, with less than a handful of sources, estimates are vague at best and only upper limits can be placed on the infrared photon density [22,23].

There is, however, a tremendous potential for cosmology and astrophysics in the study of VHE γ-ray source spectra. Primack, Bullock, Somerville and MacMinn [24–26] have shown how semi-analytical models of galaxy formation allow an estimation of how the extragalactic background light and thus the attenuation of VHE γ-rays depend on input assumptions like the history of star formation and the stellar initial mass function.

The simulations show that the extragalactic background between 1 and 10 μm, the region relevant for $\sim$ TeV γ-attenuation, is indeed sensitive to the underlying cosmology: as cold dark matter models imply early galaxy formation ($1 \leq z \leq 3$), they predict a larger extragalactic background than cold and hot dark matter models where galaxies form late ($0.2 \leq z \leq 1$).

On the other hand, at wavelengths below 1 μm and above 10 μm, the extragalactic background shows a significant dependence on the stellar initial mass function, with models having a higher fraction of high-mass stars leading to a larger extragalactic background. The observation of absorption below 1 TeV in the spectra of relatively distant blazars (z=0.5) or above 10 TeV in the spectra of nearby sources would provide a unique tool to tackle this problem.

To fully utilize these methods, it is crucial to discover more sources at different redshifts and to be able to separate external absorption from internal absorption at the source. A number of new experiments are under construction or starting data taking that are likely to address these issues by performing systematic sky searches and by detailed analysis of the spectrum of known sources over a wide range of energies.

CURRENT AND FUTURE EXPERIMENTS

The primary goals of experimental VHE astronomy are to increase the sensitivity of the observation methods, cover as large an energy range as possible, and constantly monitor the observable sky. This can only be achieved by a *combination* of instruments, based either on the Cherenkov light technique or the particle content of air showers.

Pointed Instruments

An important limitation for air Cherenkov telescopes comes from night sky background fluctuations. The energy threshold therefore roughly scales inversely with the square root of the mirror collection area. There are several approaches to overcome this problem. Whereas the MAGIC [28] collaboration is building a large 17 m diameter telescope, other groups like CELESTE [29] in France and STACEE [30] in New Mexico use already existing large mirror facilities at solar plants. As an example, STACEE uses 48 heliostat mirrors with a total reflector surface of 1785 m^2 to reflect the Cherenkov light to 3 secondary mirrors on a central tower which then

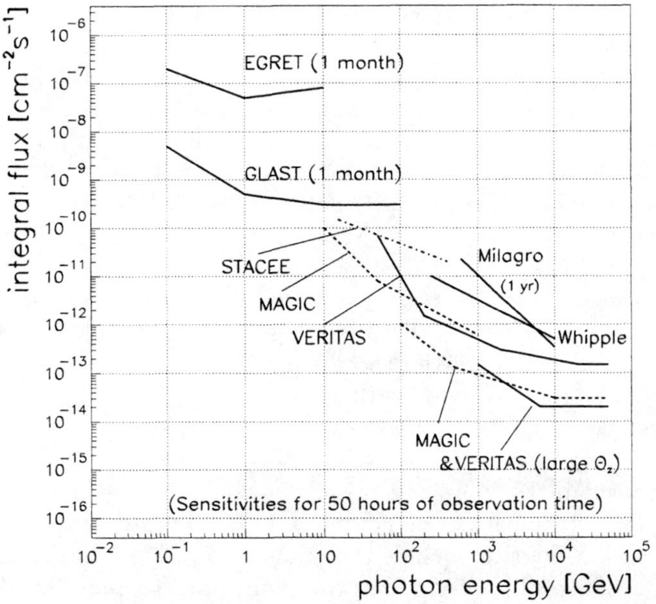

FIGURE 3. Predicted sensitivity of some proposed and operational telescopes. The numbers are based on a $5\,\sigma$ detection for the given exposure on a single source. The sensitivities for EGRET and GLAST [27], a high energy γ-ray space mission to be launched in 2005, are for 1 month, the Milagro sensitivity is for 1 year of all-sky survey.

reflect the light into a photomultiplier camera. With this technique, STACEE will work in the energy region from 20 to 250 GeV and thus bridge the energy gap between satellites and current air Cherenkov telescopes.

The VERITAS project [31] takes a different approach by building an *array* of imaging telescopes, each 10 m in diameter, placed on the corners of a hexagon with ~50 m distance. The proposed site is near Mt. Hopkins (Arizona). While this also increases the total mirror area, the stereoscopic imaging of the same shower by several telescopes makes it possible to tag the shower core and reconstruct the shower geometry more accurately. This leads to an improved angular and energy resolution and a very efficient reduction of the hadronic and muonic background. The HEGRA collaboration on La Palma is already operating an array of 5 telescopes, and first results include an almost background free signal from the Crab nebula [32].

Wide Aperture Detectors

The task of constantly monitoring a large part of the sky requires techniques that are not limited by a low duty cycle and a small field of view. The observation technique has to built on the *particle content* of air showers with all the adherent problems (see [33] for a detailed discussion of the relevant parameters). A first-generation instrument, the Milagro [34] detector located 2600 m a.s.l. near Los Alamos, New Mexico, at latitude 35.9° N, is nearing completion.

Milagro is a water Cherenkov detector of size $60 \times 80 \times 8\,\mathrm{m}^3$. Two layers of photomultiplier tubes detect the Cherenkov light produced by secondary particles entering the water of the pond. The first layer with 450 tubes on a $3 \times 3\,\mathrm{m}^2$ grid at a depth of 1.5 m allows the shower direction and thus the origin of the primary particle to be reconstructed with an (energy dependent) accuracy of about 0.5°. The second layer with 273 tubes at a depth of 7.0 m detects primarily the penetrating hadronic and muonic component of air showers. Recognizing hadronic hot spots and muons in the shower image allows a (modest) γ/hadron separation.

Results from data taken between February 1997 to April 1998 with a smaller one-layer prototype, Milagrito, include the observation of Mrk 501 in its high state of emission [35]. The full Milagro detector will start to monitor the northern hemisphere in early 1999.

As a summary, Figure 3 shows the sensitivity and the energy range of current and future satellite and ground-based telescopes. The sensitivity is based on a $5\,\sigma$ detection for the given exposure on a single source. The combination of these instruments will provide us with new data urgently needed to understand the TeV universe and will (hopefully) generate a number of exiting surprises.

ACKNOWLEDGEMENTS

It is a pleasure to thank Gaurang B. Yodh and all organizers of COSMO'98. This work is supported by the National Science Foundation (NSF) under grant number PHY-9722649.

REFERENCES

1. For recent reviews, see R.A. Ong, *Physics Reports* **305**, 93 (1998), and C.M. Hoffman, C. Sinnis, P. Fleury and M. Punch, to be published in Reviews of Modern Physics.
2. C. Fichtel et al., *Astrophys. J. Suppl. Ser.* **94**, 551 (1994).
3. D.J. Thomson et al., *Astrophys. J. Suppl. Ser.* **101**, 259 (1996).
4. M. Punch et al., *Nature* **358**, 477 (1992).
5. J. Quinn et al., *Astrophys. J.* **456**, L83 (1996).
6. M. Catanese et al., *Astrophys. J.* **501**, 616 (1998).
7. T.C. Weekes et al., *Astrophys. J.* **342**, 379 (1989).
8. T. Kifune et al., *Astrophys. J.* **438**, L91 (1995).

9. T. Yoshikoshi et al., *Astrophys. J.* **487**, L65 (1997).
10. T. Tanimori et al., *Astrophys. J.* **497**, L25 (1998).
11. M.F. Cawley and T.C. Weekes, *Exp. Astron.* **6**, 7 (1995).
12. M.F. Cawley et al., *Exp. Astron.* **1**, 173 (1990).
13. P.T. Reynolds et al., *Astrophys. J.* **404**, 206 (1993).
14. S.D. Bloom and A.P. Marscher, in *AIP Conf. Proc. 280 (CGRO)*, eds. M. Friedlander, N. Gehrels, D.J. Macomb, 1993, p.578.
15. C.D. Dermer and R. Schlickeiser, *Astrophys. J.* **416**, 458 (1993).
16. A. Konopelko et al., in *Proc. VERITAS Workshop on TeV Astrophysics of Extragalactic Sources*, eds. M. Catanese and T. Weekes, to be published in Astroparticle Phys. (1999).
17. K. Mannheim, *Astron. & Astrophys.* **269**, 67 (1993).
18. P.L. Biermann and P.A. Strittmater, *Astrophys. J.* **322**, 643 (1987).
19. J.E. McEnery, this volume.
20. K. Mannheim, *Astroparticle Phys.* **3**, 295 (1995).
21. F.W. Stecker, O.C. DeJager and M.H. Salamon, *Astrophys. J.* **390**, L49 (1992).
22. B. Funk et al., *Astroparticle Phys.* **9**, 97 (1998).
23. S. Biller et al., *Phys. Rev. Lett.* **80**, 2992 (1998).
24. D. MacMinn and J.R. Primack, *Space Science Rev.* **75**, 413 (1996).
25. R.S. Somerville and J.R. Primack, *Mon. Not. R. Astron. Soc.*, in press (1999).
26. J.R. Primack, J.S. Bullock, R.S. Somerville and D. MacMinn, in *Proc. VERITAS Workshop*, to be published in Astroparticle Phys. (1999).
27. E.D. Bloom, *Space Science Rev.* **75**, 109 (1996).
28. E. Lorenz, in *Proc. 25th Int. Cosmic Ray Conf. (Durban)* **5**, 177 (1997).
29. E. Paré, *Space Science Rev.* **75**, 127 (1996).
30. M.C. Chantell et al., *Nucl. Instr. Meth. A* **408**, 468 (1998).
31. T.C. Weekes et al., *Proposal for the Very Energetic Radiation Imaging Telescope Array System (VERITAS)*, Smithsonian Astrophysical Observatory (1996).
32. A. Daum et al., *Astroparticle Phys.* **8**, 1 (1997).
33. R.S. Miller and S. Westerhoff, *Astroparticle Phys.*, in press (1999).
34. G.B. Yodh et al., *Space Science Rev.* **75**, 199 (1996).
35. A.J. Smith, in *Proc. VERITAS Workshop*, to be published in Astroparticle Phys. (1999).

Multifrequency Observations of TeV Emitting Blazars

Julie E. McEnery

Physics Department
University of Utah, UT 84112

Abstract. The results of multifrequency observations of Mrk421 and Mrk501 during TeV γ-ray flares are presented. Flares on timescales of days were observed from each of these sources. In both cases, the X-ray and the γ-ray flux varied together. At lower energies the flare amplitude was much less. For observations of Mrk421 in April 1995 the extreme ultraviolet and optical flares lagged the γ-ray and X-ray flares by $\sim$ 1 day. In May 1996, two very intense intra-day flares were observed at TeV energies during a period of increased X-ray and optical activity. The importance of these observations within the context of non-thermal emission models of AGN is discussed.

INTRODUCTION

During the past decade the field of very high γ-ray astronomy has come of age. The launch of the EGRET satellite capable of detecting 20 MeV - 30 GeV γ-rays, and the maturing of the air Cherenkov technique sensitive at energies above 300 GeV has opened a new window on the sky. One of the most exciting discoveries has been the detection of VHE γ-rays from the blazar sub-class of AGN. To date, over 60 AGN have been detected at GeV energies by EGRET [1], while 3 have been detected at TeV energies by imaging air Cherenkov telescopes. Observations of AGN at TeV energies are particularly important. The extension of the measured spectrum up to TeV energies poses particularly stringent constraints on emission models. The very large collection area of the Cherenkov telescopes allows the detection of lower amplitude and shorter duration flares than it possible with EGRET.

Blazars are a sub class of AGN, characterised by flat radio spectra, high polarization, and exhibit variability at all energies from radio through to VHE γ-ray. These properties are believed to be a result of viewing a relativistic jet orientated close to the line of sight. The spectral energy distribution (SED) of blazars, illustrated in the paper by Westerhoff (these proceedings), consists of two bumps, the first extends from the radio through to the UV/X-ray regime, while the second bump spans X-ray to VHE γ-ray energies. To understand the mechanisms governing the γ-ray emission and the reasons determining whether a blazar is bright at TeV

CP478 , *COSMO—98*, edited by David O. Caldwell

energies or not, one must examine the emission across the entire electromagnetic spectrum.

Blazars themselves are sometimes divided into two classes, Flat Spectrum Radio Quasars (FSRQ) or BL Lacs, determined by the presence or absence of thermal features, respectively. The absence of emission lines in BL Lacs makes then hard to identify. Typically these objects are discovered by X-ray or radio surveys. Traditionally, this has resulted in a distinction between X-ray selected (XBL) and radio selected BL Lacs (RBL). The blazars detected at TeV energies are the closest blazars and are all X-ray selected BL Lacs, which are the least luminous class of blazars, with the two peaks in their SEDs shifted to higher energies relative to those of FSRQs or RBLs. Several reasons have been proposed to explain why TeV emission has been seen from only 3 blazars: (a) the shift of the two peaks of XBLs to higher energies imply that the particle spectrum may also extend to higher energies allowing for the production of higher energy photons, (b) the lack of emission lines in BL Lacs could imply a dearth of gas/dust clouds, and/or accretion disk photons which would result in less absorption of the γ-rays at the source, or (c) they are the closest blazars, therefore the γ-rays are less absorbed by pair production off the cosmic infra-red background. Extensive searches have been made for TeV emission from other AGN [6], concentrating on nearby objects, and more recently on X-ray selected BL Lacs. With the limited sample of TeV bright objects it is not possible to determine why some blazars but not others are bright at TeV energies.

One of the major roles of multifrequency observations is as a probe of non-thermal emission processes in jets of AGN. A successful model of the non-thermal emission from blazars must be able to reproduce the broadband multifrequency spectra and account for the observed changes in flux and spectral index during flares. The origin of the lower energy component in the SEDs of blazars is generally accepted to be a result of beamed, incoherent synchrotron radiation.The origin of the second peak is a subject of considerable debate. Several models have been proposed, which can loosely be divided into two classes; (a) Proton Initiated Cascade (PIC) models, in which highly relativistic protons interact with a soft photon field producing pions, $e^{\pm}$ pairs and neutrinos [11]. These particles (except for the neutrinos) develop an electromagnetic cascade until the medium becomes optically thin for the observed γ-rays. Since the high and low energy peaks in the SED are produced by different populations of relativistic particles, this class of model makes no definite predictions about temporal variability of the high energy emission or its relation to variability at lower energies. (b) A second class of models attributes the high energy emission to inverse Compton scattering of lower energy photons to γ-ray energies by the same population of electrons responsible for the synchrotron emission. These models make strong predictions on the behaviour of time lags and relative amplitude variations during flares, the detailed nature of which is determined by the source of the soft photon field, the jet geometry and the physical parameter that prompted the flaring behaviour. Multifrequency campaigns at energies from radio through γ-rays have been very successful at observing flares in TeV bright AGN, the results of these observations allow stringent constraints to be placed on important physical parameters of the jet such as magnetic field and Doppler boost factor.

OBSERVATIONS AND RESULTS

Since their discovery as TeV γ-ray sources, Mrk421 [2] and Mrk501 [3] have been routinely monitored at TeV γ-ray energies at the Whipple Observatory. The TeV γ-ray flux from each these objects is highly variable on timescales from hours to months. Dayscale flares have been observed from each of these objects during multifrequency campaigns. The results of observations of Mrk421 in April 1995 and Mrk501 in April 1997 are illustrated in figure 1. In both cases X-ray and γ-ray flares are observed to occur simultaneously, or at least with no measurable lag between the two lightcurves. The flare amplitude decreases with decreasing energy. A flare at extreme ultraviolet energies which lagged the γ-ray flare by ~ 1 day was detected in April 1995 from Mrk421. The optical observations of Mrk421 in April 1995 show a very weak flare which also lags the γ-ray data by one day. Optical observations of Mrk501 in April 1997, exhibit no simple correlation with the γ-ray data, but the emission is significantly higher than during the previous month (indicated with a dashed line).

Intraday flares at TeV energies have also been observed from both of these blazars [12,8]. One of the most extraordinary events every observed at TeV energies was the detection of two very intense hourscale flares in May 1996. The first occurred on the 7th of May with a factor of 30 increase in flux in 2.5 hours, becoming the most intense source of TeV γ-rays ever observed. The second on May 15 was of approximately 30 minutes duration, with a peak of about 14 times the average for 1996. While optical and X-ray observations were not sufficiently well sampled to look for correlated intraday variability during this period, high and variable flux was seen at both energies during this period [13,14].

DISCUSSION AND CONCLUSION

The very high energy photons observed from TeV bright blazars pose special constraints on proposed emission models. The absorption of the γ-rays by pair production both, intrinsic to the AGN and in the intergalactic medium is very important. Also in processes involving inverse Compton scattering, one must take into account the decrease in cross-section when moving from the Thomson to the Klein-Nishina regime. Both of these effects become important when $E_\gamma E_{soft} \geq m_e^2 c^4$. This has profound implications in γ-ray emission scenarios, for 5 TeV photons: the density of ambient photons with energies with energies greater than 0.1 eV must be low or the effects of $\gamma\gamma$ absorption would cause a turnover in the TeV spectrum, also if the γ-ray emission is produced by inverse Compton scattering, then the seed photons must have energies less than 0.1 eV to avoid the Klein-Nishina regime.

The effects of $\gamma\gamma$ absorption can be used to obtain a model independent limit on the Doppler factor (δ) of the jet. For a 500 GeV photon the cross-section for pair production is maximised for IR/optical photons. If one assumes that the apparent correlation between the optical and γ-ray emission imply that these photons are produced in the same location, then the requirement that the opacity be less than one implies a lower limit on the Doppler boost factor of 15.4 for Mrk421. Within the context of inverse Compton models one can also obtain limits on the magnetic field and Doppler factor. One can obtain the upper energy cutoff of the electron distribution as a function of the magnetic field (B) and Doppler factor from the observed synchrotron cutoff. The requirement that the seed photons for inverse Compton scattering be less than 0.1 eV implies for Mrk421 that $B \leq 0.01\delta$ in order to produce the observed 8 TeV γ-rays [7]. A lower limit on B can be derived by requiring that the cooling time for the electrons be smaller than the observed decay time of observed flares. The requirement that the lower limit on B is not higher than the upper limit results in a lower limit on the Doppler boost factor of 12.6 for Mrk421. A similar analysis for Mrk501 leads to $\delta \geq 2$.

Multiwavelength observations made to date do not allow a definitive distinction between emission models in the jets of AGN. However all models are becoming increasing tightly constrained. The correlation between X-ray and γ-ray flares is becoming well established, with little or no time lags between the two wavebands. The observation of correlated X-ray and TeV γ-ray emission is naturally expected from inverse Compton models. Both the X-ray and the γ-ray emission are produced by the same population of electrons, thus one expects a disturbance in the electron population to result in simultaneous flares in the inverse Compton and synchrotron spectra. The details of time lags and differences in amplitude of variability can be used as a sensitive probe of the different inverse Compton models. Recent TeV γ-ray observations indicate that the spectrum of some of these objects extends up to extremely high energies, photons of at least 18 TeV were observed from Mrk501 during a flare in 1997 [9]. This observation poses severe constraints on models based on inverse Compton scattering; in order to avoid the inefficient Klein-Nishina regime, the seed photons are required to have energies in the IR or lower, this in turn requires that the electrons be accelerated to extremely high energies. It is not clear that sufficiently efficient acceleration mechanisms exist. However, these observations also pose problems for proton models of the non-thermal emission. It is not certain the protons can be accelerated or decelerated quickly enough to explain the very short timescale variability observed in Mrk421 and Mrk501. These problems may be alleviated by variability models being currently developed to explain flares as a result of changes in the bulk motion of the blob of relativistic particle rather than changes in the particle energy distributions themselves [10].

Much has been learned about TeV bright AGN from recent multifrequency observations. The observed correlations between the different energies strongly favour inverse Compton models of γ-ray emission, although proton models cannot be ruled out. To definitively distinguish between emission models, more objects need to be detected at TeV energies and more detailed multifrequency observations during flares are required to quantify more accurately time lags and flare amplitudes over as wide a spectral range as possible.

REFERENCES

1. D.J. Thomson et al., *Astrophys. J. Suppl. Ser.* **101**, 259 (1996).
2. M. Punch et al., *Nature* **358**, 477 (1992).
3. J. Quinn et al., *Astrophys. J.* **456**, L83 (1996).
4. J.H. Buckley et al., *Astrophys. J.* **472**, L9 (1996).
5. M. Catanese et al., *Astrophys. J.* **487**, 143 (1997).
6. A.D. Kerrick et al., *Astrophys. J.* **452**, 588 (1995).
7. F.W. Samuelson et al., *Astrophys. J.* **501**, L17 (1998).
8. J.A. Gaidos et al., *Nature* **383**, 319 (1996)
9. A. Konopelko et al., in *Proc. VERITAS Workshop on TeV Astrophysics of Extragalactic Sources*, eds. M. Catanese and T. Weekes (1999).
10. C.D. Dermer et al., in *Proc. VERITAS Workshop on TeV Astrophysics of Extragalactic Sources*, eds. M. Catanese and T. Weekes (1999).
11. K. Mannheim, *Astron. & Astrophys.* **269**, 67 (1993).
12. J. Quinn et al., *Astrophys. J.* (1999) (submitted).
13. J.E. McEnery et al., in *Proc. Blazars, Black holes and Jets*, (Girona), (1996).
14. M.S. Schubnell et al., in *Proc. Blazars, Black holes and Jets*, (Girona), (1996).

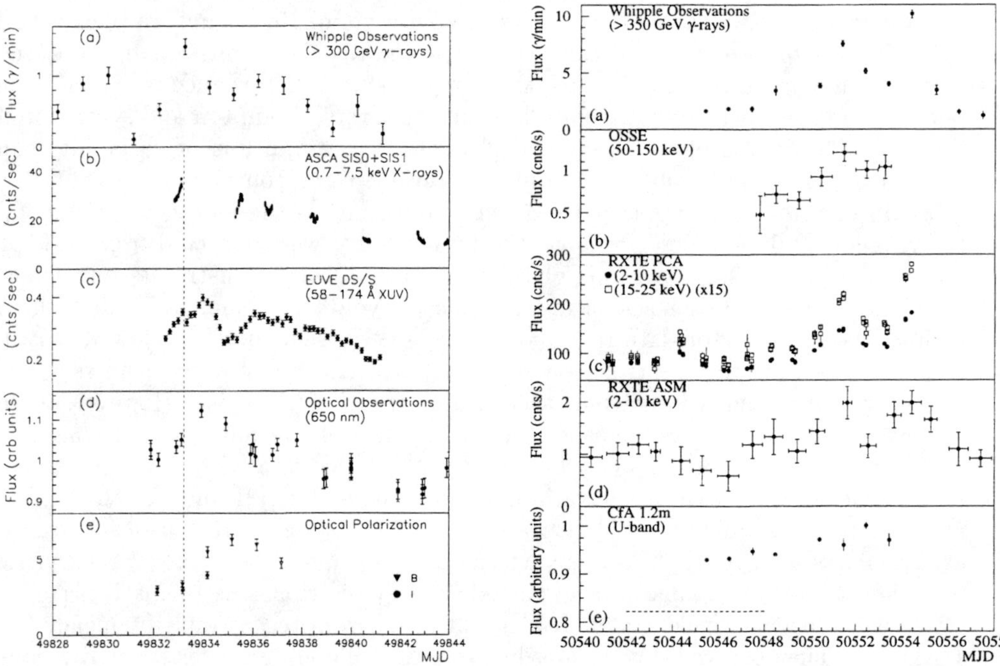

FIGURE 1. *Left:* Multi-wavelength light-curve of Mrk 421 taken in 1995 April-May [4]. *Right:* light-curves of Mrk 501 taken in 1997 April [5].

GLAST
A Partnership in Astro and Particle Physics

Gary L. Godfrey
For the GLAST Collaboration (1)

Stanford Linear Accelerator Center
Stanford, California 94309

Abstract. The Gamma Large Area Space Telescope (GLAST) is a >8000 cm^2 effective area (at >1 GeV) pair conversion telescope using silicon strip detectors that is scheduled to be launched in the year 2005. GLAST will record photons from 20 MeV to 300 GeV with a point spread function of .4° at 1 GeV. Because of its large effective area, almost 2π solid angle of view, and long experimental life (no expendables), GLAST will accumulate ~50 times more photons than EGRET (2) from any point in the sky. The collaboration presently consists of both astro and particle physicists from 28 international institutions.

INTRODUCTION

An overview of the GLAST detector is shown in Figure 1. An incoming photon is converted to an electron-positron pair in one of the thin lead foils. Downstream of each foil is a layer of x direction silicon strip detectors and a layer of y direction silicon strip detectors. The direction of the incoming photon is reconstructed using the tracks from the electron-positron pair as determined by the layers of silicon strip detectors. The energy of the electron+positron is measured in a CsI calorimeter situated below the converter/tracker section.

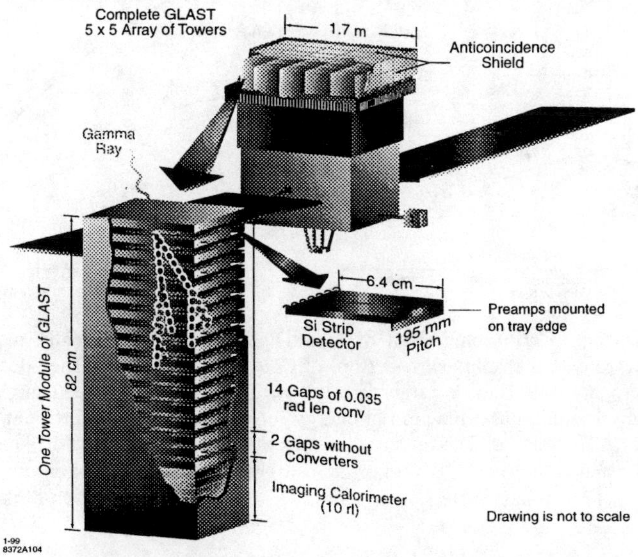

FIGURE 1. The GLAST baseline instrument. One of 25 identical towers is shown in detail.

CP478, *COSMO—98*, edited by David O. Caldwell

Surrounding the 25 towers of GLAST is a segmented-plastic-scintillator anti-coincidence detector (ACD). The combination of no ACD hit, unique gamma conversion pattern in the tracker, and electromagnetic shower pattern in the CsI will reject ~10^6 cosmics protons for each one that gets by to mimic a real gamma. The segmented ACD will help decrease a problem seen in EGRET where backsplash from real gammas in EGRET's calorimeter sometimes caused a veto in EGRET's monolithic anticoincidence dome.

From the start, the GLAST design has been guided by the use of a detailed Monte Carlo simulation. The performance of GLAST as compared with EGRET is shown in Figure 2 and summarized in Table 1. The combined result of these improvements is an ~100 times better point source sensitivity than EGRET.

Three measurements will be of interest to this conference, Particle Physics and the Early Universe.

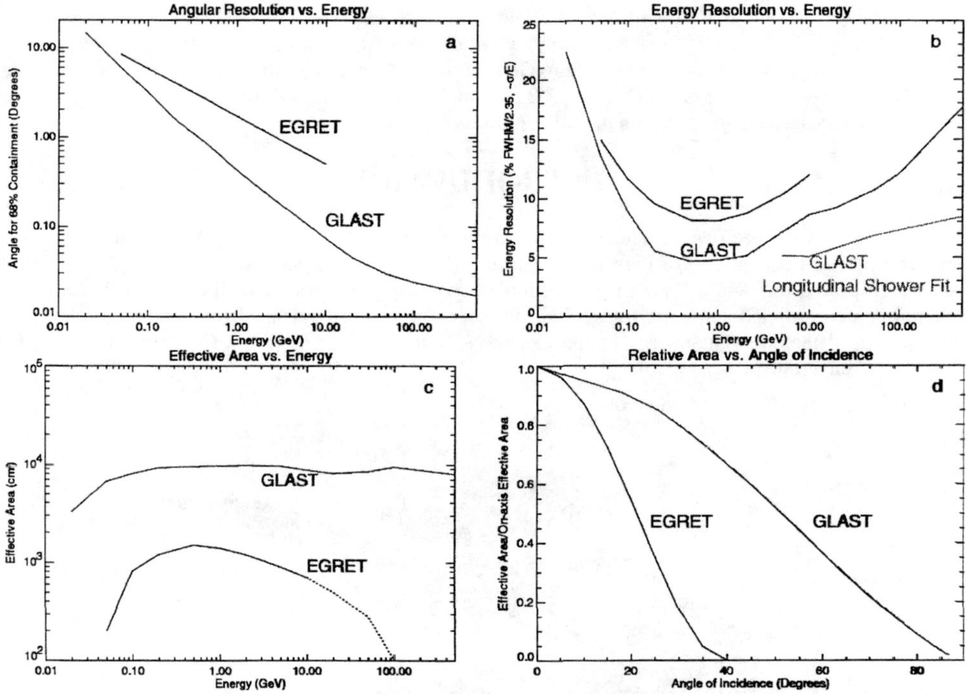

FIGURE 2. A Monte Carlo comparison of GLAST and EGRET. Normally incident photons are uniformly distributed over the zenith surface (top) of GLAST and EGRET. The GLAST curves include fiducial cuts that contain the showers transversely. a) The 68% containment angle as a function of energy. b) Energy resolution as a function of energy for photons normally incident on GLAST (10 rl, CsI) and EGRET (6 rl, NaI). c) The effective area as a function of energy for events which pass the tracker γ-signature and either the ACD veto or deposit greater than 5 GeV of energy in a single crystal in the calorimeter. d) The relative effective area for 1 GeV photons as a function of the zenith angle.

TABLE 1. GLAST baseline capabilities.

Parameter	GLAST Baseline Design
Energy Range	<.02 → 300 GeV
Single Photon Angular Resolution (on axis space angle for 68% containment)	3.1° @ 100 MeV, 0.4° @ 1 GeV, <0.1° @ 10 GeV
Point Source Sensitivity (1 yr, high gal lat, E_γ >100 MeV, 5 σ)	3.5 x 10^{-9} photons cm^{-2} sec^{-1}
Point Source Location	30 arcsec → 5 arcmin
Field of View (FWHM)	2.6 sr
Effective Area (including inefficiencies due to background rejection)	8000 cm^2 >1 GeV 4000 cm^2 @ 50 MeV >800 cm^2 @ 20 MeV
Dead Time	~20 usec
Background Rejection	Cosmic ray rejection efficiency 10^6:1
Instrument Lifetime	Expected >5 years (no consumables)

Spectral Cutoff Due to Extragalactic Background Light

The Extragalactic Background Light (EBL) was produced by galaxies undergoing starbursts during the epoch of galaxy formation. The absorption of high-energy gamma rays occurs over cosmological distance via $\gamma\gamma \rightarrow e^+e^-$ interactions with near-ultraviolet, optical and near-infrared photons that make up the EBL. The cross section for this interaction is maximized when E_{EBL}[eV]=1 TeV / (3 E_γ). Figure 3 shows how an unabsorbed AGN spectrum would appear if observed at red shifts of z=0.5 and 2 for particular EBL models (3). Determination of the EBL can provide unique information on the formation of galaxies at early epochs, and will test models for structure formation in the universe, such as those in which a neutrino mass of 5 to 10 eV plays an important role (4).

Measurement of a Possible Quantum Gravity Effect

The authors Amelino-Camelia, Ellis, Mavromatos, Nanopoulos, and Sarkar (5) have pointed out that the arrival times of high energy gamma-rays from some gamma-ray bursters (GRB), those now known to be at cosmological distances, can provide constraints on the fundamental laws of physics. If electromagnetic waves in vacuum have a dispersion with a scale of the Planck mass, then the difference in arrival times of high energy gamma-rays from a cosmological GRB sources would be easily measurable with a next generation gamma-ray telescope such as GLAST. Equation (1) gives this dispersion to lowest order.

$$c_{observed} = c_o(1+\kappa\frac{E_\gamma}{M_{Planck}}) \qquad (1)$$

Assuming equal emission times at their source, a 10 GeV gamma would arrive 30 msec later than a 100 MeV gamma from a GRB burst at z=.06 with M_{Planck}=10^{19} GeV and assuming the dimensionless constant κ=+1 . GLAST will be able to record successive gammas as close together as the trigger deadtime of 20 usec, thus making such time differences easily measurable. Such measurements would be difficult with EGRET because of its ~100 msec spark chamber deadtime (see Figure 4). Since different energy gammas may not be emitted at the same time at their source, the measurement of dispersion will entail looking for a systematic difference between the arrival time of low and high energy photons and showing that it is proportional to the GRB distance (z).

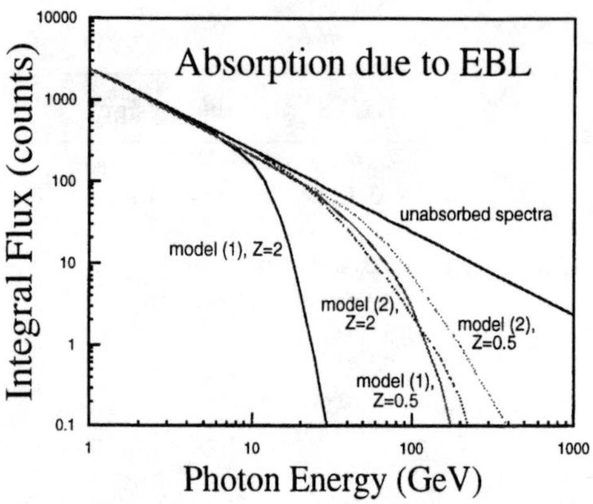

FIGURE 3. High-energy gamma-rays emitted by distant AGNs can interact with EBL photons before they have a chance to reach us. Shown above is how an unabsorbed AGN spectrum would appear if observed at redshifts of z=0.5 and 2 for two particular EBL models (based on Madau & Phinney (3)). GLAST observations of the high-energy spectra for a large sample of AGN at various redshifts will map out the EBL. The ordinate of the graph is in arbitrary units.

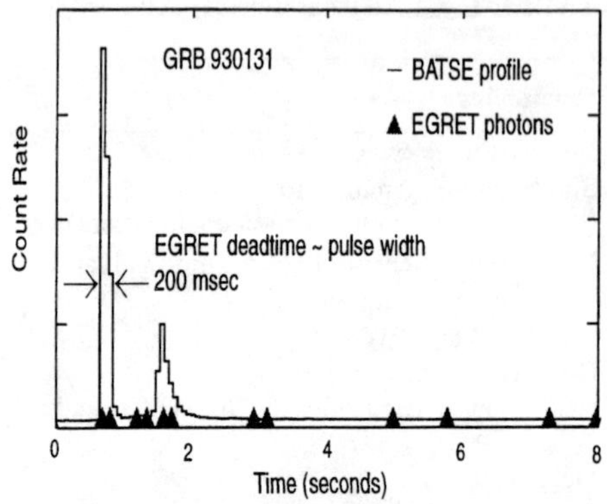

FIGURE 4. EGRET and BATSE data for Gamma Ray Burst 930131. High energy EGRET photons do occur within milliseconds of the xrays seen by BATSE. Many more high energy photons may have arrived during the BATSE spikes, but they would have been missed due to the EGRET spark chamber deadtime of ~100 msec. GLAST will have ~20 usec dead time per trigger and thus not have this severe a problem.

Neutralino Annihilation from the Galactic Center

The dark matter puzzle is currently one of the most interesting challenges confronting particle astrophysics and cosmology. The measured rotation curves of galaxies and galactic structure formation arguments are powerful evidence for dark matter in the universe. The lightest supersymmetric particle, X, is a promising candidate for this dark matter (6). It is neutral (hence the name neutralino) and stable if R parity is not violated. With the requirement that neutralinos make up most of the dark matter, current limits allow viable models in the mass interval, 30 GeV< M_X <10 TeV (7). If neutralinos make up the dark matter of our galaxy, they would have non-relativistic velocities. Hence the annihilation into the $\gamma\gamma$ and γZ final states can give rise to gamma rays with unique energies, that is, gamma-ray lines with $E_\gamma = M_X$ or $E_\gamma = M_X (1 - m_Z^2/4M_X^2)$. GLAST can search for gamma-ray lines above 30 GeV thanks to its excellent energy resolution, $\sigma_E/E \sim 1.5\%$, for gamma-rays entering at angles >56° to the normal. Tracks at these angles traverse >18 radiation lengths of CsI.

Recently, full one-loop calculations of the $XX \rightarrow \gamma\gamma$ and $XX \rightarrow \gamma Z$ annihilation processes have been performed in the minimal extension of the standard model by Bergström & Ullio (8,9). Recent n-body simulations of dark matter halos by Navarro, Frenk, & White (NFW) (10) have given indications of a universal profile where the density increases substantially near the galactic center. With this distribution, Bergström, Ullio, and Buckley (11) have calculated what would be seen for various choices of supersymmetric model parameters. Using their calculation and the GLAST Monte Carlo for the baseline instrument, Figure 5 shows that many models (each point is a model) exceed GLAST's 5 σ detection limit (the solid curves).

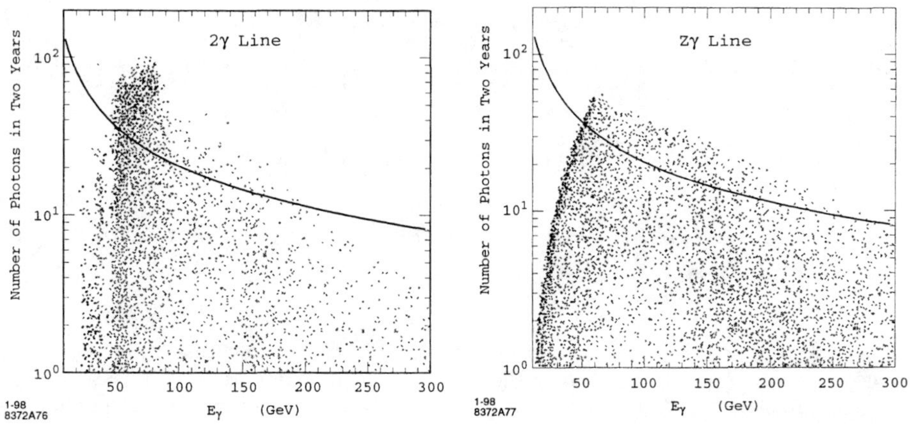

FIGURE 5. The number of events expected in GLAST for $XX \rightarrow \gamma\gamma$ and γZ from a 1-sr cone encompassing the galactic center, assuming a two-year scanning mode exposure and the Monte Carlo calculation described in the text. The NFW galaxy dark matter halo profile giving the maximal flux has been assumed. The solid lines show the number of events needed to obtain a five-sigma detection over the galactic diffuse gamma-ray background as estimated from EGRET data.

ACKNOWLEDGMENTS

This work was supported by Department of Energy contract DE-AC03-76SF00515.

REFERENCES

1. GLAST World Wide Web Page: http://www-glast.stanford.edu
GLAST Collaboration Institutions: American University, Washington, D.C. ; Boston University; Columbia University; CEA, Saclay; Ecole Polytechnique LPNHE; Hiroshima University; Institute of Physical and Chemical Research, Saitama; Institute of Space and Astronautical Science, Sagamihara; Instituto di Fisica Cosmica CNR, Milano; Kanagawa University; Lockheed-Martin Solar and Astrophysics Lab, Palo Alto; Max Planck Institut fr Extraterrestrische Physik; NASA Ames Research Center; NASA Goddard Space Flight Center; Naval Research Laboratory; Royal Institute of Technology, Stockholm; Shibaura Institute of Technology; Sonoma State University; Stanford Linear Accelerator Center; Stanford University; Stockholm University; University of California, Santa Cruz; University of Chicago; University of Rome INFN; University of Trieste INFN; University of Tokyo; University of Turin; University of Utah; University of Washington.
2. Dingus, B.L., et. al., *ApJ* **467**, 589-596 (1996).
3. Madau, P., and Phinney, E.S., *ApJ* **459**, L111 (1996); Salamon, M.H., and Stecker, F.W., submitted to *ApJ* (astro-ph/9704166).
4. Macminn, D., and Primack, J.R., astro-ph/9504032.
5. Amelino-Camelia, G., Ellis, J., Mavromatos, N.E., Nanopoulos, D.V., and Sarkar S., *Nature* **393**, 763-765 (1998), and astro-ph/9810483.
6. Weinberg, S., *Phys. Rev. Lett.* **50**, 387 (1983); Goldberg, H., *Phys. Rev. Lett.* **50**, 1419 (1983).
7. Jungman, G., Kamionkowski, M., and Griest, K., *Phys. Reports* **267**, 195 (1996).
8. Bergström, L., and Ullio, P., *Nucl. Phys.* **B504**,27(1997); Bern, Z.,et.al.,*Phys.Lett.* **B411**,86(1997).
9. Ullio, P., and Bergström, L., *Phys. Rev.* **D57**, 1962-1971 (1998).
10. Navarro, J.F., Frenk, C., and White S.D.M., *ApJ* **462**, 563 (1996).
11. Bergström, L., Ullio, P., and Buckley, J.H., *Astropart. Phys.* **9**, 137-162 (1998).

Weak Interaction Rates at High Temperatures and Densities

Jameel-Un Nabi [1] and Hans Volker Klapdor-Kleingrothaus

Max-Planck-Institut für Kernphysik, Postfach 103980, 69029 Heidelberg, Germany.

Abstract. Proton-neutron quasiparticle RPA is used for the first time to calculate weak interaction rates for *sd*- and *fp*-shell nuclei at high temperatures and densities. Particle emission processes from excited states are taken into account and is found to considerably affect the recommended β decay rates. The calculated rates take into consideration the latest experimental energy levels and ft value compilations. The calculation is done for 709 nuclei with mass number ranging from 18 to 100. The astrophysical applications of the calculated rates are highlighted.

INTRODUCTION

Weak interaction processes play a decisive role in the early stage of the core collapse of a massive star (see e.g., [1]). Electron captures on nuclei in the iron mass region, starting after the core mass exceeds the Chandrasekhar mass limit, reduce the electron pressure, thus accelerating the collapse. They also lower the electron-to-baryon ratio, Y_e, thus shifting the distribution of nuclei present in the core to more neutron-rich material. Many of the nuclei present can also β decay under these circumstances. The β decays pose a strong competition with the electron capture processes for neutron-rich nuclei due to an increase in phase space related to larger Q_β values. The electron captures and β decays also determine the physical conditions in the (quasi-) hydrostatically evolving core through their influence on the entropy per nucleon.

The weak interaction further affects the formation of heavy elements above iron via the r-process at the final stage of the supernova explosion. The weak interaction also largely determines the mass of the core, and thus the strength and fate of the shock wave formed by the supernova explosion.

Proton-neutron quasiparticle random-phase-approximation (pn-QRPA) has been shown to be a good microscopic theory for the calculation of beta decay half-lives [2–4]. Keeping in view the success of pn-QRPA theory in calculating terrestrial decay rates, we used this extended model to calculate for the first time the weak

---

[1] Talk presented by Jameel-Un Nabi

CP478 , *COSMO—98*, edited by David O. Caldwell

interaction rates in stellar matter using pn-QRPA theory. This work is the first calculation of weak rates for *fp*-shell nuclei in hot and dense matter taking into account details of the nuclear structure and presenting the rates over a wide range of temperature-density grid points and over a larger mass range. Earlier work by [5] (usually called FFN) used parameterization of the Gamow-Teller (GT) strength distribution on the basis of the independent particle model. The GT contribution was also supplemented by a contribution simulating low-lying transitions. The FFN rates were somehow then updated and extended to heavier nuclei [6]. These authors simulated the low-lying transitions by the same ft value for all nuclei, while FFN adopted specific values for individual nuclei. These calculations have naturally limitations as they lacked the important ingredient of a microscopic theory for the calculation of the GT strength. Recently a shell model diagonalization approach (SMDA) was used to calculate some supernova electron capture and β-decay rates [7,8]. However, due to the very large m-scheme dimensions involved, the GT strength distributions were calculated in truncated model spaces. These authors restricted themselves to parent excited states of a few MeV for the calculation of electron capture rates. For the calculation of β-decay rates they considered parent excited states usually up to 1 MeV and in addition, "back resonances" (the GT back resonance are states reached by the strong GT transitions in the electron capture process built on ground and excited states, see [5,6]) built on daughter states below 1 MeV.

The QRPA approach used in the present calculation, allowed the use of a very large model space up to 7 major shells (SMDA, in comparison, considered 1 major shell). We took into consideration for the first time the effect of beta-delayed particle emission processes in this type of calculations. A total of 12 different stellar weak rates for each nucleus were calculated. These include $e^{\pm}$-capture rates, $\beta^{\pm}$-rates, (anti)neutrino energy loss rates, probabilities of beta-delayed proton (neutron) emission and energy rates of beta-delayed protons (neutrons). We consider parent excitation energies up to the particle decay thresholds, S_n or S_p (after accounting for the effective Coulomb barrier which prevents a proton from being promptly emitted and the uncertainty in calculation of energy levels), and considered many 100's of initial and final states in our rate calculation. This has the effect that our calculated electron capture rates are, in general, suppressed in comparison to the corresponding rates of FFN at high temperatures and densities. The effect is more pronounced for the case of decay rates. A detailed comparison can be found in [9–11]. Our calculation of capture rates is enhanced for odd-A and odd-odd nuclei in comparison to the corresponding SMDA calculation. For even-even nuclei they are suppressed at high temperatures. For the case of decay rates our calculation is, in general, enhanced for the case of odd-A nuclei. In all other cases our calculated rates are suppressed. The degree of suppression and/or enhancement varies with temperature and density. For more quantitative conclusions we refer to [11,12]. Our results do not support the claim by [8] that for capture on odd-odd nuclei FFN placed the GT centroid at too low excitation energy. For odd-odd nuclei no experimental information is available for the GT strength distribution and [8] also did not compare the corresponding terrestrial rates of odd-odd nuclei with measured half-lives to show the reliability of their calculation. For the sake of reliability of our calculation, a comparison of all calculated terrestrial rates using the pn-QRPA theory, used in the present work, with measured half-lives, wherever possible, have been made and discussed in [3,4].

The calculated weak interaction rates for all 709 nuclei (A = 18 to 100) can be obtained as files on a magnetic tape from the authors on request. These also contain the weak interaction rates for many neutron-rich nuclei, which may play an important role even at the silicon burning stage of stellar evolution. For details of the formalism and the calculations we refer to [9,10].

In the next section we discuss some examples of the astrophysical applications of the calculated rates.

A Astrophysical applications of calculated rates

Several circumstantial arguments point to the formation of the third r-process peak at A $\sim$ 190 in 8-10 $M_\odot$ stars ([13], and the references therein). These low-mass supernovae might produce an r-process in a prompt explosion and a subsequent r-process in a neutrino driven wind. In [13] it is strongly argued that a prompt explosion of an O+Mg+Ne core may produce a reasonable r-process peak at A $\simeq$ 190, as well as r-process elements with Z $\geq$ 56. This prompt explosion was also calculated for the collapse of a O+Ne+Mg core by Hillebrandt et al. [14]. However other authors like Mayle and Wilson [15] found a delayed explosion due to late time neutrino heating for these cores.

Since these low-mass supernova stars are very interesting from astrophysical viewpoint, it is desirable to know the fate of these cores much more reliably. In their final stages of evolution the 8-10 $M_\odot$ stars undergo electron captures on ^{24}Mg and ^{20}Ne.

Miyaji and Nomoto [16] investigated ρ_{ign} (density at which oxygen deflagration starts) using the electron capture rates from the gross theory [17] and reported in their paper that it is very unlikely that oxygen burning is ignited by electron capture on ^{24}Mg and ^{24}Na ($\rho \sim 4 \times 10^9$ gcm^{-3}). However they reported that it is likely that at $\rho \sim 10^{10}$ gcm^{-3} electron capture on ^{20}Ne ignites the oxygen burning and ρ_{ign} is not low enough for complete disruption of stars.

The temperature and density of the O+Ne+Mg core at this relevant stage are, logT = 8.4 - 9.4 (in Kelvins), logρ = 9.3 - 10.4 (in g cm^{-3}). For the A = 24 systems, if convection is neglected, the electron capture process stops when all ^{24}Na is exhausted. This occurs because there is no supply of fresh material in the central region. However, if convection is considered, fresh ^{24}Mg (hence ^{24}Na) is always supplied in the convective core. As a result the entropy production continues to the higher density where gross theory electron capture rates differ from our calculated electron capture rates by two orders of magnitude.

Since the heating rate also increases in this density region, there is a possibility that the oxygen ignition begins before the electron capture on ^{24}Ne starts. To clear the situation, the core evolution calculations should be repeated with the new rates presented.

The rate of change of Y_e is important for stellar evolution. The nuclei which cause the largest changes in Y_e are neither the most abundant ones nor the ones with the strongest rates. Knowledge of both abundances *and* rates of nuclei are required for determining the most important electron capture or beta-decay nuclei for given values of ρ, T, and Y_e. The new rates presented here may revise the earlier ranked lists of nuclei [6,18,19] and may be helpful to prioritize more detailed studies on individual nuclei.

REFERENCES

1. Bethe H.A., *Rev. Mod. Phys.* **62**, 801 (1990).
2. Klapdor H.V., *Prog. Part. Nucl. Phys.* **10**, 131 (1983); *Prog. Part. Nucl. Phys.* **17**, 419 (1986); AIP Conf. Proc. **238**, 870 (1991).
3. Staudt A., Bender E., Muto K., Klapdor-Kleingrothaus H.V., *Atomic Data and Nuclear Data Tables* **44**, 79 (1990).
4. Hirsch M., Staudt A., Muto K., Klapdor-Kleingrothaus H.V., *Atomic Data and Nuclear Data Tables* **53**, 165 (1993).
5. Fuller G.M., Fowler W.A. and Newman M.J., *Ap. J. Suppl.* **42**, 447 (1980), *Ap. J. Suppl.* **48**, 279 (1982), *Ap. J.* **252**, 715 (1980).
6. Aufderheide M.B., Fushiki I., Woosley S.E., Hartmann D.H., *Ap. J. Suppl.* **91**, 389 (1994).
7. Langanke K., Martínez-Pinedo G., *Phys. Lett. B.* **436**, 19 (1998); Preprint **nucl–th/9809082** (1998).
8. Martínez-Pinedo G., Langanke K., Dean D.J., Preprint **nucl–th/9811095** (1998).
9. Nabi J.-U, and Klapdor-Kleingrothaus H.V., *Atomic Data and Nuclear Data Tables, in press* **71**, (1999).
10. Nabi J.-U, Ph.D thesis, Heidelberg University, (1999).
11. Nabi J.-U, and Klapdor-Kleingrothaus H.V., in preparation.
12. Nabi J.-U, and Klapdor-Kleingrothaus H.V., Preprint **nucl–th/9901090** (1999).
13. Wheeler J.C., Cowan J.J. and Hillebrandt W., Preprint **astro-ph/9711273** (1997).
14. Hillebrandt W., Nomoto K. and Wolff R.G., *Astr. Ap.* **133**, 175 (1984).
15. Mayle R.W. and Wilson J.R., *Ap. J.* **334**, 909 (1988).
16. Miyaji S. and Nomoto K., *Ap. J.* **318**, 307 (1987).
17. Miyaji S., Nomoto K., Yokoi K., Sugimoto D., *Publ. Astron. Soc. Japan* **32**, 303 (1980).
18. Cooperstein J. and Wambach J., *Nucl. Phys. A* **420**, 591 (1984).
19. Fuller G.M., *Ap. J.* **252**, 741 (1982).

PARTICLE PHYSICS AND COSMOLOGY

Early Inflation and Cosmology in Theories with Sub-Millimeter Dimensions*

Nima Arkani-Hamed[a], Savas Dimopoulos[b], Nemanja Kaloper[b], and John March-Russell[c]

[a] SLAC, Stanford University, Stanford, CA 94309, USA
[b] Physics Department, Stanford University, Stanford CA 94305, USA
[c] Theory Division, CERN, CH-1211, Geneva 23, Switzerland

Abstract. We discuss early cosmology in theories where the fundamental Planck mass is close to the TeV scale. In such theories the standard model fields are localized to a $(3+1)$-dimensional wall with n new transverse sub-millimeter sized spatial dimensions. The topic touched upon include: early inflation that occurs while the size of the new dimensions are still small, the spectrum and magnitude of density perturbations, the post-inflation era of contraction of our world while the internal dimensions evolve to their final "large" radius, and the production of gravitons in the bulk during these two eras. The radion moduli problem is also discussed.

Following Refs. [1–3], their has been much recent interest in the possibility that the fundamental Planck scale M_* is close to a TeV [4–9]. The weakness of gravity at long distances is then explained by n new "large" dimensions, the (present) size, b_0, of which is given by the Gauss Law relation $M_{\rm pl}^2 = (b_0)^n M_*^{n+2}$ where $M_{\rm pl} = 2 \times 10^{18}\,{\rm GeV}$. Taking $M_* \sim 1\,{\rm TeV}$ implies the case $n = 1$ is excluded since $b_0 \sim 10^{13}$ cm, while for $n \geq 2$ however, $b_0 \lesssim 1$ mm, which is the distance where our present experimental knowledge of gravitational strength forces ends. The success of the standard model (SM) up to $\sim 100\,{\rm GeV}$ implies that the SM fields can not feel these extra dimensions; they must be localized to a 3d wall, or "3-brane", in the higher dimensional space. The most attractive possibility for localizing the SM fields to the brane is to employ the D-branes that naturally occur in string theory [10,2]; this approach has the advantage of being formulated within a consistent quantum theory of gravity. However, the most important question is whether this framework is experimentally invalidated: In Ref. [3], laboratory,

*) Talk presented by Savas Dimopoulos at COSMO98. The work of SD an NK is supported in part by NSF grant PHY-9219345-004. The work of JMR is supported in part by an A.P. Sloan Foundation Fellowship.

CP478 , *COSMO—98*, edited by David O. Caldwell

astrophysical, and cosmological constraints were found not to exclude these ideas.

Here we will summarize our understanding of inflation and other features of early universe cosmology in such models [11].[†] In particular we are concerned with three very important questions: 1) How does inflation arise in these models, or is some alternative necessary? 2) How do the new spatial dimensions evolve to their current stabilized value? 3) Does either inflation or the internal evolution produce too much matter in the bulk? Additionally, is there an equivalent of the Polonyi or moduli problem?

INFLATION

The most attractive models of inflation occur while the internal dimensions are still small, far away from their final stabilized value b_0 [11].[‡] Concretely, the basic story is:

- The quantum creation of the universe takes place with the initial size of all dimensions close to the fundamental Planck scale M_*^{-1}.
- A prolonged period of inflation in a direction parallel to our brane takes place. The approximately scale-invariant nature of the observed primordial perturbation spectrum implies that, during inflation, the internal dimensions must expand more slowly than the universe on the wall. Thus we are led to consider a form of asymmetric inflationary expansion of the higher-dimensional world ($b(t) = b_I$ is essentially static). Since the internal dimensions are small the effective 4-dimensional Newton's constant is large: $G_{\mathrm{N,initial}} = 1/b_I^n M_*^{n+2} \simeq 1/M_*^2$.
- Thus the Hubble constant during this initial period of inflation can be *large* even though the energy density is quite small, $\langle V \rangle \leq \mathrm{TeV}^4$,

$$H_{\mathrm{infl}}^2 \simeq \frac{\langle V \rangle}{6 b_I^n M_*^{(n+2)}} \simeq \frac{\langle V \rangle}{M_*^2}. \tag{1}$$

Therefore inflation can be *rapid*, and moreover, the density perturbations can be large, being determined to be

$$\frac{\delta\rho}{\rho} = \frac{5}{12\pi\sqrt{2n(n-1)}} \frac{H_{\mathrm{infl}}}{M_*(M_* b_I)^{n/2} S}, \tag{2}$$

---

†) See also the companion contribution Ref. [12].

‡) In particular, we avoid all the problems [8,9] which are met if inflation occurs *after* the new dimensions reach their current, large, size. Also note that post-stabilization inflation cannot explain the age of the universe as wall-only inflation cannot begin before $t \sim H^{-1} \sim M_{\mathrm{pl}}/M_*^2 >> M_*^{-1}$, when the universe is already very large and old [9,12].

where S is a potential-dependent parameter that encapsulates both the duration of $a(t)$ inflation and the deviation of the perturbation spectrum from scale-invariant Harrison-Zeldovich. (We argue that $S \lesssim 1/50$.)

- Specifically, the deviation of the spectral index n_ρ of density perturbations $\delta\rho/\rho$ from scale invariance is given by:

$$n_\rho - 1 \simeq -\frac{n(n+2)}{2}\left(S^2 + ST(b/b_I - 1)^2\right), \tag{3}$$

where T is another potential-dependent parameter. Thus to have sufficient scale invariance we need just mildly tune $S, T \lesssim 1/50$, or alternatively $S \lesssim 10^{-3}$ and $T \sim 1$. Similarly, the number of efolds is given by

$$N_e \simeq \frac{1}{S+T}\left(\frac{2\sqrt{T}\tan^{-1}(1/\sqrt{S})}{\sqrt{S}} - \log(1+1/S) + 2\log(1+1/\sqrt{T})\right) \tag{4}$$

which for $N_e \gtrsim 100$ is actually slightly more stringent than (3).

- By going to an effective 4d theory on the wall where $b(t)$ appears as a Brans-Dicke like field (in Einstein frame), it is possible to see that the conditions on S and T are nothing but *the usual slow-roll conditions* for a scalar-gravity theory.

- In the minimal approach, the inflaton field is just the moduli describing the size of the new dimensions (the radion field of [6]), the role of the inflationary potential being played by the stabilizing potential of this internal space. In the case of a wall-localized inflaton, this early inflation might even result from the electroweak phase transition, in which case the inflaton is the Higgs. Actually, an important remark in this regard is that when the internal dimensions are small $b_I \sim M_*^{-1}$ the distinction between on-the-wall and off-the-wall physics is not meaningful: *e.g.*, the inflationary features in $V(b)$ at small b could be *due* to Higgs physics on the wall.

EVOLUTION TOWARDS STABILIZATION POINT[§]

Under quite general conditions, the inflationary era is followed by an epoch where the scale factor of our brane-universe undergoes a slow *contraction* while the internal dimensions expand towards their final stabilized value. Even with the inclusion of a potential for b, it is possible to exactly solve for the evolution during this epoch, which generalizes the usual vacuum Kasner solutions. The history can be summarized as:

[§] This section contains material in addition to that presented at COSMO98.

- Wall inflation ends, and simultaneously the radion starts to evolve to its' minimum at b_0. Almost immediately our scale factor $a(t)$ begins to contract. Both a and b have determined (subluminal) power-law dependence on time.

- At generic values of b away from the stabilization point the potential $V(b)$ can be approximated by $V = Wb^{-p}$ (V is the effective 4d potential, $[W] = 4-p$).[¶] Then the asymptotic form of the exact solutions places an upper bound on the amount of contraction of the brane as a function of b:

$$\frac{a_f}{a_i} \leq \left(\frac{b_i}{b_f}\right)^{\zeta} \tag{5}$$

where the parameter ζ is given by ($\Delta = 6n - 4np - n^2 - p^2$)

$$\zeta = \frac{n(n+p-2)}{2(2n+p)} \quad \text{for } \Delta > 0, \qquad \zeta = \frac{3n - \sqrt{3n(n+2)}}{6} \quad \text{for } \Delta < 0. \tag{6}$$

- Remarkably this epoch of contraction *ends*. There are two generic possibilities for how this can happen: The first involves the reheating of the wall. Contraction of $a(t)$ stops and reverses when ρ_{wall} satisfies

$$\rho_{wall} = n(n-1)M_*^{n+2}b^n H_b^2 \equiv \rho_b \tag{7}$$

Here ρ_b is the radion (kinetic) energy. Interestingly there is a model-independent form of such reheating that results from the primordial ρ_{wall} left over from the inflationary epoch. If there is sufficient $a(t)$ contraction, this blue-shifting de Sitter phase remnant ρ_{wall} can become comparable to ρ_b *before* b reaches the stabilization point, b_0. It is also possible that a form of reheating takes place on the wall which is totally unconnected with the contraction of $a(t)$, but that again leads to $\rho_{wall} \geq \rho_b$. Possibilities in this class include the decay of some metastable state on the wall, or the collision of some other brane with our brane. After the contraction of $a(t)$ reverses, the radion and wall-localized energy densities scale together until the stabilization point is reached. The second case is where $b(t)$ reaches the stabilization point before $\rho_{wall} = \rho_b$. By going to the Einstein frame it is possible to show that once stabilization of b occurs, the period of $a(t)$ contraction automatically ceases.

- The total amount of contraction of $a(t)$ is bounded and varies between at most 7 efoldings in the case of $n = 2$ to at most 12 efoldings when $n = 6$.

¶) An additional logarithmic dependence of V on b is quite possible and harmless.

BULK GRAVITON PRODUCTION, AND MODULI PROBLEM?

Naively one would worry that during the era of radion evolution the bulk becomes full of (dangerous) gravitons. There are actually two slightly different issues here: i) The energy density in Kaluza-Klein (KK) excitations of the graviton in the bulk, and, ii) the energy density in the (would-be) *zero mode* of the bulk graviton, namely the radion. The constraints on the energy density in KK excitations is actually more severe than that on the radion energy density, which just comes from overclosure. The reason for this is the diffuse gamma-ray background constraint arising from KK decays [3]. Again an executive summary of the relevant arguments is:

- A time-dependent gravitational field can produce particles from the vacuum. Including this, the equation for the effective 4d KK energy density becomes

$$\dot{\rho}_{KK} + 3H_a\rho_{KK} + H_b\rho_{KK} = H^5. \tag{8}$$

- From this, and the expressions for $a(t)$ and $b(t)$ during the contraction phase, one can show that the *dominant contribution to bulk KK graviton production arises from early times.* In fact $\rho_{KK,\text{final}}$ differs from the final energy density of the blue-shifted wall-localized radiation at the end of the epoch of contraction only in that it is further suppressed by a factor of $(a_f/a_i)(b_i/b_f)$, which comes from the fact that the KK gravitons are red-shifted by the *bulk* expansion, and only concentrated and not blue-shifted by the wall contraction. Thus we get

$$\frac{\rho_{\text{KK,f}}}{\rho_{\text{wall,f}}} \leq \left(\frac{a_f b_i}{a_i b_f}\right) \lesssim \left(10^{1-30/n}\right)^{1+\zeta}, \tag{9}$$

  using the conservative estimates $b_i \sim 10M_*^{-1}$ and $b_f = b_0$. Evaluating this in, for example, the case of the simple Kasner contraction with exponents given in (6) leads to ρ_{KK}/ρ_{wall} varying between 3×10^{-17} for $n = 2$ to 1×10^{-8} for $n = 6$.

- This shows that the effective temperature of the KK gravitons is well below the diffuse gamma-ray bound, even before any dilution necessary to solve the radion moduli problem. It also demonstrates that the vast majority of the energy in the bulk is in the motion of the zero mode radion ρ_b, rather than in the bulk KK modes. This is simply because that $\rho_b \simeq \rho_{wall}$ is the natural circumstance at the end of the epoch of contraction.

We now turn to the radion moduli problem:

- Finally around the stabilization point b_0 the radion field starts to oscillate freely. Since this energy density scales as $1/a^3$, and the wall-to-radion energy densities are initially comparable at the start of the oscillation era, the radion energy eventually dominates the total energy density.

- The most serious question that early universe cosmology presents in the world-as-a-brane scenario is how do we dilute this energy in radion oscillations to an acceptable level. The radion is long-lived, its' decay width back to light wall states being given by[||]

$$\Gamma_\varphi \simeq \frac{m_\varphi^3}{M_{\rm pl}^2}. \tag{10}$$

We thus require some dilution in the radion energy density, either by a short period of late inflation followed by reheating, or by a delayed reheating after ρ_b has sufficiently red-shifted. The amount of dilution of the radion energy density that we require is relatively modest, given roughly by 10^{-7}, so that only about 5 efolds of late inflation are needed.

CONCLUSIONS

We have argued that early inflation when the internal dimensions are still small can successfully accomplish all that is required of inflation, including generation of suitable $\delta\rho/\rho$ without the unpleasant introduction of very light or fine-tuned wall fields. Remarkably the era of post-inflation brane-contraction is harmless, and automatically ends via a "Big Bounce". During the phase of $b(t)$ evolution to the stabilization point, the production of bulk gravitons by the time-varying metric remains completely suppressed, ensuring that the bulk is very cold at, and after, the stabilization of the internal dimensions. The primary remaining issue is the radion moduli problem, which is no more severe than in gauge-mediated supersymmetry breaking models. Overall, then, early universe cosmology in these models is quite interesting!

REFERENCES

1. N. Arkani-Hamed, S. Dimopoulos and G. Dvali, *Phys. Lett.* **B429**, 263 (1998).
2. I. Antoniadis, N. Arkani-Hamed, S. Dimopoulos and G. Dvali, *Phys. Lett.* **B436**, 257 (1998).
3. N. Arkani-Hamed, S. Dimopoulos and G. Dvali, hep-ph/9807344.
4. I. Antoniadis, *Phys. Lett.* **B246** (1990) 377; I. Antoniadis and K. Benakli, *Phys. Lett.* **B326** (1994) 69; I. Antoniadis, K. Benakli and M. Quiros, *Phys. Lett.* **B331** (1994) 313.
5. P. Horava and E. Witten, *Nucl. Phys.* **B460** (1996) 506; E. Witten *Nucl. Phys.* **B471** (1996) 135; J.D. Lykken, *Phys. Rev.* **D54** (1996) 3693; E. Caceres, V.S. Kaplunovsky and I.M.Mandelberg, *Nucl. Phys.* **B493** (1997) 73.

[||] This is increased if there are many branes in the bulk. The "brane crystallization" mechanism of stabilization requires $N_{\rm wall} \simeq (M_{\rm pl}/M_*)^{2(n-2)/n}$ branes in the bulk. If each of these have $O(1)$ light modes then the total decay width to all branes is greatly enhanced.

6. N. Arkani-Hamed, S. Dimopoulos and J. March-Russell, hep-th/9809124.
7. K. Dienes, E. Dudas and T. Gherghetta, *Phys. Lett.* **B436**, 55 (1998); *Nucl. Phys.* **B537** (1999) 47; G. Shiu and S.H. Tye, *Phys. Rev.* **D58**, 106007 (1998); C. Bachas, hep-ph/9807415; R. Sundrum, hep-ph/9805471 and hep-ph/9807348; P. Argyres, S. Dimopoulos and J. March-Russell, hep-th/9808138; Z. Kakushadze and S.-H. Tye, hep-th/9809147; K. Dienes, *et al.*, hep-ph/9809406; K. Benakli, hep-ph/9809582; L. Randall and R. Sundrum, hep-th/9810155; G.F. Giudice, R. Rattazzi and J.D. Wells, hep-ph/9811291; S. Nussinov and R. Shrock, hep-ph/9811323; E.A. Mirabelli, M. Perelstein and M.E. Peskin, hep-ph/9811337; T. Han, J.D. Lykken and R. Zhang, hep-ph/9811350; N. Arkani-Hamed and S. Dimopoulos, hep-ph/9811353; J.L. Hewett, hep-ph/9811356; Z. Berezhiani and G. Dvali, hep-ph/9811378; K.R. Dienes, E. Dudas and T. Gherghetta, hep-ph/9811428; N. Arkani-Hamed, *et al.*, hep-ph/9811448; I. Antoniadis and C. Bachas, hep-th/9812093; Z. Kakushadze, hep-th/9812163; G. Dvali and S.H. Tye, hep-ph/9812483.
8. K. Benakli and S. Davidson, hep-ph/9810280; D. Lyth, hep-ph/9810320.
9. N. Kaloper and A. Linde, hep-th/9811141.
10. See for example: J. Polchinski, *TASI lectures on D-branes*, hep-th/9611050; C. Bachas, *Lectures on D-branes*, hep-th/9806199.
11. N. Arkani-Hamed, S. Dimopoulos, N. Kaloper, and J. March-Russell, hep-ph/9803224.
12. N. Kaloper, talk at COSMO98, and contribution to this volume.

Recent Progress in Baryogenesis

James M. Cline

Department of Physics, McGill University
Montréal, Québec H3A 2T8 Canada

Abstract. I give a synopsis, specifically aimed at nonexperts, of some of the recent developments in electroweak baryogenesis. The focus of the talk is on the present status of electroweak baryogenesis in supersymmetric models, since this is a plausible and realistic possibility that is currently being probed by experimental searches for the Higgs boson and the top squark. The question of whether it is viable to have a period of color-breaking during the electroweak phase transition is also discussed.

BARYON ASYMMETRY OF THE UNIVERSE: REVIEW

The baryons within the observed universe appear to consist essentially only of matter and not antimatter. There are no regions of gamma ray emission which would correspond to the collision between a galaxy made of normal matter with one made from antimatter, for instance. Moreover, the theory of Big Bang nucleosynthesis gives predictions for the abundances of primordial helium (both ^{4}He and ^{3}He), deuterium and lithium, if the ratio of baryons to photons in the universe is in the range

$$2 \lesssim \eta_{10} \lesssim 6; \qquad \eta_{10} \equiv \frac{n_B}{n_\gamma} \times 10^{10} .$$

If we do live in a universe with equal quantities of matter and antimatter, then the two must somehow be separated on distance scales greater than the present Hubble length, that is, the size of the presently observable universe. The simplest assumption is that indeed the whole universe has the same preponderance of matter over antimatter. It is a deep mystery of cosmology why the abundance of baryons should be this peculiar number, 10^{-10}, relative to the number of photons.

The mystery is intensified by the fact that the most natural initial condition of the universe, at the time of the big bang or shortly thereafter, *is* to have equal numbers of baryons and antibaryons,

$$n_B = n_{\bar{B}} ,$$

implying that the net baryon number of the universe was zero. There are several reasons for believing this.

CP478 , *COSMO—98*, edited by David O. Caldwell

1. Baryon (B) and Lepton (L) conservation are *accidental* symmetries of the Standard Model (SM): there is no good reason for them to be exact. For example, a dimension 9 operator consisting of 6 right-handed quark fields in a color singlet state,

$$\Lambda^{-5}(u\,d\,d)^2$$

is allowed by the gauge symmetries, and would lead to neutron-antineutron oscillations at some rate suppressed by the large mass scale Λ. At sufficiently high temperatures $T \sim \Lambda$ in the early universe, however, the effects of such a baryon number violating operator would be unsuppressed.

2. *Sphalerons*, present within the SM itself, violate B and L (but not $B - L$). These are the lowest energy field configurations with Chern-Simons number 1/2, intermediate between neighboring N-vacua of the $SU(2)$ electroweak gauge theory. (A sphaleron can be thought of as the $t = 0$ slice of an instanton, such as occur in QCD. In QCD it is chirality, rather than $B + L$, which is anomalously violated.) Local transitions which go from one N-vacuum to a neighboring one must pass through a sphaleron-like configuration. Because of the triangle anomaly in the baryon and lepton currents, each such transition is accompanied by 9 quarks and 3 leptons. At zero temperature, the energy barrier between the N-vacua, which is the sphaleron energy, is near 10 TeV, and the tunneling rate for the anomalous transition is so slow as to be entirely irrelevant. But at temperatures above that of the electroweak phase transition, $\sim$ 100 GeV, this energy barrier disappears, and sphaleron transitions are fast compared to the Hubble expansion rate in the early universe.

3. Grand Unified Theories also violate B and L through heavy gauge boson (X) vertices of the form Xqe or $X\bar{q}q$. The presence of both such interactions prevents one from assigning a conserved B or L number to the X boson.

Because of these sources of B violation, we would expect that any initial B or L asymmetry that might have been present initially would be quickly wiped out, giving

$$B = L = 0$$

as the effective initial condition at the high temperatures of the very early universe. Therefore something must have happened between now and then to produce the observed baryon asymmetry.

It was realized by Sakharov in 1967 that three things are needed to spontaneously generate the baryon asymmetry: baryon number violating interactions, CP (particle-antiparticle symmetry) violation, and loss of thermal equilibrium for the B-violating interactions. The first two features are present in the SM by virtue of sphalerons and the phase in the CKM matrix for the quarks. This source of CP violation is however too weakly coupled to the mechanism of baryon production to produce a large enough asymmetry. The third condition is unfulfilled in the SM model: the transition between the symmetric and the broken phase of the electroweak theory is so continuous that it is not a phase transition at all. Sphaleron interactions, although they eventually drop out of thermal equilibrium, do so too gradually to allow the universe to go from a state of $B = 0$ to one of nonzero B.

We therefore need new physics beyond the SM for baryogenesis. Many particle physicists believe that supersymmetry is the most natural direction in which to look for such new physics, and I will also take that point of view. Moreover, I will confine my remarks to electroweak baryogenesis, even though much interesting work on nonelectroweak baryogenesis scenarios has been done in the last year [1].

ELECTROWEAK BARYOGENESIS IN THE MSSM

Electroweak baryogenesis within the minimal supersymmetric standard model (MSSM) has evolved far beyond the status of being just a rough, qualitative theory. Rather, it has been the subject of intense, highly quantitative scrutiny, thanks to the fact that it is so much in the realm of currently testable physics. The way it must work is rather well-defined. First, the electroweak phase transition must be first order, which requires the Higgs field potential to have a barrier between the symmetric phase minimum at $H = 0$ and the true vacuum state with $H \neq 0$, as shown in figure 1. Under these conditions, the phase transition proceeds by the nucleation of spherical bubbles containing the new $H \neq 0$ phase within. The bubbles quickly expand and fill the universe with the broken phase.

Between the initial nucleation and the completion of the phase transition, particles in the high-T plasma are encountering the expanding bubble walls (figure 2). Most particles are massless outside the bubble and massive within, so the bubble wall behaves like a quantum mechanical potential which can scatter the particles. In general there is partial reflection and transmission of particles at the wall, with left-handed particles reflecting into right-handed ones, since spin is conserved, and vice versa.

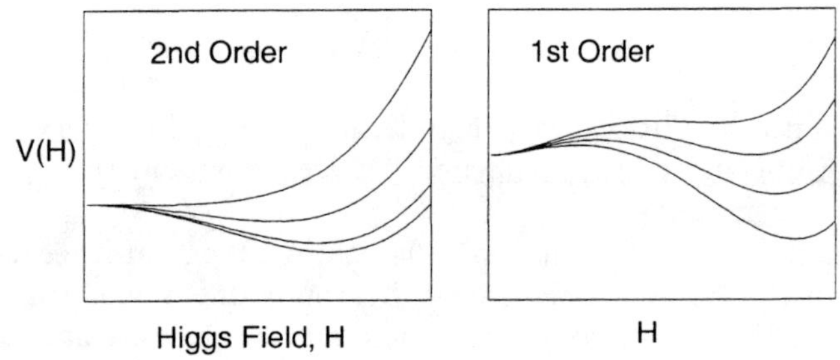

FIGURE 1. Higgs potential near a second order (left) or a first order (right) phase transition, for a series of temperatures.

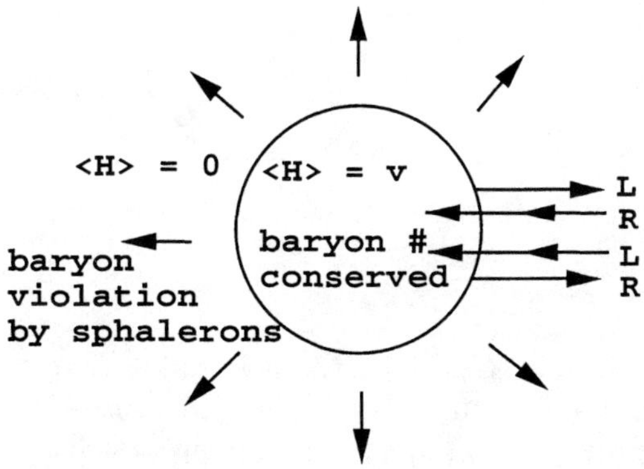

FIGURE 2. The expanding bubble wall during a first order electroweak phase transition.

It is in the reflection process that CP violation is important. If CP is violated on the wall, then particles and antiparticles can have unequal reflection probabilities. The same is true for left-handed and right-handed fermions. Thus an excess of left-handed quarks versus antiquarks can build up in front of the wall (compensated by an equal and opposite excess of right-handed quarks so that net baryon number is still zero at this point).

The asymmetry between q_l and $\bar{q}_l$ increases the free energy of the plasma locally, and sphalerons try to minimize this energy by destroying the q_l-$\bar{q}_l$ asymmetry to some extent, redistributing it among the other species of quarks and leptons. This biases the sphalerons to preferentially create an excess of baryons or antibaryons, which resides in front of the wall for a time, but eventually falls inside the bubble and becomes the baryon asymmetry of the universe (BAU) that is observed today. However the BAU can survive inside the bubbles only if sphaleron interactions are essentially shut off. This is because the CP asymmetry is not operative inside the bubbles (since chirality is not a good quantum number in the broken phase) to insure that sphalerons act preferentially to make only baryons or antibaryons. If sphalerons are not turned off, they will erase whatever BAU is created.

The above idea is highly constrained. For example, it does not appear feasible to alter the basic mechanism by replacing bubbles with other field configurations such as cosmic strings, despite the fact that this possibility has been widely discussed in the literature. We have recently shown [2] that the cosmic string scenario, when scrutinized more carefully, underproduces the baryon asymmetry by 10 orders of magnitude. The reason is essentially that CP violation at the string walls is proportional to the string velocity squared, v^2, while the density of strings in the network scales like v^{-2}. It was previously assumed that the strength of CP violation and the string density could be independently varied.

However, the most important constraint on electroweak baryogenesis is the condition for making the sphaleron interactions slow enough to preserve the BAU, which turns out to be

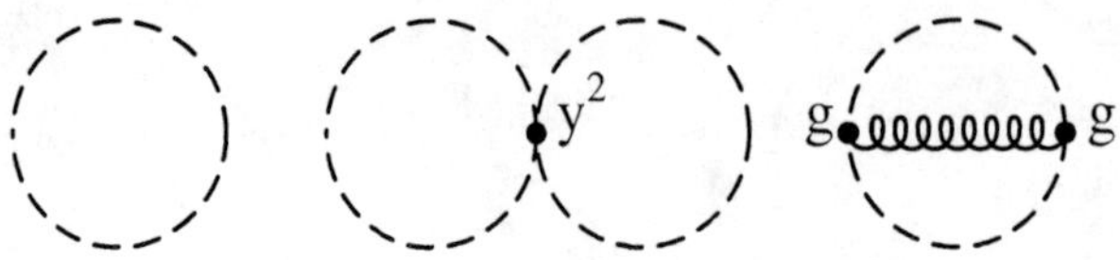

FIGURE 3. Some of the virtual squark diagrams contributing to $V(H)$.

$$\langle H \rangle \geq T \tag{1}$$

inside the bubbles, where $\langle H \rangle$ is the the field value that minimzes the potential energy. Although impossible to achieve in the SM, it *is* possible in the MSSM, provided that the right-handed top squark and the lightest Higgs boson are sufficiently light. (A light left-handed stop is disfavored by precision electroweak considerations, namely the rho or T parameter.) The significance of the top squark is its large Yukawa coupling y to the Higgs field, whose potential $V(H)$ controls the phase transition. Top squark loops such as those shown in figure 3 contribute cubic terms of the form $-y^3TH^3$ to $V(H)$, which make it possible to fulfill eq. (1). This can be understood in terms of the standard expression for the free energy of a relativistic gas of bosons, represented by the one-loop diagram in figure 3, which when expanded in powers of the boson mass over temperature contains a term proportional to m^3T. Since the mass of the squark depends on the Higgs field, this increases the magnitude of the negative H^3 term in $V(H)$.

We have studied the EWPT in the MSSM using the two-loop effective potential, and varying the parameters of the MSSM randomly over many thousands of values to search for those which give a strong enough phase transition [3]. Our results are in reasonable agreement with several similar studies [4]- [9], in predicting the mass ranges for the Higgs boson and the stop to be

$$\begin{aligned} 85 \text{ GeV} &< m_h < 107 - 116 \text{ GeV}\,; \\ 120 \text{ GeV} &< m_{\tilde{t}_R} < 172 \text{ GeV}\,. \end{aligned} \tag{2}$$

The lower bound on m_h is from the latest $L3$ (LEP) experimental limit [10] (the MSSM version of this bound is weaker than in the SM), and the upper bound is a function of the heavy stop mass, because its radiative corrections to m_h grow logarithmically with $m_{\tilde{t}}$.

One might hope that the chances of observing such a light stop at the Tevatron would be good. But experimental limits on $m_{\tilde{t}_R}$ invariably depend on other unknown MSSM parameters that determine which decay channels are open to $\tilde{t}_R$. For example, channels like $\tilde{t}_r \to c\tilde{\chi}^0$ are closed if the neutralino is heavier than $\tilde{t}_r$. Similarly, the baryon production mechanism at the bubble wall prefers values of the μ parameter and Wino mass m_2 in the range $|\mu| \sim m_2 \sim 100$ GeV, which would inspire hopes for the imminent discovery of the chargino [11,12]. This range of values is in fact excluded by chargino searches for $\mu > 0$, but not for $\mu < 0$.

For these reasons, the most foolproof experimental test of electroweak baryogenesis is the search for the light Higgs boson. The remaining window (2) is rapdily being closed by the LEP2 run at CERN. It is a pity, though, that LEP2 will terminate with a limit of only $m_h < 107$ GeV, leaving a small unexcluded window of Higgs mass values. However it is possible that a better theoretical determination of the Higgs mass, using the renormalization group, could help to close this window.

COULD COLOR HAVE BEEN BRIEFLY BROKEN?

One of the interesting possibilities that has been suggested in connection with electroweak baryogenesis with a light stop is that the SU(3) gauge group of QCD was temporarily broken. In order to get a sufficiently light right-handed stop, its bare mass-squared parameter must have been *negative*, so that the field-dependent stop mass has the form

$$m^2_{\tilde{t}_r} = -\widetilde{m}^2_U + y^2|H|^2 + \frac{g_s^2}{6}|\tilde{t}_R|^2$$

This makes the symmetric vacuum unstable toward condensation of the stop field in some random direction in color space. If $-\widetilde{m}^2_U$ is sufficiently negative, it can be energetically preferable to have a period of color breaking before the normal electroweak vacuum state takes over. In this case one would have the sequence

$$(H,\ \tilde{t}_R) \;=\; (0,\ 0) \underset{1}{\rightarrow} (0,\ v_{\tilde{t}}) \underset{2}{\rightarrow} (v_h,\ 0)$$

of phase transitions, which would change our view of cosmological history in a very interesting way. For example, the second transition tends to be very strong, which is favorable for baryogenesis. But there is an energy barrier impeding this second stage of the transition, due to a positive term in the potential

$$y^2|H|^2|\tilde{t}_R|^2$$

whose presence is mandated by supersymmetry. It could happen that the rate of tunneling from the color-broken to the electroweak phase is so small that it will never happen in the history of the universe. Guy Moore has made the following conjecture: if $m_{\tilde{t}_r}$ is ever small enough for transition 1 to take place, then the universe gets stuck in the color broken phase and never completes transition 2. Although preliminary studies of this question have been done, it deserves a more careful treatment.

We have undertaken such a study, by constructing the full two-loop effective potential $V(H,\ \tilde{t}_R)$ for the Higgs and stop fields, and computing the nucleation rate for the most likely bubbles interpolating between the color-broken and electroweak phases [13]. This involves finding the path in the $(H,\ \tilde{t}_R)$ field space along which the bubble evolves, which gives the lowest bubble energy, hence the fastest rate of transitions. The field equations with boundary conditions $(0,\ v_{\tilde{t}})$ and $(v_h,\ 0)$ at the respective ends must be solved along this path. One needs a value of bubble energy over temperature smaller than $E/T \cong 180$ to get a tunneling rate per unit volume, $\sim T^4 e^{-E/T}$, that is competitive with the Hubble rate per Hubble volume, $\mathcal{H}^4$. That is, E/T must be less than $4\ln(T/\mathcal{H})$. However, we find that even when all MSSM parameters are adjusted to the values that are optimal for tunneling, the exponent E/T is too large by an order of magnitude. It therefore appears that one must go beyond the MSSM in order to make color-breaking a real possibility just before the electroweak phase transition.

FOR THE FUTURE

Although the electroweak phase transition in the MSSM is now well understood, the details of how baryons are produced at the bubble wall are still controversial. This is a highly complex phenomenon involving quantum reflection or classical forces acting on the particles near the wall, while they are simultaneously being scattered by other particles in the plasma. It is likely that the Quantum Boltzmann Equation will be needed to put this part of the theory on a more rigorous footing.

REFERENCES

1. C. Kolda and J. March-Russell, hep-ph/9802358 (1998); B.A. Campbell, M.K. Gaillard, H. Murayama, and K.A. Olive, Nucl. Phys. B538 (1999) 351; K. Enqvist and J. McDonald, Phys. Rev. Lett. 81 (1998) 3071; E.W. Kolb, A. Riotto and I.I. Tkachev, Phys. Lett. B423 (1998) 348; K. Benakli and S. Davidson, hep-ph/9810280 (1998); R. Brustein and D.H. Oaknin, hep-ph/9809365 (1998); R. Brandenberger, I. Halperin and A. Zhitnitsky, hep-ph/9808471 (1998).
2. J.M. Cline, J.R. Espinosa, G.D. Moore and A. Riotto, hep-ph/9810261 (1998), to appear in Phys. Rev. D
3. J.M. Cline and G.D. Moore, Phys. Rev. Lett. 81 (1998) 3317.
4. J.R. Espinosa, Nucl. Phys. B475 (1996) 273
5. D. Bödeker, P. John, M. Laine and M.G. Schmidt, Nucl. Phys. B497 (1997) 387.
6. J.R. Espinosa and B. De Carlos, Nucl. Phys. B503 (1997) 24
7. M. Carena, M. Quiros and C.E.M. Wagner, Nucl. Phys. B524, (1998) 3.
8. M. Losada, Nucl. Phys. B537 (1999) 3.
9. M. Laine and K. Rummukainen, Phys. Rev. Lett. 80 (1998) 5259; Nucl. Phys. B535 (1998) 423.
10. R. Clare, L3 Report to LEPC, 12 November 1998 (unpublished)
11. M. Carena, M. Quiros, A. Riotto, I. Vilja and C.E.M. Wagner, Nucl. Phys. B503 (1997) 387.
12. J.M. Cline, M. Joyce and K. Kainulainen, Phys. Lett. B417 (1998) 79.
13. J.M. Cline, G.D. Moore and G. Servant, preprint McGill-99/01 (1999)

Baryogenesis and CP Violating Domain Walls in QCD

A. Zhitnitsky*, R. Brandenberger†,* and I. Halperin*

*Department of Physics and Astronomy, University of British Columbia, Vancouver, BC, V6T 1Z1, CANADA
†Department of Physics, Brown University, Providence, RI 02912, USA

Abstract. The analysis of the θ dependence in QCD within an effective Lagrangian approach shows that the vacuum energy $E_{vac}(\theta)$ is generically a multi-valued function of θ. This implies the existence of metastable vacua and unstable CP violating domain walls interpolating between them. Such a vacuum structure may lead to a new baryogenesis mechanism which takes place at the QCD scale in the early universe. We emphasise that the suggested ideas for baryogenesis can be, in principle, experimentally tested at RHIC.

Introduction

Recently, there has been progress in understanding the general properties of the θ dependence in QCD [1] based on a new development [2] in supersymmetric (SUSY) theories. The crucial point is the realization that the vacuum energy $E_{vac}(\theta)$ is generically a multi-valued function, which leads to the existence of extra local minima with higher energy. Taking this and other related phenomena into account may lead to the new effects in the evolution of the early universe during the QCD epoch.

What do we know about $E_{vac}(\theta)$? There are a few exact results. 1). We know the exact Ward Identity (WI): $\frac{\partial^2 E_{vac}(\theta)}{\partial\theta^2} = -\frac{m_q}{N_f}\langle 0|\bar{\Psi}\Psi|0\rangle$, which implies that at small θ the variation of the vacuum energy with θ is fixed: $E(\theta) - E(0) \sim f_\pi^2 m_\pi^2\theta^2$. As it should be, in the chiral limit the vacuum energy does not depend on θ. 2). The vacuum energy $E_{vac}(\theta)$ is a periodic function of θ, i.e. $E_{vac}(\theta \to 2\pi + \theta) = E_{vac}(\theta)$. This periodicity is a direct consequence of the quantization of the topological charge in QCD: $|\theta\rangle = \sum_n e^{in\theta}|n\rangle$, $|\theta + 2\pi\rangle \equiv |\theta\rangle$, where $|n\rangle$ is the winding state, n is an integer. We note that the vacuum energy in gluodynamics (QCD without quarks) is also a 2π periodic function. 3). In gluodynamics with the large number of colors N_c the physics should depend on θ through the combination θ/N_c in order for the U(1) problem to be solved. 4). The anomalous WI's require [3] that Goldstone

CP478, *COSMO—98*, edited by David O. Caldwell

bosons which are described by the unitary matrix U_{ij} should appear (apart from the mass term) only in the following combination with θ: $\theta - iTr\,\log U$. This is the only allowed combination for arbitrary N_c and N_f. This fact is a consequence of the transformation properties of the Goldstone fields $U \rightarrow \exp i\alpha U$ and the θ parameter $\theta \rightarrow \theta + N_f\alpha$ under $U(1)$ chiral rotations.

Based on the results outlined above, and also on large N_c arguments, Di Vecchia, Veneziano and Witten (VVW) suggested in [3] an effective lagrangian approach in order to analyze the θ dependence of $E_{vac}(\theta)$. In spite of the great simplicity and attractiveness of the VVW solution, this scenario can not be quite complete. The reason is that gluodynamics should also obey the 2π periodicity law in θ. This is not satisfied in the VVW solution. In addition, the approach of Ref. [3] does not constrain in any way the topological charge to integer values. To understand what is missing in the approach of Ref. [3], it is instructive to consider the problem of θ dependence in supersymmetric (SUSY) models.

Lessons from SUSY Theories and θ Dependence in QCD

For supersymmetric gluodynamics, there are N_c different vacua which are labeled by the θ angle and the discrete parameter $k = 0, 1, ...N_c - 1$ such that the gluino condensate depends on these parameters as $\langle\lambda\lambda\rangle \sim \exp\left(\frac{i\theta+2\pi k}{N_c}\right)$. Therefore, when θ varies continuously from 0 to 2π, the N_c distinct and disconnected Bloch type vacua undergo a cyclic permutation. The same picture holds when quarks are added. Therefore, in SUSY theories we have the property described in the Introduction: the θ dependence comes through the combination θ/N_c, but the physics is 2π periodic due to the existence of the additional states. In SUSY models they are degenerate, in non-supersymmetric theories we expect that they have different energies.

The above features of the vacuum structure can be explored within an effective Lagrangian for SUSY gluodynamics constructed a long time ago by Veneziano and Yankielowicz (VY) [4]. However, as was recently noted by Kovner and Shifman [2], the multi-valuedness of the VY effective potential makes it necessary to sum over all its branches in the partition function. This prescription restores the single-valuedness (and boundness from below, in the non-supersymmetric case) of the partition function. In terms of the original theory, it means the summation over all topological classes $Z \sim \sum_n \int DA \exp\left[-S_0 + iQ(\theta + 2\pi n)\right]$ which imposes quantization of the topological charge Q in integer units.

Now we adopt this prescription to QCD and describe the effective potential for QCD [1] which allows us to analyze the θ dependence of the ground state. The Goldstone fields are described by the unitary matrix U_{ij} corresponding to the γ_5 phases of the chiral condensate: $\langle\bar{\Psi}_L^i \Psi_R^j\rangle = -|\langle\bar{\Psi}_L \Psi_R\rangle| U_{ij}$. In terms of U the effective chiral potential is periodic in θ and takes the form

$$E_{vac}(\theta, U) = -\lim_{V\rightarrow\infty} \frac{1}{V} \log\{\sum_l \exp[VE\cos\Phi + \frac{1}{2}VTr(MU + M^+U^+)]\}, \qquad (1)$$

where V is the 4-volume and the phase Φ is

$$\Phi = -\frac{q}{p}\theta + i\frac{q}{p}\log DetU + \frac{2\pi}{p}l, \quad l = 0, 1.., p-1. \tag{2}$$

All dimensional parameters of this potential are expressed in terms of the QCD vacuum condensates, and are well known numerically: $M = diag(m_i|\langle\bar{\Psi}^i\Psi^i\rangle|)$; the constant E is related to the QCD gluon condensate $E = \langle b\alpha_s/(32\pi)G^2\rangle$.

Since Eq.(1) correctly reproduces the VVW effective chiral lagrangian [3] in the large N_c limit and reproduces the anomalous conformal and chiral Ward identities of QCD, one can argue that it represents the anomalous effective Lagrangian realizing the broken conformal and chiral symmetries of QCD.

We refer to the original papers for a detailed discussion of the properties of the effective potential (1). Here we only point out that in general $E_{vac}(\theta)$ is a multivalued function with cusp singularities at $\theta = \frac{(2k+1)\pi}{q}$ where one solution switches to another. The number of additional metastable vacuum states is very sensitive to the integer parameters p, q (2). These numbers are related to a discrete symmetry which is a remnant of the anomaly and can be found only by explicit dynamical calculations. In large N_c limit, $q/p \sim 1/N_c$ in order for the $U(1)$ problem to be solved.

Important for our baryogenesis mechanism is that the effective potential (1) leads to the existence of domain walls [5] which may furthermore acquire a fractional baryon charge when coupled to baryons. On the other hand, the fermions can stabilize the domain walls, yielding compact objects, not necessarily of spherical form, to be called *B-shells* in what follows.

Baryogenesis Mechanism

Now we are ready to formulate [6] an explicit mechanism which hopefully can explain the observed value of n_B/s and which takes place at the QCD scale without the need to introduce any new physics. Our mechanism, however, is strictly speaking not a baryogenesis, but rather a charge separation mechanism, since the total baryon current is exactly conserved. All three famous Sakharov criteria can be satisfied at the QCD scale in the presence of domain walls separating different vacuum states: 1) the domain walls are clearly out of equilibrium since the vacua are non-degenerate; 2) CP violation is unsuppressed due to the non-zero phase difference of the chiral condensate across the wall; 3) The baryon charge separation can take place due to the topological nature of the domain walls.

At the QCD phase transition at the temperature $T_c \simeq 200\,MeV$, the chiral condensate forms. Because of the presence of nearly degenerate states, a network of domain walls will arise immediately after T_c. If $\theta = 0$, there are two degenerate metastable states $|B\rangle$ and $|C\rangle$ above the true vacuum of lowest energy $|A\rangle$. A CP transformation exchanges the states $|B\rangle$ and $|C\rangle$. The baryon charge of an $(A|B)$ wall will be opposite to that of an $(A|C)$ wall, and no baryon asymmetry can be produced in this case. However, when $\theta \neq 0$ at temperatures close to T_c the situation is very different. (We expect that, in general, at $T \sim T_c$ the axion field has not yet settled into its ground state, and thus $\theta(T_c) \sim 1$ coherently over the entire present Hubble radius.). Then there is a splitting $\Delta E \sim \theta m_q$ between the energies of the lowest metastable vacua, which translates into a splitting $\Delta M \sim \theta M$ between the masses M of B-shells with negative and positive baryon charges. When the B-shells - the remnants of unstable domain walls - reach thermal equilibrium, the relative densities of negative and positive B-shells will be fixed by the Boltzmann formula, and one type of B-shells will thus be very strongly suppressed. At this moment the anti-baryon charge will be stored in the walls of the B-shells while the compensating baryon charge is left in the bulk where we live. The B-shells are invisible to us and contribute to the cold dark matter of the Universe. A more detailed discussion can be found in [6].

So far we have discussed baryogenesis by the charge separation involving the baryon charge induced on the wall. In addition, one could think of a competing mechanism, in which the charge separation proceeds due to different interactions of particles and anti-particles with a CP-breaking domain wall, similarly to what happens in electroweak baryogenesis. However, in contrast to the electroweak s-cenario, the anti-baryon charge would not be 'eaten' by sphalerons, but instead be 'hidden' inside the blobs of the metastable vacuum. The latter might be additionally stabilized in this case by the fermi pressure of anti-baryons trapped inside during the evolution of the walls. The latter mechanism could either supplement the former one, or alternatively be the sole mechanism of baryogenesis if the CP violating domain walls would not carry a baryon charge. The question which of these two mechanisms is more important is a problem for future.

Furthermore, many other important questions remain to be answered in this scenario: Are these B-shells absolutely stable? What is the average size of the object? What is the interaction of a B-shell with normal matter? What are the observational constraints on such kinds of objects? We hope to answer some of these questions in the near future. We emphasise that the suggested ideas can be, in principle, experimentally tested at RHIC.

REFERENCES

1. I.Halperin and A.Zhitnitsky, *Phys. Rev.* **D58**, 054016 (1998), hep-ph/9711398; *Phys. Rev. Lett.* **81**, 4071 (1998); *Phys. Lett.* **B440**, 77 (1998).
2. A. Kovner and M. Shifman, Phys. Rev. **D56**, 2396 (1997).
3. E. Witten, Ann. Phys.128, 363 (1980);
   P. Di Vecchia and G. Veneziano, *Nucl.Phys.* **B171**, 253 (1980).
4. G. Veneziano and S. Yankielowicz, *Phys. Lett.* **B113**, 231 (1982).
5. T. Fugleberg, I. Halperin and A. Zhitnitsky, *Phys. Rev.* **D**, (1999), to be published, hep-ph/9808469.
6. R. Brandenberger, I. Halperin and A. Zhitnitsky, hep-ph/9808471.

Sphalerons and CP-Violation

Jackie Grant[1] and Mark Hindmarsh[2]

Centre for Theoretical Physics, University of Sussex, Brighton BN1 9QJ U.K.

Abstract. We report here on the effect that CP-violation from a two Higgs doublet potential has on the sphaleron, a field configuration which mediates baryon and lepton number violation in electroweak theories. We parametrized the Higgs potential in terms of physical Higgs masses and mixing angles, one of which is CP-violating. From considering boundedness of potential we were able to derive the allowed range of the CP-violating mixing angle for given Higgs masses. We found the sphaleron and its barrier in the pure SU(2) theory in the presence of CP-violation by extending the usual spherically symmetric ansatz. From the sphaleron solution we used the method of gradient flow to find the static minimum energy path between vacua of the theory with different Chern-Simons number. We determined the sphaleron energy as a function of CP-violating mixing angle and showed that it can increase by 10-20% [1].

Any theory explaining the observed asymmetry in baryon number requires an amount of CP-violation. With this in mind we investigated the effect CP-violation has on the sphaleron energy, its Chern-Simons number (n_{CS}), and the corresponding static minimum energy path. This is important because the requirement that any baryon asymmetry created during a first order electroweak phase transition not be washed out puts an upper bound on the lightest Higgs mass. This bound is sensitive to the energy of the sphaleron configuration [2], and is order 105 GeV for the MSSM [3].

We considered the bosonic sector of the electroweak Lagrangian with two Higgs doublets in the limit that $\theta_w = 0$. The potential we chose was the most general two doublet renormalizable potential with real expectation values and the symmetry $\phi_1 \to \phi_1$ and $\phi_2 \to -\phi_2$, which is imposed on dimension four terms to reduce the effect of flavour changing neutral currents [4]. Such a potential has nine independent real couplings, one of which (χ_2) determines the degree of CP violation.

$$V(\phi_1, \phi_2) = (\lambda_1 + \lambda_3)\left(\phi_1^\dagger\phi_1 - \frac{v_1^2}{2}\right)^2 + (\lambda_2 + \lambda_3)\left(\phi_2^\dagger\phi_2 - \frac{v_2^2}{2}\right)^2$$
$$+2\lambda_3\left(\phi_1^\dagger\phi_1 - \frac{v_1^2}{2}\right)\left(\phi_2^\dagger\phi_2 - \frac{v_2^2}{2}\right) + \lambda_4\left[\phi_1^\dagger\phi_1\phi_2^\dagger\phi_2 - \mathrm{Re}^2(\phi_1^\dagger\phi_2) - \mathrm{Im}^2(\phi_1^\dagger\phi_2)\right]$$

1) j.j.grant@sussex.ac.uk
2) m.b.hindmarsh@sussex.ac.uk

CP478, *COSMO—98*, edited by David O. Caldwell

$$+\lambda_+ \left[\left(\mathrm{Re}(\phi_1^\dagger \phi_2) - \frac{v_1 v_2}{2} \right)^2 + \mathrm{Im}^2(\phi_1^\dagger \phi_2) \right] + \chi_1 \left[\left(\mathrm{Re}(\phi_1^\dagger \phi_2) - \frac{v_1 v_2}{2} \right)^2 - \mathrm{Im}^2(\phi_1^\dagger \phi_2) \right]$$
$$+2\chi_2 \left(\mathrm{Re}(\phi_1^\dagger \phi_2) - \frac{v_1 v_2}{2} \right) \mathrm{Im}(\phi_1^\dagger \phi_2) \tag{1}$$

Although χ_2 controls the amount of CP-violation, it is preferable to associate the CP-violation with something more physical: the mixing angle between CP-even and odd Higgs particles. To achieve this we inverted the mass-squared matrix of the Higgs particles to get the four masses of the three neutral Higgses and the conjugate pair of charged Higgses, and the three mixing angles between the neutral Higgses as functions of the couplings. Inverting these relations allowed us to fix eight of our couplings by choosing these four masses, three mixing angles and knowing the vacuum expectation value ($v = \sqrt{v_1^2 + v_2^2}$). We then had only one of the nine couplings to fix by hand (which we chose as λ_3). Of the three neutral Higgses, in the absence of any mixing two are CP-even and one is CP-odd (this is the axion when massless), the quantity that gives us the degree of CP-violation is then the mixing angle θ_{CP} between the CP-even and odd neutral higgses. This θ_{CP} we chose as our only and full measure of CP-violation; when $\chi_2 = 0$; $\theta_{CP} = 0$ and visa versa.

We now have a potential with masses, mixings and the vacuum expectation value as (all bar one of) our couplings, and a physical measure, θ_{CP}, of the only CP-violation in our theory. Before we proceed however we had to find the conditions for which our input parameters (masses, mixings, v, and λ_3) gave a bounded potential and a stable vacuum. Vacuum stability is automatically ensured by starting with positive squares for the physical masses. Boundedness of potential requires that the eigenvalues of

$$\frac{\partial^2 V(\phi_1, \phi_2)}{\partial x_i \partial x_j} \tag{2}$$

are all positive, where $i, j = 1, 2, 3, 4$; $x_1 = \phi_1^\dagger \phi_1$, $x_2 = \phi_2^\dagger \phi_2$, $x_3 = \mathrm{Re}(\phi_1^\dagger \phi_2)$, and $x_4 = \mathrm{Im}(\phi_1^\dagger \phi_2)$. This gave us four conditions on couplings (and therefore on masses and mixings) which were difficult to satisfy. In fact, we were unable to find values of masses and mixings that gave a bounded potential for the full range of θ_{CP}. Large ranges of θ_{CP} favoured two nearly degenerate light ($\sim$100 GeV) neutral Higgses, a heavy third neutral Higgs, and light ($\leq$ 100 GeV) charged Higgses.

This proceedure gives a bounded potential with a stable vacuum, with all but one of the couplings determined in terms of the masses, mixings, and the vacuum expectation value. Given this potential it is now straightforward to find the sphaleron solution. We extended the usual SU(2) spherically symetric ansatz of [5] in the manner of [6] so that for the two Higgs doublets we have

$$\phi_\alpha = \frac{v_\alpha}{\sqrt{2}} (F_\alpha(r) + i G_\alpha(r) \hat{x}^a \sigma^a) \begin{bmatrix} 0 \\ 1 \end{bmatrix} \tag{3}$$

where $\alpha = 1, 2$ and $F_\alpha(r)$ and $G_\alpha(r)$ are complex. The ansatz with this extended form of the Higgs doublets does not generally give a spherically symmetric Lagrangian density. If however

$$F_\alpha(r) = \lambda(r) G_\alpha(r) \tag{4}$$

then the terms in the Lagrangian density that break spherical symmetry vanish and the two Higgs doublets are between them a function of five variables, one more than for the ansatz with no CP-violation. We used this condition as a check on the consistency of our numerical results. One important consequence of a spherically symmetric ansatz is that the mass of the charged Higgs fields decouples from the Lagrangian. This means that our results are independent of the mass of the charged Higgses. However the boundedness of potential is sensitive to this mass.

With this ansatz and a suitable initial guess for the fields, we used a numerical relaxation method to relaxed to the sphaleron. The configuration had n_{CS} of 1/2. Having found the sphaleron we used the method of gradient flow [7] to find the static minimum energy path connecting the two vacua. Gradient flow forces the fields of the theory to evolve with a velocity orthogonal to the contours of equal static energy, $\mathcal{E}_s$.

$$\frac{\partial f}{\partial t} = -\kappa \frac{\delta \mathcal{E}_s}{\delta f} \tag{5}$$

where κ is a friction term, and f are the fields of the theory. This method had one problem. To measure the n_{CS} it is useful to remain in the radial gauge, and as all the fields have an equation of evolution, gradient flow does not keep the gauge. To compensate for this we carried out a gauge transformation after each step to return ourselves to the radial gauge. We obtained a barrier that was symmetric about n_{CS} of 1/2, with no unusual features for the extra doublet and the CP-violation. This is especially interesting as with CP-violation there is no obvious reason for this symmetry.

We were also able to determine the effect that CP-violation has on the energy of the sphaleron. To do this, we chose values for physical parameters: the masses and mixing angles, and used v=246 GeV to get the couplings in our potential (one of which we still had to fix). We then checked for what range of θ_{CP} (with all other input parameters fixed) we had a bounded potential. We then found the sphaleron energy as a function of this range of θ_{CP}.

For demonstration purposes, it is useful to have a bounded potential over as large a range of θ_{CP} as possible. To do this we were forced to set the mass of the charged Higgs particles to zero. The value of the charged Higgs mass decoupled for our ansatz, and so this does not effect the sphaleron energy. Had we chosen as our charged Higgs mass 65 GeV, we would only have a bounded potential for $0.7 \leq \theta_{CP} \leq 1.5$, $1.6 \leq \theta_{CP} \leq 2.4$, $\theta_{CP} = 0$, and $\theta_{CP} = \pi$. The reason our potential can be bounded for large charged Higgs mass with no CP-violation is because with physical parameter $\theta_{CP} \neq 0$ or π the couplings $\tan\beta$ ($=v_2/v_1$) and

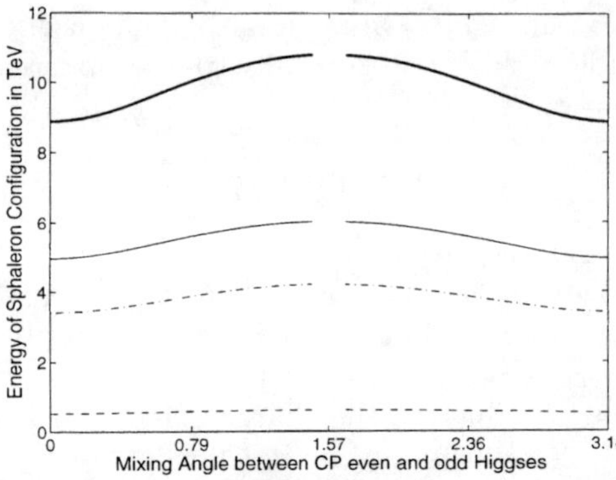

FIGURE 1. The sphaleron energy as a function of CP-violating mixing angle, θ_{CP}, for neutral Higgs masses of 100 GeV, 105 GeV, and 500 GeV, and CP-conserving mixing angles of $\frac{\pi}{4}$, and $-\frac{\pi}{4}$, and λ_3=0. The reason the energy is symmetric about $\pi/2$ is because we chose equal magnitudes for the CP-conserving mixing angles. The solid, dot-dashed and dashed lines are the contributions to the energy from the gauge field, the mixed or covariant derivative, and potential terms respectively.

χ_2 are determined entirely from masses, mixings and v, and the one coupling to be fixed can be chosen arbitrarily as any of the other couplings. With $\theta_{CP} = 0$ or π, $\chi_2 = 0$, and $\tan\beta = 0/0$. $\tan\beta$ has become a free parameter. With $\tan\beta$ a free parameter the condition of boundedness of potential is easier to satisfy.

The CP-violation is in the Higgs potential only, and since the Higgs potential contributes about 10% of the total energy, it is conceivable that the total energy does not vary much with θ_{CP}. However the CP-violation effects all the contributions to the energy and as a result the total energy is very sensitive to the degree of CP-violation in the system. In conclusion we find that although CP-violation does not effect the n_{CS} of the sphaleron. The sphaleron energy is sensitive to CP-violation from the Higgs sector. This suggests that the upper bound on the lightest Higgs to prevent washout of the baryon and lepton asymmetry is increased in the presence of CP-violation.

REFERENCES

1. J. Grant, and M. Hindmarsh, hep-ph/9811289.
2. V.A. Kuzmin, V.A. Rubakov, and M.E. Shaposhnikov Phys. Lett. B **155** 036 (1985).
3. J.M. Cline,and G.D. Moore, hep-ph/9806354.
4. J.F. Gunion, H.E. Haber, G. Kane, and S. Dawson, "The Higgs Hunter's Guide" (Addison-Wesley, Redwood City, 1990).
5. B. Ratra and L.G. Yaffe, Phys. Lett. B **205** 057 (1988).
6. B. Kastening, R.D. Peccei, and X. Zhang, Phys. Lett. B **266** 413 (1991).
7. N. Manton, in "Formation and Interactions of Topological Defects", NATO ASI Series, eds. R. Brandenberger and Davis, (Plenum, New York, 1995).

Neutrino Degeneracy and Chemically Inhomogeneous Universe

Alexandre D. Dolgov

Teoretisk Astrofysik Center
Juliane Maries Vej 30, DK-2100, Copenhagen, Denmark [1]

Abstract.
A possibility that the universe may be strongly inhomogeneous chemically, while very smooth energetically, is considered. A possible mechanism which could lead to such a picture is a large and inhomogeneous lepton asymmetry. The model may explain possibly observed variation of primordial deuterium by an order of magnitude and predicts helium-rich and possibly helium-poor regions where the helium mass fraction differs from the observed 25% by a factor of 2 in either direction. Helium variation can be observed by a different dumping rate of CMB angular spectrum at small angles at different patches of the sky.

It is usually assumed that the universe is well described by a homogeneous and isotropic matter distribution. Of course at the present day the matter is quite clumpy and the mass (or energy) density contrast is large at galactic and even at cluster scales. Averaged over larger scales the matter distribution is probably very smooth. The observed clumpiness evolved from initially small density fluctuations due to gravitational instability. The smoothness of the young universe is strongly confirmed by very small angular fluctuations of the temperature of the cosmic microwave (CMB) radiation. Though the energetic homogeneity of the universe is well verified up to red shift $z = 10^3$, corresponding to the last scattering of CMB, its chemical homogeneity remains an assumption, maybe quite natural, but still an assumption. In what follows I will consider the question if a big chemical inhomogeneity at cosmologically large scales could be compatible with the observed smooth mass or energy distribution in the universe. We discuss a possible mechanism which may give rise to a large variation of element abundances in the early universe, preserving cosmological homogeneity, $\delta\rho/\rho \ll 1$, and possible manifestations of this variation of abundances. To a large extent my talk is based on our paper with B. Pagel [1].

The starting point of this investigation from the observational side was the reported measurements [2] - [8] of the deuterium abundance in Lyman-limit absorp-

---

[1] Also: ITEP, Bol. Cheremushkinskaya 25, Moscow 113259, Russia.

CP478, *COSMO—98*, edited by David O. Caldwell

tion line systems with red-shifts $0.48 < z < 3.5$ on the line of sight to quasars. Surprisingly some groups have claimed a very high value for the deuterium-to-hydrogen ratio, D/H $\approx$ (a few) $\times 10^{-4}$. Though these results are subject to different kinds of criticism [4,9,10], they put forward a very important question: what is known about element abundances at very large distances?

From theoretical point of view it is an interesting challenge to find if there exist (not too unnatural) cosmological scenarios consistent with the observed smoothness of the universe but predicting large abundance variations. An example of such mechanism was proposed in ref. [11] (see also [12]), where a model of leptogenesis was considered which, first, gave a large lepton asymmetry, which could be close to unity, and, second, this asymmetry might strongly change on astronomically large scales, l_L. The magnitude of the latter depends on the unknown parameters of the model and can easily be in the mega-giga parsec range. The model is based on the Affleck-Dine [13] scenario of baryogenesis but in contrast to the original one it gives rise to a large (and varying) lepton asymmetry and to a small baryonic one. There are two more models [14]- [16] in the recent literature, where a large lepton asymmetry together with a small baryonic one is advocated, though without any significant spatial variation.

If we assume that the variation of deuterium abundance by approximately an order of magnitude is indeed real, then the characteristic scale l_L should be smaller than a gigaparsec. The lower bound on this scale may be much smaller. It can in principle be determined by measurements of the abundances of light elements at large distances in our neighborhood, say, $z \geq 0.05$.

A variation of deuterium abundance may be also explained by a variation of the cosmic baryon-to-photon ratio. This possibility was explored in refs. [17,18]. The isocurvature fluctuations on large scales, $l > 100$ Mpc, which are necessary to create the observed variation of deuterium, are excluded [18] by the smallness of angular fluctuations of the cosmic microwave background radiation (CMB). Variations of baryonic number density on much smaller scales, $M \sim 10^5 M_\odot$, are not in conflict with the observed smoothness of CMB and in principle can explain the data subject to a potential conflict with the primordial ^{7}Li abundance [17].

Exactly the same criticism of creating too large fluctuations in CMB temperature is applicable to a simple version of the model with a varying lepton asymmetry. One can check that the necessary value of the chemical potential of electron neutrinos ξ_{ν_e} should be close to -1 to explain the possibly observed variation of deuterium by roughly an order of magnitude. Such a change in ξ_{ν_e} with respect to zero value, assumed in our part of the world, would induce a variation in total energy density during the RD stage at a per cent level, which is excluded by the smoothness of CMB. However, this objection can be avoided if there is a conspiracy between different leptonic chemical potentials such that in different spatial regions they have the same values but with interchange of electronic, muonic and/or tauonic chemical potentials. Since the abundances of light elements are much more sensitive to the magnitude of the electron neutrino chemical potential than to those of muon and tauon neutrinos, the variation of ξ_{ν_e} (accompanied by corresponding variations of

ξ_{ν_μ} and ξ_{ν_τ}) would lead to a strong variation in the abundance of deuterium and other light elements. The equality of, say, ξ_{ν_e} at one space point to ξ_{ν_μ} at another point looks like a very artificial fine-tuning, but this may be rather naturally realized due the lepton flavor symmetry, $e \leftrightarrow \mu \leftrightarrow \tau$.

If lepton asymmetry changes at large distances, then not only deuterium but also ^4He would not remain constant in space. Playing with the nucleosynthesis code [19] one can check that in the deuterium rich regions the mass fraction of helium could be larger than 50% (twice larger than in our neighborhood). There may also exist the so called mirror regions with a positive and large chemical potential of electronic neutrinos. In such regions abundances of both deuterium and helium would be much smaller that those observed nearby. For more detail see ref. [1]. Surprisingly nothing is known about helium abundance at large distances. All accurate measurements of ^{4}He based on emission lines known to us were done at most at $z = 0.045$ corresponding to a distance of $140h^{-1}$ Mpc [20], whereas helium line and continuum absorption measurements made at high redshifts give the abundance merely within "a factor of a few" owing to uncertain ionization corrections [21]. In the regions with a large fraction of ^4He one would expect bluer stars with a shorter life-time, though the structure formation there may be inhibited due to a less efficient cooling. In the helium poor regions the effects may be opposite. As was noted by A. Kusenko at this Conference, a study of supernovae may help to put limits on abundance variation. This problem definitely deserves further and more detailed investigation.

Lepton conspiracy, mentioned above, would diminish energy density fluctuations in first approximation. However there are some more subtle effects which could be either dangerous for the model or observable in CMB. The first one is related to the binding energy of ^4He (7 MeV per nucleon). Since the mass fraction of ^{4}He may change by a factor of 2 in deuterium- (and helium-) rich regions (from 25% to more than 50%), this means that the variation in baryonic energy density may be as large as $2 \cdot 10^{-3}$. Rescaling the estimates of ref. [18] one can find [1] for the fluctuations of the CMB temperature: $\delta T/T \approx 10^{-5}\,(R_{hor}/10\lambda)$, where λ is the wavelength of the fluctuation and R_{hor} is the present day horizon size. The restriction on the amplitude of temperature fluctuations would be satisfied if $\lambda > 200 - 300\text{Mpc}/h_{100}$ ($h_{100} = H/100$ km/sec/Mpc). Surprisingly direct astrophysical effects of such big fluctuations of the helium mass fraction at distances above 100 Mpc cannot be observed presently, at least the evident simple ones.

Another effect which would induce energy inhomogeneities, is the heating of neutrinos by e^+e^--annihilation at $T \lesssim 1$ MeV, when neutrinos practically decoupled from plasma, and the corresponding cooling of photons. (For the most recent and precise calculations of this cooling see ref. [23]). The efficiency of the cooling depends upon the chemical potential of neutrinos and would create fluctuations in CMB temperature at the level of $2 \cdot 10^{-5}$ [1].

A variation of mass fraction of primordial ^4He could be observed in the future high precision measurements of CMB anisotropies at small angular scales [22]. There are two possible effects, first, a slight difference in recombination temperature which logarithmically depends on hydrogen-to-photon ratio, and second, a strong suppression of high multipoles with an increase of R_p. The latter is related to the earlier helium recombination with respect to hydrogen and correspondingly to a smaller number of free electrons at the moment of hydrogen recombination. This in turn results in an increase of the mean free path of photons in the primeval plasma and in a stronger Silk damping [24]. The position and the magnitude of the first acoustic peak remains practically unchanged [22].

This effect seems to be very promising for obtaining a bound on or an observation of a possible variation of primordial helium mass fraction. If this is the case then the amplitude of high multipoles at different directions on the sky would be quite different. The impact of the possible variation of primordial abundances on the angular spectrum of CMB anisotropy at low l is more model dependent. It may have a peak corresponding to the characteristic scale $R > 200 - 300$ Mpc or a plateau, which would mimic the effect of the hot dark matter.

Acknowledgments The work of A.D. was supported by Danmarks Grundforskningsfond through its funding of the Theoretical Astrophysical Center.

REFERENCES

1. A.D. Dolgov and B.E.J. Pagel, *New Astronomy* (to be published); astro-ph/9711202.
2. D. Tytler, X-M. Fan, and S. Burles, Nature **381** (1996) 207.
3. M. Rugers and C.J. Hogan, *Astrophys. J.* **469**, L1 (1996).
4. D. Tytler, S. Burles, and D. Kirkman, astro-ph/9612121.
5. A. Songalia, E.J. Wampler, and L.L. Cowie, *Nature* **385**, 137 (1997).
6. S. Burles and D. Tytler, astro-ph/9712109;*Astrophys. J.* (in press).
7. J.K. Webb, R.F. Carswell, K.M.Lanzetta, et al. *Nature*, **388**, 250 (1997).
8. D. Tytler, S. Burles, L. Lu, et al, astro-ph/9810217.
9. S.A. Levshakov, W.H. Kegel, and F. Takahara, astro-ph/9801108, submitted to *A&A Letters* (1998); astro-ph/9802110, submitted to *Astrophys. J. Letters.* (1998).
10. S.A. Levshakov, D. Tytler, and S. Burles, astro-ph/9812114.
11. A.D.Dolgov and D.K.Kirilova, *J.Moscow Phys.Soc.*, **1**, 217 (1991).
12. A.D.Dolgov, *Phys.Repts.* **222**, No. 6 (1992).
13. I. Affleck and M. Dine, *Nucl. Phys.* **B249**, 361 (1985).
14. R. Foot, M.J. Thompson, and R.R. Volkas, *Phys. Rev.* D **53**, 5349 (1996);
15. R. Foot and R.R. Volkas, Phys. Rev. D **55**, 5147 (1997).
16. A. Casas, W.Y. Cheng, and G. Gelmini, hep-ph/9709289.
17. K. Jedamzik and G. Fuller,*Astrophys. J.* **452**, 33 (1995)
18. C.J. Copi, K.A. Olive, and D.N. Schramm, astro-ph/9606156.
19. L. Kawano, Fermilab-Pub-92/04-A (1992).
20. Y.I. Izotov, T.X. Thuan, and V.A. Lipovetsky, *Astrophys. J. Suppl.* **108**, 1 (1997)
21. C.J. Hogan, S.F. Anderson, and M.H. Rugers, *Astr. J.* **113**, 1495 (1997).
22. W. Hu, D. Scott, N. Sugiyama, and M. White, *Phys. Rev.* D **52**, 5498 (1995).
23. A.D. Dolgov, S.H. Hansen, and D.V. Semikoz, *Nucl. Phys.* B **524**, 621 (1998).
24. J. Silk, *Astrophys. J.*, **151**, 459 (1968).

Cosmology with a time-varying speed of light

Andreas Albrecht*

*Department of Physics
One Shields Ave.
University of California
Davis CA 95616

Abstract. Cosmic inflation is the only known mechanism with the potential to explain the very special initial conditions which are required at the early stages of the evolution of our universe. This article outlines my work with Joao Magueijo which attempts to construct an alternative mechanism based on a time-varying speed of light.

INTRODUCTION

As we reconstruct the past history of the universe, we learn that the universe as we see it today must have evolved from a special state that was extremely flat and homogeneous. Such a state is highly unstable toward both the formation of inhomogeneities and evolution away from flatness, because the influence of gravity tends to drive matter away from a flat homogeneous state. Until the advent of inflation, cosmologists had no way of explaining how the universe could have started out so precisely balanced in a state of such high instability.

Inflation addresses this issue by changing the story at very early times. During an early period of inflation, the matter is placed in a peculiar potential dominated state where the effects of gravity are very different. During inflation, flatness is an attractor, and the deviations (or perturbations) from homogeneity destined for our observable universe can be calculated. With a suitable tuning of model parameters, the perturbations can be given a sufficiently small amplitude and even naturally acquire a spectrum which gives good agreement with observations.

The theory of cosmic perturbations has for some time now benefited from the existence of alternative models. The presence of alternatives has made it possible to systematically evaluate each alternative, and has even helped us to discover the most fundamental way in which inflationary theory could be falsified [1]. So far, things are looking very good for an inflationary origin for the cosmic perturbations.

But inflation offers us much more than a theory of perturbations. It also is supposed to explain the origin of the flatness and overall smoothness of the universe.

CP478 , *COSMO—98*, edited by David O. Caldwell

In this role, inflation theory has faced no serious competition. While this fact in itself might be seen as a success, it would certainly be much more gratifying if the significant place in the theoretical landscape currently occupied by inflation could be earned by doing better than some serious contenders. After all, inflation is just the first idea we have had to explain these cosmic puzzles.

The above reasoning has motivated me to search for alternatives for some time now. So far everything that I have tried has ended up being just another version of standard inflation, once it was forced into a workable form. This experience encourages the view that inflation might be *the* unique mechanism by which the initial condition problems can be addressed, and also explains the extreme nature of the idea I outline below.

The idea I present here starts with a very simple observation. A statement of the unusual nature of the initial conditions in standard cosmology usually carries with it a description of the "horizon problem". Basically, in the standard big bang model any mechanism which operates in the early universe and attempts to "set up" the correct initial conditions would have to act causally, because what we observe today is composed of many causally disconnected regions in the early universe (see Figure 1). Inflation solves this problem because a period of superluminal expansion radically changes the causality structure of the universe. Another way of changing the causality structure is to have light travel faster in the early universe (Figure 2).

Joao Magueijo and I have pursued this idea, to the extent of setting up a phenomenological model of how physics might look with a time varying speed of

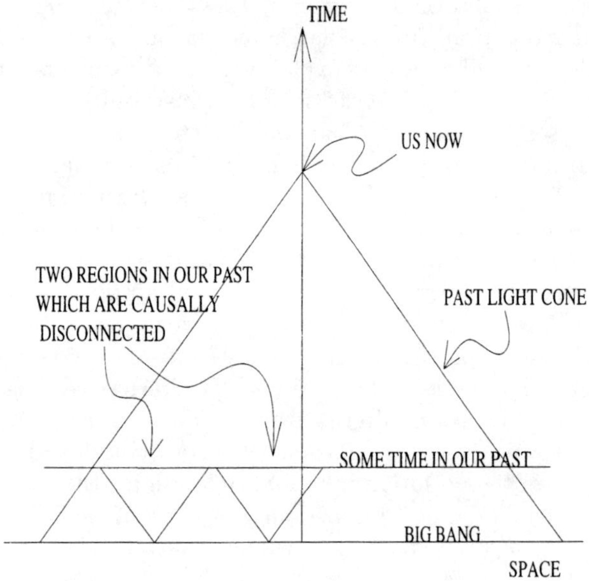

FIGURE 1. The horizon problem in standard cosmology.

light (VSL) [2]. Interestingly, we have found that our model exhibits energy non-conservation of just the sort that can fill in energy deficits and pull down energy peaks to produce a flat homogeneous universe from a wide range of initial conditions.

Of course our model also breaks Lorentz invariance, which may seem unreasonable to may physicists. In defense of choosing this radical direction, let me comment that many theorists are prejudiced *against* the idea that the spacetime continuum is really a continuum down to arbitrarily small scales. Any deviation from a true continuum would necessarily break Lorentz invariance. In particular many ideas about our 3+1 dimensional world that are coming from superstring theory (and its offspring) suggest that properties of this (3+1D) manifold are emergent as a low-energy limit of something quite different. If the VSL picture really takes root, the picture I describe below could well be a phenomenology of the dynamics of the universe as the 3+1 manifold we inhabit emerges from very different physics governing high energies. In this picture the speed of light varies in the early universe and then holds constant, so that the standard cosmology can proceed. This much is in the same spirit as inflation, where a period of unusual physics is placed in the early universe, without changing the standard physical picture which does an excellent job of explaining many aspects of the universe.

I should mention that the idea of using a varying speed of light to explain initial conditions first appears in print in a paper by Moffat [3]. His paper takes the idea in a somewhat different direction than we have. Also, subsequent work by Barrow

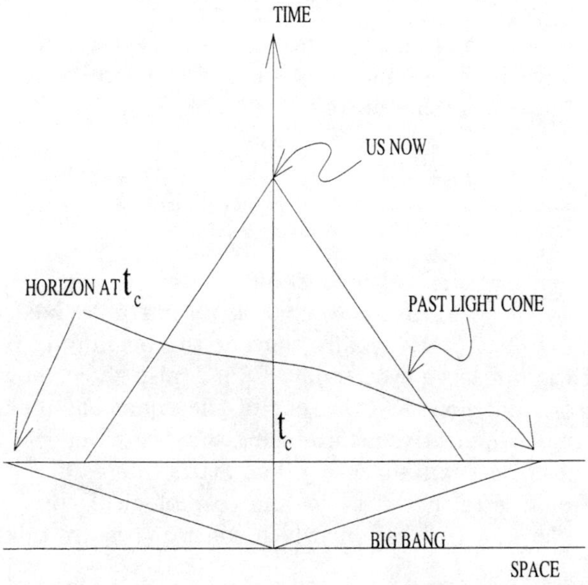

FIGURE 2. The horizon problem solved by a time-varying speed of light.

and Magueijo has taken VSL in a variety of different directions. I will mention this briefly in the final section.

OUR PRESCRIPTION

To pursue the idea of VSL, Magueijo and I have used the following simple prescription: VSL models necessarily have preferred frame, because Lorentz invariance is broken. We assume that in that special frame, the Lagrangian is the same as usual, with the substitution $c \to c(t)$. We also assume that the dynamics of c do not affect the curvature, so that the Riemann tensor and the Ricci scalar are to be computed (in the preferred frame) with c held fixed.

This scheme is spelled out in [2]. There we carefully discuss the question of why this scheme is *not* simply ordinary physics under a strange reparameterization, and thus why what I describe below can not be viewed as simply an unusual way of describing inflation.

COSMOLOGICAL EQUATIONS

Under our VSL scheme energy is not conserved when c is varying. In a cosmology which is Robertson-Walker in the preferred frame we get

$$\dot{\rho} + 3\frac{\dot{a}}{a}\left(\rho + \frac{p}{c^2}\right) = -\rho\frac{\dot{G}}{G} + \frac{3Kc^2}{4\pi G a^2}\frac{\dot{c}}{c}. \tag{1}$$

This equation is the usual equation for energy conservation when c and G are constant, and I have included the term in $\dot{G}$ for future reference. To observe the effect on the flatness of the universe, we look at the evolution of $\epsilon \equiv \Omega - 1$ ($\Omega \equiv \rho/\rho_c$):

$$\dot{\epsilon} = (1+\epsilon)\epsilon\frac{\dot{a}}{a}(1+3w) + 2\frac{\dot{c}}{c}\epsilon \tag{2}$$

where we have taken $p = w\rho c^2$ with constant w. Here we can see how in standard big bang cosmology ($w > -1/3$, $\dot{c} = 0$) $\epsilon = 0$ is an unstable fixed point, leading to the need to tune ϵ to extremely small values at the beginning, in order to match a value of ϵ which is not large even today. Eqn. 2 also shows how inflation makes $\epsilon = 0$ an attractor, and how $\dot{G}$ drops out of the equation, making at least this version of a varying G ineffective at producing the desired effect. One can also see how negative values of $\dot{c}/c$ will make $\epsilon = 0$ an attractor.

There is also an interesting effect on the cosmological constant. In order to discuss this, we must be careful about which constant we are talking about.

$$S = \int dx^4 \sqrt{-g}\left(\frac{c^4(R + 2\Lambda_1)}{16\pi G} + \mathcal{L}_M + \mathcal{L}_{\Lambda_2}\right) \tag{3}$$

Equation 3 shows the action in the preferred frame, where $\mathcal{L}_M$ is the matter fields Lagrangian. The term in Λ_1 is a geometrical cosmological constant, as first introduced by Einstein. The term in Λ_2 represents the vacuum energy density of the quantum fields [4]. VSL is only able to affect Λ_1, and we simply call this Λ in what follows. If we define $\epsilon_\Lambda = \rho_\Lambda/\rho_m$ where $\rho_\Lambda = \frac{\Lambda c^2}{8\pi G}$ we find

$$\dot{\epsilon}_\Lambda = \epsilon_\Lambda\Big(3\frac{\dot{a}}{a}(1+w) + 2\frac{\dot{c}}{c}\frac{1+\epsilon_\Lambda}{1+\epsilon}\Big). \tag{4}$$

Ordinary cosmology has a Λ problem in the sense that Λ rapidly comes to dominate, and must be tuned initially in order not to be super-dominant today. Here we can see that inflation, with $-1 \leq w \leq -1/3$, cannot provide the necessary tuning, while VSL ($\dot{c} < 0$) can have that effect. This, by the way also helps illustrate how VSL is not physically equivalent to inflation.

We have also considered the evolution of perturbations. We have found that for a sudden change in c, the density contrast Δ obeys

$$\frac{\Delta^+}{\Delta^-} = \frac{c^+}{c^-}. \tag{5}$$

For the large variation in c required to produce a flat universe, we have found that the universe has all perturbations reduced to unobservable levels. This leave a blank slate which requires something like the defect models of cosmic structure formation to provide perturbations [5,6].

SCENARIO BUILDING

So far I have discussed the machinery of VSL, but how can this be turned into a cosmological scenario? One can start the discussion by considering a sudden transition between c^- and c^+. Let us assume for a moment that before the transition we have a flat FRW universe. We find under these circumstances the temperature obeys

$$T^+/T^- = c^-/c^+. \tag{6}$$

If we want $T^+ \approx T_{Planck}$ and require $c^-/c^+ > 10^{60}$ to solve the flatness problem, on is starting with a very cold T^-. Immediately, one can see that fine tuning is required to have a cold flat universe before the transition, and this scenario does not make much sense. Interestingly, unlike inflation Equation 2 shows that VSL can create energy even in a empty open ($\epsilon = -1$) universe. It might be more interesting to build scenarios based on that starting point. Just as inflation has seen the scenario-building change radially over the years, we feel there is a lot to be learned about how to implement VSL before we understand the best scheme to use. What we have so far is a very interesting mechanism that can move the universe toward a flat homogeneous state.

DISCUSSION AND CONCLUSIONS

Since our paper there have been a number of other publications on VSL. One new direction pursued by Barrow and Magueijo [7,8] looks at a possible power-law $c(t)$ that could have c varying even today. They have found interesting attractor solutions which keeps Ω_Λ at a constant fraction of the total Ω, but they have their work cut out for them understanding primordial nucleosynthesis in that model. Also, Moffat has further explored the idea of spontaneous breaking of Lorentz symmetry [9].

Probably the greatest problem with the VSL idea is that we have no fundamental picture of what makes c vary. The phenomenological treatment in [2] does not address this question. What we can say is our work shows that there are interesting cosmological rewards for considering a time varying speed of light, and with that motivation it may be possible to make some interesting discoveries.

Despite this problem, it is already possible to falsify at least the fast-transition version of VSL. Since the perturbations turn out to be infinitesimal after the transition, structure formation must be left to active models which have their own characteristic signatures that differentiate them from inflation [10]. If the microwave background comes out with characteristic inflationary features, these VSL models will be ruled out.

This work was supported in part by UC Davis. I would like to thank Joao Magueijo for a fruitful collaboration, and Richard Garavuso for his comments on this manuscript.

REFERENCES

1. A. Albrecht, How to falsify scenarios with primordial fluctuations from inflation, in Critical dialogues in cosmology, ed. Neil Turok, PUP Princeton.
2. A. Albrecht and J. Magueijo, PRD 59, 43516 (1999)
3. J. Moffat, International Journal of Physics D, Vol. 2, No. 3 (1993) 351-365; Foundations of Physics, Vol. 23 (1993) 411.
4. S. Weinberg, Theories of the cosmological constant, in Critical dialogues in cosmology, ed. Neil Turok, PUP Princeton.
5. A. Vilenkin and E.P.S. Shellard, Topological defects in Cosmology, CUP, (1996).
6. R.A. Battye, J.Robinson, A. Albrecht Phys.Rev.Lett. 80 (1998) 4847-4850
7. J. D. Barrow PRD 59, 43515 (1999)
8. J. D. Barrow, J. Magueijo "Solving the Flatness and Quasi-flatness Problems in Brans-Dicke Cosmologies with a Varying Light Speed" astro-ph/9901049, Phys.Lett. B447 (1999) 246, Phys. Lett. B443 (1998) 104
9. J. W. Moffat "Varying Light Velocity as a Solution to the Problems in Cosmology" astro-ph/9811390
10. A. Albrecht, D. Coulson, P. Ferreira, J. Magueijo Phys.Rev.Lett. 76 (1996) 1413-1416

Energy-Momentum Tensor in an Expanding Universe

Salman Habib, Carmen Molina-París[1], and Emil Mottola

Los Alamos National Laboratory, Theoretical Division T8, Los Alamos, New Mexico 87545

Abstract.
We study a quantum scalar field in a de Sitter universe. The initial state of the system is of adiabatic order four, and not necessarily de Sitter invariant. We define an adiabatic number basis in which the energy-momentum tensor has a quasi-classical form in terms of the number of particles present. Numerical results are presented for a massive field with $\xi = 1/6$ (conformal coupling) and $\xi = 0$ (minimal coupling).

INTRODUCTION AND MOTIVATION

In this paper we want to justify the interest in studying the back-reaction of quantum fields in a classical gravitational background [1]. The purpose is to present the general time-dependent initial value problem for a FRW cosmology (which is suitable for implementation on a computer). The basic ingredients are: (1) an initial adiabatic state for the quantum field; (2) an adiabatic number basis in which the energy-momentum tensor admits a quasi-classical interpretation in terms of the average number of particles present at a given time. The motivation is the studies of time-dependent problems in field theories in flat spacetime such as scalar QED and $\lambda\phi^4$ theory [2].

QUANTUM SCALAR FIELD IN DE SITTER

The metric for a de Sitter spacetime with closed spatial sections is [3]

$$g_{tt} = -1, \;\; g_{ij} = a^2(t)h_{ij}(\vec{x}), \;\; a(t) = Z^{-1}\cosh(Zt) \;, \tag{1}$$

with $h_{ij}(\vec{x})$ the metric of a 3-d sphere and $a(t)$ the scale factor.

The wave equation for the quantum scalar field is

$$[-\Box + m^2 + \xi R]\Phi(t, \vec{x}) = 0 \;. \tag{2}$$

[1]) Based on a talk presented by Carmen Molina-París at Cosmo 98, Asilomar, Monterey, 15th-20th of November 1998.

CP478 , *COSMO—98*, edited by David O. Caldwell

The field Φ can be decomposed as

$$\Phi(t,\vec{x}) = \frac{1}{[a(t)]^{3/2}} \sum_{\vec{k}} a_{\vec{k}} f_k(t) \mathcal{Y}_{\vec{k}}(\vec{x}) + a^\dagger_{\vec{k}} f^*_k(t) \mathcal{Y}^*_{\vec{k}}(\vec{x}) \ , \tag{3}$$

so that the mode functions $f_k(t)$ satisfy

$$\ddot{f}_k(t) + \Omega^2_k(t) f_k(t) = 0 \ . \tag{4}$$

We want to look for solutions of semi-classical nature (WKB)

$$\tilde{f}_k(t) = \frac{1}{(2W_k)^{1/2}} \exp\left[-i\int_{t_0}^t dt' W_k(t')\right] = \frac{1}{(2W_k)^{1/2}} \exp\left[-i\theta_k(t)\right] \ , \tag{5}$$

as they are useful for both: (1) the adiabatic regularization scheme [4], and (2) the definition of particle number [5]. Particle number at time t can be defined with respect to a time dependent basis which permits a semi-classical correspondence by means of a time-dependent Bogoliubov transformation

$$f_k \stackrel{\text{def}}{=} \alpha_k \tilde{f}_k + \beta_k \tilde{f}^*_k \ , \tag{6}$$

$$\dot{f}_k \stackrel{\text{def}}{=} \left(-iW_k + \frac{V_k}{2}\right) \alpha_k \tilde{f}_k + \left(iW_k + \frac{V_k}{2}\right) \beta_k \tilde{f}^*_k \ . \tag{7}$$

The adiabatic bilinears needed in this basis are [2]

$$\mathcal{N}_k(t) = |\beta_k|^2, \ \mathcal{R}_k(t) = \Re\left(\alpha_k \beta^*_k e^{-2i\theta_k}\right), \ \mathcal{I}_k(t) = \Im\left(\alpha_k \beta^*_k e^{-2i\theta_k}\right) \ . \tag{8}$$

ENERGY-MOMENTUM TENSOR

The classical energy-momentum tensor is [3]

$$T_{\mu\nu} = (1-2\xi)(\partial_\mu \Phi)(\partial_\nu \Phi) + \left(2\xi - \frac{1}{2}\right) g_{\mu\nu} g^{\alpha\beta} (\partial_\alpha \Phi)(\partial_\beta \Phi) \tag{9}$$

$$- 2\xi \Phi \nabla_\mu \nabla_\nu \Phi + 2\xi g_{\mu\nu} \Phi \Box \Phi + \xi G_{\mu\nu} \Phi^2 - \frac{1}{2} m^2 g_{\mu\nu} \Phi^2 \ . \tag{10}$$

The initial state of the system is given by [6]

$$\alpha_k(t_0) = 1, \ \beta_k(t_0) = 0 \ \Rightarrow \ \mathcal{N}_k(t_0) = \mathcal{R}_k(t_0) = \mathcal{I}_k(t_0) = 0 \ . \tag{11}$$

It can be shown [6] that by making use of the adiabatic bilinears the expectation value of the energy in the initial state of the system is

$$\langle \epsilon \rangle = \frac{1}{4\pi^2 a^3} \sum_{k=1}^{+\infty} k^2 \left[\epsilon^{\mathcal{N}}_k \left(\mathcal{N}_k + \frac{1}{2}\right) + \epsilon^{\mathcal{R}}_k \mathcal{R}_k + \epsilon^{\mathcal{I}}_k \mathcal{I}_k\right] \ . \tag{12}$$

The vacuum polarization term is proportional to 1/2, and the particle contribution is described by the adiabatic bilinears.

We choose W_k and V_k such that the UV-divergences of $T_{\mu\nu}$ [4] are contained in the vacuum term, that is

$$W_k = W_k^{(4)}, \; V_k = -\frac{\dot{W}_k^{(2)}}{W_k^{(2)}} \quad \text{for} \quad {}^{(1)}H_{\mu\nu} = 0 \, . \tag{13}$$

In a de Sitter spacetime as ${}^{(1)}H_{\mu\nu} = 0$, all the divergences have been matched. In a FRW universe, the remaining fourth adiabatic order divergences are logarithmic and can be handled by rescaling the bare coupling of this tensor in the back-reaction calculation [7].

PARTICLE PRODUCTION AND HORIZON CROSSING

Suppose we look at the evolution of the mode $f_k(t)$, with comoving momentum k and physical momentum $k_{\text{phy}} = k/a(t)$. De Sitter spacetime has a length scale given by $H^{-1} = a/\dot{a}$. It is reasonable to expect that particle production occurs at horizon crossing, that is $k_{\text{phy}} \propto H(t)$. The crossing time t_k is the moment at which the mode $f_k(\text{t})$ crosses the horizon. There are two crossing times for each mode f_k (at $\pm\, t_k$), due to the time-reversal invariance of de Sitter. The following three situations can arise:

1. $t_0 < -t_k$ The mode enters the horizon at $-t_k$, then freezes inside the horizon, and exits at time t_k, to reach its late time regime.
2. $-t_k < t_0 < t_k$ The mode is born inside the horizon, and it exits at time t_k, to reach its late time regime.
3. $t_0 > t_k$ The mode is born outside the horizon, and will never get inside it.

CONCLUSIONS AND WORK IN PROGRESS

We have studied the time-dependent initial value problem for a quantum field in a FRW cosmology. We have defined an adiabatic particle number basis in which the UV-divergences of $\langle T_{\mu\nu} \rangle$ are in the vacuum polarization term. $\langle T_{\mu\nu} \rangle_{\text{R}}$ admits a quasi-classical interpretation in terms of the number of particles present at a given time, and its first time derivative. We have theoretical and numerical covariant conservation of $\langle T_{\mu\nu} \rangle_{\text{R}}$. For a de Sitter universe, particle production takes place at horizon crossing, in agreement with the numerical simulations. The energy, pressure and trace are not red-shifted away due to the expansion of the universe. Back-reaction effects can be important.

Work in progress includes applications to other FRW universes, the case of an interacting field $\frac{\lambda}{4!}\phi^4$, back-reaction [taking into account the dynamics of the scale factor $a(t)$], and applications to inflation and early density perturbations.

NUMERICAL RESULTS FOR $T_0 = 0$

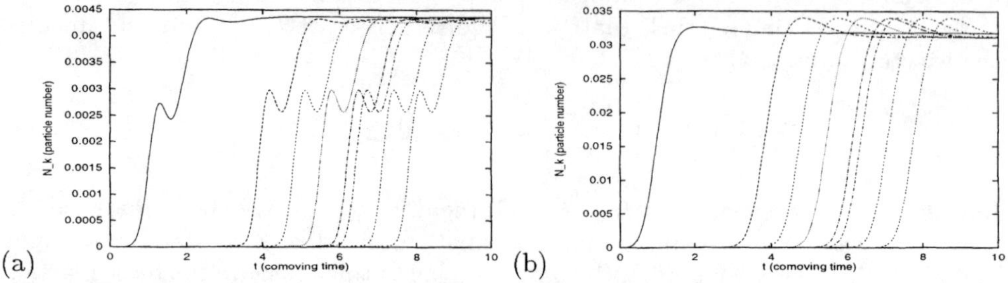

FIGURE 1. Particle number for (a) m=1.0, ξ=1/6, and (b) m=1.6, ξ=0, and k=1, 20, 50, 100, 200, 250, 500, 1000

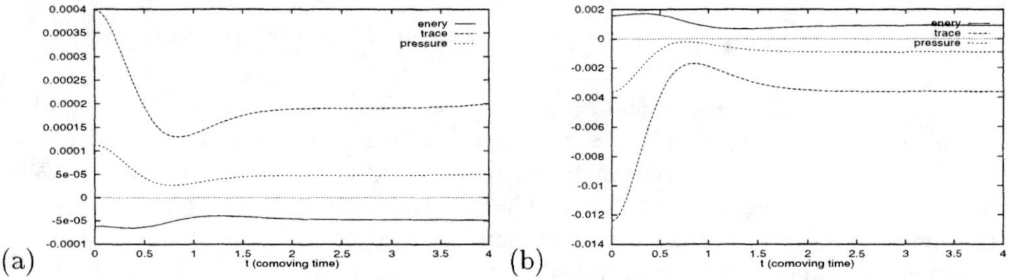

FIGURE 2. Energy, trace and pressure for (a) $m = 1.0$ and $\xi = 1/6$, and (b) $m = 1.6$ and $\xi = 0$

ACKNOWLEDGMENTS

It is a pleasure to acknowledge very helpful discussions with Paul Anderson who suggested to us a new method of subtraction which explicitly maintains covariant conservation of the energy-momentum tensor. This research is supported by the Department of Energy under contract W-7405-ENG-36.

REFERENCES

1. W.-M. Suen and P. R. Anderson, *Phys. Rev.* D **35**, 2940 (1987).
2. F. Cooper, S. Habib, Y. Kluger, and E. Mottola, *Phys. Rev.* D **55**, 6471 (1997).
3. N. D. Birrell and P. C. W. Davies, *Quantum Fields in Curved Space*, (Cambridge University Press, England, 1982).
4. T. S. Bunch, *J. Phys. A: Math. Gen.* **13**, 1297 (1980).
   Paul R. Anderson and Leonard Parker, *Phys. Rev.* D **36**, 2963 (1987).
5. Leonard Parker, *Phys. Rev.* **183**, 1057 (1969).
6. P. R. Anderson, S. Habib, C. Molina-París, and E. Mottola, *in preparation.*
7. Leonard Parker and Jonathan Z. Simon, *Phys. Rev.* D **47**, 1339 (1993).

Dirichlet Solitons in Field Theories

Mark Trodden

Particle Astrophysics Theory Group, Department of Physics, Case Western Reserve University, 10900 Euclid Avenue, Cleveland, OH 44106-7079, USA.

Abstract. I briefly describe a new class of soliton configurations in field theories. These consist of topological defects which can end when they intersect other defects of equal or higher dimensionality. Such configurations may be termed "Dirichlet topological defects", in analogy with the D-branes of string theory. I provide a specific example - cosmic strings that terminate on domain walls - and discuss some new directions for this work, including an interesting and qualitatively different extension to supersymmetric theories.

The study of topological soliton solutions to classical field theories has led to many important new ideas in particle physics and cosmology. In particle physics, a common feature of the recent progress in both supersymmetry (SUSY) and string theory has been the discovery of *dualities*. These dualities map the calculationally difficult limit of one theory into the (hopefully) calculationally easier limit of another theory. More precisely, dualities often interchange the roles of the fundamental and solitonic degrees of freedom. For the case of Seiberg-Witten dualities in SUSY field theories, the relevant solitons are monopoles, so that the electric and magnetic degrees of freedom are interchanged. For string theories, the relevant objects are the D-branes; extended configurations on which fundamental strings can end. In both cases, the study of solitons in the theories has led to a better understanding of how particle physics works.

In cosmology, solitons, or *defects*, play two main roles. First, one may use cosmology to constrain candidate particle physics theories. Examples of this include the requirement that symmetry breaking schemes not allow magnetic monopoles in the early universe, and vorton constraints on theories admitting superconducting cosmic strings. Second, the dynamics, interactions, and microphysics of topological defects can provide explanations for cosmological problems. Examples of this include the use of cosmic strings and textures as seeds for the large scale structure, and the idea that topological solitons might play a role in the generation of the baryon asymmetry of the universe.

Topological defects are solitonic solutions whose stability is guaranteed by a topological conservation law. When a symmetry group G is spontaneously broken

CP478 , *COSMO—98*, edited by David O. Caldwell

to a subgroup H, the types of defects supported depend on the homotopy properties of the vacuum manifold, $\mathcal{M} = G/H$. In a $(d+1)$-dimensional spacetime, p-dimensional defects ($p < d$) exist if the homotopy group $\pi_{d-p-1}(\mathcal{M})$ is nontrivial. (For reviews see [1].) In addition to these basic defects, there are various composite solutions which combine two of the types, generally when a $(p-1)$-dimensional defect serves as the boundary of a p-dimensional defect. Such configurations have interesting cosmological applications, such as in the Langacker-Pi mechanism for solving the monopole problem [2].

In this brief review I will describe a new class of topological defects [6] which is complementary to those mentioned above. These consist of field configurations in which one type of topological defect can terminate when intersecting other defects of equal or higher dimensionality. Such configurations may be termed "Dirichlet topological defects", in analogy with the D-branes of string theory. The latter are extended objects on which fundamental strings can end [7]. The models considered here are ordinary field theories, which support topological solitons which resemble these objects in fundamental string theory. Here I will only have space to provide the specific example of cosmic strings that terminate on domain walls, and to comment on future directions for investigation.

Strings arise most simply from the breakdown of U(1) symmetries. Therefore consider two complex fields $\psi_i = \rho_i e^{i\theta_i}$, $(i = 1, 2)$, and a single real scalar ϕ, transforming under two U(1) and one $\mathbf{Z}_2$ symmetries in the following way

$$\begin{aligned} \mathbf{Z}_2 : &\quad \{\phi \to -\phi \ , \ \psi_1 \leftrightarrow \psi_2\} \ , \\ \mathrm{U}(1)_1 : &\quad \psi_1 \to e^{-i\omega_1}\psi_1 \ , \\ \mathrm{U}(1)_2 : &\quad \psi_2 \to e^{-i\omega_2}\psi_2 \ . \end{aligned} \tag{1}$$

The two U(1)'s may be taken to be either global or gauge symmetries. In the latter case, ω_1 and ω_2 are functions of spacetime, and there are two gauge fields $A_\mu^{(1)}$, $A_\mu^{(2)}$, with the usual transformation properties, and associated covariant derivatives.

Write a general, renormalizable potential in the convenient form

$$\begin{aligned} V(\phi, \psi_1, \psi_2) = \lambda_\phi(\phi^2 - \tilde{v}^2)^2 &+ \lambda_\psi \left[|\psi_1|^2 + |\psi_2|^2 - \tilde{w}^2 + g(\phi^2 - \tilde{v}^2)\right]^2 \\ &+ h|\psi_1|^2|\psi_2|^2 - \mu\phi(|\psi_1|^2 - |\psi_2|^2) \ , \end{aligned} \tag{2}$$

where $v = \langle|\phi|\rangle$ is the root of the cubic equation

$$8\lambda_\phi\lambda_\psi v^3 + 6\lambda_\psi g\mu v^2 - (8\lambda_\phi\lambda_\psi\tilde{v}^2 + \mu^2)v - 2\lambda_\psi(g\tilde{v}^2 + \tilde{w}^2)\mu = 0 \tag{3}$$

that reduces to $\tilde{v}$ at $\mu = 0$, and w is given by

$$w = \left(\tilde{w}^2 + g(\tilde{v}^2 - v^2) + \frac{\mu v}{2\lambda_\psi}\right)^{1/2} \ . \tag{4}$$

In the vacuum the real scalar ϕ takes the vev $\pm v$ and there may exist domain walls separating these two values. When $\langle\phi\rangle = +v$, the vacuum has $|\psi_1| = v$ and $\psi_2 = 0$, while when $\langle\phi\rangle = -v$ the vacuum has $|\psi_2| = v$ and $\psi_1 = 0$.

In this model, therefore, the unbroken symmetry group in the true vacuum is U(1), and the vacuum manifold is $\mathcal{M} = [\mathrm{U}(1) \times \mathrm{U}(1) \times \mathbf{Z}_2]/\mathrm{U}(1) = S^1 \times \mathbf{Z}_2$, admitting walls and strings. When $\langle\phi\rangle = +v$, the complex field ψ_1 can form cosmic strings with winding number n, around which θ_1 will change by $2\pi n$. Such a string ends if it intersects a D-wall, since $\langle\psi_1\rangle = 0$ on the other side. Analogous statements hold for the ψ_2 field when $\langle\phi\rangle = -v$.

In the core of a string the corresponding U(1) symmetry is restored. In the gauge case, therefore, the gauge bosons associated with, for example, $\mathrm{U}(1)_1$ are massless both in the core of a ψ_1-string on the $\langle\phi\rangle = v$ side of the D-wall, and anywhere on the $\langle\phi\rangle = -v$ side of the D-wall. As usual, outside the ψ_1-string the gauge field is pure gauge, such that it cancels the gradient energy of the scalars by enforcing the vanishing of the covariant derivative. The gauge field is thus given by $A^{(1)}_\mu = -\partial_\mu\theta_1$. Consequently, there is magnetic flux through the string (which I take to have winding number n), given by $\Phi^{(1)} = -n\pi$. This flux flows through the string until it hits the wall; on the other side of the wall the symmetry is unbroken everywhere, and the magnetic field describes a monopole configuration emanating from the point where the string intersects the wall.

Configurations of the this type, with strings ending on walls, have recently been discussed in the context of supersymmetric QCD [8]. There, the string consists of non-Abelian flux, and the wall separates different chiral vacua, with shifted values of the QCD θ-parameter. The intersections of strings and domain walls can be thought of as quarks. The structures of these QCD configurations and the scalar field models discussed here are obviously very similar, and the relationship between them deserves further investigation. (One difference is that the flux in the strings considered in [8] does not propagate freely on the other side of the wall, as the symmetry is still broken there; rather, it is confined to the wall itself. It should not be difficult to extend models of the type considered in this paper to include such situations.)

I have described a class of topological defects in classical field theories in (3+1) dimensions, consisting of Dirichlet defects on which fundamental defects of lower dimension can terminate. While the search for models supporting these configurations is inspired by the appearance of D-branes in string theory, there are important differences between the two sets of objects. In all of the theories considered, the basic degrees of freedom are scalar fields and gauge fields, out of which all of the higher-dimensional objects are constructed. Gravity and supersymmetry are not included (although there are no obstacles to the appropriate generalizations [9]). Furthermore, the specific dependence of D-brane energy on the string coupling constant is not a feature of our models, and the Ramond-Ramond gauge fields to which D-branes couple are absent. Nevertheless, it may be interesting to compare the dynamical behavior of Dirichlet defects to that of D-branes in string theory, and search for models in which the similarities between the two systems are even stronger.

One obvious direction in which to generalize the models considered here is to consider q-dimensional defects ending on p-dimensional D-defects in d spatial dimensions. (There are a variety of such objects in string theory and M-theory, with configurations governed by charge conservation.) A number of interesting issues arise in this case, especially for gauge symmetries. For example, to make topological defects of dimension q in d spatial dimensions requires that $\pi_{d-q-1}(\mathcal{M})$ be nontrivial, for example by breaking SO$(d-p)$ to SO$(d-p-1)$ (for which $\mathcal{M} = S^{d-p-1}$). In such a model, the unbroken symmetry group SO$(d-p-1)$ is non-Abelian for $p \leq d-4$ and the low-energy gauge theory is expected to be strongly coupled, and the resulting defects to be confined.

Finally, as with any species of topological defect, it is also natural to ask what the cosmological consequences of the formation of these objects in the early universe might be.

ACKNOWLEDGMENTS

I would like to thanks my collaborators, Sean Carroll and Simeon Hellerman for a lot of fun working on these topics. Thanks also to the organizers, for a great meeting. This work was supported by the Department of Energy (D.O.E.).

REFERENCES

1. A. Vilenkin and E.P.S. Shellard, *Cosmic Strings and Other Topological Defects*, Cambridge University Press (1994); M.B. Hindmarsh and T.W.B. Kibble, *Rept. Prog. Phys.* **58**, 477 (1995).
2. P. Langacker and S.-Y. Pi, *Phys. Rev. Lett.* **45**, 1 (1980).
3. F.A. Bais, *Phys. Lett.* **98B**, 437 (1981); A. Vilenkin, *Nucl. Phys.* **B196**, 240 (1982); M.B. Hindmarsh and T.W.B. Kibble, *Phys. Rev. Lett.* **55**, 2398 (1985).
4. G. Dvali, H. Liu, and T. Vachaspati, *Phys. Rev. Lett.* **80**, 2281 (1998).
5. I. Chuang, R. Durrer, N. Turok, and B. Yurke, *Science* **251**, 1336 (1991); A. Sornborger, S.M. Carroll, and T. Pyne, *Phys. Rev. D* **55**, 6454 (1997).
6. S. Carroll and M. Trodden, *Phys. Rev.* **D57**, 5189 (1998).
7. P. Horava, *Nucl. Phys.* **B327**, 461 (1989); *Phys. Lett.* **B231** 351 (1989); J. Dai, R.G. Leigh, and J. Polchinski, *Mod. Phys. Lett.* **A4**, 2073 (1989); J. Polchinski, *Phys. Rev. Lett.* **75**, 4724 (1995), hep-th/9510017; J. Polchinski, S. Chaudhuri, and C.V. Johnson, "Notes on D-branes", hep-th/9602052; J. Polchinski, "TASI Lectures on D-branes", hep-th/9611050.
8. E. Witten, **Nucl. Phys. B507**, 658 (1997); S.-J. Rey, to appear.
9. S.C. Davis, A.-C. Davis and M. Trodden, *Phys. Lett.* **B405**, 257 (1997); *Phys. Rev.* **D57**, 5184 (1998); J.R. Morris, *Phys. Rev.* **D52**, 1096 (1995); *Phys. Rev.* **D53**, 2078 (1996); *Phys. Rev.* **D56**, 2378 (1997).

Coupling Constant Unification within Supersymmetric Left-Right SO(10)

F. Astorga

Instituto de Física y Matemáticas,
Universidad Michoacana de San Nicolás de Hidalgo,
Apdo. Postal 2-82,
58041 Morelia, Michoacán, México

Abstract. Some results of a second order unification analysis within SUSY L-R SO(10) are presented here. This analysis is performed considering two intermediate scales between M_z and the unification scale.

INTRODUCTION

Supersymmetry has found a place in unification. On one hand, the supersymmetric extension of the Standard Model (SUSY-SM) embedded in SU(5) [1] has proved to overcome the failure of plain SU(5) in achieving unification in the light of recent data on the values of the coupling constants at the mass of the Z (M_z). On the other hand, the partial solution provided by this symmetry to the gauge hierarchy problem makes it natural to consider it in higher-rank groups like SO(10).

The inclusion of supersymmetry in SO(10) [2] enriches the already attractive structure of this model and allows extra intermediate scales corresponding to the breaking of the associated symmetries on the way down to the Standard Model. In this paper we consider two intermediate scales between M_z and the Grand Unification scale M_G: M_R is the scale at which a supersymmetric $SU(3)_c \times SU(2)_L \times SU(2)_R \times U(1)_{B-L}$ group coming from SO(10) breaks to a SUSY Standard Model $SU(3)_c \times SU(2)_L \times U(1)_Y$ group, and M_s, where SUSY is broken to leave the ordinary SM. Both are thought as 'effective' scales around which a mass spectrum of the new particles involved should distribute.

Unification is described in terms of the convergence of the couplings of the strong (α_s), weak (α_2) and electromagnetic (α_1) interactions. The evolution of these couplings as functions of an energy scale μ that is normally identified with a typical energy of the process involved, is given by the following equation coming from the Renormalization Group Equation (RGE) [3]:

CP478, *COSMO—98*, edited by David O. Caldwell

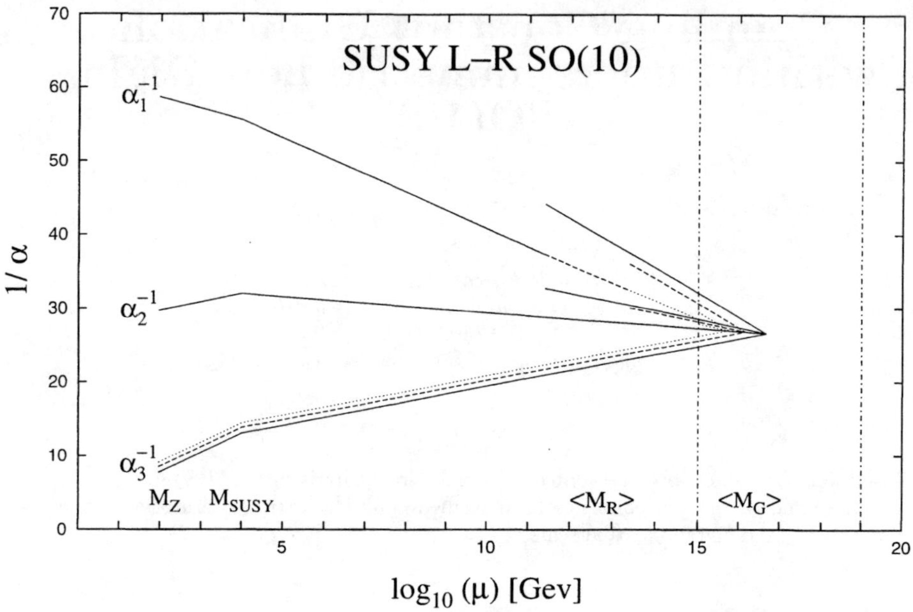

FIGURE 1. Evolution of the inverse of the couplings at second order of approximation and considering two intermediate scales. This shows the running of the couplings for three different α_{0s}initial conditions. The two vertical dot-dash lines show the constraints on the unification scale imposed by proton decay experimental limits and the Planck mass.

$$\mu\frac{\partial\alpha_i}{\partial\mu} = \beta(\alpha_i) = -\frac{2}{4\pi}\beta_i{\alpha_i}^2 - \frac{2}{(4\pi)^2}\sum_{j=1}^{3}\beta_{ij}{\alpha_i}^2\alpha_j - \ldots \qquad (i = 1, 2, 3) \tag{1}$$

where β is an unknown function which can be expanded in a power series of the couplings.

The initial conditions on the values of the couplings are obtained from experimental data coming, among other experiments, from LEP. The increasing precision on these values makes them a fine probe to test the unification hypothesis. Here we use the values $\alpha_{01} = 0.017045 \pm 0.000036$, $\alpha_{02} = 0.03365 \pm 0.00022$ and take $\alpha_{0s} = 0.118 \pm 0.003$ as an average value for the strong coupling [4].

SECOND ORDER ANALYSIS

At this order of approximation, we have to include the cubic term in the α-expansion shown in equation (1). This gives us a coupled system of differential equations for the evolution of the couplings whose solution requires a numerical procedure. This second order analysis is carried out using a Runge-Kutta-Merson integration method [5]. Partial results of this second order analysis are shown in Figure 1.

CONCLUSIONS

In general terms it can be said that, as far as unification is concerned, a two-intermediate-scale model could be better equipped to fit data than a model with only one scale, both because the physically interesting aspects of the scales involved, and because of the mobility of these in a two-scale approach. This SUSY L-R SO(10) fufils equally well all unification constraints, as SUSY SU(5) and L-R SO(10) do. This shows the wide variety of unification scenarios that are still allowed.

ACKNOWLEDGEMENTS

I would like to thank CIC of Universidad Michoacana and SNI for financial support in the form of a research grant.

REFERENCES

1. Amaldi, U., de Boer, W., and Fürstenau, H., *Phys. Lett.* **B260**, 447 (1991).
2. Lee, Dae-Gyu, and Mohapatra, R.N., *Phys. Rev.* **D52**, 4125 (1995).
3. Stückelberg, E.C.G., and Peterman, A., *Helvetica Physica Acta* **26**, 499 (1953).
4. Schmelling, M., *Proceedings of the 28th International Conference on High Energy Physics*, edited by Z. Ajduk and A.K. Wroblewski, World Scientific, 1997, p. 91.
5. Press, H.W., Flannery, B.P., Teukolsky, S.A., and Vetterling, W.T., *Numerical Recipes, The Art of Scientific Computing*, Cambridge University Press, 1987, p. 550.

SUSY and Symmetry Nonrestoration at High Temperature

Borut Bajc

J. Stefan Institute, 1001 Ljubljana, Slovenia

Abstract. The status of internal symmetry breaking at high temperature in supersymmetric models is shortly reviewed. This possibility could solve some well known cosmological problems, such as the domain wall, monopole and false vacuum problems.

INTRODUCTION

The fascinating idea of symmetry nonrestoration at high temperature has been proposed long ago [1,2], but is still a subject of ongoing research (see for example [3,4] for recent reviews and list of references). Apart from being interesting by itself the phenomenon of symmetry nonrestoration could also naturally solve some cosmological problems. At first sight such a possibility seems forbiden in supersymmetric models due to the existence of a no-go theorem [5]. I will show how this difficulty can be circumvented.

Why symmetry nonrestoration at high temperature? As it is well known, when the group G of a higher symmetry phase is simple and the lower symmetry H has a $U(1)$ factor, a phase transition at the grandunified scale between them causes a huge overproduction of monopoles, which leads to a cosmological catastrophe, called the monopole problem. Similarly, the domain wall problem could arise during a phase transition to a disconnected vacuum manifold.

Such cosmological problems can be trivially solved if one finds that there were no phase transtions, i.e. that the grandunified group G (for example $SU(5)$) was spontaneously broken already at high temperature $T > M_X$ to its low-energy subgroup H (the SM group $SU(3)_c \times SU(2)_L \times U(1)_Y$) [6,7]. Another possible solution is that the low-energy subgroup H does not contain a $U(1)$ factor below M_X [8]. In such a scenario electromagnetism would be spontaneously broken during a period of the history of the universe and would eventually get restored later. Anyway, both in the case that G is not restored and in the case that H does not contain a $U(1)$ factor, we have symmetry breaking at high T, a phenomenon which is counterintuitive but extremely interesting.

Of course symmetry nonrestoration does not mean that inflation is not needed

CP478, *COSMO—98*, edited by David O. Caldwell

any more. After all it has to solve the usual horizon problem, homogeneity, etc. The only suggestion here is that inflation does not solve the monopole problem. Only in this way there is a hope that some monopole could one day be detected and thus give a clear experimental signature of a GUT. In fact if inflation solves the monopole problem, it pushes the monopole number essentially to zero, while thermal production even in the case of symmetry nonrestoration (and thus no phase transition) can give a sensible but nondangerous monopole density today [9,6]. More important, inflation needs more than just GUT, while symmetry nonrestoration could be obtained at least in principle from minimal models. We will see later that in SUSY such a solution does not only exist, but it is also very natural.

Why SUSY? The historical reason to consider seriously supersymmetric models is the solution to the hierarchy problem; on top of that the minimal $SU(5)$ SUSY GUT predicts gauge coupling unification, which cannot be achieved in its nonsupersymmetric version [10]. Two special reasons make the study of gauge symmetry nonrestoration in SUSY models at high temperature especially interesting.

The first one is the appearence of a new cosmological problem [11]. It comes from the fact that SUSY models have (at $T = 0$) often more than one vacuum, which are degenerate and disconnected. For example the minimal $SU(5)$ with one adjoint has three degenerate vacua: the one with $SU(5)$ symmetry (call it vacuum 1), one with $SU(5)$ broken to $SU(3) \times SU(2) \times U(1)$ (vacuum 2) and a third one with the symmetry $SU(4) \times U(1)$ (vacuum 3). Obviously, at low energy (but above the scale M_W) our universe was in vacuum 2. If one assumes as usual that at high T the GUT symmetry was restored, when going from $T \gg M_X$ to $T \ll M_X$ the ground state would remain trapped in the wrong vacuum with $SU(5)$ unbroken. The point is that the barrier between two vacua is huge (of order M_X^4), so that tunneling is essentially impossible. Again, the problem could be solved if also at $T \gg M_X$ the GUT symmetry was spontaneously broken to the SM gauge group.

The second special reason to make SUSY models interesting is that apparently the program of symmetry nonrestoration does not work. There exists in fact a no-go theorem [5]: any internal symmetry in a renormalizable SUSY model gets restored at high enough temperatures. But, as is usually the case with no-go theorems, it is possible to evade them changing the conditions under which they are valid. In the following I will describe two such possibilities.

LARGE EXTERNAL CHARGE DENSITY

The above no-go theorem considers only cases with zero charge density, so it can not be applied here. The essential idea was developed for ordinary models long ago, but it was realized only in [12] that it can be applied to SUSY models as well.

It is based on the fact that for large charge densities, it is energetically more favourable for a given system to put the charge in a (nontrivial) vacuum than in thermal excitations. Formally, a negative mass term $-\mu^2|\phi|^2$ (μ is the chemical potential) has to be added to the effective potential, which drives the vev $|\phi|$ to a nonzero value for charge densities above some critical value. Both the charge density of the universe and its critical value scale as T^3, so the condition for an internal symmetry to be broken remains invariant as long as the charge is really conserved.

The most attractive way of implementing the above scenario is to have a large lepton number in the universe [13,14]. Notice that a large lepton number in the universe is experimentally allowed. From primordial nucleosynthesis constraints and galaxy formation one gets an approximate limit $n_L \leq 70T^3$ for $T \gg M_W$ [15], which is an order of magnitude more than the critical density found in [14]. Also, such a large lepton number is consistent with a small baryon number ($\approx 10^{-9}T^3$) since sphalerons are not operative when $SU(2)_L$ is spontaneously broken (similarly to the usual $T = 0$ tunneling, it is exponentially suppressed). So the picture is consistent, but unfortunately one Higgs only in the SM is not enough to break completely the whole gauge group, leaving the system with electromagnetic gauge invariance at any temperature. Thus the SM itself does not possess the solution to the monopole problem. This difficulty can be however easily circumvented in the MSSM, where several boson fields can get nonzero vevs breaking partially or completely the gauge group and thus solve the monopole problem.

In some type of SUSY GUTs the role of the lepton symmetry can be played by an R-symmetry [16]. A large R-charge would be eventually washed out at a later stage of the evolution of the universe due to supersymmetry breaking, so that contrary to the case with the lepton charge no signal of any R-charge would remain today. Nevertheless such a scenario would postpone the creation of monopoles to a nondangerous era.

FLAT DIRECTIONS AT HIGH TEMPERATURE

Recently a new idea for symmetry nonrestoration in SUSY models was proposed [17,18]. At $T = 0$ it is very common in supersymmetric theories to have flat directions. The point is that by definition they can be very softly coupled to the rest of the world (for example with higher dimensional nonrenormalizable terms), so that it is not difficult to find them out of thermal equilibrium at high temperature. This is exactly the case which was not considered by the no-go theorem: in [5] all the fields were assumed to be in thermal equilibrium, i.e. to develop a positive mass term $+T^2|\phi|^2$ in the potential.

Since there is no one-loop contribution to the mass term for the flat direction field, one must consider the two-loop diagrams. The bosonic ones are nothing else than the "butterfly" diagrams of [19]. However, in this case the sign of such a mass term can be negative, leading to $< \phi > \neq 0$ at high T: this means symmetry breaking at high T, which could be a possible solution to the cosmological problems of monopoles, domain walls and wrong vacuum. The condition for a negative sign to the mass term of the flat direction can be written as a constraint to the non canonical Kähler potential. All this is very natural and not at all difficult to achieve.

The essential point is that the temperature provides the necessary breaking of supersymmetry needed to give large vevs to the would have been flat directions.

ACKNOWLEDGMENTS

It is a pleasure to thank all the people that taught me the physics I have been describing in this short paper: Goran Senjanović, who introduced me in this beautiful and interesting subject, Toni Riotto, Alejandra Melfo and Gia Dvali. I thank the organizers for this inspiring conference in a nice environment and for financial support. This work and the long trip to California were supported by the Ministry of Science and Technology of Slovenia.

REFERENCES

1. Weinberg S., *Phys. Rev.* **D9**, 3357 (1974).
2. Mohapatra R.N., and Senjanović G., *Phys. Rev. Lett.* **42**, 1651 (1979); *Phys. Rev.* **D20**, 3390 (1979).
3. Senjanović G., *COSMO 97*, ed. L. Roszkowski, Singapore: World Scientific, 1998, pp. 437-445., hep-ph/9805361.
4. Bajc B., *COSMO 97*, ed. L. Roszkowski, Singapore: World Scientific, 1998, pp. 467-469., hep-ph/9805352.
5. Haber H., *Phys. Rev.* **D 46**, 1317 (1982); Mangano M., *Phys. Lett.* **147B**, 307 (1984).
6. Dvali G., Melfo A., Senjanović G., *Phys. Rev. Lett.* **75**, 4559 (1995); Salomonson P., Skagertan B.S., Stern A., *Phys. Lett.* **B151**, 243 (1985).
7. Dvali G., Senjanović G., *Phys. Rev. Lett.* **74**, 5178 (1995).
8. Langancker P., Pi S.-Y., *Phys. Rev. Lett.* **45**, 1 (1980).
9. Turner M., *Phys. Lett.* **115B**, 95 (1982).
10. Einhorn M.B., Jones D.R.T., *Nucl. Phys.* **B196**, 475 (1982); Marciano W.J., Senjanović G., *Phys. Rev.* **D25**, 3092 (1982).
11. Weinberg S., *Phys. Rev. Lett.* **48**, 1776 (1982).
12. Riotto A., Senjanović G., *Phys. Rev. Lett.* **79**, 349 (1997).
13. Linde A.D., *Phys. Rev.* **D14**, 3345 (1976); Liu J., Segre G., *Phys. Lett.* **B338**, 259 (1994).
14. Bajc B., Riotto A., Senjanović G., *Phys. Rev. Lett.* **81**, 1355-1358 (1998).
15. Kang H.-S., Steigman G., *Nucl. Phys.* **B372**, 494 (1992), and references therein.
16. Bajc B., Riotto A., Senjanović G., *Mod. Phys. Lett.* **A13**, 2955-2964 (1998).
17. Dvali G., Krauss L., preprint CWRU-P31-98, hep-ph/9811298.
18. Bajc B., Senjanović G., preprint IJS-TP-98-24, hep-ph/9811321.
19. Bajc B., Melfo A., Senjanović G., *Phys. Lett.* **B387**, 796 (1996); Bajc B., Senjanović G., *Nucl. Phys. Proc. Suppl.* **52A**, 246-250 (1997).

SPINS AND CHARGES UNIFY

NORMA MANKOČ BORŠTNIK

Department of Physics, University of Ljubljana, Jadranska 19
J. Stefan Institute, Jamova 39 Ljubljana , 1 111, Slovenia

Abstract. In a space of d Grassmann coordinates two types of generators of Lorentz transformations, one of spinorial and the other of vectorial character, define the representations of the group $SO(1,d-1)$ and of the subgroups $SO(1,3), SU(3), SU(2)$, $U(1)$ for fermions and bosons, respectively, unifying all the internal degrees of freedom. When accordingly all interactions are unified, the Yang-Mills fields, the Higgs fields and the Yukawa couplings appear as a part of a gravitational field. The theory suggests four families of quarks and leptons. The equal number of fermions and bosons assures supersymmetry.

Introduction. The Standard electroweak model, which has besides free parameters to be determined by the experiment also several assumptions, does not say: i) Why $SU(3) \times SU(2) \times U(1)$ is the input symmetry of the model? ii) Why fermions are left handed $SU(2)$ doublets and right handed $SU(2)$ singlets, that is, why at all the weak charge is connected with the spin? iii) Why charges of spinors are in the fundamental representations of the corresponding groups, while charges of gauge bosons are in the adjoint representations of the same groups? iv) Where do the generations come from? v) Where do the Higgs bosons and the Yukawa couplings come from?

Embedding the charge groups into the group $SU(5)$ fixes uniquely the $Y(U(1))$ charge, which is as well the free parameter of the Standard model, but leaves the connection between the handedness and the charges undetermined. To connect handedness and charges, spins and charges have to unify.

I am presenting the approach which unifies spins and charges and answers the four of the above open questions of the Standard model.

Coordinate Grassmann space and internal degrees of freedom of fermions and bosons. We define a d-dimensional Grassmann space of real anticommuting coordinates $\{\theta^a\}$, $a = 0,1,2,3,5,6,...,d$, satisfying the anticommutation relations $\theta^a\theta^b + \theta^b\theta^a := \{\theta^a, \theta^b\} = 0$. Linear transformation actions on vectors $(\alpha\theta^a + \beta x^a)$, which leave forms $(\alpha\theta^a + \beta x^a)(\alpha\theta^b + \beta x^b)\eta_{ab}$ invariant, form the Lorentz transformations $L^a{}_c L^b{}_d \eta_{ab} = \eta_{cd}$.

A linear space spanned over a Grassmann coordinate space of d coordinates has the dimension 2^d. If monomials $\theta^{\alpha_1}\theta^{\alpha_2}....\theta^{\alpha_n}$ are taken as a set of basic vectors

CP478 , *COSMO—98*, edited by David O. Caldwell

with $\alpha_i \neq \alpha_j$,**half of the vectors have an even** (those with an even n) and **half of the vectors have an odd** (those with an odd n) **Grassmann character**. Any vector in this space may be represented as a linear superposition of monomials $f(\theta) = \alpha_0 + \sum_{i=1}^{d} \alpha_{a_1 a_2..a_i} \theta^{a_1}\theta^{a_2}....\theta^{a_i}, \quad a_k < a_{k+1}$, where constants $\alpha_0, \alpha_{a_1 a_2..a_i}$ are complex numbers. We define the generalized commutation relations (which follow from the corresponding Poisson brackets [1]): $\{A, B\} := AB-(-1)^{n_{AB}}BA$, fulfilling the relation[1] $\{A, B\} = (-1)^{n_{AB}+1}\{B, A\}$,

We define the following linear operators [1]

$$p^\theta{}_a := -i\overrightarrow{\partial}^\theta{}_a, \quad \tilde{a}^a := i(p^{\theta a} - i\theta^a), \quad \tilde{\tilde{a}}^a := -(p^{\theta a} + i\theta^a),$$

$$S^{ab} := (\theta^a p^{\theta b} - \theta^b p^{\theta a}), \quad \tilde{S}^{ab} := -\frac{i}{4}[\tilde{a}^a, \tilde{a}^b], \quad \tilde{\tilde{S}}^{ab} := -\frac{i}{4}[\tilde{\tilde{a}}^a, \tilde{\tilde{a}}^b], \tag{2.1}$$

with $[A, B] := AB - BA$ and $S^{ab} = \tilde{S}^{ab} + \tilde{\tilde{S}}^{ab}$ and find accordingly

$$\{p^{\theta a}, p^{\theta b}\} = 0 = \{\theta^a, \theta^b\}, \quad \{p^{\theta a}, \theta^b\} = -i\eta^{ab}, \quad \{\tilde{a}^a, \tilde{a}^b\} = 2\eta^{ab} = \{\tilde{\tilde{a}}^a, \tilde{\tilde{a}}^b\},$$

$$\{\tilde{a}^a, \tilde{\tilde{a}}^b\} = 0, \quad \{\tilde{S}^{ab}, \tilde{\tilde{S}}^{cd}\} = 0 = \{\tilde{S}^{ab}, \ \tilde{\tilde{a}}^c\} = \{\tilde{a}^a, \tilde{\tilde{S}}^{bc}\}. \tag{2.2}$$

We see that θ^a and $p^{\theta a}$ **form a Grassmann odd Heisenberg algebra, while** $\tilde{a}^a$ and $\tilde{\tilde{a}}^a$ **form the Clifford algebra.** Either S^{ab} or $\tilde{S}^{ab}$ or $\tilde{\tilde{S}}^{ab}$ fulfil the Lie algebra of the Lorentz group $SO(1, d-1)$ in the d-dimensional Grassmann space: $\{M^{ab}, M^{cd}\} = -i(M^{ad}\eta^{bc} + M^{bc}\eta^{ab} - M^{ac}\eta^{bd} - M^{bd}\eta^{ac})$, with M^{ab} equal either to S^{ab} or to $\tilde{S}^{ab}$ or to $\tilde{\tilde{S}}^{ab}$. We shall make use of $\tilde{a}^a$, $\tilde{S}^{ab}$ only.

The algebra of the group $SO(1, d-1)$ contains [1] n subalgebras defined by operators $\tau^{Ai}, A = 1, n; i = 1, n_A$, where n_A is the number of elements of each subalgebra, with the properties

$$[\tau^{Ai}, \tau^{Bj}] = i\delta^{AB} f^{Aijk}\tau^{Ak}, \tag{2.3}$$

if operators τ^{Ai} can be expressed as linear superpositions of operators M^{ab} $\tau^{Ai} = c^{Ai}{}_{ab}M^{ab}, \quad c^{Ai}{}_{ab} = -c^{Ai}{}_{ba}, \quad A = 1, n, \quad i = 1, n_A, \quad a, b = 0, d$. Here f^{Aijk} are structure constants of the (A) subgroup with n_A operators. **According to the two types of operators M^{ab} there are two types of operators τ^{Ai} defining subalgebras of spinorial and vectorial character**, respectively.

By solving the eigenvalue problem [1]

$$< \theta|\tilde{A}_i|\tilde{\varphi} >= \tilde{\alpha}_i < \theta|\tilde{\varphi} >, \quad < \theta|A_i|\varphi >= \alpha_i < \theta|\varphi >, \quad i = \{1, r\}, \tag{2.4}$$

where $\tilde{A}_i$ and A_i stand for commuting operators of the subgroups $SO(1, 3), SU(3), SU(2)$ and $U(1)$ of spinorial and vectorial character, respectively, we find that **the operators $\tilde{S}^{ab}$ define** the fundamental or **the spinorial representations of the Lorentz group and subgroups** , **while** S^{ab} **define the**

[1]) Here n_{AB} is equal to 1 if A and B have Grassmann odd character and equal to 0 otherwise.

vectorial representations of the Lorentz group $SO(1,d-1)$ and subgroups. We find the representations of these subgroups expressed as polynomials of θ^a. In the subspace of $\theta^0,..,\theta^3$ we find four times two spinors, suggesting that **there are four families of quarks and leptons**. We also find a scalar, a pseudoscalar and (two) three vectors and (two) four vectors. In the subspace of $SU(2)$ we find in the spinorial sector two doublets and four singlets (of either even or odd Grassmann character), while in the vectorial sector we find five singlets, one triplet (of Grassmann even character) and four doublets (of Grassmann odd character). We find that **the operator** $\tilde{S}^{mh}$, if $m = 0,1,2,3$ and h belongs to the space of the subgroup $SU(2)$, **transforms the left handed $SU(2)$ doublets to right handed $SU(2)$ singlets, connecting left handed weak doublets and right handed weak singlets into the same multiplet**. The corresponding operators connecting the spin and the $SU(3)$ charges transform left handed triplets into right handed antitriplets or antisinglets. In the space of $SU(3)$ subgroup we find in the spinorial sector triplets and singlets, in the vectorial sector we find singlets and octets and we find also triplets.

The representations, which are the direct product of the representations of the $SO(1,3), SU(3)$ and $SU(2)$ subgroups, with well defined Y charge, carry all the quantum numbers, needed in the Standard model to describe four families of quarks and leptons (left handed weak doublets in the same multiplet with the right handed weak singlet), all Yang-Mills fields and the Higgs scalar. We also find vectorial representations, which are the $SU(3)$ triplets, but we don't find spinorial representations, which are the $SU(3)$ octets, which means that in spite of the fact that the supersymmetry is guaranteed due to equal number of spinorial and vectorial degrees of freedom, **the ordinary supersymmetry in this approach is not possible.**

Lagrange function for a particle in gauge fields. For a free particle which lives in a d-dimensional ordinary space of commuting and in a d-dimensional Grassmann space of anticommuting coordinates $X^a \equiv \{x^a, \theta^a\}$ and has its geodesics parametrized by an ordinary parameter (τ) and a Grassmann odd parameter(ξ) ($X^a = x^a(\tau) + \varepsilon\xi\theta^a(\tau)$), we define the dynamics by choosing the action [1] $I = \frac{1}{2}\int d\tau d\xi E E^i_A \partial_i X^a E^j_B \partial_j X^b \eta_{ab}\eta^{AB}$, where $\partial_i := (\partial_\tau, \overrightarrow{\partial}_\xi), \tau^i = (\tau, \xi)$, while E^i_A determines a metric on a two dimensional superspace τ^i , $E = det(E^i_A)$. We choose $\eta_{AA} = 0, \eta_{12} = 1 = \eta_{21}$, while η_{ab} is the Minkowski metric. The action is invariant under the Lorentz transformations of supercoordinates: $X'^a = L^a{}_b X^b$ and is locally supersymmetric.

The dynamics of a point particle in gauge fields, the gravitational and the Yang-Mills fields, is obtained by transforming vectors from a freely falling to an external coordinate system [1] by supervielbeins $X_\mu = \mathbf{e}^a{}_\mu X^a$, $\mathbf{e}^a{}_\mu = e^a{}_\mu + \varepsilon\xi\theta^b e^a{}_{\mu b}$. The index a refers to a freely falling coordinate system (a Lorentz index), the index μ refers to an external coordinate system (an Einstein index). Rewriting the action for a freely falling particle in terms of external coordinates, the action follows, which defines the two canonical momenta of the system:

$$p_{0\mu} = p_\mu - \frac{1}{2}\tilde{S}^{ab} e_{a\mu b} = p_\mu - \frac{1}{2}\tilde{S}^{ab}\omega_{ab\mu}\ , \ \omega_{ab\mu} = \frac{1}{2}(e_{a\mu b} - e_{b\mu a}), \ \text{with} \ e_{a\mu b} = f^{\nu}{}_{a,\mu} e_{b\nu}$$

$$\text{and} \ p^\theta_\mu = -i\theta_a e^a{}_\mu, \ \tilde{a}_a = i(p^\theta_a - i\theta_a), \tilde{\tilde{a}}_a = 0. \tag{3.1}$$

The last equation brings the Clifford algebra into the theory. One can find the two constraints $p_0^\mu p_{0\mu} = 0 = p_{0\mu} f^{1\mu}{}_a \tilde{a}^a$.

For a free particle, the two constraints quantize to the Dirac-like and to the Klein-Gordon equation [1], **provided that** $\tilde{\gamma}^m = -\tilde{\tilde{a}}^0 \tilde{a}^m, \ m = 0, 1, 2, 3,$ **is recognized as the Dirac operator**:

$$\tilde{\gamma}^m p_m |\tilde{\Psi}>= 0, \ \ p^m p_m |\tilde{\Psi}>= 0. \tag{3.2}$$

To see how Yang-Mills fields enter into the theory, the Dirac-like equation has to be rewritten in terms of components of fields which determine the ordinary gravitation in the four dimensional subspace and of components, which determine gravitation in higher dimensions, assuming that the coordinates of ordinary space with indices higher than four stay compacted to unmeasurable small dimensions, so that the four dimensional subspace of Grassmann space with the generators defining the Lorentz group $SO(1,3)$ is (almost) decomposed from the rest of the Grassmann space with the generators forming the (compact) group $SO(d-4)$. Assuming the case in which the gravity in four dimensional space-time is zero: $e^m{}_\alpha = \delta^m_\alpha$ and in addition $e^h{}_\alpha = 0 = e^m{}_\sigma, \ \alpha, m \in (0,3), \ \ \sigma, h \in (5,d)$, while nonzero vielbeins $e^k{}_\sigma$ depend on θ^a and $x^\alpha, \alpha \in \{0,3\}$ only, we find

$$\tilde{\gamma}^a f^\mu{}_a p_{0\mu} = \tilde{\gamma}^\alpha (p_\alpha - \sum_{A,i} \tilde{\tau}^{Ai} A^{Ai}_\alpha), \tag{3.3}$$

with $\sum_{A,i} \tilde{\tau}^{Ai} A^{Ai}_\alpha = \frac{1}{2}\tilde{S}^{hk}\omega_{hk\alpha}, \ \ h, k = 5, 6, 7, ..d$, which **demonstrates the Dirac equation in the presence of the Yang-Mills fields** in the four dimensional subspace.

If, however, we assume that $f^\alpha_{m,\sigma}$ is also nonzero, **the additional term** $\tilde{\gamma}^h f^\sigma{}_h p_{0\mu}$ **appears** in Eq. (3.3),**which**, according to what we have said in Sect. 2, **takes care of the Yukawa couplings**.

REFERENCES

1. N. Mankoč Borštnik, *Modern Phys. Lett.* **A 10**, 587 (1995), N. Mankoč -Borštnik, *Phys.Lett.* **B 292**, 25 (1992), *Il Nuovo Cimento* **A 105**, 1461 (1992), *J. of Math. Phys.* **34**, 8 (1993), *Int. J. of Mod. Phys.* **A 9**, 1731 (1994), *J. of Math. Phys.* **36(4)** 1593 (1995), N. Mankoč -Borštnik and S. Fajfer, *Nuovo Cimento* **B 112** (1997) 1637, N. Mankoč -Borštnik and A. Borštnik, *J. Phys. G: Nucl. Part. Phys.* **24** (1998) 963, N. Mankoč Borštnik, *Modern Phys. Lett.* **A 10** 587 (1995),

THE MISSING ENERGY/MATTER OF THE UNIVERSE

Quintessence and the Rest of the World

Sean M. Carroll

Institute for Theoretical Physics, University of California,
Santa Barbara, California 93106, USA
Email: carroll@itp.ucsb.edu

Abstract. A nearly-massless, slowly-rolling scalar field ϕ may provide most of the energy density of the current universe. One potential difficulty with this idea is that couplings to ordinary matter, even if suppressed by the Planck scale, should lead to observable long-range forces and time dependence of the constants of nature. I explore the possibility that an approximate global symmetry serves to suppress such couplings even further. Such a symmetry would allow a coupling of ϕ to the pseudoscalar $F_{\mu\nu}\widetilde{F}^{\mu\nu}$ of electromagnetism, which would rotate the polarization state of radiation from distant sources. This effect is fairly well constrained, but it is conceivable that future improvements could lead to a detection of a cosmological scalar field. (Based closely on astro-ph/9806099 [1].)

Recently a number of pieces of evidence, especially studies of the Hubble diagram for Type Ia supernovae [2], have lent support to the idea that the universe is dominated by a smooth component with an effective negative pressure, leading to an accelerating expansion. While the most straightforward candidate for such a component is the cosmological constant [3], a plausible alternative is dynamical vacuum energy, or "quintessence" [4,5].

A number of models for quintessence have been put forward, the most popular of which invoke a scalar field in a very shallow potential, which until recently was overdamped in its evolution by the expansion of the universe. For generic potentials the requisite shallowness implies that excitations of the field are nearly massless, $m_\phi \equiv \sqrt{V''(\phi)/2} \leq H_0 \sim 10^{-33}$ eV. To provide the necessary energy density, the present value of the potential must be approximately the closure density of the universe, $V(\phi_0) \sim (10^{-3}\ \text{eV})^4$, so the field itself will typically be of order $\phi_0 \sim 10^{18}$ GeV $\sim M_{Pl} = (8\pi G)^{-1/2}$, where M_{Pl} is the reduced Planck mass.

The exchange of very light fields gives rise to forces of very long range, so it is interesting to consider the direct interaction of the quintessence field ϕ to ordinary matter. Although it is traditional to neglect (or set to zero) the couplings of this light scalar to the standard model, we expect that our low-energy world is described by an effective field theory obtained by integrating out degrees of freedom with momenta larger than some mass scale M, in which case it is appropriate to include

CP478 , *COSMO—98*, edited by David O. Caldwell

non-renormalizable interactions suppressed by appropriate inverse powers of M. For example, ϕ can couple to standard-model fields via interactions of the form $\beta_i(\phi/M)\mathcal{L}_i$, where β_i is a dimensionless coupling and $\mathcal{L}_i$ is any gauge-invariant dimension-four operator, such as $F_{\mu\nu}F^{\mu\nu}$ or $i\bar{\psi}\gamma^\mu D_\mu\psi$. In the absence of detailed knowledge about the structure of the theory at high energies, the couplings β_i are expected to be of order unity. The mass parameter M, meanwhile, represents the energy scale characterizing the phenomena which we have integrated out to obtain the low-energy description. We cannot specify it with precision, but it should not be higher than the Planck scale $M_{Pl} \sim 10^{18}$ GeV, where quantum gravity becomes relevant.

The scalar force mediated by ϕ will not obey the Equivalence Principle (which is only compatible with forces mediated by spin-two fields), and hence is constrained by Eötvös-type experiments. Su *et al.* [6] have found that the differential acceleration of various test bodies, in the direction of the Sun, is less than 10^{-12} times the strength of gravity. Such limits can be translated into constraints on the dimensionless couplings β_i; for example, we may calculate the charge on a test body due to a coupling $\beta_{G^2}(\phi/M)\,\mathrm{Tr}(G_{\mu\nu}G^{\mu\nu})$, where $G_{\mu\nu}$ is the field strength tensor for QCD (*cf.* [7]). Although it is difficult to compute QCD matrix elements to high precision, the Su *et al.* results can be used to place a conservative upper limit

$$|\beta_{G^2}| \leq 10^{-4}\left(\frac{M}{M_{Pl}}\right) . \tag{1}$$

Similar considerations constrain other couplings, although typically not quite as well (see *e.g.* [8]).

A related phenomenon is the time variation of the constants of nature. For the dynamical nature of ϕ to be relevant today, we expect a change in ϕ of order M_{Pl} over cosmological timescales $t_0 \sim H_0^{-1}$. In that case, a coupling such as $\beta_{F^2}(\phi/M)F_{\mu\nu}F^{\mu\nu}$ will lead to evolution of the fine-structure constant α. Various observations constrain such variation. For example, isotopic abundances in the Oklo natural reactor imply that $|\dot{\alpha}/\alpha| < 10^{-15}\ \mathrm{yr}^{-1}$ over the past two billion years [9]; this leads to the limit

$$|\beta_{F^2}| \leq 10^{-6}\left(\frac{MH_0}{\langle\dot{\phi}\rangle}\right) , \tag{2}$$

where $\langle\dot{\phi}\rangle$ is the average rate of change of ϕ in the last two billion years. (See also [10].) Again, changes in other couplings are also constrained.

There is clearly good evidence against the existence of a nearly-massless scalar field coupled to the standard model via nonrenormalizable interactions with strength of order $1/M_{Pl}$. It would be premature, however, to conclude that the idea of quintessence is ruled out, as we may consider imposing an approximate global symmetry of the form $\phi \to \phi + const$. Indeed, such symmetries are invoked in pseudo-Goldstone boson models of quintessence [5], as an explanation for the naturalness of the small mass m_ϕ: in the limit as the symmetry is exact, this mass goes to zero. This same effect could explain the small values of the dimensionless couplings β_i.

An important consequence of this viewpoint is that interactions which are invariant under $\phi \rightarrow \phi + const$ — that is, derivative couplings of ϕ — should be present with couplings β_i of order unity. The leading allowed term in the standard model is $\beta_{F\widetilde{F}}(\phi/M)F_{\mu\nu}\widetilde{F}^{\mu\nu}$, where $F_{\mu\nu}$ is the electromagnetic field strength tensor and $\widetilde{F}^{\mu\nu} = \frac{1}{2}\epsilon^{\mu\nu\rho\sigma}F_{\rho\sigma}$ is its dual. The combination $F_{\mu\nu}\widetilde{F}^{\mu\nu}$ is a total divergence, so this interaction does represent a derivative coupling, and respects the symmetry $\phi \rightarrow \phi + const$.

Such a coupling can lead to potentially observable effects: a spatially homogeneous but slowly-varying ϕ field would rotate the direction of polarization of light from distant radio sources [11]. The dispersion relation for electromagnetic radiation in the presence of a time-dependent ϕ becomes $\omega^2 = k^2 \pm (\beta_{F\widetilde{F}}/M)\dot{\phi}k$, where $+/-$ refer to right- and left-handed circularly polarized modes, respectively. If we define χ to be the angle between some fiducial direction in the plane of the sky and the polarization vector from an astrophysical source, then in the WKB limit where the wavelength of the radiation is much less than that of ϕ, the difference in group velocity for the two modes leads to a rotation $\Delta\chi = (\beta_{F\widetilde{F}}/M)\Delta\phi$.

Rotations of this type are potentially observable, as distant radio galaxies and quasars often have a well-defined relationship between their luminosity structure and polarization structure [12]; recent observations provide a stringent upper limit on any redshift-dependent changes in this relationship [13,14]. As an example we may consider the data given by Leahy [13] for $\Delta\chi$ as a function of redshift. Simply taking the mean value all of the points (for which the minimum redshift is $z = 0.425$) yields $\langle\Delta\chi\rangle = -0.6° \pm 1.5°$. This implies a bound

$$|\beta_{F\widetilde{F}}| \leq 3 \times 10^{-2}\left(\frac{M}{|\Delta\phi|}\right) , \tag{3}$$

where $\Delta\phi$ is the change in ϕ between a redshift $z = 0.425$ and today.

Does Eq. (3) constitute a good limit? We expect $\Delta\phi$ to be of order M_{Pl}, so the factor $M/|\Delta\phi|$ is likely to be less than or of order unity. However, while 3×10^{-2} is less than one, it is not remarkably less; we might imagine that dimensionless constants conspire to make $\beta_{F\widetilde{F}}$ naturally smaller than this value even without suppression by some symmetry. For example, couplings of this form can arise through triangle graphs in ordinary field theories (*i.e.*, even disregarding the possibility of exotic quantum gravitational effects); such graphs lead to $\beta_{F\widetilde{F}} = N\alpha/4\pi$, where α is the fine structure constant and N is a dimensionless factor which depends on the field content of the model. Since $\alpha/4\pi \sim 6 \times 10^{-4}$, it is by no means implausible that the interaction under consideration could exist but have evaded detection thus far. This raises the exciting possibility that improvements in the limits from radio galaxy polarization measurements could lead to a detection of quintessence. Since the relevant observed quantity is an angle, it is hard to imagine significant contamination by systematic errors, so the observation of a large number of sources can be expected to improve these limits substantially.

Alternatively, we can consider models of quintessence in which the dynamics are more complicated than those of a slowly-rolling scalar field. An example is given by models of variable-mass particles [15]. In that case, however, there is an instability on small scales due to the ability of the particles to cluster together. This possibility is prevented in simple quintessence models by the extraordinarily low mass of the scalar field, which is precisely the origin of the long-range forces that have gone undetected. It may therefore be the case that the only viable models of quintessence are those invoking approximate global symmetries.

This research was supported by the National Science Foundation under grant PHY/94-07194.

REFERENCES

1. S.M. Carroll, *Phys. Rev. Lett.* **81**, 3067 (1998), astro-ph/9806099.
2. A.G. Riess *et al.*, *Astron. Journ.* **116**, 109 (1998), astro-ph/9805201; S. Perlmutter *et al.*, *Astrophys. J.*, in press, astro-ph/9812133.
3. S.M. Carroll, W.H. Press, and E.L. Turner, *Annu. Rev. Astron. Astrophys.* **30**, 499 (1992).
4. See for example C. Wetterich, *Nucl. Phys.* **B302**, 645 (1988); B. Ratra and P.J.E. Peebles, *Phys. Rev. D* **37**, 3406 (1988); P.J.E. Peebles and B. Ratra, *Astrophys. J.* **325**, L17 (1988); K. Coble, S. Dodelson, and J.A. Frieman, *Phys. Rev. D* **55**, 1851 (1997), astro-ph/9608122; M.S. Turner and M. White, *Phys. Rev. D* **56**, 4439 (1997), astro-ph/9701138; R.R. Caldwell, R. Dave and P.J. Steinhardt, *Phys. Rev. Lett.* **80**, 1582 (1998), astro-ph/9708069.
5. C.T. Hill, D.N. Schramm, and J.N. Fry, *Comments Nucl. Part. Phys.* **19**, 25 (1989); J. Frieman, C. Hill, and R. Watkins, *Phys. Rev. D* **46**, 1226 (1992); J. Frieman, C. Hill, A. Stebbins, and I. Waga, *Phys. Rev. Lett.* **75**, 2077 (1995), astro-ph/9505060.
6. Y. Su *et al.*, *Phys. Rev. D* **50**, 3614 (1994).
7. R.D. Peccei, J. Solà and C. Wetterich, *Phys. Lett.* **B195**, 183 (1987).
8. T. Damour, in *Les Houches Summer School on Gravitation and Quantizations*, eds. J. Zinn-Justin and B. Julia, Les Houches, France, 5 Jul - 1 Aug 1992, gr-qc/9606079.
9. T. Damour and F. Dyson, *Nucl. Phys.* **B480**, 37 (1996), hep-ph/9606486.
10. J.K. Webb *et al.*, astro-ph/9803165.
11. S.M. Carroll, G.B. Field, and R. Jackiw, *Phys. Rev. D* **41**, 1231 (1990); S.M. Carroll and G.B. Field, *Phys. Rev. D* **43**, 3789 (1991).
12. D.J. Saikia and C.J. Salter, *Annu. Rev. Astron. Astrophys.* **26**, 93 (1986); J.N. Clarke, P.P. Kronberg, and M. Simard-Normandin, *Mon. Not. R. Astr. Soc.*, **190**, 205 (1980); P. Haves and R.G. Conway, *Mon. Not. R. Astr. Soc.*, **173**, 53p (1975).
13. J.P. Leahy, astro-ph/9704285.
14. J.F.C. Wardle, R.A. Perley, and M.H. Cohen, *Phys. Rev. Lett.* **79**, 1801 (1997); astro-ph/9705142.
15. G.W. Anderson and S.M. Carroll, astro-ph/9711288.

Evidence for dark matter in clusters from lensing studies

Priyamvada Natarajan

CITA, 60 St. George Street, Toronto M5S 1A7, Canada

Abstract. Gravitational lensing studies of clusters of galaxies provide the most compelling evidence for the existence of dark matter in the Universe. Here, we briefly outline recent work that provides constraints on the granularity of the dark matter distribution within clusters. The inferred re-distribution of dark matter on small-scales in clusters has implications for dark matter candidates. The future prospects of combining lensing studies of clusters with other complementary probes - X-ray data, spectroscopic data, and Sunyaev-Zeldovich temperature decrements in order to better understand the formation and evolution of clusters is outlined.

Lensing by Galaxy Clusters

Clusters of galaxies have the optimum cross-section for lensing the nearly isotropically distributed high redshift faint galaxy population. Lensing by an extended mass concentration can be understood in terms of a mathematical mapping (e.g. Blandford & Narayan 1986; Fort & Mellier 1994; Bartelmann & Narayan 1996) from the source plane onto the image plane with the properties that it conserves surface brightness and is achromatic. The deflections produced are non-linear with impact parameter and therefore produce both amplification and distortion of the background sources. There are two important effects: the isotropic magnification and the non-isotropic distortion. The isotropic magnification is caused by mass interior to the beam and is pronounced in the region of the image plane where the local surface mass density Σ is of the order of the critical surface mass density, $\Sigma_{\rm crit}$, which occurs in the dense cores of rich clusters producing multiple images and arcs ($\Sigma_{\rm crit}$ depends on the angular distance to the source and lens and hence on cosmological parameters). The anisotropic distortion of images is caused by the gradient of the two-dimensional potential and characterizes the 'weak lensing' regime, the signal being arclets (single weakly sheared images) produced even at large distances from the cluster centre.

CP478 , *COSMO—98*, edited by David O. Caldwell

Granularity of the mass distribution in clusters

Here we report the results of our studies to constrain the fraction of dark matter associated with individual galaxy halos within clusters - as a probe of the mass distribution on intermediate scales ($\simeq$ 50 kpc). Our technique involves quantifying the 'local' weak lensing induced by the dark halos around bright cluster galaxies. A measure of these small-scale perturbations can be derived from the statistical distribution of the shapes of faint background galaxies. We use the maximum-likelihood techniques developed in the theoretical discussion by Natarajan & Kneib (1995; 1997) and correlate the local perturbations with cluster galaxies of known morphology and luminosity. In this way, we provide new constraints on both the mass and the extent of dark halos in cluster galaxies. Both strong and weak lensing constraints are incorporated simultaneously. We have successfully applied these techniques to a new wide-field *Hubble Space Telescope* (*HST*) image of the rich cluster AC114 at $z = 0.31$ [NKSE 98].

Future prospects

Besides establishing the existence of dark matter, lensing studies in recent years have evolved to the level where they are begining to provide key insights into the physics of cluster formation and evolution. Listed below are a few of the emerging techniques and applications of cluster lensing that are likely to develop significantly over the next few years with the advent of large 8-10 m class telescopes and proposed instruments like the Advanced Camera for Survey on HST, AXAF and BIMA.

- **Understanding galaxy evolution in clusters**

  Probing substructure within clusters with current data, we find that the **total** mass of a fiducial L^* cluster spheroidal is primarily contained with $\sim$ 15 kpc, with some indication that the halos of the S0 population may be more truncated than those of the ellipticals. Comparisons with similar estimates for field galaxies suggests that the cluster galaxies in AC114 may possess less extensive and less massive halos [for details see NKSE 98]. Comparing the total mass in cluster galaxies within 250 kpc of the center (down to the magnitude limit of our selection criterion) to the total mass of the cluster, we estimate that approximately 11% of the mass of the cluster is bound to individual cluster galaxies, which is in very good agreement with the results from N-body simulations of galaxy evolution in clusters (Dubinski 1998; Moore et al. 1996; Governato et al. 1998). This fraction has important consequences for the rate of galaxy interactions and hence the evolution of the cluster on the whole.

- **Combining Lensing, X-ray and S-Z data**

  The prospects of combining data from complementary probes of the structure and dynamics of clusters are very promising (Zaroubi et al. 1998; Natarajan 1999). Using the detailed mass distribution constructed by lensing studies in conjunction with X-ray and S-Z data insights into the physics of the Intra-cluster Medium can be obtained thereby enabling a more comprehensive understanding of how clusters assemble and evolve.

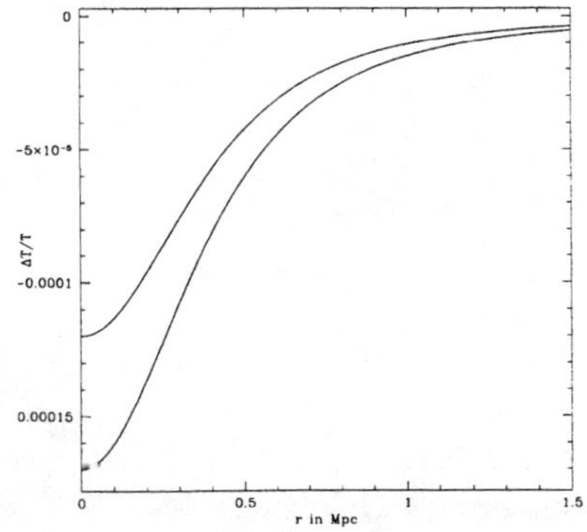

FIGURE 1. The predicted S-Z decrement as a function of radius for the cluster A2218 at $z = 0.175$. The two curves correspond to 2 different values of the assumed central X-ray temperature - 7 keV & 8 keV and have been obtained using the mass model reconstructed from gravitational lensing. Note that spherical symmetry is not assumed here, rather the inferred ellipticity of the lensing mass is used. Additionally, the electron number density distribution $n_e(r)$ is assumed to be a β model and a mean axis-ratio has been assumed to model the elongation of the cluster along the line-of-sight.

- **Constraining cosmological parameters**

  The variation of the M/L ratio with scale provides constraints on the density parameter Ω, and can be probed via gravitational lensing using both strong and weak lensing effects (Natarajan 1998) and can be compared with other methods (for a compilation of other methods see Bahcall 1998).

  The advantage of lensing is the fact that the effect is independent of the dynamical state of the lensing mass, therefore, it is not plagued by the systematics that the other techniques suffer from. With follow-up studies of galaxy scale lenses and the growing number of well-studied cluster lenses, this is emerging as a powerful tool in providing constraints on Ω.

Acknowledgments

I acknowledge my collaborators: Jean-Paul Kneib, Ian Smail and Richard Ellis for many useful and fun conversations.

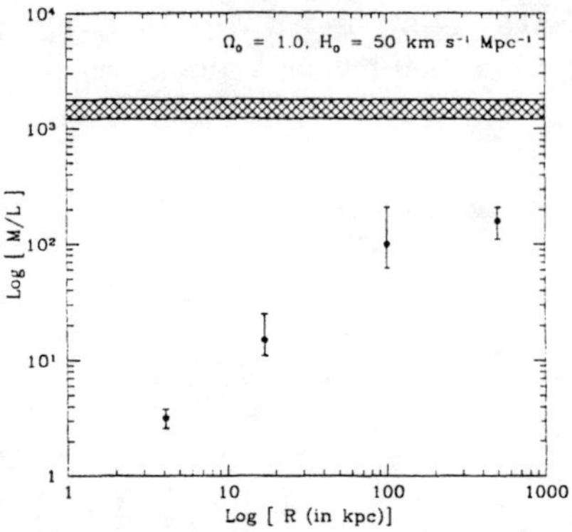

FIGURE 2. The variation of the M/L ratio with scale as probed by **gravitational lensing** methods. The point on the smallest scale (4.1 kpc) is from the M/L determined at the Einstein radius for the galaxy lens HST14176 [Hjorth & Kneib 1998]; the second point at 17 kpc is from the galaxy-galaxy lensing in the cluster in AC114 [NKSE 98]; the third from galaxy-galaxy lensing in the field [BBS 96] and the final point is the total mass-to-light ratio of the cluster AC114 computed at 500 kpc [NKSE 98]. The cross-hatched region in the plot is the global mass-to-light ratio required in order to close the Universe.

REFERENCES

1. N. Bahcall, 1998, Proc. of the Potsdam Cosmology Workshop 'Large-scale structure - tracks and traces', eds. Muller et. al., World Scientific, 1998.
2. M. Bartelmann & R. Narayan, 1996, Lectures on gravitational lensing, Jerusalem Winter School on Cosmology [astro-ph/9606001].
3. M. Birkinshaw, 1998, to appear in Phys. Reports [astro-ph/9808050].
4. R. Blandford & R. Narayan, 1986, ApJ, 310, 568.
5. T. Brainerd, R. Blandford & I.R. Smail, 1996, ApJ, 466, 623 [BBS 96].
6. B. Fort & Y. Mellier, 1994, A&AR, 5, 239.
7. B. Geiger & P. Schneider, 1998, MNRAS, 295, 497.
8. J. Hjorth & J.P. Kneib, 1998, in prep.
9. N. Kaiser & G. Squires, 1993, ApJ, 404, 441.
10. B. Moore, N. Katz & G. Lake, A. Dressler & A. Oemler, 1996, Nature, 379, 613.
11. P. Natarajan & J.P. Kneib, 1997, MNRAS, 287, 833.
12. P. Natarajan, J.P. Kneib, I.R. Smail & R.S. Ellis, 1998, ApJ, 474, 25 [NKSE98].
13. P. Natarajan, 1998, Proc. of the Potsdam Cosmology Workshop 'Large-scale structure - tracks and traces', eds. Muller et. al., World Scientific.
14. P. Natarajan & J.P. Kneib, 1995, Proc. of IAU 173, 'Astrophysical applications of gravitational lensing', eds. Hewitt & Kochanek.
15. P. Natarajan, 1999, in prep.
16. A. J. Tyson, G. Kochanski, & I.P. Dell'Antonio, 1997, ApJ, 474, 25.
17. S. Zaroubi, G. Squires, Y. Hoffmann & J. Silk, 1998, ApJ Lett, 500, 87.

Dark Matter Implications of the Microlensing Results

Kim Griest

Physics Department
University of California, San Diego
La Jolla, CA 92093
kgriest@ucsd.edu

Abstract. While gravitational microlensing is well established as an important method of discovering dark objects, the dark matter contribution of the objects thus discovered is still uncertain. We discuss the data, the problems, and the way to resolve the issues.

INTRODUCTION

Gravitational microlensing has been established as a powerful method of discovering and characterizing populations of dark objects. Our collaboration, the MACHO collaboration, has discovered over 300 events towards the Galactic bulge, but for the dark matter question which is the subject of this talk, only the dozen or so events towards the Large (LMC) and Small (SMC) Magellanic Clouds are relevant.

The microlensing surveys set out to use gravitational lensing to search for dark matter in the form of Massive Compact Halo Objects (Machos), and since we discovered the signal we were looking for, the first and simplest interpretation of these events was that they represent a substantial fraction of the dark halo of the Milky Way Galaxy. A likelihood analysis gives a most likely contribution of around $2\times10^{11} M_{\odot}$, almost independent of the halo model used [1]. This is several times the total mass in known material such as stars, dust, and gas, i.e. if this interpretation is correct, then the LMC and SMC microlensing events represent the discovery of the dominant detected component of our Galaxy. However, this interpretation is somewhat problematic, and as I will discuss, the interpretation of these events is still not settled. The main problem is that we usually cannot tell where along the line-of-sight the dark objects reside. If the lenses are not in the Galactic halo, then the estimate of the mass in lenses may be irrelevant for the dark matter problem. This is the subject of this talk.

CP478, *COSMO—98*, edited by David O. Caldwell

MICROLENSING BASICS

The large surveys teams, MACHO [2,1], EROS [3,4], and OGLE [5,6] monitor millions of stars in the LMC and SMC, which are at distances of 50 kpc and 60 kpc respectively, distances that place them inside the large dark halo of the Milky Way. The dark halo is needed to explain the observed flat rotation curve of the Milky Way, which implies that the halo objects, whatever they may be, move with a mean speed of approximately 270 km/s.

When a compact object passes directly in front of a more distant star, the compact objects acts a gravitational lens and magnifies the source star. Perfect alignment results in a circular image called the Einstein ring, of radius

$$r_E = 610 R_\odot [\frac{m}{M_\odot} \frac{L}{\mathrm{kpc}} x(1-x)]^{1/2},$$

where L is the distance to the source star, $x = D_L/L$, D_L is the distance to the lens, and m is the mass of the lens. The light in the ring image gives theoretically infinite magnification, but perfect rings are expected to be extremely rare. Imperfect alignment gives two images instead of a ring but also can magnify the source substantially. Since the observer (on Earth), the lens, and the source are all in relative motion, these stellar magnifications are transient events, lasting for a duration $\widehat{t} = 2r_E/v_\perp$, where $v_\perp$ is the transverse speed of the lens relative to the line-of-sight.

The main problem of microlensing interpretation can be seen from these formulas. During a typical event, there is only one measurable observable, the duration $\widehat{t}$, but there are four physical parameters, L, m, $v_\perp$, and x. L can be determined since the source is visible, but there is a degeneracy between the m, $v_\perp$, and x parameters. Thus on an event-by-event basis one cannot determine either the lens distance or the lens mass. By assuming a mass and velocity distribution, one can determine the typical lens mass in a statistical way, and this is what is done in obtaining a Macho halo fraction of $\sim 50\%$ and a typical Macho mass of $0.5 M_\odot$ [1]. Thus these conclusions depend upon the assumption that the events come from objects in the dark halo. An assumption of a non-halo population for these lenses can give quite a different determination of the amount of material that the dozen or so LMC events represent.

INTERPRETATION OF RESULTS

Before discussing alternative interpretations of the dozen or so LMC/SMC events, I should point out one very robust dark matter result from the microlensing surveys. Both the MACHO and EROS collaborations did intensive searches for short duration microlensing events and found none. Short duration events would be due to low mass Machos, so the non-observation of such events allows one to rule out a dark halo consisting of low mass compact objects. The most sensitive analysis to date [7] implies that objects with masses between $10^{-7}M_\odot$ and $10^{-3}M_\odot$ (Mars to Jupiter mass) make up less than 25% of the dark halo, and objects with mass m in the range $3.5 \times 10^{-7}M_\odot \leq m \leq 4.5 \times 10^{-5}M_\odot$ make up less than 10% of the dark matter. These results are not very sensitive to the halo model used and do not suffer from the interpretation problems discussed below.

The main problem with the dark matter interpretation of the LMC/SMC microlensing events is that the most likely lens mass is around $0.5M_\odot$. Objects with masses above $0.08M_\odot$ should be main sequence stars, which of course would be visible and therefore ruled out as dark matter candidates. If the mass determined from the microlensing likelyhood analysis had been in the brown dwarf range (below $0.08M_\odot$), then people probably would have accepted the dark matter interpretation. Now the question becomes, "What could these objects be?"

Many possible answers to this question have been given, but none of them are particularly compelling. For example, old white dwarf stars have masses of around $0.6M_\odot$, and are probably too dim to be seen in any extant experiment. But a star loses more than 50% of its mass on route to becoming a white dwarf, so one must ask where all that material is today. Some ideas have been given, but it seems problematic to hide that much material. Neutron stars and regular black hole remnants of an early generation of stars have similar problems. Primordial black holes formed in the early Universe would avoid these problems, but so far no mechanism that could create large numbers of these objects without substantial fine-tuning of the model parameters has been suggested. Thus no compelling candidate for dark halo Machos has appeared.

So some workers have turned to non-dark matter alternatives. However, each of these alternatives is also problematic, so for now, there is no compelling explanation for the LMC microlensing events. For example, suppose there is a small dwarf galaxy in front of the LMC that contains the lenses. The surveys monitor only about 20 square degrees on the sky, and derive the large dark matter mass by assuming that the LMC line-of-sight is typical. A small galaxy could give 6 or so events [8], in which case the surveys have only discovered a small neighboring galaxy and not the dark matter. One group [9], claims to detect stars from just such a galaxy, but several other groups dispute both the the claim of an intervening galaxy [10,11] and its ability to produce enough microlensing [12,13]. So it seems unlikely that an intervening dwarf galaxy is the explanation for the LMC events.

Another non-dark matter possibility is that the LMC is lensing stars in the LMC itself [14]. This is possible, but explicit calculations of the size of this effect using standard models of the LMC suggest that only one of the dozen or so LMC events is likely to have come from LMC self-lensing [1,15]. Of course, the LMC may be more extended and more complicated than the simple disk models used, and in my opinion, this possibility is still one of the more likely explanations. However, I should note that recent much improved neutral hydrogen maps of the LMC show it to be a very symmetrical disk, with spiral features, just as the standard LMC model would predict [16]. In summary, both the dark matter and non-dark matter interpretations have problems, and no clear answer is just now at hand.

WHAT IS NEEDED

Clearly we need to discover where the lenses lie. One way to ensure that LMC self lensing, or an intervening dwarf galaxy is not the cause of the events, is to find lensing towards the SMC. This have been done by two collaborations [17,18,23]. Unfortunately, the SMC is known to be extended along the line-of-sight, and so, in contrast to the LMC case, one expects SMC self-lensing to contribute to, or even dominate, the dark matter halo signal towards the SMC. Thus the two SMC events found do not resolve the question.

So what is needed is a clear determination of the distances to some LMC lenses. With standard point-source-point-lens (pspl) microlensing this is not possible on an event by event basis. However, there are several types of unusual microlensing cases where deviations from the pspl form occur and where it is possible to eliminate at least one of the three $(m, v_\perp, x)$ variables from the equations.

For example, if the event has a long duration, the Earth's motion around the Sun can affect the lightcurve. Since the Earth's motion is well known, the projected transverse speed can be determined, resulting in a unique relationship between the lens mass and distance. This effect is called microlensing parallax, and has been detected in events towards the bulge [19], but unfortunately not yet towards the LMC.

Similar information is obtained for non-pspl events where the finite size of the source star is resolved. This happens for the rare events where the lens comes very close to the observer-source line-of-sight, and therefore causes high magnification. This effect has also been seen only in the bulge [20].

Finally, if the source is is a binary star [21], or the lens is a binary system [22], much more complicated lightcurves can result and additional information can be obtained. The most interesting case occurs when the lens is actually a binary system. General relativity predicts a complicated caustic structure in this case – caustics being positions on the sky (roughly behind the lens) where the magnification is theoretically infinite. The finite size of the source star prevents infinite magnification from occurring, but magnifications of more than 100 can occur for the brief time it takes the caustic to pass over the source star's limb. The time for which the source is highly magnified therefore allows determination of the caustic sweep speed. For lenses in the halo, sweep speeds of $\sim$ 1000 km/s are expected, while for lenses in the LMC or SMC, speeds of $\sim$ 100 km/s are typical.

Recently the MACHO collaboration detected such an event towards the SMC [18]. Three follow-up programs (MACHO GMAN, EROS, and PLANET) were enlisted and helped detect the caustic crossing described above [18,23,24]. The duration of this crossing was such that a sweep velocity of $\sim$ 90 km/s was found – strong evidence that the lens in this case is in the SMC itself and not in the dark matter halo. Since self lensing is expected to be large for the SMC, this does not resolve the dark matter question, but a few events like this towards the LMC should help in straightening out the interpretation of LMC microlensing.

CONCLUSIONS AND DISCUSSION

In summary, with current data, the dark matter interpretation of the microlensing events is still up-in-the-air. What is most needed are determinations of the distances to some lenses causing LMC events. Even one lens distance determination of ~ 10 kpc, typical of a halo lens, would show that some of the microlensing was due to halo objects and therefore imply a substantial Macho contribution to the Milky Way dark matter. Alternately several lens distances of ~ 50 kpc, would show that LMC/LMC lensing is the dominate effect and lead to the conclusion that Machos are probably not a main contributor to the dark matter.

Fortunately, the possibility of getting these distances is improving. Several new microlensing experiments such as EROS2 [23] and OGLE2 [6], are in operation and a next generation microlensing project (NGMP) has been proposed [25]. In addition, the search for extra-solar planets using follow-up on microlensing events towards the bulge is rapidly expanding, and these experiments get the detailed photometry needed to determine the lens distances in some cases. Also needed for the future are better models of the LMC, SMC, and Milky Way, and if Macho dark matter is confirmed, ideas of what Machos are and how they could have formed will also be needed.

ACKNOWLEDGEMENTS

This work was supported in part by the U.S. Department of Energy under grant DEFG0390ER 40546, and in part by the NSF Center for Particle Astrophysics through AST 9120005.

REFERENCES

1. C. Alcock, *et al.*, ApJ 486, 697, 1997.
2. C.Alcock, *et al.*, ApJ 479, 119, 1997.
3. Renault, C., *et al.*, A&A, 324, 69 (1997).
4. Palanque-Delabrouille, N., *et al.*, A&A, 332, 1 (1998).
5. Udalski, A., *et al.*, Acta Astronomica, 44, 165 (1994).
6. Udalski, A., *et al.*, Acta Astronomica, 47, 319 (1997).
7. C. Alcock, *et al.*(MACHO + EROS collaboration), ApJ Lett. 499, L9 (1998).
8. Zhao, H., MNRAS, 294, 139 (1998).
9. Zaritsky, D., and Lin, D.N.C., AJ, 114, 2545 (1997).
10. Beaulieu, J.P. & Sackett, P.D. astro-ph/9710156 (1997).
11. C.Alcock, *et al.*, ApJ Lett. 490, L59, 1997.
12. Gould, A. ApJ 499, 728 (1998).
13. Bennett, D.P., ApJ Lett, 493, L79 (1998)
14. Sahu, K., Nature, 370, 275 (1994)
15. Gould, A., ApJ, 404, 451 (1993)

16. Kim, S., *et al.*, ApJ, 503, 674 (1998)
17. C. Alcock, *et al.*, ApJ Lett. 491, L11, 1997
18. C.Alcock, *et al.*, astroph/9807163, submitted to ApJ, 1998.
19. Afonso, C, *et al.*, astro/ph-9806380, submitted to A&A (1998).
20. C.Alcock, *et al.*, ApJ Lett. 454, L125 1995.
21. C. Alcock, *et al.*, ApJ 491, 436, 1997.
22. K. Griest and W. Hu, *Astrophysical Journal,* **397,** 362 (1992).
23. Mao, S. & Paczynski, B., ApJL 374, L37 (1991).
24. Albrow, M. D., *et al.*, astro-ph/9807086, submitted to ApJ Lett (1998).
25. Stubbs, C. private communication, 1998.

What Could the Machos Be?[1]

HongSheng Zhao

Sterrewacht Leiden, Niels Bohrweg 2, 2333CA Leiden, The Netherlands

Abstract. If the Universe has a significant baryonic dark component in the form of compact objects in galaxy halos (machos), then there is a minute chance (about 10^{-7}) that one of the Galactic machos passes sufficiently close to our line of sight to a star out of some 10^7 monitored stars in the Magellanic Clouds (MCs) that it brightens by more than 0.3 magnitude due to gravitational focusing. After a brief discussion of the current controversy over the interpretation of the observed events, i.e., whether the lensing is caused by halo white dwarfs or machos in general or by stars in various observed or hypothesized structures of the Clouds and the Galaxy, I propose a few observations to put ideas of the pro-macho camp and the pro-star camp to test. In particular, I propose a radial velocity survey towards the MCs.

Current Debate. Experimental searches for microlenses in the line of sight to the MCs by MACHO, OGLE, EROS and a number of follow-up surveys have found more than 16 candidates to the LMC and two to the SMC. Most of them are clustered within 2° of the center of the LMC (cf. upper left panel of Fig.1). The observed rate falls short of explaining the rotation curve of the Galaxy with a smooth halo of machos (Alcock et al. 1997a). Whether this is yet another major puzzle in astronomy which requires explanation by fine tuning is an open question (Adams & Laughlin 1996, Chabrier et al. 1996, Charlot & Silk 1995, Flynn et al. 1996, Graff & Freese 1996, Gates et al. 1998, Gibson & Mould 1997, Honma & Kan-Ya 1998). Another complication to the otherwise plausible conversion of the event rate to Ω_{macho} of the Universe is the inevitable background events, on top of any macho signal, coming from lensing of two stars in the LMC disc (Sahu 1994). Since disc-disc lensing is not very efficient if the LMC disc is cold and thin (Gould 1994, Wu 1994), more speculative models have been put forward by the pro-star camp to boost up the star-star lensing by hypothesizing a variety of special structures to the MCs. In particular, a connection is drawn between the unexpected rate of microlensing and the Milky Way-MCs and SMC-LMC interactions (Zhao 1998a, b, Weinberg 1998); the latter is yet another long-standing issue, rooted in the tidal vs. ram-stripping formation of the Magellenic stream, which is resolved finally by recent

[1] Based on an invited talk at COSMO-98 in Asilomar, AIP Proc. "Particle Physics and the Early Universe", ed. David Caldwell

CP478 , *COSMO—98*, edited by David O. Caldwell

observations (Putman et al. 1998 and references therein). Several observations and theoretical arguments suggest that our line of sight to the LMC passes through a 3-dimensional stellar distribution more extended in the line of sight than a simple thin disc of the LMC (Evans et al. 1998, Kunkel et al. 1997, Zaritsky et al. 1997, 1999, Zhao 1996). Others argue against additional structures other than the thin disc of the LMC (Alcock et al. 1997b, Gallart 1998, Beaulieu & Sackett 1998, Bennett 1998, Gould 1998, Johnston 1998, Ibata et al. 1999). The related issue on the efficiency of self-lensing of a tidally stretched SMC bar has also been brought up a few times (Alcock et al. 1997c, Palanque-Delabrouille et al. 1998, Zhao 1998a), but was highlighted soon after the discovery of the 98-SMC-1 caustic binary event (Afonso et al. 1998, Alcock et al. 1999, Albrow et al. 1999, Becker et al. 1998, Sahu & Sahu 1998, Udalski et al. 1998). The observations are clearly in support of the idea that the SMC has a non-equilibrium structure extended in the line of sight (Caldwell & Coulson 1986, Mathewson, Ford & Visvanathan 1986, Welch et al. 1987, Westerlund 1997). While there is still some dispute about the interpretation of the two binary lens candidates LMC-9 (Bennett et al. 1996, Zhao 1998a) and 98-SMC-1 (Honma 1999, Kerins & Evans 1999), the consensus seems to be that self-lensing in the LMC and SMC is significant if not dominant. In general, models of the lens and source distribution to the LMC fall in two broad classes.

Objects with motion decoupled from the Clouds. These are dark objects or stars of origin independent of the MCs. They generally move with a velocity quite different from the MCs. Besides the Galactic dark halo, they include the thick disc, and maybe a segment of the warp of our Galaxy (Evans et al. 1998) or even a Sagittarius-like undigested substructure in the halo which by chance is in the foreground or background of the MCs (Zhao 1996). Recent observations, including RR Lyrae sample in the MACHO sample, are not in favor of chance alignment models. Stars in the warp would have small heliocentric velocities as they participate in the Galactic rotation just like the Sun. The hypothesized substructure should be a cold feature in the velocity range between $-300\,\mathrm{km\,s^{-1}}$ to $300\,\mathrm{km\,s^{-1}}$.

Objects co-moving with the Clouds. These are generally stars which are one way or another generated by dynamical processes in the formation and evolution of the MCs or their progenitor. A predicted orbital and tidal disruption history of the Ancient Magellanic Galaxy (AMG) is shown (the two right panels of Fig. 2) in a model for the progenitor galaxy (Zhao 1998b). The lower left is a plot of the end-result of an N-body simulation (kindly made available by Lance Gardiner) of the LMC-SMC-Milky Way interaction, which creates a rich variety of stellar and gaseous substructures shown here. Zhao (1996, 1998a,b) suggested lensing of several such substructures, loosely bound or completely unbound to the MCs. These include a completely unbound grand tidal arm of the MCs which extends to Ursa Minor and Draco, a localized common halo bridging the MCs, a tidally strongly deformed SMC, a polar ring structure or a tidal halo of the LMC (cf. Fig. 2). The LMC disc might also not be perfectly thin and flat. Similar to the Galactic disc, which has a strong warp at the edge of the disc due to perhaps tidal interaction with the LMC or the Sagittarius dwarf galaxy, the gas-rich LMC disc might sport a thickened disc (Weinberg 1998) due to repeated harassment of the SMC and the Galaxy. All these models imply tidal interaction and could match observations, such as, the Magellanic Stream and its leading arm, proper motions of Ursa Minor and Draco, distributions of gas and stellar associations in the Magellanic Bridge, and a possible large line of sight extent of the SMC and a polar ring structure seen in the velocity distribution of LMC carbon stars (Kunkel et al. 1997). Weinberg's model is also motivated by a diffuse structure seen in star count data of USNO2 which seems to end near the tidal radius of the LMC.

These extra stars surrounding the Clouds are still highly speculative with highly uncertain geometries. It is debatable whether their structure is better described as a thickened disc, or a warp, a flare, a polar ring, a tidal halo etc.. But the common feature is that they circulate around the Galaxy together with the Clouds. In this sense they can be called co-moving objects, with a proper motion and radial velocity typically within, say $100\,\mathrm{km\,s^{-1}}$, of MCs. For example, the Vertical Red Clump feature identified by Zaritsky et al. (1997, 1999) in their color magnitude diagram of the LMC might be stars a few kpc in front of the LMC disc. If this interpretation

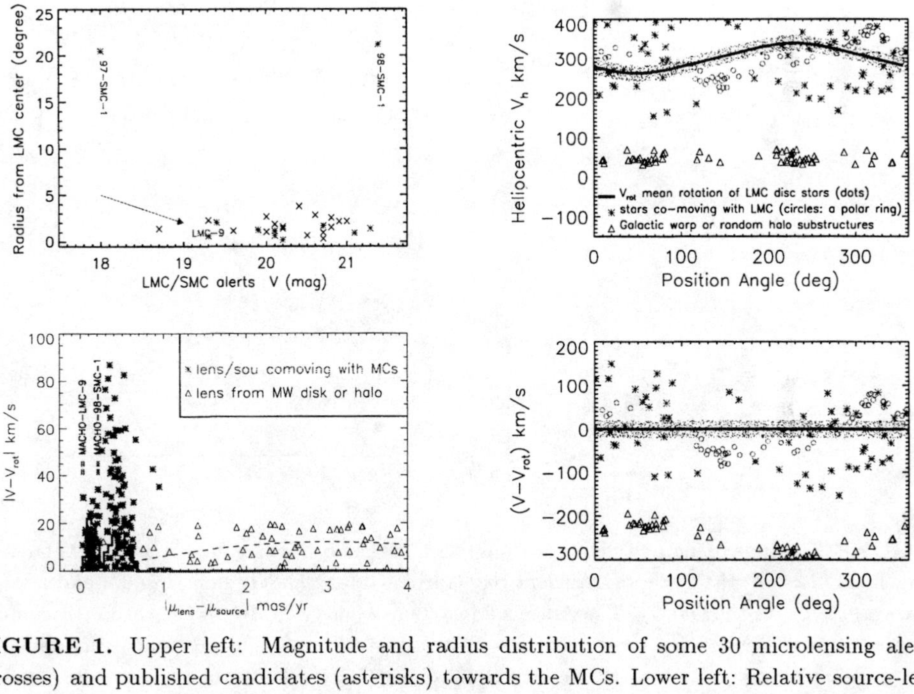

FIGURE 1. Upper left: Magnitude and radius distribution of some 30 microlensing alerts (crosses) and published candidates (asterisks) towards the MCs. Lower left: Relative source-lens proper motion vs. the offset in radial velocity of lensed source from average LMC disc stars in the same field. Upper right: The radial velocity of stars in the general direction of the LMC as a function of the position angle. Lower right: Distribution of the same objects after removing the mean rotation of the LMC disc (the sinusoidal curve).

holds up against countering arguments, the VRC objects would belong to the co-moving class rather than an independent stream since they move with a velocity very similar to the LMC.

These models typically predict a lens-source separation > 1 kpc and events of diameter crossing time > 30days for a $0.1 M_{\odot}$ stellar lens, and require a total mass in tidal substructures above 10% of that of the LMC, or comparable to that of the SMC to produce significant lensing. The challenge is to keep a large amount of unbound material near the Clouds, unless they are born recently (say due to a recent SMC-LMC encounter) or they remain loosely bound. For example, part of the common halo or bridge in the simulation shown in Fig. 2 is loosely bound to the MCs. It contains about $10^9 M_{\odot}$ in a mix of gas and stars with a velocity dispersion of about $60\,\mathrm{km\,s^{-1}}$ (Gardiner et al. 1994, 1996).

Smoking Guns for the two broad classes of models. Star-star lensing leaves detectable traces in the kinematical and spatial distribution of the lensed

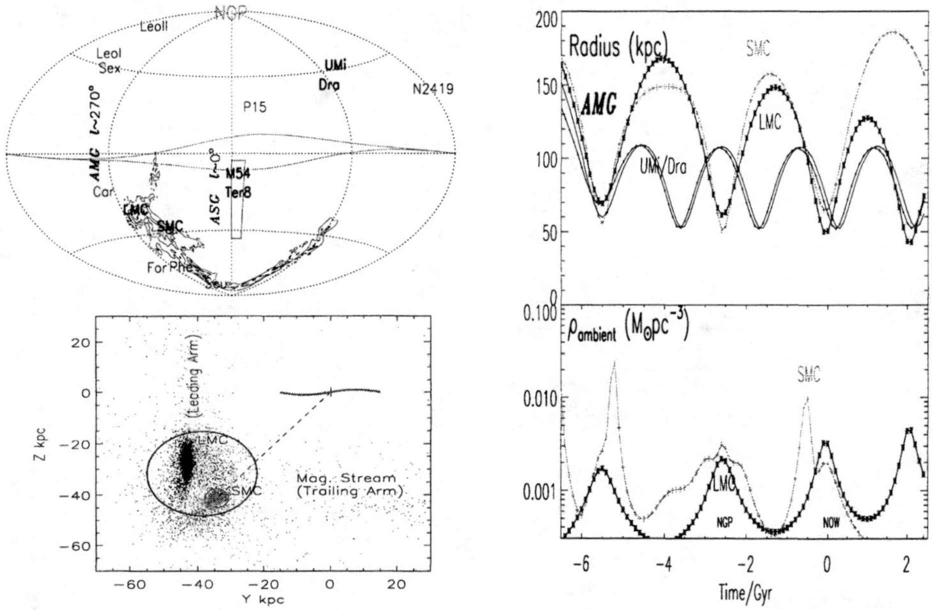

FIGURE 2. Upper left: Aitoff projections of a few halo dwarf galaxies and globulars, the Magellanic Stream, the disrupted Ancient Sagittarius Galaxy (ASG, along $l \sim 0^o$) and a sketch of a warped disc of the Galaxy. The Ancient Magellanic Galaxy (AMG) is the suspected common progenitor of objects on the great circle $l \sim 270^o$. Lower left: Simulated tidal structures along the Magellanic Stream and in the vicinity of LMC/SMC. The Sun-Galactic center line is along $Y = Z = 0$. Upper right: Predicted evolution of the distance from the Galactic center of the LMC, SMC, Ursa Minor and Draco for the past 6.5 Gyrs under the assumption that they come from a common progenitor. Lower right: Predicted history of tidal shocks (measured in ambient density) on the MCs, including a very recent one on the SMC.

sources. This is illustrated schematically by the right two panels of Fig. 1. The narrow sinusoidal band of the PA vs. V_r plot is indicative of the rotation of the kinematically cold (dispersion less than $20\,\mathrm{km\,s^{-1}}$) young thin disc component; an older, thicker disc would also rotate but with a larger dispersion. Overplotted are the predicted radial velocities of extra stars coming from either a decoupled component (triangles), which is clearly offset from the LMC, or a substructure comoving with the LMC (asterisks), which would generally show as a thicker band with either little dependence on the PA, or dependence different from the rotating disc (e.g., stars on the polar ring of Kunkel et al.). The co-moving material can be identified after subtracting the rotation (the lower right panel); they are markedly offset from the narrow distribution of the LMC thin disc stars. Here I propose a number of ways to resolve the substructures in the line of sight.

I. Self-lensing of stars comoving with the Clouds would induce a gradient of the event rate per survey star per year across the LMC disc since it is modulated by the structure of the LMC, whose width and density vary with the line of sight. The typical event time scale varies similarly. On the other hand if the lenses come from a smooth macho halo or any extended smooth structure decoupled from the MCs, it should be nearly the same for all lines of sight to the LMC disc. A tricky point here is that the dark halo could have fine structures, maybe a Sgr-like stream made of dark machos, which could induce a gradient as well.

II. Self-lensing prefers sources at the backside or behind the LMC disc. This is because lensing is most efficient if the source is located a few kpc behind a dense screen of stars, here the LMC disc. As a result, one should find a slight bias of lensed sources towards fainter magnitude than average stars in the survey, which are mostly LMC disc stars. Together with the spatial bias in (i) self-lensing should induce a bias in the magnitude vs. radius relation (maybe in the direction of the arrow in upper left panel of Fig. 1). No such bias is expected for macho models. Again there could be ambiguity if the stars at the backside of the LMC disc are systematically younger or older, hence intrinsically brighter or fainter.

III. Furthermore, these lensed sources behind the LMC disc are kinematically different from average stars in the LMC. This is perhaps the strongest signal of self-lensing since radial velocities can be measured so accurately that we can look for a small offset. There are two complementing effects here. First a radial velocity survey should pick up a small fraction of outliers of the rotation curve of the LMC disc (cf. right panels of Fig. 1), which may belong to some puffed-up distribution surrounding the LMC disc. Second the outliers at the backside of the LMC are more likely picked as lensed sources, with a velocity set apart from the rotation speed of the LMC disc *in the same field* by typically more than $20\,\mathrm{km\,s^{-1}}$ (cf. the lower left panel of Fig. 1). In contrast, if all events come from LMC disc stars being lensed by foreground Milky Way machos or disc stars, then the lensed sources would follow the motions of average stars in the cold, rotating disc.

IV. A continuation and expansion of the MACHO survey hopefully can yield a small sample of exotic events (e.g., when a source star passes the caustics of a binary lens) for which we can often tease out the relative lens-source parallax or proper motion. Their distribution would be a direct probe of the dynamics and structure of the lens population. This effect can be integrated with the radial velocity bias to set part various lens populations (cf. lower left panel of Fig. 1). The potential has clearly been demonstrated by LMC-9 and 98-SMC-1, where the observed small relative proper motion also seems to favor lensing by stars co-moving with the LMC (asterisks and p.m. histogram) than lensing by foreground populations in the Milky Way (triangles and the broad Gaussian curve with a p.m. dispersion of 3mas/yr, typical for halo machos or Galactic thick disc stars). I expect a handful of such events in the next five years with an extrapolation of current surveys, while the Next Generation Microlensing Survey (NGMS, Stubbs 1998) holds the promise of a similar number of events each year. Again whether a sample of caustic crossing binary events is a fair sample depends on whether we expect the fraction of machos in close binaries to be the same as that of stars.

Detection or non-detection of these subtle effects would be the key to resolve the controversy of the Galactic dark matter. The radial velocity survey is by far the most promising approach since it is not limited to the rare high amplification (caustic) events, it is not essential to take spectra of the source star during lensing, and the effect is big and easy to detect. While the strength of each of the "smoking guns" are still subject to details of our assumptions of, e.g., the star formation history of the LMC and intrinsic properties of machos, a complementary studies of lensing-induced systematic bias in radial velocity, distance, proper motion and spatial distributions is likely to give a robust conclusion on the nature of Galactic dark matter. More detailed simulations of the velocity distributions of various structures will be reported elsewhere, together with discussions of optimal strategies of carrying out the proposed observations to lift current degeneracies in interpreting the microlensing data. Finally studies of these effects can all be integrated in the NGMS. Tests of these effects could be first done towards the Galactic bulge where self-lensing of the Galactic bar surely plays an important role. New constraints will also come from microlensing surveys towards M31 (Crotts & Tomaney 1996, Ansari et al. 1997). A more distant but exciting prospect is to complement ground-based microlensing light curves with astrometric follow-up observations with the SIM satellite (2005-2010). Furthermore, GAIA (under study for launching 2009-2014) could in principle detect microlensing on the basis of astrometric shift alone without relying on the alerting system from the ground. The relative proper motion and parallax of the lens and the source will be the definitive test of the lens location and kinematics.

The author thanks Wyn Evans, Ken Freeman, Puragra Guhathakurta, Rodrigo Ibata, Konrad Kuijken, Joel Primack, Penny Sackett and Chris Stubbs for helpful discussions, and Dennis Zaritsky for a critical reading.

REFERENCES

1. Adams, F. C., & Laughlin, G. 1996, ApJ, 468, 586

2. Afonso, C. et al. (EROS), 1998, A&A, 337, 17
3. Alcock, C. et al. (MACHO), 1997a, ApJ, 486, 697
4. Alcock C. et al. (MACHO), 1997b, ApJ, 490, L59
5. Alcock C. et al. (MACHO), 1997c, ApJ, 491, L11
6. Alcock, C. et al. (MACHO/GMAN), 1999, ApJ submitted (astro-ph/9807163)
7. Albrow, M. D. et al. (PLANET), 1999, ApJ in press (astro-ph/9807086)
8. Ansari, R. et al. (EROS), 1996, A&A, 314, 94.
9. Crotts, A. P. S., Tomaney, A. B., 1996, ApJ, 473, 87
10. Beaulieu, J.-P. & Sackett, P. 1998, AJ 116, 209
11. Bennett, D. et al. 1996, Nucl. Phys. B, Proc. Suppl., 51B, 131 (astro-ph/9606012)
12. Bennett, D. 1998, ApJ, 493, L79
13. Caldwell J.A.R. & Coulson I.M. 1986, MNRAS, 218, 223
14. Chabrier, G., Segretain, L., & Méra, D. 1996, ApJ, 486, L21
15. Charlot, S., & Silk, J. 1995, ApJ, 445, 124 F
16. Evans, W. et al. 1998, ApJ, 501, L45
17. Flynn, C., Gould, A., & Bahcall, J. N. 1996, ApJ, 466, L55
18. Gallart, C. 1998, ApJ, 495, 43
19. Gardiner, L. T. & Noguchi, M., 1996, MNRAS, 278, 191
20. Gardiner, L. T., Sawa, T., Fujimoto, M., 1994, MNRAS, 266, 567
21. Gates, E. et al. 1998, ApJ, L145
22. Gibson, B. K., & Mould, J. R. 1997, ApJ, 482, 98
23. Gould, A. 1994, ApJ, 421, L75
24. Gould, A. 1998, ApJ, 499, 728
25. Graff, D. S., & Freese, K. 1996, ApJ, 456, L49
26. Honma, M 1999, ApJ 511, L29
27. Honma, M & Ka-Ya, Y. ApJ 503, L139
28. Ibata, R., Lewis, G. & Beaulieu, J.-P. 1999, ApJ, 509, L291
29. Johnston, K. 1998, ApJ, 495, 297
30. Kerins, E. & Evans, N.W. 1999, ApJ, in press
31. Kunkel, W. et al. 1997, ApJ, 488, L129
32. Palanque-Delabrouille, N. et al. (EROS), 1998, A&A, 332, 1 (astro-ph/9710194)
33. Putman, M. et al. 1998, Nature, 394, 752
34. Sahu, A. 1994, Nature, 370, 275
35. Sahu, K. C. & Sahu, M. S., 1998, ApJ, 508, L147
36. Stubbs, C. 1998, astro-ph/9810488
37. Udalski, A. et al. (OGLE), 1998, Acta Astronomica, 48, 383
38. Weinberg, M. 1998, 3rd Stromlo Symp. ed. B. Gibson et al. (astro-ph/9811204)
39. Welch, D. L. et al. 1987, ApJ, 321, 162
40. Westerlund, B. E., 1997, *The Magellanic Clouds*, Cambridge Univ. Press
41. Wu, X.P. 1994, ApJ, 435, 66
42. Zaritsky, D. & Lin, D. 1997, AJ, 114, 2545
43. Zaritsky, D. et al. 1999, AJ, in press
44. Zhao, H.S. 1996, astro-ph/9606166
45. Zhao, H.S. 1998a, MNRAS, 294, 139
46. Zhao, H.S. 1998b, ApJ, 500, L49

Dark matter from Affleck-Dine baryogenesis

Alexander Kusenko

Department of Physics and Astronomy, University of California, Los Angeles, CA 90095-1547

Abstract. Fragmentation of the Affleck-Dine condensate into Q-balls could fill the Universe with dark matter either in the form of stable baryonic balls, or LSP produced from the decay of unstable Q-balls. The dark matter and the ordinary matter in the Universe may share the same origin.

The Affleck-Dine (AD) scenario for baryogenesis [1] has several appealing features. First, the requisite scalar degrees of freedom with a non-zero baryon number are necessarily present in theories with low-energy supersymmetry. Second, the scalar potential of the MSSM and other supersymmetric generalizations of the Standard Model has numerous flat directions. At the end of inflation, the scalar fields develop large expectation values along these flat directions setting the stage for the AD scenario. The most economical supersymmetric extension of the Standard Model, MSSM, with some appropriate supersymmetry breaking [2], is already sufficient for the AD baryogenesis [3]. Finally, the observed value of the baryon asymmetry, $\eta_B \sim 10^{-10}$, can easily be accommodated in this scenario.

Supersymmetry also predicts the existence of non-topological solitons, Q-balls [4], with non-zero baryon and lepton numbers [5]. The interior of these Q-balls has the same field structure [6] as the AD condensate; and, in fact, very large Q-balls can form via the fragmentation of the scalar condensate in the early Universe [7,8].

Depending on their size and the mode of supersymmetry breaking, the baryonic balls formed in the process of AD baryogenesis can be stable or unstable. If they are stable, they can presently exist as a form of dark matter [7] with distinctive observational signatures [9]. If they are unstable [10], the LSP produced from their decay can contribute to dark matter [8]. In both cases, the ordinary matter and the dark matter come from the same origin, the primordial scalar condensate, and the amounts of dark matter (Ω_{DM}) and the ordinary matter (Ω_M) are related [11,12]. One may hope to understand, therefore, why Ω_M and Ω_{DM} are the same order of magnitude. This equality is inevitably fortuitous in theories that assume the dark matter formed as a result of a freeze-out or other process unrelated to baryogenesis (see, however, Ref. [13]). If, however, both components arise from AD baryogenesis, it is reasonable to ask whether a relation $\Omega_M \sim \Omega_{DM}$ can have a natural

CP478, *COSMO—98*, edited by David O. Caldwell

explanation [11,12].

At the end of inflation, the scalar fields acquire large expectation values along the flat directions and begin to roll towards their potential minima. Under very generic conditions, such a motion of the scalar condensate can become unstable with respect to small coordinate-dependent perturbations [7]. For an adiabatically slow motion of the initially homogeneous condensate $\phi(x,t) = \rho(t)\exp(i\Omega(t))$, the instability sets in when the second derivative of the potential, $U''(\phi(x,t)) \equiv U''(\rho(t))$, is smaller than $(d\Omega(t)/dt)^2$. We note in passing that for many potentials, in particular, those that arise as a result of gauge-mediated supersymmetry breaking, $U''(\phi(x,t))$ can be negative for some values of ϕ, in which case the condition $U''(\phi(x,t)) < (d\Omega(t)/dt)^2$ is satisfied automatically. If the expansion of the universe occurs at a rate that allows this instability to grow, the condensate becomes inhomogeneous and breaks up into Q-balls.

Q-balls produced from the shattered AD condensate can further evolve by absorbing baryons from or releasing baryons into the surrounding plasma [12,14].

Stable baryonic Q-balls can be the dark matter. Their signatures [9] in different detectors are characterized by a straight track with a large energy deposition, $\sim$ 100 GeV/cm, and no attenuation throughout the detector volume. The dark-matter Q-balls would have typical galactic velocities $\sim 10^{-3}c$. Assuming an order-one contribution to energy density of the Universe, one can set limits on their flux and, hence, their baryon number Q_B. The present limits require that $Q_B > 10^{22}$ [9,15]. The predicted range for Q_B consistent with $\Omega_M \sim \Omega_{DM}$ is $Q_B = 10^{26\pm2}$ [12].

The entire cosmologically interesting range of dark-matter Q-balls [7,9,12] can be explored by a detector with a surface area of several square kilometers. Since the required sensitivity is extremely low (thanks to the huge energy release expected from the passage of a superball), it is conceivable that a relatively inexpensive dedicated experiment could perform an exhaustive search and ultimately discover or rule out the stable Q-balls as dark-matter particles.

Baryonic Q-balls can absorb protons and neutrons, a process accompanied by a release of energy. This is because a stable baryonic ball has a smaller mass than a collection of neutrons with the same total baryon number. As a result, primordial Q-balls can accumulate in the center of a neutron star, grow through an absorption of neutrons, and eventually reduce the neutron star mass below the limit of its stability, which is about $0.2M_\odot$. At that point gravity is no longer strong enough to keep the neutrons from decaying, and the star explodes [16]. The lifetime of a neutron star is of order $(m/200\text{GeV})^5$ Gyr, where m is the scale associated with supersymmetry breaking. Perhaps, these mini-supernova explosions [17], which release of order 10^{52} erg, can be observable; they may also be related to (some) gamma-ray bursts.

The cosmological limits on dark mater impose some constraints on the form of the potential along a flat direction responsible for baryogenesis [18]. Thanks to the sensitivity of the Q-ball properties to the high-energy scales (*cf.* [19]), one may hope to learn about the higher-dimensional operators due to the Planck-scale physics, as well as the supersymmetry-breaking terms that lift the flat directions [18].

Enqvist and McDonald have suggested that the unstable SUSY Q-balls can also play a role in the production of dark matter [8,11,20–22]. The decay of a SUSY Q-ball into quarks is accompanied by the production of the lightest supersymmetric particles [8] that contribute to dark matter. This scenario offers an explanation of why $\Omega_{M} \sim \Omega_{DM}$ [11] because the dark-matter neutralinos are produced non-thermally with a number density that is related to the density of baryons. This scenario fits naturally in cosmology with a D-term inflation [23] and a reheating temperature below $10^{4\pm1}$ GeV [20].

Formation and decay of the unstable SUSY Q-balls could cause the isocurvature fluctuations that may be observable by MAP and PLANK [21].

The scenarios described above were studied in the context of MSSM. Of course, some additional scalar fields with global charges that often appear in the non-minimal SUSY models can also form stable Q-balls [24]. The non-abelian Q-balls [25], as well as the Q-balls associated with the conservation of a gauge charge [26], may also play an important role in cosmology.

To summarize, the Affleck-Dine baryogenesis in the MSSM can produce both the matter nucleons and the lumps of the scalar condensate, Q-balls. The stable Q-balls, as well as the products of the decay of the unstable Q-balls can contribute to dark matter. These possibilities are illustrated by the following diagram:

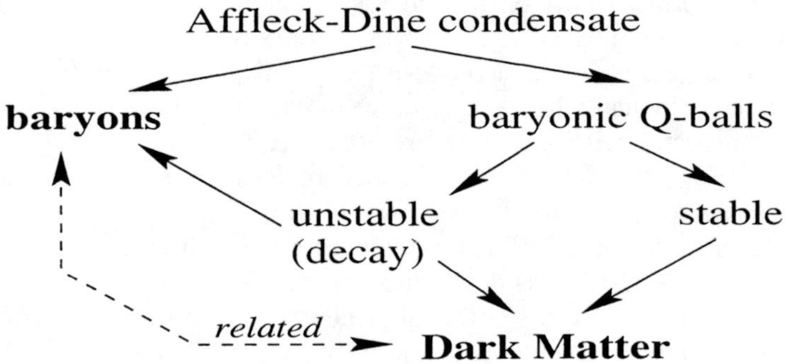

Formation of Q-balls can have profound implications for cosmology[1]. Since the fragmentation of a coherent scalar condensate [7,8] is the only conceivable mechanism that could lead to the formation of very large Q-balls, their observation would seem to speak unambiguously in favor of such process having taken place. Thus, although non-observation of Q-balls cannot rule out the AD baryogenesis, their observation would be a definite confirmation thereof. In addition, the decay of unstable SUSY Q-balls could produce the LSP dark matter at late times and out of equilibrium.

REFERENCES

1. I. Affleck and M. Dine, Nucl. Phys. **B 249** (1985) 361.

[1] For a more detailed review, see, *e. g.*, Ref. [27].

2. M. Dine, L. Randall, and S. Thomas, Phys. Rev. Lett. **75** (1995) 398.
3. M. Dine, L. Randall, and S. Thomas, Nucl. Phys. **B 458** (1996) 291; J. A. Casas and G. B. Gelmini, Phys. Lett. **B410** (1997) 36; B. A. Campbell, M. K. Gaillard, H. Murayama, and K. A. Olive, Nucl.Phys. **B538** (1999) 351.
4. G. Rosen, J. Math. Phys. **9** (1968) 996; *ibid.* **9** (1968) 999; R. Friedberg, T. D. Lee, and A. Sirlin, Phys. Rev. **D13** (1976) 2739; S. Coleman, Nucl. Phys. **B262**(1985) 263; A. Kusenko, Phys. Lett. **B 404** (1997) 285.
5. A. Kusenko, Phys. Lett. **B 405** (1997) 108.
6. A. Kusenko, M. Shaposhnikov, and P. Tinyakov, Pisma Zh. Eksp. Teor. Fiz. **67** (1998) 229 [JETP Lett. **67** (1998) 247] (hep-th/9801041).
7. A. Kusenko and M. Shaposhnikov, Phys. Lett. **B 418** (1998) 46.
8. K. Enqvist and J. McDonald, Phys. Lett. **B 425** (1998) 309.
9. A. Kusenko, V. A. Kuzmin, M. Shaposhnikov, and P. G. Tinyakov, Phys. Rev. Lett. **80** (1998) 3185.
10. A. Cohen, S. Coleman, H. Georgi and A. Manohar, Nucl. Phys. **B272** (1986) 301.
11. K. Enqvist and J. McDonald, Nucl. Phys. **B538** (1999) 321.
12. M. Laine and M. Shaposhnikov, Nucl. Phys. **B532** (1998) 376.
13. D. B. Kaplan, Phys. Rev. Lett. **68** (1992) 741.
14. J. A. Frieman, G. B. Gelmini, M. Gleiser and E. W. Kolb, Phys. Rev. Lett. **60** (1988) 2101; K. Griest, E. W. Kolb and A. Maassarotti, Phys. Rev. **D40** (1989) 3529; J. Ellis, J. Hagelin, D. V. Nanopoulos and K. Tamvakis, Phys. Lett. **B125**, 275 (1983); K. Griest and E. W. Kolb, Phys. Rev. **D40** (1989) 3231; J. A. Frieman, A. V. Olinto, M. Gleiser and C. Alcock, Phys. Rev. **D40** (1989) 3241; A. Kusenko, Phys. Lett. **B 406** (1997) 26.
15. I. A. Belolaptikov *et al.*, astro-ph/9802223.
16. A. Kusenko, M. Shaposhnikov, P. G. Tinyakov, and I. I. Tkachev, Phys.Lett. **B 423** (1998) 104.
17. M. Colpi, S. L. Shapiro, and S. A. Teukolsky, Astrophys. J. 414 (1993) 717; K. Sumiyoshi, S. Yamada, H. Suzuki, and W. Hillebrandt, astro-ph/9707230.
18. M. Axenides, E. G. Floratos, G. K. Leontaris, and N. D. Tracas, hep-ph/9811371.
19. G. Dvali, A. Kusenko, and M. Shaposhnikov, Phys. Lett. **B 417** (1998) 99.
20. K. Enqvist and J. McDonald, Phys. Rev. Lett. **81** (1998) 3071; hep-ph/9806213.
21. K. Enqvist and J. McDonald, hep-ph/9811412.
22. K. Enqvist and J. McDonald, Phys. Lett. **B440** (1998) 59.
23. P. Binétruy and G. Dvali, Phys. Lett. **B388** (1996) 241; E. Halyo, Phys. Lett. **B387** (1996) 43; C. Kolda and J. March-Russell, hep-ph/9802358.
24. D. A. Demir, hep-ph/9810453.
25. A. M. Safian, S. Coleman, and M. Axenides, Nucl. Phys. **B297** (1988) 498; M. Axenides, E. Floratos, and A. Kehagias, hep-ph/9810230.
26. K. Lee, J. A. Stein-Schabes, R. Watkins and L. M. Widrow, Phys. Rev. **D39** (1989) 1665; T. Shiromizu, Phys. Rev. D58 (1998) 107301; T. Shiromizu, T. Uesugi, M. Aoki, hep-ph/9811420.
27. A. Kusenko, Invited talk at 2nd International Conference on Dark Matter in Astro and Particle Physics (DARK98), Heidelberg, Germany, 20-25 Jul 1998; to appear in Proceedings; hep-ph/9808276.

Non-Baryonic Dark Matter – A Theoretical Perspective

Leszek Roszkowski*

Department of Physics, Lancaster University, Lancaster LA1 4YB, England

Abstract. I review axions, neutralinos, axinos, gravitinos and super-massive Wimpzillas as dark matter candidates.

INTRODUCTION

Some twenty years after the dark matter (DM) problem was taken seriously by particle theorists, we still don't know the exact nature of the hypothetical non-luminous material which presumably fills all space and forms extended halos around galaxies and their clusters [1]. While there may well be more than one type of DM, arguments from large structures make us believe that a large, and presumably dominant, fraction of DM in the Universe is made of some massive particles which at the time of entering matter dominance would be already non-relativistic, or *cold*. From the particle physics point of view, cold DM (CDM) could be made of some particles which would generically be called weakly interacting massive particles (WIMPs).

WIMPs do not necessarily have to interact only via weak interactions *per se*. One expects that they should preferably be electrically and color neutral, and therefore be, as it is often stated, "non-baryonic". Otherwise, they would dissipate their kinetic energy. (For baryonic DM, see the talks by K. Griest [2] on MACHOs and by Kusenko on Q-balls [3].)

Among WIMPs, there exist several interesting candidates for CDM which are well-motivated by the underlying particle physics. The neutralino of supersymmetry (SUSY) is considered by many a "front-runner" by being perhaps the most "genuine" WIMP. The axion is another well-motivated candidate. But by no means should one forget about other possibilities. While some old picks (sneutrinos and neutrinos with mass in the GeV range) are now ruled out, axinos and gravitinos have recently been revamped as possibilities for CDM. A new type of super-heavy Wimpzilla has also been proposed. In this talk I will briefly summarize main results and review recent developments in the field of WIMP and WIMP-like DM.

Neutrinos, the only WIMPs that are actually known to exist, are not considered

CP478, *COSMO—98*, edited by David O. Caldwell

particularly attractive as DM candidates. It has long been believed that their mass is probably very tiny, as suggested by favoured solutions to the solar and atmospheric neutrino problems, which would make them hot, rather than cold DM. This picture has recently been given strong support by first direct evidence from Superkamiokande for neutrinos' mass [4]. While the new data only gives the $\mu - \tau$ neutrino (mass)^2 difference of $2.2 \times 10^{-3}\,\mathrm{eV}^2$, it it very unlikely that there would exist two massive neutrinos with cosmologically relevant mass of 5 to 40 eV and such a tiny mass difference.

Current estimates of the lower bound on the age of the Universe lead to $\Omega h^2 < 0.25$. Recent results from high-redshift supernovae type Ia imply $\Omega_{\mathrm{matter}} \simeq 0.3$. The Hubble parameter is now constrained to 0.65 ± 0.1. Since $\Omega_{\mathrm{baryon}} h^2 \lesssim 0.015$, one obtains $\Omega_{\mathrm{CDM}} h^2 \lesssim 0.15$ or so. Assuming that CDM accounts for most of matter in galactic halos, one obtains a very rough estimate $\Omega_{\mathrm{CDM}} h^2 \gtrsim 0.025$.

AXIONS

Axions are spin-zero particles which are predicted by the Peccei-Quinn (PQ) solution [5] to the strong CP problem. As it is well known, the Lagrangian of QCD allows for a PC-violating term $\frac{\alpha_s}{8\pi}\bar{\theta}G\widetilde{G}$, where G is the gluon field strength. This term would contribute about $5 \times 10^{-16}\bar{\theta}\,\mathrm{e\,cm}$ to the electric dipole moment of the neutron, thus violating the current experimental bound by some ten orders of magnitude. In order to explain the required strong suppression in the value of $\bar{\theta}$, Peccei and Quinn postulated a new global $U(1)$ symmetry which would be spontaneously broken at some scale f_a. The pseudogoldstone boson associated with this scenario is the axion [6]. Because of a QCD chiral anomaly, the axion acquires mass $m_a \approx \Lambda^2_{QCD}/f_a$ where f_a is *a priori* an arbitrary parameter.

Axions are also relevant cosmologically. A variety of astrophysical and cosmological constraints have now narrowed the range of axion mass to $10^{-6}\,\mathrm{eV} \lesssim m_a \lesssim 10^{-3}\,\mathrm{eV}$ in a broad range of axion models. Rather remarkably, in this mass range, axion relic abundance is of order one, making them a possible DM candidate. Because they are produced out of thermal equilibrium, axions quickly become non-relativistic and are cold relics.

Axions are currently being searched for in microwave cavities immersed in a strong magnetic field, as reviewed in a separate talk by Sadoulet [7].

NEUTRALINOS

The DM candidate which has attracted perhaps the most attention from both the theoretical and experimental communities is the neutralino. It is a neutral Majorana particle, the lightest of the mass eigenstates of the fermionic partners of the gauge and Higgs bosons: the bino, wino and the higgsinos. If it is the lightest SUSY particle (LSP), it is massive and stable, due to assumed R-parity. A perfect candidate for a WIMP! There is much literature devoted to the neutralino as DM, including a number of excellent reviews (see, *e.g.*, Ref. [8]). Here I will only summarize the most essential results and comment on recent developments and updates.

Neutralino properties as DM and ensuing implications for SUSY spectra are quite model dependent but certain general conclusions can be drawn. Two benchmark models are normally considered. One is the supersymmetrized Standard Model (MSSM) with a minimum number of GUT assumptions. The other, much more predictive, model is the Constrained MSSM (CMSSM), or an effective minimal supergravity model. In the CMSSM one assumes that gaugino and scalar (sfermion and Higgs) soft SUSY-breaking masses unify to $m_{1/2}$ and m_0, respectively, at a GUT scale $M_{\rm GUT} \simeq 2 \times 10^{16}\,{\rm GeV}$. Masses of all the superpartners at the Fermi scale are then obtained by running RGE's down from $M_{\rm GUT}$ to m_Z. Higgs masses are determined through the condition of electroweak symmetry breaking (EWSB).

It is now generally accepted that the neutralino as DM candidate should most naturally contain a dominant bino component [9]. The (admittedly rough and subjective although commonly used) argument in the MSSM is that of naturalness: no superpartner masses are expected to significantly exceed 1 TeV. This, because of GUT-related relations among the masses of the gluino, the wino and the bino, $M_1 \simeq 0.5 M_2$ and $M_2 \simeq 0.3 m_{\widetilde{g}}$, implies $m_\chi \lesssim 150\,{\rm GeV}$. Of course this bound is only indicative but it gives us some idea for the expected range of m_χ. Another implication is that, because of the structure of the neutralino mass matrix, higgsino-like neutralinos are also strongly disfavored (the region $|\mu| \ll M_2$) [9].

Remarkably, in the CMSSM, the bino-like neutralino typically automatically *comes out* to be the LSP in a very large part of the SUSY parameter space [10,11]. This happens mostly as a result of imposing EWSB which typically produces $|\mu| \gg M_2$. Because of this property, and the fact that, roughly, $\Omega_\chi h^2 \sim m_0^4/m_\chi^2$, one is often able to put a *cosmological* ($\Omega_\chi h^2 < 0.25$) upper bound on SUSY mass parameters. Remarkably, the bound in the ball-park of 1 TeV, as generally expected! There are some loop-holes in this attractive picture (which, to me, is a powerful illustration of unity of particle physics and cosmology). For large $\tan\beta$, the bound disappears in a broad range around $m_A/2$ (half of the mass of the pseudoscalar Higgs) where the annihilation cross-section becomes large. Another recently pointed out effect is the neutralino's co-annihilation with the next lightest $\widetilde{\tau}_R$ in the region of $m_{1/2} \gg m_0$ [12].

Neutralinos and other SUSY particles have been constrained by searches at LEP and elsewhere. Mass limits are somewhat model dependent but are currently around 30 GeV (MSSM) to 50 GeV (CMSSM), as reviewed by J. Ellis [12]. I also refer to his talk for an explanation why higgsino-like LSPs are basically ruled out in a class of CMSSM-like models (with somewhat relaxed assumptions about unification of scalar masses).

Detection of (SUSY) WIMPs follows two broad avenues, as reviewed in separate talks [7,13]. Here I would like to make some comments about possible evidence for

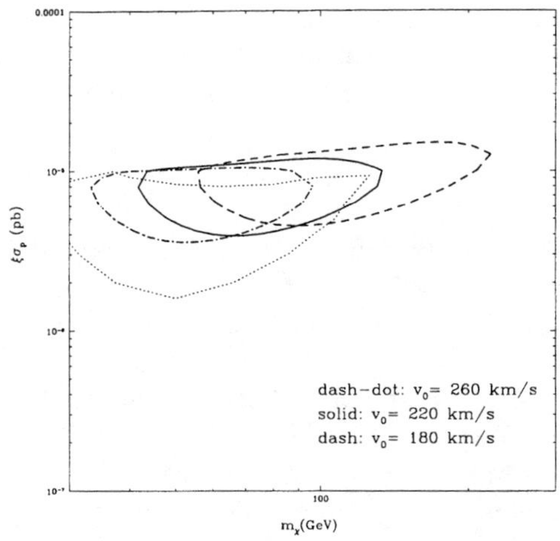

FIGURE 1. Contours of the function $\kappa = 15$ defined in the text for different values of the peak of the halo WIMP velocity distribution. Denoted by dots is the $2-\sigma$ region selected by DAMA assuming $v_0 = 220\, km/s$.

a WIMP signal in annual modulation. A superposition of the motion of the Earth around the Sun with that of the Sun around the center of the Milky Way leads to a small but sizeable (a few per cent) periodic variation in the effective velocity of halo WIMPs and therefore also in the detection rates [14]. The rate should reach its peak on the 2nd of June. Based on 14,962 day × kg of data, DAMA has recently reported on a possible signal in their NaI setup [15], which also confirmed DAMA's earlier indication [16] based on 4,549 day × kg of data. DAMA used the maximum likelihood method to compute in the k-th energy bin the most probable value of $S_k = S_{0,k} + S_{t,k} \cos(\omega(t - t_0))$, where ($S_{0,k}$) $S_{t,k}$ are time (in-)dependent components in the notation of Ref. [15]. The measured values of $S_{0,k}$ and $S_{t,k}$ are fitted with two parameters $\xi\sigma_p$ and $m_{\rm WIMP}$, where $\xi = \rho_{\rm WIMP}/(0.3\, {\rm GeV}/cm^3)$ and $\rho_{\rm WIMP}$ is the local density of WIMPs.

The derived ranges $m_{\rm WIMP} = 59\, {\rm GeV}^{+17}_{-14}\, {\rm GeV}$ and $\xi\sigma_p = 7.0^{+0.4}_{-1.2} \times 10^{-6} pb$ (at 99.6% CL), are in the ball-park of what one could expect from a genuine WIMP signal. We should be looking with attention for other dedicated experiments with similar sensitivity, like CDMS or UKDMC, to soon falsify the effect. Here I would like to note that the region selected by DAMA is probably too restrictive. The effect is rather sensitive to assumptions about the structure of the halo. In the analysis performed by DAMA only one value of the peak of the WIMP (Maxwellian) velocity distribution $v_0 = 220\, km/sec$ in the halo was assumed and only for one model of the halo. Since the Galactic halo has not been directly observed and there still is a considerable disagreement about a correct model of the halo, quoted error bars for v_0 and the local halo density should, in my opinion, be treated only as estimates (if not guesstimates). Varying v_0 within a reasonable range leads to a significant enlargement of the selected region. This can be seen in Fig. 1. We plot the function κ defined as

$$\kappa = \Sigma_{\text{energy bins}} \left[\left(S_{0,k} - S_{0,k}^{exp} \right)^2 / \sigma_{0,k}^2 + \left(S_{t,k} - S_{t,k}^{exp} \right)^2 / \sigma_{t,k}^2 \right], \qquad (1)$$

where the $S_{.,.}$s and $S_{.,.}^{exp}$s are the calculated and measured values and the σ's are the experimental error bars given in Table 2 of Ref. [15]. We minimize κ to determine $\xi\sigma_p$ and m_{WIMP} for a given value of v_0. This is plotted in Fig. 1. One can see that the region of $\kappa < 15$ which broadly matches the $2-\sigma$ region of DAMA strongly depends on v_0. Allowing for different halo models has a further significant effect [17]. (See also Ref. [18].)

Assuming that the effect reported by DAMA is caused by a genuine WIMP signal, it is interesting to ask what ranges of SUSY parameters it corresponds to. As reported here by Arnowitt [19] and Fornengo [20], it is indeed possible to find such SUSY configurations which could reproduce the signal but only for large enough $\tan\beta \gtrsim 10$. For smaller $\tan\beta$ the mass of the pseudoscalar contributing to the WIMP-nucleon cross-section becomes too small. The corresponding values of $\Omega_\chi h^2$ are typically rather small, below 0.06 (MSSM) or 0.02 (CMSSM), although larger values can also be found. I expect these conclusions to hold even if one neglects some arbitrariness in enlarging the experimental region of DAMA [20], which I don't find justified [17], and in deciding which value of *global* $\Omega_\chi h^2$ corresponds to the local WIMP density [20]. The effect of varying v_0 and admitting different halo models will have a significant impact on broadening the experimental region and therefore also on the allowed configurations of SUSY masses and couplings [17].

AXINOS

Each of the two DM candidates described above resulted from an attractive, albeit yet unconfirmed, idea in particle physics: the PQ symmetry and SUSY. Taken together, these predict an axino, the fermionic partner of the axion. Similarly to the axion, the axino couples to ordinary matter with a very tiny coupling proportional to $1/f_a$ where $10^{(9\div10)}\,\text{GeV} \lesssim f_a \lesssim 10^{12}\,\text{GeV}$.

It is plausible to consider the axino as the LSP since its mass is basically a free parameter which can only be determined in specific models. As we have seen above, the neutralino has been accepted in the literature as a "canonical" candidate for the LSP and an attractive dark matter candidate. But with current LEP bounds between 30 and 50 GeV, it becomes increasingly plausible that there may well be another SUSY particle which will be lighter than the neutralino, and therefore a candidate for the LSP and dark matter.

Primordial axinos decouple from the thermal soup very early, around $T \simeq f_a$, similarly to the axions. The early study of Ragagopal, *et al.*, [21] concluded that, in order to satisfy $\Omega h^2 < 1$, the primordial axino had to be light ($\lesssim 2$ keV),

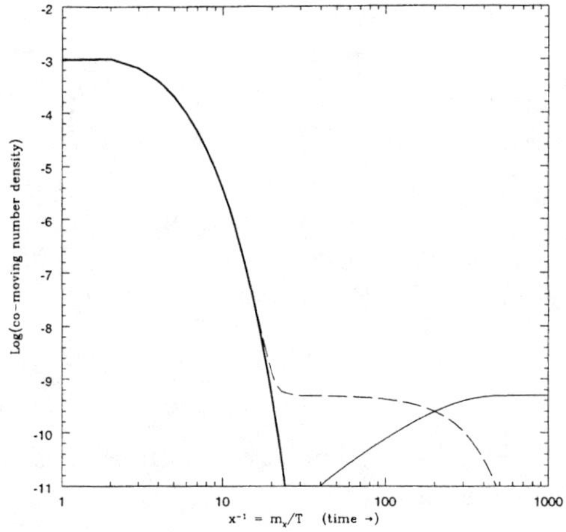

FIGURE 2. A schematic behavior of the co-moving number density: the thermal equilibrium (thick solid), NLSP neutralino (dash) and LSP axino (thin solid).

corresponding to warm dark matter, unless inflation would be invoked to dilute their abundance. In either case, one did not end up with axino as cold DM.

It has recently been shown [22] that the axino can be a plausible *cold* dark matter candidate after all, and that its relic density can naturally be of order the critical density. The axino can be produced as a non-thermal relic in the decays of heavier SUSY particles. Because its coupling is so much weaker, superparticles first cascade decay to the next lightest SUSY partner (NLSP) for which the most natural candidate would be the neutralino. The neutralino then freezes out from thermal equilibrium at $T_f \simeq m_\chi/20$. If it were the LSP, its co-moving number density after freeze-out would remain nearly constant. In the scenario of [22], the neutralino, after decoupling from the thermal equilibrium, subsequently decays into the axino via, *e.g.*, the process

$$\chi \rightarrow \widetilde{a}\gamma \tag{2}$$

as shown in Fig. 2. This process was already considered early on in Ref. [23] (see also [21]) in the limit of a photino NLSP and only for both the photino and axino masses assumed to be very low, $m_{\widetilde{\gamma}} \leq 1\,\mathrm{GeV}$ and $m_{\widetilde{a}} \leq 300\,\mathrm{eV}$, the former case now excluded by experiment. In that case, the photino lifetime was typically much larger than 1 second thus normally causing destruction of primordial deuterium from Big Bang nucleosynthesis (BBN) by the energetic photon of (2). Avoiding this led to lower bounds on the mass of the photino, as a function of f_a, in the MeV range [23].

Because both the NLSP neutralino and the CKR axino are both heavy (GeV mass range), the decay (2) is now typically very fast. In the theoretically most favored case of a nearly pure bino [9,11], the neutralino lifetime can be written as

$$\tau \simeq 2.12 \times 10^{-1}\text{sec} \left(\frac{1}{NC_{aYY}}\right)^2 \left(\frac{f_a}{10^{11}\,\text{GeV}}\right)^2 \left(\frac{50\,\text{GeV}}{m_\chi}\right)^3 \tag{3}$$

where the factor NC_{aYY} is of order one. One can see that it is not difficult to ensure that the decay takes place well before 1 second in order for avoid problems with destroying successful predictions of Big Bang nucleosynthesis. The axino number density is equal to that of the NLSP neutralino. Therefore its relic abundance is

$$\Omega_{\tilde{a}} h^2 = \left(\frac{m_{\tilde{a}}}{m_\chi}\right) \Omega_\chi h^2. \tag{4}$$

The axinos are initially relativistic but, by the time of matter dominance they become redshifted by the expansion and become cold DM.

GRAVITINOS

Another old DM candidate has recently been re-analyzed. In the context of supergravity, there exists the fermionic partner of the graviton. Just like the graviton, it couples to ordinary matter only gravitationally with the strength $1/M_{\text{P}}$ where $M_{\text{P}} = 2.4 \times 10^{18}\,\text{GeV}$. Gravitino's mass is model dependent but is expected to be of order the SUSY breaking scale $M_{SUSY} \lesssim 1\,\text{TeV}$.

The story of primordially produced gravitinos is analogous to the one of axinos. Along with gravitons, they decouple at $T \sim M_{\text{P}}$. If they were the LSPs, they would "overclose the Universe", unless either $m_{\tilde{G}} \lesssim 2\,\text{keV}$ or inflation followed to dilute their density. In order for subsequent reheating not to re-populate them, one requires $T_{\text{reh}} \lesssim 10^9\,\text{GeV}$.

Recently, a scenario for producing thermal gravitinos in the context of leptogenesis has been considered [24]. Decays of heavy ($\sim 10^{16}\,\text{GeV}$) Majorana neutrinos violate L which, via $B-L$, leads to baryon asymmetry consistent with the observed values $n_B/s \sim 10^{-9}$. For this picture to work, and assuming hierarchical neutrino masses, the baryogenesis temperature of $T_B \sim 10^{10}\,\text{GeV}$, and therefore T_{reh} at least this large, are required. Gravitinos are then produced in the thermal bath mainly via two-body processes involving gluinos. Their resulting relic abundance is given by [24]

$$\Omega_{\tilde{G}} h^2 = 0.60 \left(\frac{T_B}{10^{10}\,\text{GeV}}\right) \left(\frac{100\,\text{GeV}}{m_{\tilde{G}}}\right) \left(\frac{m_{\tilde{g}}(\mu)}{1\,\text{TeV}}\right)^2 \tag{5}$$

where $\mu \sim 100\,\text{GeV}$. One can see that for plausible values of $T_B, m_{\tilde{G}}$ and $m_{\tilde{g}}$ one obtains $\Omega_{\tilde{G}} h^2 \sim 1$.

One still has to make sure that the gravitinos are not produced in large numbers in out-of-equilibrium decays of the NLSP which, because of gravitino's tiny couplings, would take place long after BBN. There are ways to satisfy this, for example, when the NLSP is a higgsino in the mass range between m_W and some 300 GeV for which Ωh^2 is small enough (< 0.008). We note here that this requirement should be easily satisfied also for a bino-like NLSPs if one, or more, of the scalar leptons and squarks (except perhaps for sneutrinos and stops) is sufficiently light to suppress bino's relic abundance.

SUPER-HEAVY WIMPS: WIMPZILLAS

We have seen that plausible DM candidates need not be WIMPs, nor do they have to be produced thermally (axinos). Other possible candidates exist. For example, in a stringy scenario, these may be some class of moduli [25]. An interesting class of non-thermal super-heavy relics has recently been suggested [26]. (Thermally-produced WIMPs of mass above some 500 TeV would give $\Omega h^2 > 1$ [27].) Dubbed Wimpzillas, such relics could be as heavy as $M_{\rm GUT}$. Moreover, they could even carry electric or color charge. One has to make sure that Wimpzillas do not annihilate efficiently enough to be in chemical equilibrium at any time. This is basically guaranteed because their number density must be very tiny in order not to "overclose" the Universe [26].

Several possible mechanisms for generating Wimpzilla-like candidates have been suggested. They could for example be produced as a result of "freezing out" quantum fluctuations at the end of inflation or by gravitational effects. First-hand description of several of them can be found in these Proceedings [26,28].

It is remarkable that the Universe, and our halo, may well be filled with such obese relics which, despite possibly large interactions with ordinary species, would be gentle enough to remain in the background. At the end, the frightening Wimpzilla may reveal itself as a "monster with a human face".

SUMMARY

Who is the WIMP? The key to answering this question will ultimately be in the hands of experimentalists. Some attractive candidates (axion, neutralino) will be either discovered or basically ruled out during the next decade. Axinos and gravitinos, and perhaps also Wimpzillas, may well have to wait for future generations of dark matter enthusiasts.

ACKNOWLEDGEMENTS

I am greatly indebted to David Caldwell and other members of the Local Organizing Committee for setting up an inspiring meeting in one of the most beautiful corners in the world, in the spirit of COSMO workshops.

REFERENCES

1. See, *e.g.*, M.S. Turner, in the Proceedings.
2. K. Griest, in the Proceedings.
3. A. Kusenko, in the Proceedings.
4. Super-Kamiokande Collaboration, talk by Y. Suzuki at *Neutrino-98*, Takayama, Japan, June 1998; Super-Kamiokande Collaboration, Y. Fukuda, *et al.*, Phys. Lett. **B433**, 9 (1998); Phys. Lett. **B436**, 33 (1998); Phys. Rev. Lett. **81**, 1562 (1998).
5. R. D. Peccei and H. R. Quinn, Phys. Rev. Lett. **38**, 1440 (1977); Phys. Rev. **D16**, 1791 (1977).
6. S. Weinberg, Phys. Rev. Lett. **40**, 223 (1978); F. Wilczek, Phys. Rev. Lett. **40**, 279 (1978).
7. B. Sadoulet, in the Proceedings.
8. G. Jungman, M. Kamionkowski, K. Griest, Phys. Rep. **267**, 195 (1996).
9. L. Roszkowski, Phys. Lett. **B 262**, 59 (1991); see also J. Ellis, D.V. Nanopoulos, L. Roszkowski, and D.N. Schramm, Phys. Lett. **B245**, 251 (1990).
10. P. Nath and R. Arnowitt, Phys. Lett. **B289**, 368 (1992).
11. R.G. Roberts and L. Roszkowski, Phys. Lett. **B309**, 329 (1993); G.L. Kane, C. Kolda, L. Roszkowski, and J. Wells, Phys. Rev. **D49**, 6173 (1994).
12. J. Ellis, in the Proceedings and references therein.
13. L. Bergström, in the Proceedings.
14. A. Drukier, *et al.*, Phys. Rev. **D33**, 3495 (1986); K. Freese, *et al.*, Phys. Rev. **D37**, 3388 (1988).
15. R. Bernabei, *et al.* (The DAMA Collaboration), ROM2F/98/34 (August 1998).
16. R. Bernabei, *et al.* (The DAMA Collaboration), Phys. Lett. **B424**, 195 (1998).
17. M. Brhlik and L. Roszkowski, to be published.
18. N. Fornengo, talk at IDM-98, Buxton, UK, September '98, hep-ph/9812210.
19. R. Arnowitt, in the Proceedings; R. Arnowitt and P. Nath, hep-ph/9902237.
20. N. Fornengo, in the Proceedings and references therein.
21. K. Ragagopal, M.S. Turner, and F. Wilczek, Nucl. Phys. **B358**, 447 (1991).
22. L. Covi, L. Roszkowski and J.E. Kim, LANCS-TH/9824 (December 1998), submitted to Phys. Rev. Lett.
23. J.E. Kim, A. Masiero, and D.V. Nanopoulos, Phys. Lett. **B139**, 346 (1984).
24. M. Bolz, W. Buchmüller, and M. Plümacher, Phys. Lett. **B443**, 209 (1998).
25. R. Brustein, in the Proceedings; R. Brustein and M. Hadad, hep-ph/9810526.
26. E. Kolb, in the Proceedings; D.J.H. Chung, E.W. Kolb, and A. Riotto, Phys. Rev. Lett. **81**, 4048 (1998).
27. K. Griest and M. Kamionkowski, Phys. Rev. Lett. **64**, 615 (1990).
28. I. Tkachev, in the Proceedings; V. Kuzmin and I. Tkachev, hep-ph/9809547.

Limits on Sparticle Dark Matter

John Ellis*

* *Theoretical Physics Division, CERN*
CH - 1211 Geneva 23

Abstract. Arguments are given that the lightest supersymmetric particle should be a neutralino χ. Minimizing the fine tuning of the gauge hierarchy favours $\Omega_\chi h^2 \sim 0.1$. There are important constraints on the parameter space os the MSSM from the stability of the electroweak vacuum. Co-annihilation with the next-to-lightest supersymmetric particle is potentially significant. Incorporating the latest accelerator constraints from LEP and elsewhere, we find that 50 GeV $\lesssim m_\chi \lesssim$ 600 GeV and $\tan\beta \gtrsim 2.5$, if soft supersymmetry breaking parameters are assumed to be universal.

THE LIGHTEST SUPERSYMMETRIC PARTICLE IN THE MSSM

The motivation for supersymmetry at an accessible energy is provided by the gauge hierarchy problem [1], namely that of understanding why $m_W \ll m_P$, the only candidate for a fundamental mass scale in physics. Alternatively and equivalently, one may ask why $G_F \sim g^2/m_W^2 \gg G_N = 1/m_P^2$, where M_P is the Planck mass, expected to be the fundamental gravitational mass scale. Or one may ask why the Coulomb potential inside an atom is so much larger than the Newton potential, which is equivalent to why $e^2 = \mathcal{O}(1) \gg m_p m_e/m_P^2$, where $m_{p,e}$ are the proton and electron masses.

One might think it would be sufficient to choose the bare mass parameters: $m_W \ll m_P$. However, one must then contend with quantum corrections, which are quadratically divergent:

$$\delta m_{H,W}^2 = \mathcal{O}\left(\frac{\alpha}{\pi}\right) \Lambda^2 \tag{1}$$

which is much larger than m_W, if the cutoff Λ representing the appearance of new

CP478 , *COSMO—98*, edited by David O. Caldwell

physics is taken to be $\mathcal{O}(m_P)$. This means that one must fine-tune the bare mass parameter so that it is almost exactly cancelled by the quantum correction (1) in order to obtain a small physical value of m_W. This seems unnatural, and the alternative is to introduce new physics at the TeV scale, so that the correction (1) is naturally small.

At one stage, it was proposed that this new physics might correspond to the Higgs boson being composite [2]. However, calculable scenarios of this type are inconsistent with the precision electroweak data from LEP and elsewhere. The alternative is to postulate approximate supersymmetry [3], whose pairs of bosons and fermions produce naturally cancelling quantum corrections:

$$\delta m_W^2 = \mathcal{O}\left(\frac{\alpha}{\pi}\right) |m_B^2 - m_F^2| \tag{2}$$

that are naturally small: $\delta m_W^2 \lesssim m_W^2$ if

$$|m_B^2 - m_F^2| \lesssim 1\text{TeV}^2. \tag{3}$$

There are many other possible motivations for supersymmetry, but this is the only one that gives reason to expect that it might be accessible to the current generation of accelerators and in the range expected for a cold dark matter particle.

The minimal supersymmetric extension of the Standard Model (MSSM) has the same gauge interactions as the Standard Model, and the Yukawa interactions are very similar:

$$\lambda_d Q D^c H + \lambda_\ell L E^c H + \lambda_u Q U^c \bar{H} + \mu \bar{H} H \tag{4}$$

where the capital letters denote supermultiplets with the same quantum numbers as the left-handed fermions of the Standard Model. The couplings $\lambda_{d,\ell,u}$ give masses to down quarks, leptons and up quarks respectively, via distinct Higgs fields H and $\bar{H}$, which are required in order to cancel triangle anomalies. The new parameter in (4) is the bilinear coupling μ between these Higgs fields, that plays a significant rôle in the description of the lightest supersymmetric particle, as we see below. The gauge quantum numbers do not forbid the appearance of additional couplings

$$\lambda L L E^c + \lambda' L Q D^c + \lambda U^c D^c D^c \tag{5}$$

but these violate lepton or baryon number, and we assume they are absent. One significant aspect of the MSSM is that the quartic scalar interactions are determined, leading to important constraints on the Higgs mass, as we also see below.

Supersymmetry must be broken, since supersymmetric partner particles do not have identical masses, and this is usually parametrized by scalar mass parameters

$m_{0_i}^2|\phi_i|^2$, gaugino masses $\frac{1}{2}M_a\tilde{V}_a \cdot \tilde{V}_a$ and trilinear scalar couplings $A_{ijk}\lambda_{ijk}\phi_i\phi_j\phi_k$. These are commonly supposed to be inputs from some high-energy physics such as supergravity or string theory. It is often hypothesized that these inputs are universal: $m_{0_i} \equiv m_0, M_a \equiv M_{1/2}, A_{ijk} \equiv A$, but these assumptions are not strongly motivated by any fundamental theory. The physical sparticle mass parameters are then renormalized in a calculable way:

$$m_{0_i}^2 = m_0^2 + C_i m_{1/2}^2 \,, \quad M_a = \left(\frac{\alpha_a}{\alpha_{GUT}}\right) \; m_{1/2} \tag{6}$$

where the C_i are calculable coefficients [4] and MSSM phenomenology is then parametrized by $\mu, m_0, m_{1/2}, A$ and $\tan\beta$ (the ratio of Higgs v.e.v.'s).

Precision electroweak data from LEP and elsewhere provide two qualitative indications in favour of supersymmetry. One is that the inferred magnitude of quantum corrections favour a relatively light Higgs boson [5]

$$m_h = 66^{+74}_{-39} \pm 10 \text{ GeV} \tag{7}$$

which is highly consistent with the value predicted in the MSSM: $m_h \lesssim 150$ GeV [6] as a result of the constrained quartic couplings. (On the other hand, composite Higgs models predicted an effective Higgs mass $\gtrsim 1$ TeV and other unseen quantum corrections.) The other indication in favour of low-energy supersymmetry is provided by measurements of the gauge couplings at LEP, that correspond to $\sin^2\theta_W \simeq 0.231$ in agreement with the predictions of supersymmetric GUTs with sparticles weighing about 1 TeV, but in disagreement with non-supersymmetric GUTs that predict $\sin^2\theta_W \sim 0.21$ to 0.22 [7]. Neither of these arguments provides an accurate estimate of the sparticle mass scales, however, since they are both only logarithmically sensitive to m_0 and/or $m_{1/2}$.

The lightest supersymmetric particle (LSP) is expected to be stable in the MSSM, and hence should be present in the Universe today as a cosmological relic from the Big Bang [8]. This is a consequence of a multiplicatively-conserved quantum number called R parity, which is related to baryon number, lepton number and spin:

$$R = (-1)^{3B+L+2S} \tag{8}$$

It is easy to check that $R = +1$ for all Standard Model particles and $R = -1$ for all their supersymmetric partners. The interactions (5) would violate R, but not a Majorana neutrino mass term or the other interactions in $SU(5)$ or $SO(10)$ GUTs. There are three important consequences of R conservation: (i) sparticles are always produced in pairs, e.g., $pp \to \tilde{q}\tilde{g}X$, $e^+e^- \to \tilde{\mu}^+\tilde{\mu}^-$, (ii) heavier sparticles decay into lighter sparticles, e.g., $\tilde{q} \to q\tilde{g}$, $\tilde{\mu} \to \mu\tilde{\gamma}$, and (iii) the LSP is stable because it has no legal decay mode.

If such a supersymmetric relic particle had either electric charge or strong interactions, it would have condensed along with ordinary baryonic matter during the formation of astrophysical structures, and should be present in the Universe today in anomalous heavy isotopes. These have not been seen in studies of H, He, Be, Li, O, C, Na, B and F isotopes at levels ranging from 10^{-11} to 10^{-29} [9], which are far below the calculated relic abundances from the Big Bang:

$$\frac{n_{relic}}{n_p} \gtrsim 10^{-6} \text{ to } 10^{-10} \tag{9}$$

for relics with electromagnetic or strong interactions. Except possibly for very heavy relics, one would expect these primordial relic particles to condense into galaxies, stars and planets, along with ordinary bayonic material, and hence show up as an anaomalous heavy isotope of one or more of the elements studied. There would also be a 'cosmic rain' of such relics [10], but this would presumably not be the dominant source of such particles on earth. The conflict with (9) is sufficiently acute that the lightest supersymmetric relic must presumably be electromagnetically neutral and weakly interacting [8]. In particular, I believe that the possibility of a stable gluino can be excluded. This leaves as scandidates for cold dark matter a sneutrino $\tilde{\nu}$ with spin 0, some neutralino mixture of $\tilde{\gamma}/\tilde{H}^0/\tilde{Z}$ with spin 1/2, and the gravitino $\tilde{G}$ with spin 3/2.

LEP searches for invisible Z^0 decays require $m_{\tilde{\nu}} \gtrsim 43$ GeV [11], and searches for the interactions of relic particles with nuclei then enforce $m_{\tilde{\nu}} \gtrsim$ few TeV [12], so we exclude this possibility for the LSP. The possibility of a gravitino $\tilde{G}$ LSP has attracted renewed interest recently with the revival of gauge-mediated models of supersymmetry breaking [13], and could constitute warm dark matter if $m_{\tilde{G}} \simeq$ 1 keV. In this talk, however, I concentrate on the $\tilde{\gamma}/\tilde{H}^0/\tilde{Z}^0$ neutralino combination χ, which is the best supersymmetric candidate for cold dark matter.

The neutralinos and charginos may be characterized at the tree level by three parameters: $m_{1/2}$, μ and tanβ. The lightest neutralino χ simplifies in the limit $m_{1/2} \rightarrow 0$ where it becomes essentially a pure photino $\tilde{\gamma}$, or $\mu \rightarrow 0$ where it becomes essentially a pure higgsino $\tilde{H}$. These possibilities are excluded, however, by LEP and the FNAL Tevatron collider [11]. From the point of view of astrophysics and cosmology, it is encouraging that there are generic domains of the remaining parameter space where $\Omega_\chi h^2 \simeq 0.1$ to 1, in particular in regions where χ is approximately a $U(1)$ gaugino $\tilde{B}$, as seen in Fig. 1 [14].

Purely experimental searches at LEP enforce $m_\chi \gtrsim 30$ GeV, as seen in Fig. 2 [15]. This bound can be strengthened by making various theoretical assumptions, such as the universality of scalar masses m_{0_i}, including in the Higgs sector, the cosmological dark matter requirement that $\Omega_\chi h^2 \leq 0.3$ and the astrophysical preference that $\Omega_\chi h^2 \geq 0.1$. Taken together as in Fig. 3, we see that they enforce

$$m_\chi \gtrsim 42 \text{ GeV} \tag{10}$$

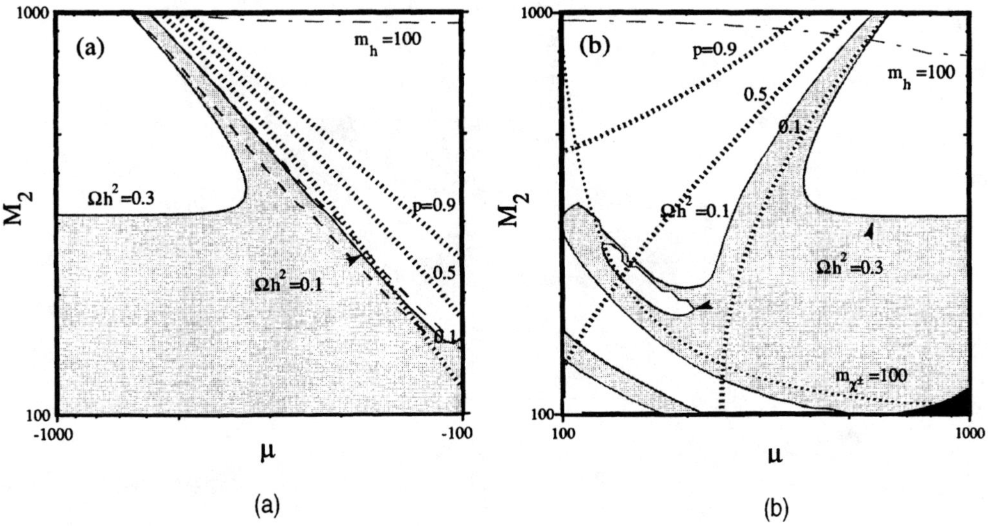

FIGURE 1. Regions of the (μ, M_2) plane in which the supersymmetric relic density may lie within the interesting range $0.1 \le \Omega h^2 \le 0.3$ [14].

and LEP should eventually be able to establish or exclude m_χ up to about 50 GeV. As seen in Fig. 4, LEP has already explored almost all the parameter space available for a Higgsino-like LSP, and this possibility will also be thoroughly explored by LEP [15].

WHAT IS THE "NATURAL" RELIC LSP DENSITY?

Should one be concerned that no sparticles have yet been seen by either LEP or the FNAL Tevatron collider? One way to quantify this is via the amount of fine-tuning of the input parameters required to obtain the physical value of m_W [16]:

$$\Delta_o = Max_i \mid \frac{a_i}{m_W} \frac{\partial m_W}{\partial a_i} \mid \tag{11}$$

where a_i is a generic supergravity input parameter. As seen in Fig. 5, the LEP exclusions impose [17]

$$\Delta_o \gtrsim 8 \tag{12}$$

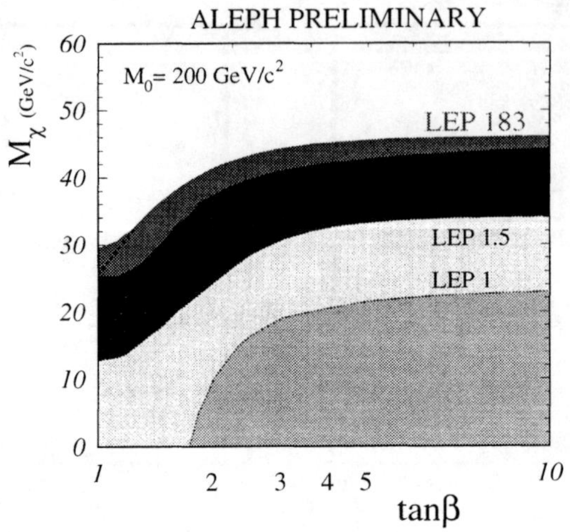

FIGURE 2. Experimental lower limit on the lightest neutralino mass, inferred from unsuccessful chargino and neutralino searches at LEP [15].

Although fine-tuning is a matter of taste, this is perhaps not large enough to be alarming, and could in any case be reduced significantly if a suitable theoretical relation between some input parameters is postulated [17].

It is interesting to note that the amount of fine tuning Δ_o is minimized when $\Omega_\chi h^2 \sim 0.1$ as preferred astrophysically, as seen in Fig. 6 [18]. This means that solving the gauge hierarchy problem naturally leads to a relic neutralino density in the range of interest to astrophysics and cosmology. I am unaware of any analogous argument for the neutrino or the axion.

IS OUR ELECTROWEAK VACUUM STABLE?

For certain ranges of the MSSM parameters, our present electroweak vacuum is unstable against the development of vev's for $\tilde{q}$ and $\tilde{l}$ fields, leading to vacua that would break charge and colour conservation. Among the dangerous possibilities are flat directions of the effective potential in which combinations such as $L_iQ_3D_3$, H_2L_i, LLE, H_2L acquire vev's. Avoiding these vacua imposes constraints that depend on the soft supersymmetry breaking parameters: they are weakest for $A \simeq m_{1/2}$. Figure 7 illustrates some of the resulting constraints in

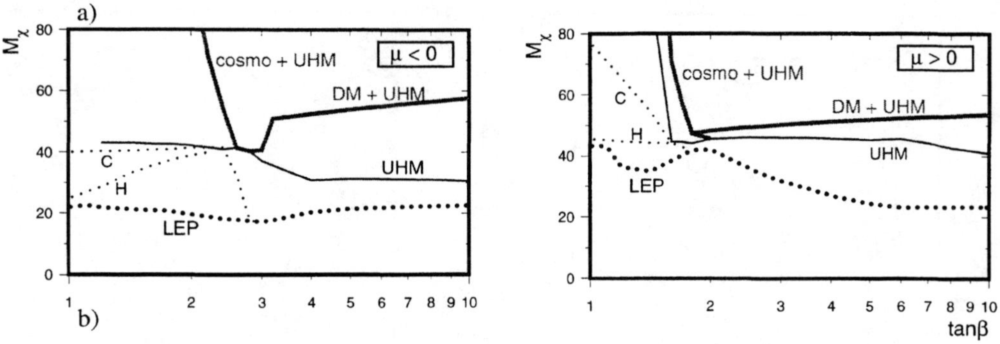

FIGURE 3. Theoretical lower limits on the lightest neutralino mass, obtained by using the unsuccessful Higgs searches (H), the cosmological upper limit on the relic density (C), the assumption that all input scalar masses are universal, including those of the Higgs multiplets (UHM), and combining this with the cosmological upper (cosmo) and astrophysical lower (DM) limits on the cold dark matter density [11].

the $(m_{1/2}, m_0)$ plane, for different values of $\tan\beta$ and signs of μ [19]. We see that they cut out large parts of the plane, particularly for low m_0. In combination with cosmology, they tend to rule out large values of $m_{1/2}$, but this aspect needs to be considered in conjunction with the effects of co-annihilation, that are discussed in the next section.

CO-ANNIHILATION EFFECTS ON THE RELIC DENSITY

As m_χ increases, the LSP annihilation cross-section decreases and hence its relic number and mass density increase. How heavy could the LSP be? Until recently, the limit given was $m_\chi \lesssim 300$ GeV [20]. However, it has now been pointed out that there are regions of the MSSM parameter space where co-annihilations of the χ with the stau slepton $\tilde{\tau}$ could be important, as seen in Fig.8 [21]. These co-annihilations would suppress Ω_χ, allowing a heavier neutralino mass, and we now find that [21]

$$m_\chi \lesssim 600\,\mathrm{GeV} \tag{13}$$

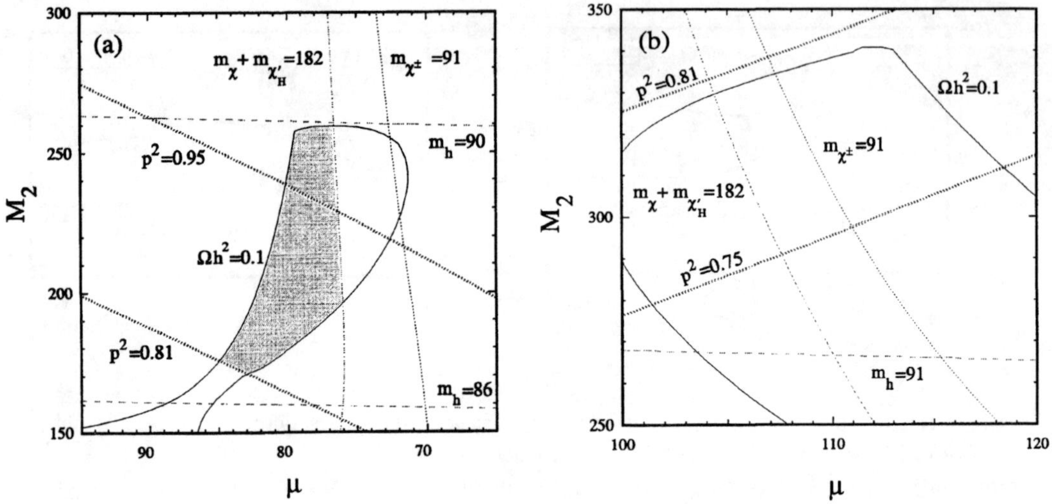

FIGURE 4. The regions of the (μ, M_2) plane where the lightest supersymmetric particle may still be a Higgsino, taking into account the indicated LEP constraints [14]. The Higgsino purity is indicated by p^2.

is possible if we require $\Omega_\chi h^2 \leq 0.3$. In the past, it was thought that all the cosmologically-preferred region of MSSM parameter space could be explored by the LHC [22], as seen in Fig. 9, but it now seems possible that there may be a delicate region close to the upper bound (13). This point requires further study.

CURRENT LEP CONSTRAINTS

The LEP constraints on MSSM particles have recently been updated [15], constraining the parameter space and hence the LSP. The large luminosity accumulated during 1998 has enabled the lower limit on the chargino mass to be increased essentially to the beam energy: $m_{\chi^\pm} \gtrsim 95$ GeV, except in the deep Higgsino region, where the limit decreases to about 90 GeV because of the small mass difference between the chargino and the LSP, which reduces the efficiency for detecting the $\chi^\pm$ decay products. There are also useful limits on associated neutralino production $e^+e^- \to \chi\chi'$, which further constrain the LSP. Without further theoretical assumptions, the purely experimental lower limit on the neutralino mass has become

$$m_\chi \gtrsim 32 \text{ GeV} \tag{14}$$

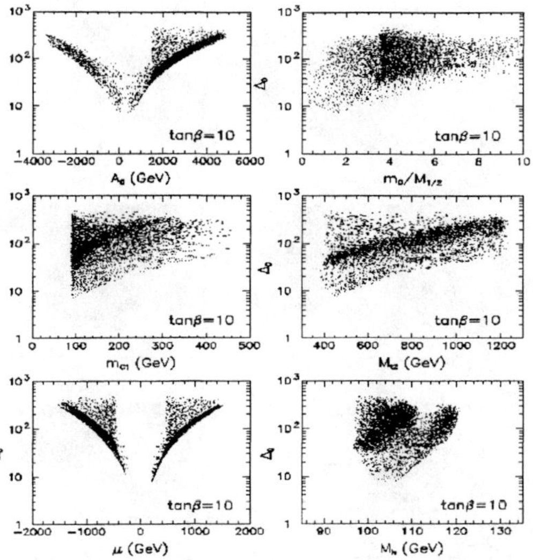

FIGURE 5. The fine-tuning price Δ_0 imposed by LEP for tanβ = 10, as a function of model parameters [17].

for large values of m_0 whatever the value of $\tan\beta$, decreasing to a minimum of 28 GeV for small m_0.

There are other new LEP limits that come into play with supplementary theoretical assumptions. These include a lower limit on the slepton mass, assuming universality ($m_{\tilde{l}} \equiv m_{\tilde{e}} = m_{\tilde{\mu}} = m_{\tilde{\tau}}$):

$$m_{\tilde{l}} > 90 \text{ GeV} \tag{15}$$

for $m_{\tilde{l}} - m_\chi \gtrsim 5$ GeV. There is also a new lower limit

$$m_{\tilde{t}} > 85 \text{ GeV} \tag{16}$$

assuming the dominance of $\tilde{t} \to c\chi$ decay, for $m_{\tilde{t}} - m_\chi \gtrsim 10$ GeV. Most important, however, is the new lower limit on the mass of the lightest Higgs boson in the MSSM. The L3 collaboration reports

$$m_h > 95.5 \text{ GeV} \tag{17}$$

for $\tan\beta \lesssim 3$. Combining all four LEP experiments, the lower limit (17) would probably be increased to 98 GeV, corresponding to the kinematic limit $\sqrt{s}$ (= 189 GeV) - m_Z.

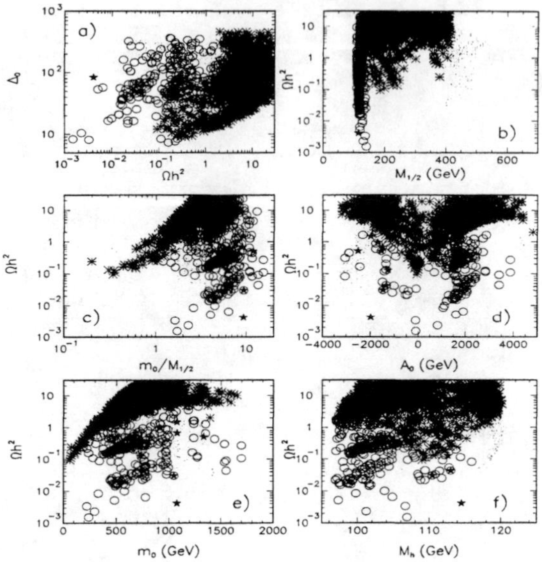

FIGURE 6. The correlation between the fine-tuning price Δ_0 and the relic density Ωh^2, showing dependences on model parameters [18].

The MSSM Higgs and other limits now appear to effectively exclude the possibility of Higgsino dark matter. Moreover, for $\mu < 0$, we now find $\tan\beta \gtrsim 3.0$, whereas a slightly smaller value is allowable if $\mu > 0$. For values of $\tan\beta$ close to these lower limits, the lower limit on m_χ increases sharply, qualitatively as in Fig. 6 but now shifted to the right. The valley in Fig. 6a for $\mu < 0$ is now filled in, so, pending a more complete evaluation, we estimate that

$$m_\chi \gtrsim 50 \text{ GeV} \tag{18}$$

for either sign of μ.

SUMMARY

We have seen that current experimental constraints impose $m_\chi \gtrsim 50$ GeV if universal soft supersymmetry breaking mass parameters are assumed, and that $m_\chi \lesssim 600$ GeV if we require $\Omega_\chi h^2 \leq 0.3$. Values of m_χ close to the lower limit may be explored by forthcoming runs of LEP in 1999 and 2000: the searches for the Higgs boson will be particularly interesting to follow. Thereafter, Run II of the Tevatron collider has the best accelerator chances to find supersymmetry, until the LHC comes along.

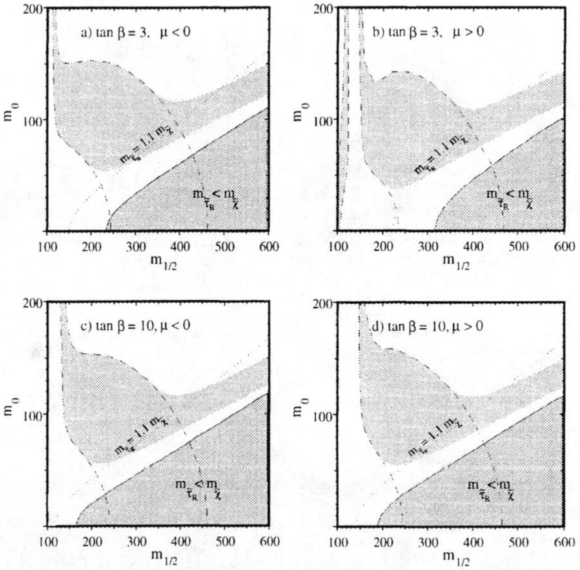

FIGURE 7. The change in the domain of parameter space allowed by the requirements $0.1 \leq \Omega h^2 \leq 0.3$ after (shaded region) and before (dashed lines) including $\tilde{\tau}$ co-annihilation [20].

In the mean time, non-accelerator searches looking directly for LSP-nucleus scattering or indirectly at LSP annihilation products will be offering stiff competition. There is already one direct search that does not claim not to have observed LSP-nucleus scattering [23]. The possible signal would correspond to a domain of MSSSM parameter space close to the present limits. The LSP interpretation of the signal is not yet generally accepted, since a complete annual modulation cycle has not yet been reported. However, healthy scepticism should not obscure the fact that it is consistent with the limits on sparticle dark matter reported here. Time only will tell whether accelerator or non-accelerator experiments will win the race to discover supersymmetry.

REFERENCES

1. L. Maiani, *Proc. Summer School on Particle Physics*, Gif-sur-Yvette, 1979 (IN2P3, Paris, 1980) p. 3;
   G 't Hooft, in: G 't Hooft et al., eds., *Recent Developments in Field Theories* (Plenum Press, New York, 1980);
   E. Witten, *Nucl.Phys.* B188 513 (1981);
   R.K. Kaul, *Phys.Lett.* 109B 19 (1982).
2. For a review, see: E. Farhi and L. Susskind, *Phys.Rep.* 74C, 277 (1981);
   S. Dimopoulos and L. Susskind, *Nucl.Phys.* B155, 237 (1979);

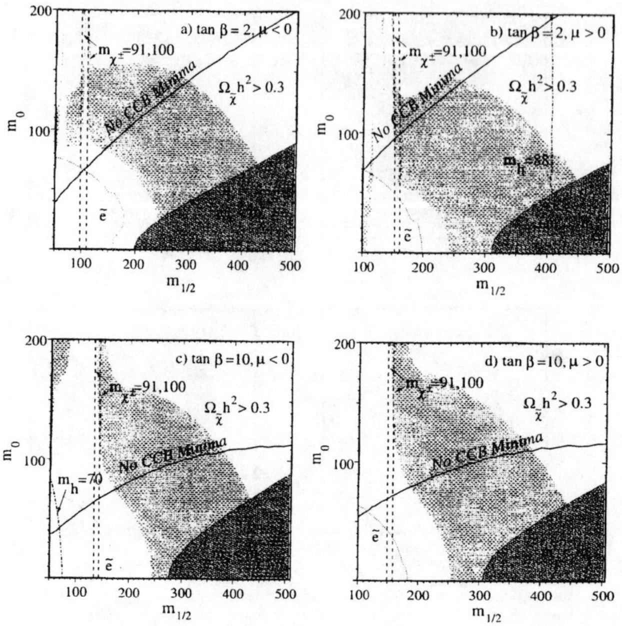

FIGURE 8. The combined cosmological and experimental constraints on the constrained MSSM, for $\tan\beta = 2, 10$ and both $\mu < 0$ and $\mu > 0$. The dashed contours represent current and future LEP chargino bounds, dotted contours are slepton bounds, and dot-dashed contours are Higgs bounds. The light-shaded region gives $0.1 < \Omega_\chi h^2 < 0.3$. Below the solid contour, CCB minima are present in the LLE, LH_2 direction. We have chosen $A_0 = -m_{1/2}$ to minimise the area containing CCB minima.

E. Eichten and K. Lane, *Phys.Lett.* B90, 125 (1980);
M.E. Peskin and T. Takeuchi, *Phys.Rev.* D46, 381 (1992);
G. Altarelli, R. Barbieri and S. Jadach, *Nucl.Phys.* B369, 3 (1992).

3. P. Fayet and S. Ferrara, *Phys.Rep.* 32, 251 (1977);
H.E. Haber and G.L. Kane, *Phys.Rep.* 117, 75 (1985).
4. See, e.g., C. Kounnas, A.B. Lahanas, D.V. Nanopoulos and M. Quirós, *Phys.Lett.* 132B, 95 (1983).
5. M. Grünewald and D. Karlen, talks at International Conference on High-Energy Physics, Vancouver 1998,
`http://www.cern.ch/LEPEWWG/misc`.
6. Y. Okada, M. Yamaguchi and T. Yanagida, *Progr.Theor.Phys.* 85, 1 (1991);
J. Ellis, G. Ridolfi and F. Zwirner, *Phys.Lett.* B257, 83 (1991), *Phys.Lett.* B262, 477(1991);
H.E. Haber and R. Hempfling, *Phys.Rev.Lett.* 66, 1815 (1991);
R. Barbieri, M. Frigeni and F. Caravaglios, *Phys.Lett.* B258, 167 (1991);
Y. Okada, M. Yamaguchi and T. Yanagida, *Phys.Lett.* B262, 54 (1991).
7. J. Ellis, S. Kelley and D.V. Nanopoulos, *Phys.Lett.* B249, 441 (1990) and *Phys.Lett.* B260, 131 (1991);

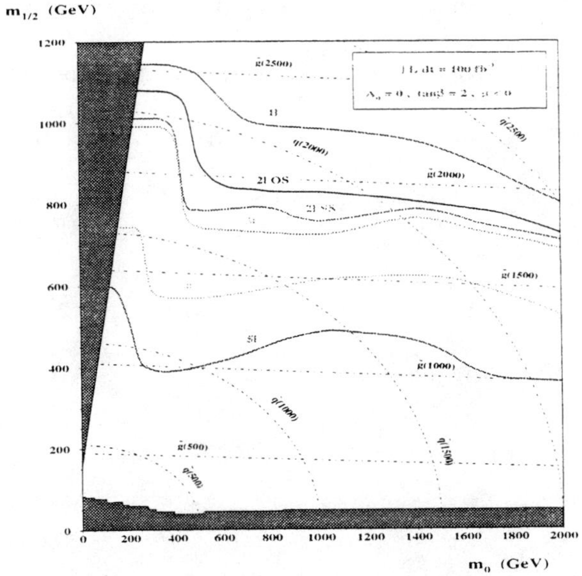

FIGURE 9. The region of the $(m_0, m_{1/2})$ plane accessible to sparticle searches at the LHC [22].

U. Amaldi, W. de Boer and H. Furstenau, *Phys.Lett.* B260, 447 (1991);
P. Langacker and M. Luo, *Phys.Rev.* D44, 817 (1991).

8. J. Ellis, J.S. Hagelin, D.V. Nanopoulos, K.A. Olive and M. Srednicki, *Nucl.Phys.* B238, 453 (1984).
9. J. Rich, M. Spiro and J. Lloyd-Owen, *Phys.Rep.* 151, 239 (1987);
P.F. Smith, *Contemp.Phys.* 29, 159 (1998);
T.K. Hemmick et al., *Phys.Rev.* D41, 2074 (1990).
10. R.N. Mohapatra and S. Nussinov, *Phys.Rev.* D57, 1940 (1998).
11. J. Ellis, T. Falk, K. Olive and M. Schmitt, *Phys.Lett.* **B388** (1996) 97 and *Phys.Lett.* B413, 355 (1997).
12. H.V. Klapdor-Kleingrothaus and Y. Ramachers, *Eur.Phys.J.* A3, 85 (1998).
13. G. Giudice and R. Rattazzi, hep-ph/9801271.
14. J. Ellis, T. Falk, K. Olive and M. Schmitt, *Phys.Lett.* B413, 355 (1997).
15. LEP Experiments Committee meeting, Nov. 12th, 1998,
`http:www.cern.ch/Committees/LEPC/minutes/LEPC50.html`
This source provides some preliminary updates of this and other experimental plots from papers contributed to the International Conference on High-Energy Physics, Vancouver 1998:
`http://ichep.triumf.ca/main.asp`.
16. J. Ellis, K. Enqvist, D.V. Nanopoulos and F. Zwirner, *Mod.Phys.Lett.* A1, 57 (1986);
G.F. Giudice and R. Barbieri, *Nucl.Phys.* B306, 63 (1988).
17. P.H. Chankowski, J. Ellis and S. Pokorski, *Phys.Lett.* B423, 327 (1998);
P.H. Chankowski, J. Ellis, M. Olechowski and S. Pokorski, hep-ph/9808275.

18. P.H. Chankowski, J. Ellis, K.A. Olive and S. Pokorski, hep-ph/9811284.
19. S. Abel and T. Falk, hep-ph/9810297.
20. K.A. Olive and M. Srednicki, *Phys.Lett.* B230, 78 (1989) and *Nucl.Phys.* B355, 208 (1991);
    K. Griest, M. Kamionkowski and M.S. Turner, *Phys.Rev.* D41, 3565 (1990).
21. J. Ellis, T. Falk and K.A. Olive, hep-ph/9810360.
22. S. Abdullin and F. Charles, hep-ph/9811402.
23. DAMA Collaboration, R. Bernabei et al., Preprint INFN-AE-98-20 (1998).

Cosmological Consequences of Supergravity Models

R. Arnowitt[1] and Pran Nath[2]

[1]*Center for Theoretical Physics, and Department of Physics, Texas A & M University, College Station, TX 77843-4242 and*
[2]*Department of Physics, Northeastern University, Boston, MA 02115*

Abstract

Gravity mediated supergravity models with R-parity invariance give rise naturally to a dark matter candidate, the lightest neutralino. Models with both universal soft breaking (mSUGRA) and with non-universal soft breaking (in the Higgs and third generation sectors) are shown to be consistent with the recent DAMA data suggesting the existence of an annual modulation effect. This data, if confirmed, would imply significant constraints on SUSY predictions at accelerators.

1 INTRODUCTION

There is currently much interaction between particle physics and cosmology on the nature of cold dark matter (CDM). Thus, supersymmetric theories with R-parity invariance possess a natural candidate for CDM, i.e. the lightest neutralino ($\tilde{\chi}_1^0$) which over most of the parameter space is the lightest supersymmetric particle (LSP) and hence absolutely stable.

Astronomical measurements show that dark matter plus some type of cosmological vacuum energy constitute most of the matter in the universe. One defines $\Omega_i = \rho_i/\rho_c$, where ρ_i is the mass density of type "i" and $\rho_c = 3H^2/8\pi G_N$ (G_N is Newton's constant and the Hubble constant H is parameterized by $H = (100kms^{-1}Mpc^{-1})h$ with $0.50 \lesssim h \lesssim 0.75$). Measurements give for the matter density $\Omega_m \simeq 0.3$ and for the cosmological constant $\Omega_\Lambda \simeq 0.6$ [1], while nucleosynthesis arguments imply for baryonic matter that $\Omega_B \simeq 0.05$. Thus, there appears to be a large amount of exotic CDM.

We analyze here the CDM question within the framework of supergravity models with R-parity invariance where supersymmetry (SUSY) breaking occurs at a high scale in a "hidden" sector and is transmitted by gravity to the physical sector [2], and discuss the following items: (1) The dependence on the SUSY parameters of the relic density of neutralinos $\Omega_{\chi_1^0}h^2$. (2) The direct detection of local (Milky Way) $\tilde{\chi}_1^0$ from their scattering by quarks in a nuclear target, and the SUSY parameter dependence of the proton cross section $\sigma_{\chi_1^0-p}$. (3) Comparison of the above theoretical results with the DAMA data [3] which suggests the observation of an annual modulation effect.

2 SUPERGRAVITY MODELS

Supergravity models depend on three functions of the scalar components ϕ_i of the chiral multiplets (χ_L, ϕ_i): the gauge kinetic function $f_{\alpha\beta}(\phi_i, \phi_i^+)$ (which enters in the gauge bosom kinetic energies), the Kahler potential $K(\phi_i, \phi_i^+)$ (which enters in the chiral multiplet kinetic energy and elsewhere), and the superpotential $W(\phi_i)$ which is broken into a physical sector part

CP478 , *COSMO—98*, edited by David O. Caldwell

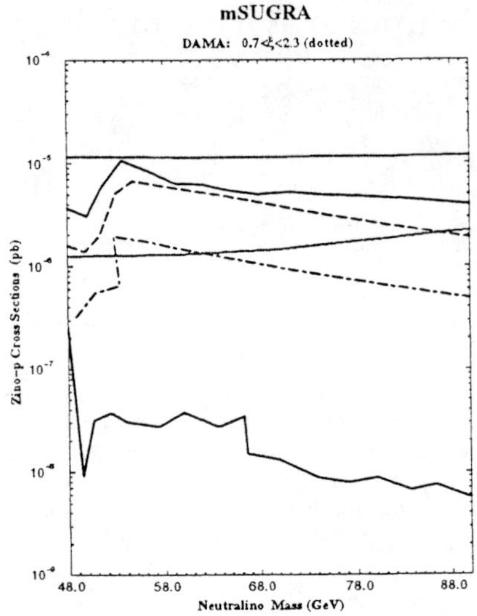

Figure 1: Maximum and minimum $\sigma_{\chi_1^0-p}$ for mSUGRA for $\tan\beta \leq 30$ (solid), $\tan\beta \leq 20$ (dashed), and $\tan\beta \leq 10$ (dot-dashed). The horizontal curves are the DAMA 95% C.L. bounds combined with the uncertainty in ξ ($0.7 \leq \xi \leq 2.3$).

and a hidden sector part: $W = W_{phys} + W_{hid}$. In gravity mediated models, supersymmetry is assumed broken in the hidden sector at a high scale e.g. $\gtrsim M_G$ (the GUT scale), the breaking being transmitted gravitationally to the physical sector. The Kahler potential governs the nature of the soft breaking terms. Thus, if the fields in K giving rise to SUSY breaking couple universally to the different generations, one obtains the minimal model (mSUGRA) with universal soft breaking depending on m_0 (squark mass), A_0 (cubic soft breaking mass), $\tan\beta =< H_2 > / < H_1 >, m_{1/2}$ and the sign of μ (the Higgs mixing parameter). (We assume here that the soft breaking parameters are all real.) Non-universal coupling leads to more complicated models. We maintain universality here in the first two generations (to supress FCNC) and in the gaugino mass, but allow arbitrary soft breaking masses in the Higgs and third generation masses at M_G. We parameterize these as follows: $m_{H_1}^2 = m_0^2(1+\delta_1)$; $m_{H_2}^2 = m_0^2(1+\delta_2)$; $m_{q_L}^2 = m_0^2(1+\delta_3)$; $m_{u_R}^2 = m_0^2(1+\delta_4)$; $m_{e_R}^2 = m_0^2(1+\delta_5)$; $m_{d_R}^2 = m_0^2(1+\delta_6)$; $m_{\ell_L}^2 = m_0^2(1+\delta_7)$ where $q_L = (\tilde{t}_L, \tilde{b}_L)$, $u_R \equiv \tilde{t}_R$ etc. and δ_i are the deviations from the universal value m_0^2. In addition, there are now three A_0 parameters, $A_{0t}, A_{0b}, A_{0\tau}$. If supersymmetry breaks above the GUT scale, group symmetry then significantly reduces the number of parameters. Thus, for any group containing an SU(5) subgroup with matter in usual the 10 and $\bar{5}$ representations, one has that $\delta_3 = \delta_4 = \delta_5$, $\delta_6 = \delta_7$ and $A_{0b} = A_{0\tau}$, significantly reducing the number of parameters.

One of the important quantities that is sensitive to non-universal effects is the μ parameter. Thus, for low and intermediate $\tan\beta$ one has for the m_0^2 dependence[4]:

$$\mu^2 = \frac{t^2}{t^2-1}[\{\frac{1-3D_0}{2} + \frac{1}{t^2}\} + \{\frac{1-D_0}{2}(\delta_3+\delta_4) - \frac{1+D_0}{2}\delta_2 + \frac{1}{t^2}\delta_1\}]\, m_0^2 \qquad (1)$$

where $t = \tan\beta$ and $D_0 \cong 1 - (m_t/200\sin\beta)^2$. Thus, different choices of the signs of δ_i effect the size of μ^2.

While the MAP and Planck Survey or satellites will determine the different types of matter in the universe to high accuracy, the current measurements still have large errors. In the following, we will assume $0.05 \leq \Omega_{CDM}h^2 \leq 0.30$ which extends over the current range of measurements.

The $\tilde{\chi}_1^0$ created at the time of the Big Bang annihilate in the early universe through s-channel Z and h (Higgs) poles and t-channel squark and slepton poles. For light $\tilde{\chi}_1^0$ i.e. $m_{\chi_1^0} \lesssim (50-55)\,GeV$, rapid annihilation through the former process allow (and sometimes require) m_0 to become large so that the lower bound of $\Omega_{\chi_1^0}h^2$ is not violated. However, for $m_{\chi_1^0} \gtrsim (50-55)\,GeV$, the s-channel poles give small contributions and m_0 must be small, i.e. $m_0 \lesssim 150\,GeV$ in order that the t-channel processes give sufficient annihilation so that the upper bound of $\Omega_{\chi_1^0}h^2$ not be violated.

The above results affect what may be expected in detection experiments of local $\tilde{\chi}_1^0$ dark matter in the halo of the Milky Way. We consider here direct detection of $\tilde{\chi}_1^0$ from their scattering by quarks in a nuclear target. The $\tilde{\chi}_1^0 - q$ cross section $\sigma_{\chi_1^0-q}$ proceeds through s-channel squark poles and t-channel Z and h, H^0 (Higgs) poles. For $m_{\chi_1^0} \lesssim 50\,GeV$ we saw that m_0 could be large, making the squarks heavy, and hence $\sigma_{\chi_1^0-q}$ is small. As $m_{\chi_1^0}$ rises past $55\,GeV$, m_0 becomes small and $\sigma_{\chi_1^0-q}$ rises. In general, $\sigma_{\chi_1^0-q}$ will subsequently decrease as $m_{\chi_1^0}$ becomes heavier. Thus, we expect that $\sigma_{\chi_1^0-q}$ will rise to a maximum at $m_{\chi_1^0} \cong (50-55)\,GeV$, and then gradually become smaller for large $m_{\chi_1^0}$, a behavior that we will see explicitly below.

We also note here that $\sigma_{\chi_1^0-p}$ is sensitive to two other SUSY parameters. Thus, $\sigma_{\chi_1^0-p}$ increases as $\tan\beta$ increases and increases as μ^2 decreases.

As the Earth goes around the Sun (with velocity 30 km/s) and the Sun moves around the Galaxy (with velocity 232 km/s), the Earth's relative motion to dark matter in the Galaxy will modulate annually about 7% with a period of one year, reaching a maximum at June 2, and a minimum at December 1. One expects therefore a corresponding change in the scattering event rates seen by a terrestial detector. The DAMA NaI detector in the Gran Sasso National Laboratory has seen indications of such an effect based on 19,511 kg days of data [3]. Based on this data, the DAMA experiment find a wimp mass of $M_w = (59^{+17}_{-14})\,GeV$ and proton cross section of $\xi\sigma_{w-p} = (7.0^{+.04}_{-1.2}) \times 10^{-6}\,pb$ where $\xi = \rho_{\ell oc}/(0.3\,GeV\,cm^{-3})$ and $\rho_{\ell oc}$ is the local Milky Way wimp density. Current estimates are $0.7 \leq \xi \leq 2.3$.

We first compare these experimental results with what is theoretically expected from mSUGRA models for m_0, $m_{\tilde{g}}$ (gluino mass) $\leq 1\,TeV$, $\mid A_t/m_0 \mid \leq 7$. Fig. 1 shows the maximum and minimum cross sections for three values of $\tan\beta$. The maximum values of $\sigma_{\chi_1^0-p}$ generally coincides with the maximum allowed $\tan\beta$ (the minimum curves for each $\tan\beta$ choice are the same). We see, as expected from the discussions above, that the theoretical curves peak around $m_{\chi_1^0} \cong 55\,GeV$, and then gradually fall off. Fig. 1 shows that the DAMA data is consistent with mSUGRA expectations for $\tan\beta \gtrsim 8$.

As discussed above, one expects $\sigma_{\chi_1^0-p}$ to increase as μ^2 decreases. From Eq. (4), one sees that this can occur for non-universal soft breaking with the choices δ_1, δ_3, $\delta_4 < 0$, $\delta_2 > 0$. This can be seen in Fig. 2. The $\tan\beta = 10$ maximum curve of Fig. 2 lies significantly above the $\tan\beta = 10$ (dot-dashed) curve of Fig. 1. For non-universal soft breaking, theory is consistent with the DAMA data for $\tan\beta \gtrsim 6$.

3 CONCLUSIONS

SUGRA models with R-parity invariance can treat cosmology and accelerator physics in a unified way, and dark matter detection experiments are now beginning to achieve a sensitivity

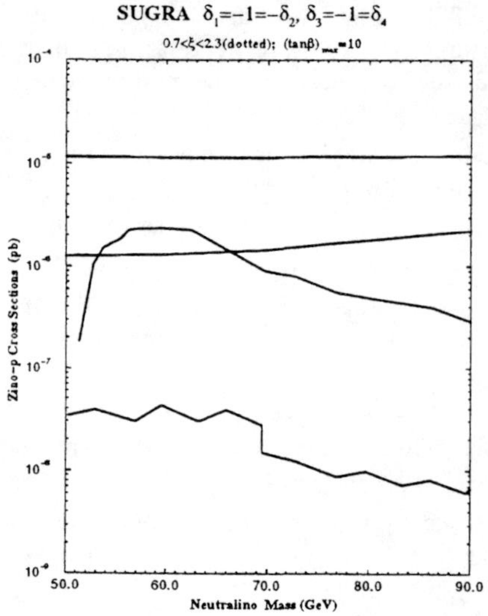

Figure 2: Maximum and minimum $\sigma_{\chi_1^0-p}$ for the case $\delta_1 = -1 = -\delta_2$, $\delta_3 = -1 = \delta_4$ with $\tan\beta \leq 10$. The horizontal curves are as in Fig. 1.

that could influence what one might expect at accelerators. While the DAMA data is yet to be confirmed, we have seen that SUGRA models can be consistent with such results. Should it turn out that dark matter detectors show that $m_{\chi_1^0} \simeq 60\,GeV$, then one would expect approximately that $m_{\chi_2^0} \cong m_{\chi_1^\pm} \simeq (110-120)\,GeV$, $m_h \cong (100-120)\,GeV$, $m_{\tilde{g}} \simeq (400-450)\,GeV$, and $\tan\beta \gtrsim 6-8$. Thus a good deal of the SUSY particle spectrum would be approximately determined, and these predictions could be tested at the upgraded Tevatron and the LHC.

Acknowledgements

This research was supported in part by NSF grants PHY-9722090 and PHY-96020274.

References

[1] For a review see M. Turner, Astro-ph/9901109 (1999).

[2] A. H. Chamseddine, R. Arnowitt and P. Nath, Phys. Rev. Lett. **49**, 970 (1982). For reviews see P. Nath, R. Arnowitt and A. H. Chamseddine, Applied N = 1 Supergravity (World Scientific, Singapore 1984); H. P. Nilles, Phys. Rep. **110**, 1 (1984); R. Arnowitt and P. Nath, Proc. of VII J. A. Swieca Summer School (World Scientific, Singapore 1994).

[3] R. Bernabei et al., INFN/AE-98/34 (1998).

[4] P. Nath and R. Arnowitt, Phys. Rev. **D56**, 2820 (1997).

Relic Neutralinos and Dark Matter [1]

Nicolao Fornengo

Dipartimento di Fisica Teorica, Università di Torino
and INFN - Sezione di Torino, via P. Giuria 1, 10125 Torino, Italy
E-mail: fornengo@to.infn.it

Abstract. We discuss the relevance for supersymmetric theories of the further indication of a possible modulation effect in WIMP direct detection recently reported by the DAMA/NaI Collaboration. The main results of our analysis show that the annual modulation data are widely compatible with an interpretation in terms of a relic neutralino as the major component of dark matter in the Universe, both in the MSSM and in SUGRA schemes. We also discuss the ensuing implications for searches of supersymmetry at accelerators and for indirect searches of neutralino dark matter.

Direct detection of relic WIMPs has a typical signature which consists in the annual modulation of the detection rate [1]. This effect is a consequence of the orbital motion of the Earth around the Sun, which induces a time dependence in the differential detection rate.

Over the last year, the DAMA/NaI Collaboration reported on the analyses of their data collected during two periods of data taking [2,3]. The data have been analysed by employing a maximum likelihood method, which allows to test the hypothesis of the presence of a yearly modulated signal against a time–independent background, by properly considering the energy and time behaviour expected for a recoil of massive WIMPs. By combining the data of the two periods of data taking, the hypothesis of presence of modulation is statistically favoured at 99.6% C.L., and pins down a 2–σ C.L. region in the plane $\xi\sigma_{\rm scalar}^{\rm (nucleon)}$ - m_χ, where m_χ is the WIMP mass, $\sigma_{\rm scalar}^{\rm (nucleon)}$ is the WIMP–nucleon scalar elastic cross section and $\xi = \rho_\chi/\rho_l$ is the fractional amount of local WIMP density ρ_χ with respect to the total local dark matter density ρ_l. This region is shown in Fig.1(a) as a closed dashed curve, for $\rho_l = 0.3$ GeV cm^{-3}. These results refer to a halo model with $v_{\rm rms} = 270$ Km s^{-1}, $v_{\rm esc} = 650$ Km s^{-1} and $v_\odot = 232$ Km s^{-1} [4]. A variation of these velocity parameters inside their 2–σ ranges of uncertainty affects the maximum likelihood value of m_χ and $\xi\sigma_{\rm scalar}^{\rm (nucleon)}$ by roughly $\pm$ 30% and by $\sim \pm$ 12%, respectively.

In this note, which is based on Refs. [4,5], we derive the theoretical implications of the experimental data of Refs. [2,3], assuming that the indication of the possible annual modulation is due to relic neutralinos. We select and discuss the explorability at accelerators of the relevant susy configurations in two different frameworks: the low energy minimal supersymmetric extension of the standard model (MSSM)

[1] Report on the work done in collaboration with A. Bottino, F. Donato and S. Scopel.

CP478, *COSMO—98*, edited by David O. Caldwell

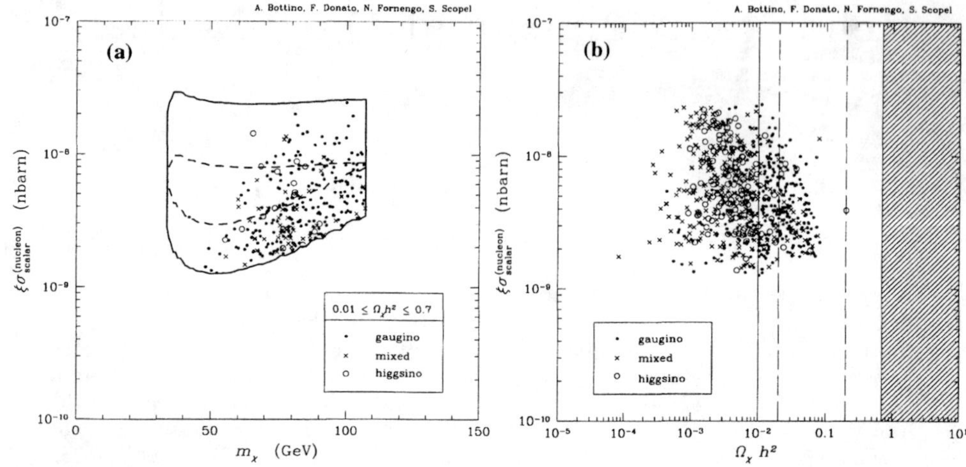

FIGURE 1. (a) Cosmologically relevant (i.e. $0.01 \leq \Omega_\chi h^2 \leq 0.7$) **MSSM** configurations compatible with the annual modulation data in the plane m_χ–$\xi\sigma_{\rm scalar}^{\rm (nucleon)}$. The dashed contour line delimits the 2–σ C.L. region obtained by the DAMA/NaI Collaboration for $\rho_l = 0.3$ GeV cm^{-3}. The solid contour line is obtained from the dashed line by accounting for the uncertainty range of ρ_l; (b) **MSSM** configurations compatible with the annual modulation data in the plane $\Omega_\chi h^2$ – $\xi\sigma_{\rm scalar}^{\rm (nucleon)}$. The two solid lines delimit the $\Omega_\chi h^2$–range of cosmological interest. The two dashed lines delimit the most appealing interval for $\Omega_\chi h^2$. The hatched area is excluded by cosmology.

and supergravity–inspired schemes (SUGRA models), both with strict unification at the GUT scale and with possible deviations from universality in the Higgs sector. For other discussions on the interpretations of the DAMA/NaI modulation data, see Ref. [6].

In order to discuss which sector of the susy parameter space is selected by the DAMA data, we first enlarge the region delimited by the 2–σ C.L. dashed contour line of Fig.1(a) in order to account for the uncertainty in the value of ρ_l: 0.1 GeV cm^{-3} $\leq \rho_l \leq$ 0.7 GeV cm^{-3}. The susy configurations are then selected by the requirement that m_χ and $\xi\sigma_{\rm scalar}^{\rm (nucleon)}$, evaluated in the MSSM and SUGRA schemes, lie inside the ensuing region, delimited by a solid contour in Fig. 1(a). The quantity ξ has been related, as usually, to the neutralino relic abundance $\Omega_\chi h^2$ as [4]: $\xi = \rho_\chi/\rho_l$. Here $(\Omega h^2)_{\rm min} = 0.01$ [4].

By varying the susy parameters inside the ranges reported in Refs. [4], we find that a large portion of the modulation region is covered by susy configurations, both in the MSSM and in SUGRA schemes. One example is shown if Fig. 1(a) for the MSSM and for a cosmologically relevant neutralino. This figure shows that *the annual modulation region is largely compatible with a relic neutralino as the major component of dark matter.*

The neutralino relic abundance is plotted versus $\xi\sigma_{\rm scalar}^{\rm (nucleon)}$ in Fig. 1(b), for the MSSM. We notice that a large fraction of the neutralino relic abundance falls into the more restricted range $0.02 \lesssim \Omega_{\rm CDM}h^2 \lesssim 0.2$, quite appealing as indicated from

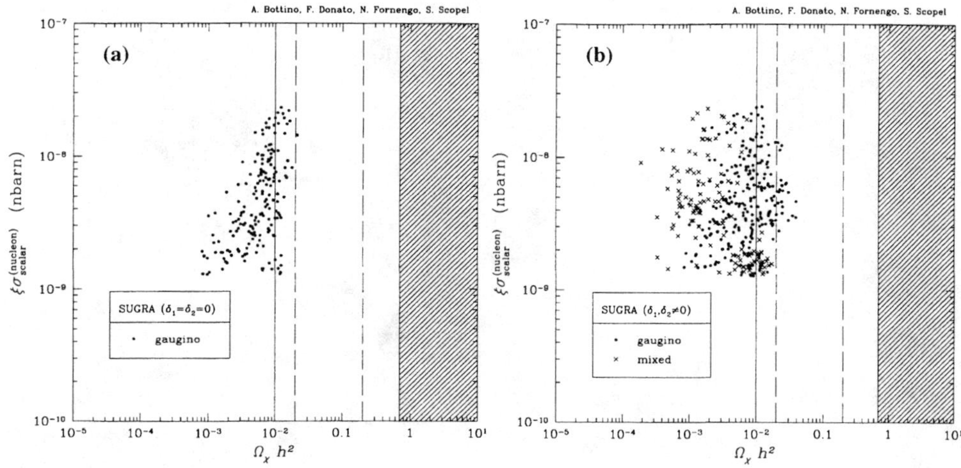

FIGURE 2. Scatter plot of **SUGRA** configurations compatible with the annual modulation data in the plane $\Omega_\chi h^2 - \xi\sigma_{\rm scalar}^{\rm (nucleon)}$. (a): universal SUGRA models. (b): SUGRA models with deviations from universality in the Higgs sector. Notations are as in Fig. 1(b).

recent analyses on the value of the matter content of the Universe [4].

In the case of SUGRA models, the situation is shown in Fig. 2, where we notice that slightly lower values of $\Omega_\chi h^2$ are compatible with the annual modulation data. Nevertheless, also in SUGRA schemes, a neutralino which could account for the major component of CDM is compatible with the annual modulation data.

Supersymmetric configurations compatible with the annual modulation data can be probed by searches at accelerators. This result is shown in Fig.3. Section (a) shows the relevant configurations in terms of m_h and $\tan\beta$. The scatter plot refers to the MSSM, while the SUGRA configurations fall inside the closed thick solid lines. In the case of MSSM, a correlation between $\tan\beta$ and m_h is quite apparent. In the case of SUGRA models, only large values of $\tan\beta$ ($\tan\beta \gtrsim 10$) are selected (for a discussion of these points, see Ref. [4]). From Fig.3(a) we notice that a good number of susy configurations are explorable at LEP2, while others will require experimental investigation at a high luminosity Fermilab Tevatron. In Fig.3(b) we display the same configurations in the plane $m_\chi - \tan\beta$. Since the reach of LEP2 extends only up to the dashed vertical line, at $m_\chi \simeq 50$ GeV, the exploration of the whole interesting region will require Tevatron upgrades or LHC.

CONCLUSIONS

In this note, we have analysed in terms of relic neutralinos the total sample of new and former DAMA/NaI data [2,3], which provide the indication of a possible annual modulation effect in the rate for WIMP direct detection. The remarkable result of our analysis is that *the annual modulation data are widely compatible with a relic neutralino making up the major part of dark matter in the Universe*, both in the low energy MSSM and in SUGRA schemes.

We have also investigated the possibility of exploring at accelerators the susy

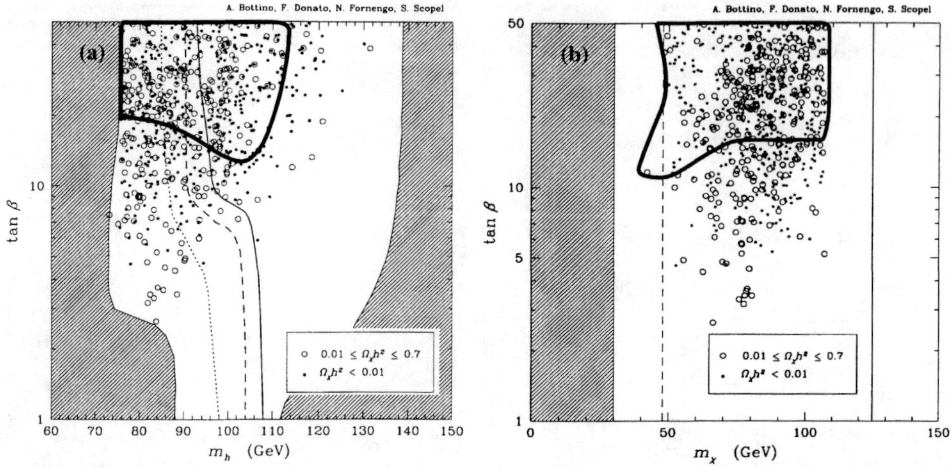

FIGURE 3. Explorability at accelerators of the **MSSM** (scatter plot) and **SUGRA** (inside the closed thick solid line) configurations compatible with the annual modulation data. (a) Plane m_h – $\tan\beta$. The hatched region on the left (right) is excluded by LEP183 (by theory). The dotted and the dashed curves are the reach of LEP2 at 192 GeV and 200 GeV, respectively. The solid line is the 95% C.L. bound reachable at LEP2, in case of non discovery of a neutral Higgs boson. (b) Plane m_χ – $\tan\beta$. The hatched region on the left is excluded by LEP. The dashed and the solid lines denote the reach of LEP2 and TeV33, respectively.

configurations which are compatible with the annual modulation data. We have shown that an analysis of the main features of these susy configurations is within the reach of present or planned experimental set–ups. The same configurations can also be largely probed by means of WIMP indirect detection, namely a neutrino flux from neutralino annihilation in the Earth or the Sun and the cosmic ray antiproton flux due to neutralino annihilation in the galactic halo. For these analyses we refer to Ref. [5].

REFERENCES

1. A.K. Drukier, K. Freese and D.N. Spergel, *Phys. Rev.* D **33**, 3495 (1986); K. Freese, J. Frieman and A. Gould, *Phys. Rev.* D **37**, 3388 (1988).
2. R. Bernabei et al., *Phys. Lett.* B **424**, 195 (1998).
3. R. Bernabei et al., preprint ROM2F/98/27, August 1998; preprint ROM2F/98/34 and INFN/AE-98/20, August 1998, http://mercury.lngs.infn.it/lngs/preprint/preprint.html.
4. A. Bottino, F. Donato, N. Fornengo and S. Scopel, hep–ph/9808456, to appear in *Phys. Rev.* D; and hep–ph/9808459, to appear in *Phys. Rev.* D.
5. A. Bottino, F. Donato, N. Fornengo and S. Scopel, hep–ph/9809239, to appear in *Astropart. Phys.*
6. R. Arnowitt, these Proceedings.

Lifetime Limit of LSP from Cosmological Light Elements[1]

Jihn E. Kim

School of Physics, Korea Institute for Advanced Study, 207-43 Cheongryangri-dong, Seoul 130-012, Korea, and Department of Physics, Seoul National University, Seoul 151-742, Korea

Abstract. From the present cosmic abundance of the light elements, one can obtain a lifetime bounds of LSP. From the consideration of deuterium, we obtain $\tau_\chi \geq 10^6$ s [1].

INTRODUCTION

More than 20 years have passed since the Lee-Weinberg bound on the stable heavy neutrino was proposed [2]. It applies to an absolutely stable particle which needs a conserved quantum number. The decaying particle cosmology was propsed around the same time [3]. But the decaying particle cosmology has been extensively applied for the gravitino [4]. In this talk, we focus on the decay of the lightest supersymmetric particle (LSP), χ. For χ to decay, the R-parity must be broken.

In supergravity, there can be nonrenormalizable interactions for the R-parity violation, but we neglect these compared to the renormalizable ones. There can be also R-violating bilinear terms, of the form $L_i H_2$, which can be rotated away at the superpotential level. However, in the presence of soft terms some bilinear terms cannot be rotated away. But for decay processes, this subtle point is not important. Thus we consider the R-violating trilinear terms,

$$W = \frac{1}{2}\lambda_{ijk} L^i L^j E^{ck} + \lambda'_{ijk} L^i Q^j D^{ck} + \frac{1}{2}\lambda'' D^{ci} D^{cj} D^{ck} \tag{1}$$

which are allowed by (i) supersymmetry and (ii) gauge symmetry. Therefore, it is reasonable to consider the phenomenology of R-parity violation. With the above superpotential present, the lepton number and/or baryon number are broken. That is the reason for requiring the $R = (-1)^{3B+L+2S}$ conservation. Anyway, with the R-parity violation, we expect neutrino oscillation, proton decay, and other abnormal

[1] This work is supported in part by KOSEF, MOE through BSRI 98-2468, and Korea Research Foundation.

CP478, *COSMO—98*, edited by David O. Caldwell

processes. From laboratory experiments, the single bounds on the couplings are not very strong, e.g. $\lambda_{121} < 0.05, \lambda_{122} < 0.05, \lambda'_{111} < 0.001, \lambda'_{112} < 0.02, \lambda''_{112} < 10^{-6}$, and $\lambda''_{113} < 10^{-5}$. However, a combined bound from lower limit of the proton decay lifetime is very strong

$$\lambda'\lambda'' < 10^{-24}. \tag{2}$$

If a universal strength for the R-violating couplings are assumed, then these couplings are very small. For example, if a singlet scalar field ϕ carrying odd R-parity develops a VEV, $<\phi>\equiv \epsilon M_P$, then the R-parity is spontaneously broken. At low energy, these effects are represented by λ, λ', and λ'' couplings which carry negative R-parity. These may arise from nonrenormalizable interactions containing ϕ field. In this case, these couplings can be of the same order.

Thus, to allow reasonably large R-violating couplings, one usually fobid λ' or λ'' completely.

In general, laboratory experiments give upper bounds on the couplings. However, the cosmological bounds depend on the region of couplings. It is schematically shown in Fig. 1. In this talk, I am interested in the region $\tau_\chi \sim 10^{-6}$ s.

THE LIFETIME OF LSP

The neutralino is a mixture of the neutral gauginos and neutral Higgsinos,

$$\chi = N_1\tilde{B} + N_2\tilde{W} + N_3\tilde{H}_1^0 + N_4\tilde{H}_2^0. \tag{3}$$

The decay can occur, for example, through $\chi \to Q+\tilde{q}$ where the virtual $\tilde{q}$ transforms to $L + D^c$. For this process, the interaction is given by

$$-\frac{\lambda'_{ijk}}{M_{\tilde{f}}^2}\left(gN_2 + \frac{1}{3}g_1N_1\right)\int d^4\theta U^{aj}\chi E^i D_a^{ck} + \text{h.c.} \tag{4}$$

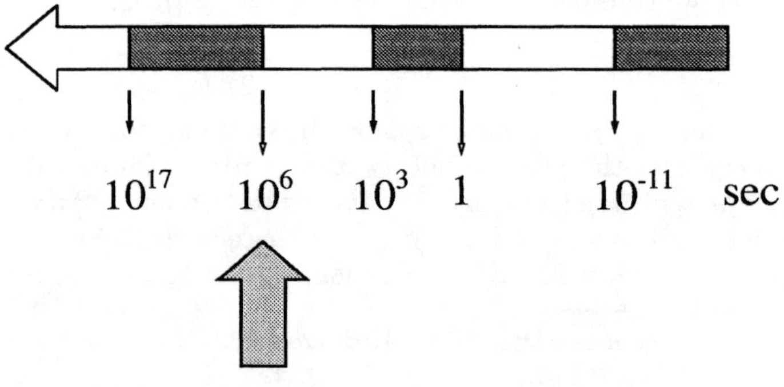

FIGURE 1. A schematic view of allowed windows for the χ lifetime. Our interest is near the point marked by the grey upward arrow.

For the $\chi \to u^c + d + e^+$ decay, there exist 9 diagrams. The possible diagrams increase very rapidly if one considers colored final states. Kinematically excluding tt^c final states, there are 1,080 diagrams. Thus, we assume the *universal sfermion mass* to simplify the expression. Then the χ decay rate is given by

$$\Gamma_{\text{tot}} = \frac{g^2}{256\pi^3}\frac{m_\chi^5}{M_{\tilde{f}}^4} \cdot \{\sum |\lambda_{ijk}|^2 \left(\tfrac{1}{8}N_2^2 + \tfrac{3}{8}N_1^2 \tan^2\theta_W\right) + \sum_{j\neq 3} |\lambda'_{ijk}|^2 \left(\tfrac{3}{4}N_2^2 + \tfrac{7}{12}N_1^2 \tan^2\theta_W\right)$$
$$+ \sum_{ik} |\lambda'_{i3k}|^2 \left(\tfrac{3}{8}N_2^2 + \tfrac{7}{24}N_1^2 \tan^2\theta_W - \tfrac{1}{2}N_1 N_2 \tan\theta_W\right) + \sum_{k\neq 3} |\lambda''_{ijk}|^2 N_1^2 \tan^2\theta_W\} \quad (5)$$

where the overall coefficient in front of the curly bracket is $1.8 \times 10^{15}\ \text{s}^{-1}$ for $M_{\tilde{f}} = 1$ TeV and $m_\chi = 30$ GeV.

COSMOLOGICAL BOUND ON τ_χ

The decoupling temperature T_D

The last moment when the heavy particles are in equilibrium with photons is the decoupling temperature T_D. For the neutralino, it is determined by a competition between the universe expansion rate and χ destruction rate. The destruction rate is a function of sfermion mass for the t-channel process $\chi + \chi \to f + f^c$. If there exists a significant Higgsino component, i.e. $(N_1^2 + N_2^2)/|N_{3\text{ or }4}^2)| > (M_Z^2/M_{\tilde{f}}^2)^2$, then the s-channel Z exchange diagram dominates. It is the easiest way to estimate the decoupling temperature. However, we also include the sfermion exchange diagrams in the numerical estimation. Nevertheless, the decoupling temperature does not depend on these parameters very much. Let us introduce a decoupling factor d

$$d = \frac{m_\chi}{T_D}. \quad (6)$$

The decoupling factor d is given for various parameter sets in Ref. [1]. For example, $M_{\tilde{f}} = 300$ GeV, $m_\chi = 60$ GeV, $\tan\beta = 10$, and the gaugino dominated neutralino ($N_1 = 0.9 - 1$, $N_2 = 0.1 - 0.5$, and $N_3 = N_4 = 0$) give $d = 21 - 22$. For a wide range of parameters, d is in the range 15–30.

In particular, we note the following. Higgsino dominated χ gives a little bit smaller decoupling temperature and hence a lower χ density. The t and u channel processes are important for $N_{3,4} > 0.1$ if $M_{\tilde{f}} > 300$ GeV. However, the decoupling temperature is rather insensitive to $\tan\beta$. In the numerical analyses, we further assumed the relations $m_{A,h_1,h_2}/m_\chi = 1/3$ and $\Gamma_{A,h_1,h_2} = 1/500$. We note, however, that the cross section and the decoupling temperatures are not sensitive to these assumptions.

In the next section, we are interested in relatively light neutralino case, which excludes the possibility of producing t, W, Z, H in the $\chi\chi$ annihilation.

The χ lifetime bound

If χ decays after 1 s, it affects the nucleosynthesis in a way of destructing the already manufactured light elements. Note that our χ lifetime is $10^{-15}/|\lambda|^2$ s and the cosmic time scale in the radiation dominated era is $t = 0.3N^{-1/2}M_P/T^2$.

Requiring that $E_\chi <$ (energy of relativistic particles), we obtain $\tau_\chi < 4.8 \times 10^{10}$ (GeV$/m_\chi)^2$ s, which gives a $|\lambda|$ bound of order 10^{-13}.

A more stringent bound comes from the dissociation of light elements, in particular the D dissociation [4,5]. The χ decay products degrade very rapidly by scattering with background photons. We are interested in the photons with $E > 2.225$ MeV. But e^+e^- production from scattering on the background radiation turned out to be important also [5]. In this analysis, another important temperature parameter T_* is introduced. At T_*, the rates for the Compton scattering and the e^+e^- production are comparable. The critical photon energy E_* corresponding to T_* is given by $E_*T_* = (1/50)$ MeV^2. Then we note that 10^{-9} (with respect to the photon number) photons with $\omega > 25T_*$ can be used to produce e^+e^-. In this case, the e^+e^- scatter off the background radiation to transfer the energy to photons which can dissociate D. Therefore, the maximum allowable lifetime $\tau_{\rm max}$ is determined by the condition that if it decayed later than $\tau_{\rm max}$, the probability for photon to scatter with e is negligible, namely it mostly scatter off with D to dissociate it. Therefore, χ must decay before $\tau_{\rm max}$. The equation for $\tau_{\rm max}$ is

$$2\frac{n_\chi}{n_e}\frac{\epsilon}{\epsilon_0}\int \sum(\epsilon_0, T)e^{-t/\tau_{\rm max}}\frac{dt}{\tau_{\rm max}} = 1 \tag{7}$$

where

$$\sum(\epsilon, T) = \int_{Q_D}^{\omega_1}\frac{\sigma_D}{\sigma_{KN}+\sigma_{pp}}2\left(1-\frac{\omega}{\omega_m}\right)\frac{d\omega}{\omega_m}. \tag{8}$$

Here ϵ is the photon energy in the fraction of χ mass, ω is the photon frequency, σ_D, σ_{KN}, and σ_{pp} are the deuteron dissociation cross section, Klein-Nishina cross section, and pp production cross section, respectively. The maximum scattered photon energy is $12\gamma T$.

We calculated $\tau_{\rm max}$ from the above formula to obtain

$$\tau_{\rm max} = (2.2 - 2.5) \times 10^6 \text{ s}. \tag{9}$$

For $\delta_B \simeq 5\times 10^{-10}$ and 60 GeV photino-like (bino-like) neutralino, the combination of R-violating couplings is bounded as

$$\begin{aligned}&\sum 0.12(0.05)|\lambda_{ijk}|^2 + \sum 0.31(0.07)|\lambda'_{ijk}(j\neq 3)|^2 + \sum 0.04(0.04)|\lambda'_{i3k}|^2\\ &+\sum 0.23(0.12)|\lambda''_{ijk}(i<j, k\neq 3)|^2 > 7.7\times 10^{-24}\end{aligned} \tag{10}$$

CONCLUSION

Thus the sum of the R-violating couplings is bounded from below to a few times 10^{-12}, the details of which depends on the nature of the neutralino. We studied the lifetime region around 10^6 s in Fig. 1. Of course, our bound is most meaningful if the R-violating couplings are of the same order. It is interesting to note that there is a allowed lifetime window between 10^3 s and 10^6 s.

REFERENCES

1. J. E. Kim, B. Kyae and J. D. Park, preprint hep-ph/9810503.
2. B. W. Lee and S. Weinberg, Phys. Rev. Lett. **39**, 165 (1977).
3. D. A. Dicus, E. W. Kolb and V. L. Teplitz, Phys. Rev. Lett. **39**, 168 (1977) and **39**, 973 (1997)(E).
4. J. Ellis, J. E. Kim and D. V. Nanopoulos, Phys. Lett. **B145**, 181 (1984); J. Ellis, G. B. Gelmini, J. L. Lopez, D. V. Nanopoulos, and S. Sarkar, Nucl. Phys. **B373**, 399 (1992).
5. D. Lindley, Astrophys. J. **294**, 1 (1985). See, also, S. Sarkar, Rep. Prog. Phys. **59**, 1493 (1996).

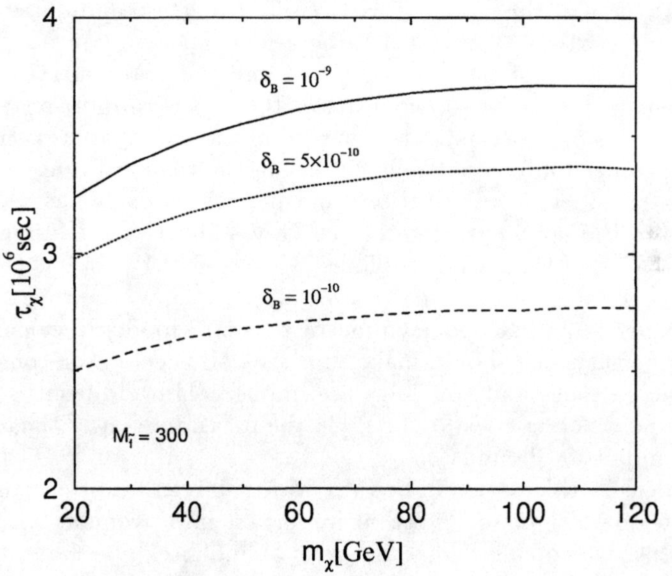

FIGURE 2. The maximum allowable lifetime of χ for $M_{\tilde{f}} = 300$ GeV, $N_1 = 0.99$, $N_2 = 0.14$, and $N_3 = N_4 = 0$.

Indirect Detection of Neutralino Dark Matter

L. Bergström

Department of Physics, Stockholm University, Box 6730, SE-113 85 Stockholm, Sweden
E-mail: lbe@physto.se

Abstract. Dark matter detection experiments are getting ever closer to the sensitivity needed to detect the primary particle physics candidates for nonbaryonic dark matter. In fact, some weak indications of possibly exotic signals already exist. Indirect detection methods include searching for antimatter and gamma rays, in particular gamma ray lines, in cosmic rays and high-energy neutrinos from the centre of the Earth or Sun caused by accretion and annihilation of dark matter particles. A review is given of recent progress, both on the theoretical and experimental sides.

I INTRODUCTION

The mystery of the dark matter is one of the outstanding questions in standard cosmology [1]. One of the favoured particle dark matter candidates is the lightest supersymmetric particle χ, assumed to be a neutralino, i.e. a mixture of the supersymmetric partners of the photon, the Z and the two neutral CP-even Higgs bosons present in the minimal extension of the supersymmetric standard model. The attractiveness of this candidate stems from the fact that its generic couplings and mass range naturally give a relic density in the required range to explain halo dark matter. Besides, its motivation from particle physics has recently become stronger due to the apparent need for 100 GeV - 10 TeV scale supersymmetry to achieve unification of the gauge couplings in view of LEP results. (For a thorough review of supersymmetric dark matter, see [2].)

When using the minimal supersymmetric standard model in calculations of relic dark matter density, one should make sure that all accelerator constraints on supersymmetric particles and couplings are imposed. In addition to significant restrictions on parameters given by LEP [3], the measurement of the $b \to s\gamma$ process is providing important bounds.

The relic density calculation in the MSSM for a given set of parameters is nowadays accurate to 10 % or so. A recent important improvement is the inclusion of coannihilations, which can change the relic abundance by a large factor in some instances [4].

CP478 , *COSMO—98*, edited by David O. Caldwell

II INDIRECT DETECTION TECHNIQUES

In principle, all stable particles produced in annihilation processes in the halo can serve as signatures of a neutralino dark matter candidate. However, electrons and protons are much too abundant in the ordinary cosmic rays to be useful. Much lower background fluxes, and therefore greater potential for detection of an additional component, are present in the case of positrons and antiprotons. The problem is still that the signal does not have a distinct spectral signature (and magnetic fields in the galaxy cause a diffusion and complete isotropisation of the interstellar flux). The featureless energy distribution of positrons and antiprotons is due to the fact that the primary annihilation processes are into quarks, heavy leptons, gauge bosons and Higgs particles, whereas the positrons and antiprotons are secondary or tertiary decay products. The reason for the small direct coupling to electron-positron pairs is the Majorana nature of the neutralino coupled with the fact that the annihilation takes place essentially at rest. Selection rules then force the coupling to fermions to be proportional to the fermion mass.

There has recently been a new balloon-borne detection experiment [6], with increased sensitivity to eventual positrons from neutralino annihilation, and an excess of positrons over that expected from ordinary sources has been found. However, since there are many other possibilities to create positrons by astrophysical sources, e.g. in nearby supernova remnants, the interpretation is not yet conclusive. However, a new theoretical calculation has shown that the experimental situation is compatible with the presence of a neutralino-induced component, if the dark matter is clumpy so as to enhance the signal [8]. (Remember that such a non-homogeneous distribution of dark matter is in fact what is expected in many cold dark matter formation scenarios [7].)

Antiprotons could for some supersymmetric parameters constitute a useful signal [9], but even with the upcoming space experiments [10,11] it will be quite difficult to disentangle a low-energy signal from the smooth cosmic-ray induced background. For kinematical reasons, antiprotons created by pair-production in cosmic ray collisions with interstellar gas and dust are born with relatively high energy, whereas antiprotons from neutralino annihilation populate also the sub-100 MeV energy band. A problem that plagues estimates of the signal strength of both positrons and antiprotons is, however, the uncertainty of the galactic propagation model and solar wind modulation. Also, secondary $\bar{p}$ interactions in the interstellar medium and $\bar{p}$ production on helium nuclei also give low-energy antiprotons which could mask an eventual signal [15].

Many supersymmetric models, compatible with all accelerator constraints, give predictions for the antiproton flux [12,21] which are in agreement with very recent data from the BESS collaboration [14]. The question remains, however, whether or not cosmic-ray induced secondary antiprotons can fully explain the results. The most complete theoretical calculations [15,16] indicate that the band of uncertainties of the conventional flux is large enough to encompass the new data. This is shown in Fig. 1 (a), where the present BESS data [14] are compared to the default background prediction of [15]. In Fig. 1 (b) a 24 % reduction of the secondary flux (well within the band of model uncertainties) is combined with the contribution from one of the supersymmetric models of [15] to give an equally good fit to the data.

Obviously, it will not be easy to tell these two cases apart, even with the high-quality data which can soon be expected [10].

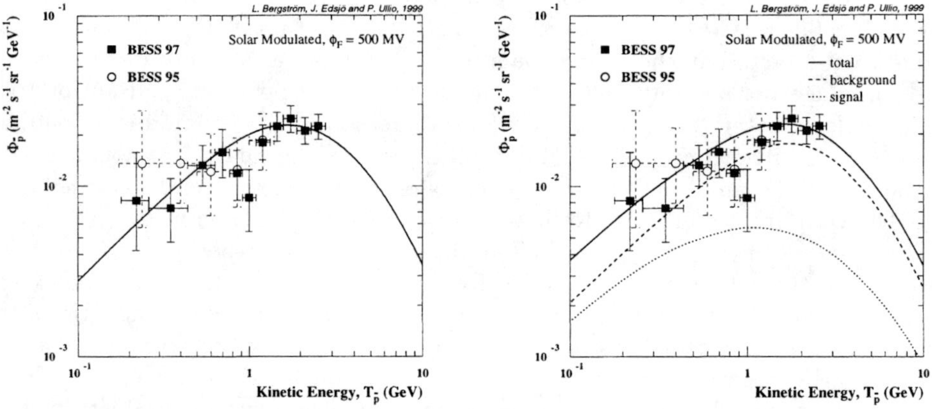

FIGURE 1. (a) Cosmic-ray antiproton flux as a function of $\bar{p}$ kinetic energy. Shown are the BESS 1995 and 1997 data points [14]. The solid line represents the default cosmic-ray induced background flux according to the analysis in [15]. (b) Example of a composite spectrum consisting of the reference background $\bar{p}$ flux in (a) reduced by 24 % with the addition of the predicted flux from annihilating dark matter neutralinos of one of the MSSM models of [15].

Even allowing for large systematic effects, the measured antiproton flux gives, however, rather stringent limits on the lifetime of hypothetical R-parity violating decaying neutralinos [17].

A Methods with distinct experimental signature

With these problems of positrons and antiprotons, one would expect that problems of gamma rays and neutrinos are similar, if they only arise from secondary decays in the annihilation process. For instance, the gamma ray spectrum arising from the fragmentation of fermion and gauge boson final states is quite featureless and gives the bulk of the gammas at low energy where the cosmic gamma ray background is severe. Also, the density of neutralinos in the halo is not large enough to give a measurable flux of secondary neutrinos, unless the dark matter halo is very clumpy. However, neutrinos can escape from the centre of the Sun or Earth, where neutralinos may have been gravitationally trapped and therefore their density enhanced. Also, gamma rays may result from loop-induced annihilations [18,23] $\chi\chi \to \gamma\gamma$ or $\chi\chi \to Z\gamma$.

The rates of these processes are difficult to estimate because of uncertainties in the supersymmetric parameters, cross sections and halo density profile. However, in contrast to the other proposed detection methods they have the virtue of giving very distinct, "smoking gun" signals: high-energy neutrinos from the centre of the Earth or Sun, or monoenergetic photons with $E_\gamma = m_\chi$ or $E_\gamma = m_\chi(1 - m_Z^2/4m_\chi^2)$ from the halo. In fact, also continuum gammas may give a good signature, since the average angular distribution should follow the dark matter distribution of the halo, in contrast to the background (cosmic-ray induced gamma ray caused mainly by protons hitting interstellar hydrogen) which should emanate mainly from the disk. This possibility has received increased attention recently due to some indications from EGRET data [19] that an excess of GeV-scale gamma-rays appear to come from the halo and could be caused by neutralino annihilation [13,21,22]. However, since there are other explanations possible (such as π^0 production from cosmic rays having escaped the disk), it is plausible that other independent evidence (such as spectral features - ideally a line) has to be provided before a detection of a dark matter-induced signal can be claimed.

B Gamma ray lines

The detection probability of a gamma line signal depends on the very poorly known density profile of the dark matter halo.

To illustrate this point, let us consider the characteristic angular dependence of the gamma-ray intensity from neutralino annihilation in the galactic halo. Annihilation of neutralinos in an isothermal halo with core radius a leads to a gamma-ray flux of

$$\frac{d\mathcal{F}}{d\Omega} \simeq (2 \times 10^{-11}\mathrm{cm}^{-2}\mathrm{s}^{-1}\mathrm{sr}^{-1}) \times \frac{(\sigma_{\gamma\gamma}v)_{29}(\rho_\chi^{0.3})^2}{(m_\chi/\,10\,\mathrm{GeV})^2}\left(\frac{R}{8.5\ \mathrm{kpc}}\right)J(\Psi) \tag{1}$$

where $(\sigma_{\gamma\gamma}v)_{29}$ is the annihilation rate in units of $10^{-29}\,\mathrm{cm}^3\,\mathrm{s}^{-1}$, $\rho_\chi^{0.3}$ is the local neutralino halo density in units of 0.3 GeV cm^{-3} and R is the distance to the galactic center. The integral $J(\Psi)$ is given by

$$J(\Psi) = \frac{1}{R\rho_0^2}\int_{\mathrm{line-of-sight}} \rho^2(\ell)d\ell(\Psi), \tag{2}$$

and is evidently very sensitive to local density variations along the line-of-sight path of integration.

We remind of the fact that since the neutralino velocities in the halo are of the order of 10^{-3} of the velocity of light, the annihilation can be considered to be at rest. The resulting gamma ray spectrum is a line at $E_\gamma = m_\chi$ of relative linewidth 10^{-3} which in favourable cases will stand out against background. The process $\chi\chi \to Z\gamma$ is treated analogously and has a similar rate [23].

To compute $J(\Psi)$, a model of the dark matter halo has to be chosen. Recently, N-body simulations have given a clue to the final halo profile obtained by hierarchical clustering in a CDM scenario [24]. It turns out that the universal halo profile found in these simulations has a rather significant enhancement $\propto 1/r$ near the halo centre. If applicable to the Milky Way, this would lead to a much enhanced annihilation rate towards the galactic centre, and also to a very characteristic angular dependence of the line signal. This would be very beneficial when discriminating against the galactic and extragalactic γ ray background, and Air Cherenkov Telescopes (ACTs) would be eminently suited to look for these signals, if the energy resolution is at the $10 - 20$ % level.

In Fig. 2, we show the gamma ray line flux given in a scan of supersymmetric models consistent will all experimental bounds (including $b \to s\gamma$), assuming an effective value of 10^3 for the average of $J(\Psi)$ over the 10^{-3} steradians that typically an Air Cherenkov Telescope (ACT) would cover. (See [20] for details.)

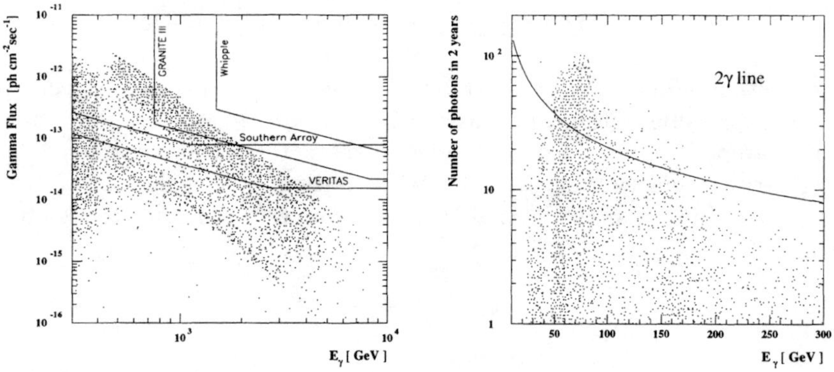

FIGURE 2. Results for the gamma ray line flux in an extensive scan of supersymmetric parameter space in the MSSM [20]. Shown in the figure to the left is the number of events versus photon energy in an Air Cherenkov Telescope of area $5 \cdot 10^4$ m^2 viewing the galactic centre for one year. The halo profile of [24] for the dark matter has been assumed. In the figure to the right the 5-σ discovery limit of the GLAST satellite is shown assuming a 2-year exposure and the same halo parameters.

It can be seen that the models which give the highest rates should be within reach of the new generation of ACTs presently being constructed. These will have an effective area of almost 10^5 m^2, a threshold of some tens of GeV and an energy resolution approaching 10 %. In favourable cases, especially at the low m_χ end, also a smaller area detector with better energy resolution and wider angular acceptance such as the proposed GLAST satellite [25] could reach discovery potential, as also shown in the Figure.

C Indirect detection through neutrinos

Another promising indirect detection method is to use neutrinos from annihilations of neutralinos accumulated in the centre of the Sun or Earth. This will be a field of extensive experimental investigations in view of the new neutrino telescopes (AMANDA, BAIKAL, NESTOR, ANTARES) in operation or under construction.

To illustrate the potential of neutrino telescopes for discovery of dark matter through neutrinos from the Earth or Sun, we present the results of a full calculation [26]. In Fig. 3 it can be seen that a neutrino telescope of area around 1 km^2, which is a size currently being discussed, would have discovery potential for a range of supersymmetric models.

If a signal were established, one can use the angular spread caused by the radial distribution of neutralinos (in the Earth) and by the energy-dependent mismatch between the direction of the muon and that of the neutrino (for both the Sun and the Earth) to get a rather good estimate of the neutralino mass [28]. If muon energy can also be measured, one can do even better [29].

It should again be noted that all indirect detection signals depend strongly, and in partly different ways, on the distribution of dark matter in the galactic halo. This was investigated in detail in [13], where also a comparison between the various methods (including direct detection) was performed.

Acknowledgements

I wish to thank my collaborators, in particular Joakim Edsjö, Paolo Gondolo and Piero Ullio, for many helpful discussions. This work has been supported in part by the Swedish Natural Science Research Council (NFR).

REFERENCES

1. L. Bergström and A. Goobar, *Cosmology and Particle Astrophysics*, Wiley/Praxis (Chichester), 1999.
2. G. Jungman, M. Kamionkowski and K. Griest, Phys. Rep. **267**, 195 (1996).
3. See, e.g., P. Abreu et al. (DELPHI Collaboration), Phys. Rev. Lett. **71**, 674 (1993); Phys. Rev. Lett. **74**, 2885 (1995).
4. J. Edsjö and P. Gondolo, Phys. Rev. D **56**, 1879 (1997).
5. A. Bottino et al., Phys. Lett. B **402**, 113 (1997).
6. S.W. Barwick et al., Phys. Rev. Lett. **75**, 390 (1995); Astrophys. J. **482**, L191 (1997).
7. B. Moore et al., Ap. J. Lett. **499**, L5 (1998).
8. E.A. Baltz and J. Edsjö, Phys. Rev. D **59**, 023511, (1999).
9. G. Jungman and M.Kamionkowski, Phys. Rev. D **49**, (1994) 2316; A. Bottino, C. Favero, N. Fornengo, G. Mignola, Astropart. Phys. **3**, 77 (1995); A. Bottino, F. Donato, N. Fornengo and P. Salati, Phys. Rev. D **58**, 123503 (1998).
10. AMS experiment, S. Ahlen et al., Nucl. Instrum. Meth. **A350**, 351 (1994).

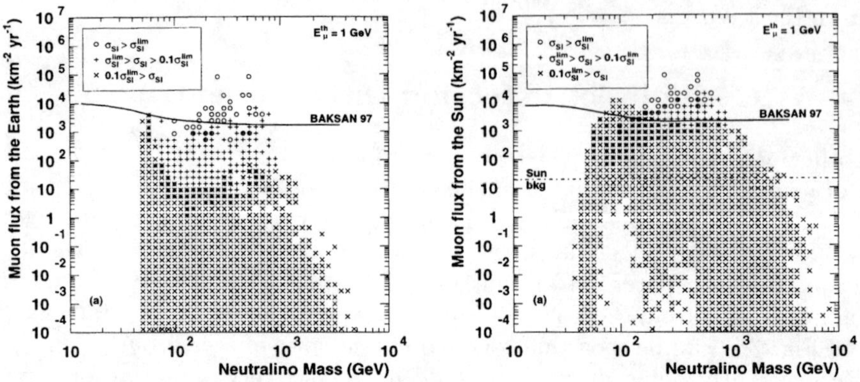

FIGURE 3. The indirect detection rates from neutralino annihilations in the Earth (left) and the Sun (right) versus the neutralino mass. The horizontal line is the Baksan limit [27]. For details, see [26].

11. O. Adriani et al. (PAMELA Collaboration) LNF-95-051 (1995).
12. A. Bottino et al., hep-ph/9809239.
13. L. Bergström, J. Edsjö, P. Gondolo and P. Ullio, Phys. Rev. D **59**, 043506 (1999).
14. S. Orito, talk given at International Conference on High-Energy Physics, Vancouver, 1998.
15. L. Bergström, J, Edsjö and P. Ullio, astro-ph/9902012 (1999).
16. M. Simon, A. Molnar and S. Roesler, Ap. J. **499**, 250 (1998).
17. E.A. Baltz and P. Gondolo, Phys. Rev. D **57**, 2969 (1998).
18. L. Bergström and H. Snellman, Phys. Rev. D **37**, (1988) 3737; A. Bouquet, P. Salati and J. Silk, Phys. Rev. D **40**, 3168 (1989); G. Jungman and M. Kamionkowski, Phys. Rev. D **51**, 3121 (1995); L. Bergström and P. Ullio, Nucl. Phys. B **504**, 27 (1997).
19. D. Dixon et al., New Astron. **3**, 539 (1998).
20. L. Bergström, J. Buckley and P. Ullio, Astropart. Phys. **9**, 137 (1998).
21. L. Bergström, J. Edsjö and P. Ullio, Phys. Rev. D **58**, 083507 (1998).
22. P. Gondolo, astro-ph/9807347.
23. L. Bergström and J. Kaplan, Astropart, Phys. **2**, 261 (1994); P. Ullio and L. Bergström, Phys. Rev. D. **57**, 1962 (1998).
24. J.F. Navarro, C.S. Frenk and S.D.M. White, Astrophys. J. **462**, 563 (1996).
25. GLAST Collaboration, http://www.glast.stanford.edu.
26. L. Bergström, J. Edsjö and P. Gondolo, Phys. Rev. D **55**, 1765 (1997); Phys. Rev. D **58**, 103519 (1998).
27. M.M. Boliev et al (Baksan telescope), Nucl. Phys. B (Proc. Suppl.) **48**, 83 (1996), TAUP 95, Toledo, Spain, 17–21 September 1996, Editor: A. Morales.
28. J. Edsjö and P. Gondolo, Phys. Lett. B **357**, (1995) 595.
29. L.Bergström, J.Edsjö and M.Kamionkowski, Astropart. Phys. **7**, 147 (1997).

Indirect Detection of Neutralino Annihilation from Three-body Channels

Xuelei Chen and Marc Kamionkowski

Department of Physics, Columbia University,
538 West 120th Street New York, NY 10027, USA

I INTRODUCTION

The neutralino is one of the most promising dark matter candidates [1]. It could be detected indirectly by observation of energetic neutrinos from neutralino annihilation in the Sun and/or Earth. Energetic neutrinos are produced by decays of neutralino annihilation products. These neutrinos can be detected by neutrino detectors such as AMANDA and super-Kamiokande [2], which observe the upward muons produced by the charged-current interactions in the rock below the detector. For the models which can be tested by the current or next generation of detectors, an equilibrium of accumulation and annihilation is reached, so the annihilation rate is determined by the capture rate in the Sun or Earth. The event rate is proportional to the second moment of the neutrino energy spectrum, so it is this neutrino energy moment weighted by the corresponding branching ratio that determines the detection rate.

By now, the cross sections for annihilation have been calculated for all two-body final states that arise at tree level. Roughly speaking, among the two-body channels the $b\bar{b}$ and $\tau^+\tau^-$ final states usually dominate for $m_\chi < m_W$. Neutralinos that are mostly higgsino annihilate primarily to gauge bosons if $m_\chi > m_W$, because there is no s-wave suppression mechanism for this channel. Neutralinos that are mostly gaugino continue to annihilate primarily to $b\bar{b}$ pairs until the neutralino mass exceeds the top-quark mass, after which the $t\bar{t}$ final state dominates, as the cross section for annihilation to fermions is proportional to the square of the fermion mass.

Three-body final states arise only at higher order in perturbation theory and are therefore usually negligible. However, some two-body channels easily dominate the cross section when they are open because of their large couplings; for example the W^+W^- for the higgsinos and $t\bar{t}$ for gauginos. This suggests that their corresponding three-body final states can be important just below these thresholds. Moreover, the neutrinos produced in these three-body final states are generally much more energetic than those produced in b and τ decays, Recently, we calculated the s-wave cross section for the processes $\chi\chi \to W^+W^{-*} \to Wf\bar{f}'$ and $\chi\chi \to t\bar{t}^* \to tW^-\bar{b}$,

CP478 , *COSMO—98*, edited by David O. Caldwell

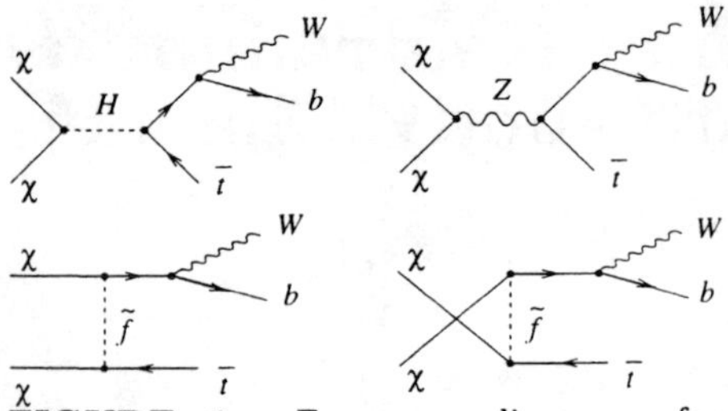

FIGURE 1. Feynman diagrams for $\chi\chi \to tt^*$.

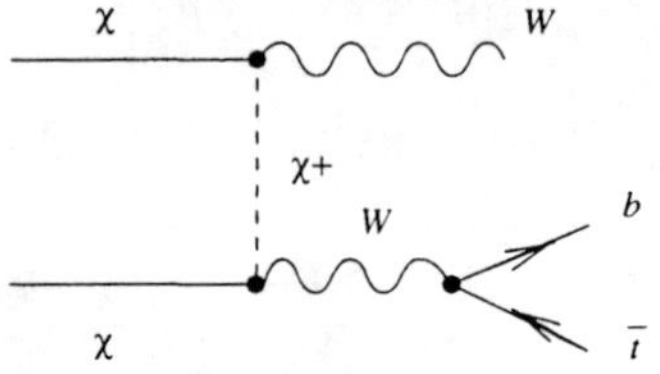

FIGURE 2. neglected diagrams

and their charge conjugates in the $v_{\rm rel} \to 0$ limit [3].

II CALCULATION

Below the top pair threshold, the neutralino may annihilate via a virtual top quark: $\chi\chi \to tt^* \to tWb$. The Feynman diagrams for this process are shown in Fig. 1. Like annihilation to $t\bar{t}$ pairs, annihilation to this three-body final state takes place via s-channel exchange of Z^0 and A^0 (pseudo-scalar Higgs bosons) and t- and u-channel exchange of squarks. Although there are additional diagrams for this process, such as those shown in Fig. 2, these are negligible for the gaugino because of the small coupling. On the other hand, for higgsinos the gauge boson channel dominates and the tt^* would be unimportant anyway. In the $v_{\rm rel} \to 0$ limit

$$\sigma v_{\rm rel} = \frac{N_c}{128\pi^3} \int_{x_{6min}}^{x_{6max}} dx_6 \int_{x_{4min}}^{x_{4max}} dx_4 \frac{1}{4} |\mathcal{M}|^2 , \tag{1}$$

the amplitude is given by $M = M_t - M_u + M_s$.

The three-body cross section for a series of typical models are shown in Fig. 3. In these models the neutralino is primarily gaugino. As expected, the three-body cross section approaches the two-body value above the top threshold (we take m_t = 180 GeV). Below the top mass it is non-zero but drops quickly. The flux of upward muons is proportional to the second moment of the neutrino energy spectrum weighted by branching ratios, which is given by

$$B_F \langle Nz^2 \rangle = \frac{3}{128\pi^3} \int dx_4 dx_6 |\mathcal{M}|^2 \left(\langle Nz^2 \rangle_t x_4^2 + \langle Nz^2 \rangle_W x_5^2 + \langle Nz^2 \rangle_b x_6^2 \right), \tag{2}$$

in the three-body case. Although the three-body cross section is small except just below the $t\bar{t}$ threshold, its contribution to the the second moment, $B_F \langle Nz^2 \rangle$, may be important, as illustrated in Fig. 4. This is because the $\langle Nz^2 \rangle$ for top quarks and W bosons is significantly larger than that for the light fermions [4].

Below the W^+W^- threshold, the neutralino can annihilate to a real W and a virtual W. The W-bosons then decay independently into a fermion pair $f\bar{f}'$, which can be $\tau\nu$, $\mu\nu$, $e\nu$, cs, or ud. About 10% of these decay into a muon (or anti-muon) and a muon anti-neutrino (or neutrino).

The WW^* calculation is similar to the tWb calculation. In the $v \to 0$ limit, only chargino exchange in the t and u channels shown in Fig. 5 are important. Neutrinos can be produced either by the virtual W^* or by decay of the real W. The neutrinos produced by decay of the muon or other fermions can be neglected, and the real W boson has a probability $\Gamma_{W\to\mu\nu}$ to decay to a muon neutrino. On the other hand this W boson is produced in all $\chi\chi \to Wf\bar{f}'$ channels, and each of these channels has approximately the same cross section, so the contribution has a weight of $n_{chan}\Gamma_{W\to\mu\nu} \approx 1$. The cross section and second moment for annihilation in the Earth are shown in Fig. 6 and Fig. 7. Below $m_W/2$, the four-body channel might become significant, and may smooth the jump in much the same fashion as the three-body channel does near the two-body channel threshold, but we will not consider it here.

III CONCLUSIONS

The contributions of these three-body channels are important only in a limited region of parameter space. However, they may produce a large effect. In fact, our calculation shows that although the cross sections of these annihilations are significant only just below the two-body channel threshold, due to the high energy of the neutrinos they produce, they can enhance the neutrino signal by many times and actually dominate the neutrino signal far below the two-body threshold. Furthermore, the regions in question may be of particular interest. For example, motivated by collider data, Kane and Wells proposed a light higgsino [5], and recent DAMA results [6] suggest a WIMP candidate with $m_x \simeq 60$ GeV (but see [7]). There are also arguments that the neutralino should be primarily gaugino with a mass

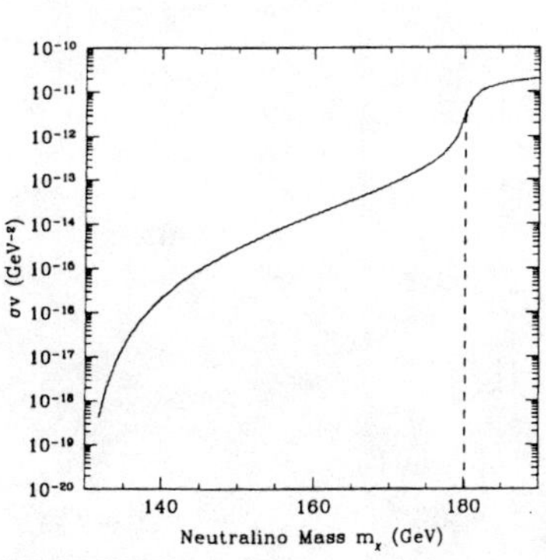

FIGURE 3. total annihilation cross section.

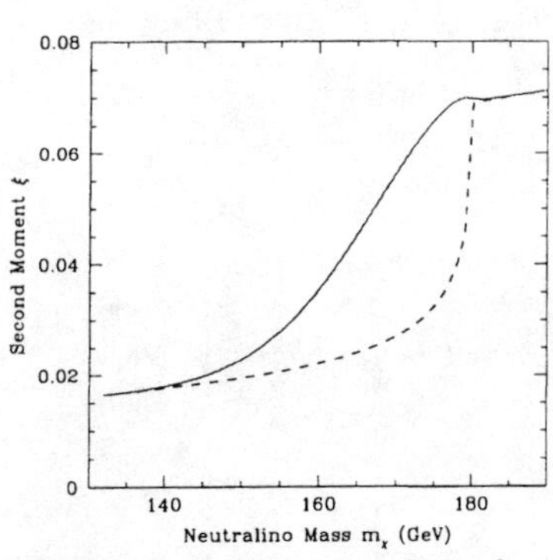

FIGURE 4. Detection Rate for neutrinos from the core of Earth.

Dashed curves show two-body only result, solid curves includes three-body contribution.

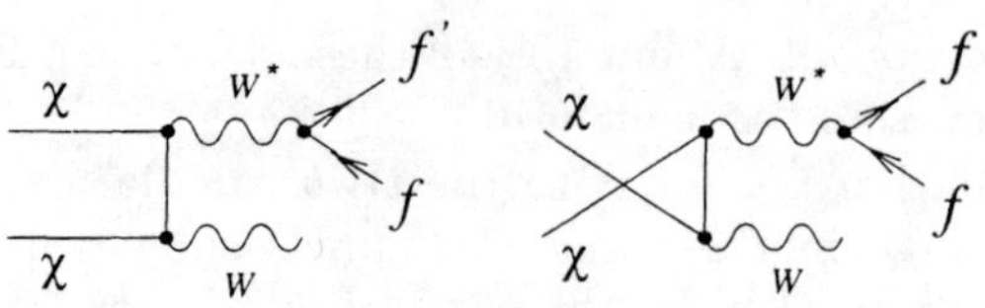

FIGURE 5. Feynman Diagrams for $\chi\chi \to WW^*$

somewhere below but near the top-quark mass [8].

There are many parameters in the minimal supersymmetric model. The results shown in Figs. 4 and 7 are of course model dependent, and these effects might be more or less important in models with different parameters.

ACKNOWLEDGMENTS

We thank G. Jungman for help with the neutdriver code. This work was supported by D.O.E. Contract No. DEFG02-92-ER 40699, NASA NAG5-3091, and the Alfred P. Sloan Foundation.

REFERENCES

1. G. Jungman, M. Kamionkowski and K. Griest, *Phys. Rept.* **267** (1996) 195.
2. For a review, see, e.g. the talks of L. Roszkowski and L. Bergstrom in this volume.
3. X. Chen and M. Kamionkowski, *J. High Energy Phys.* **07** (1998) 001.
4. G. Jungman and M. Kamionkowski, *Phys. Rev.* **D 51** (1995) 328.
5. G. L. Kane and J. D. Wells, *Phys. Rev. Lett.* **76** (1996) 4458; K. Freese and M. Kamionkowski, *Phys. Rev.* **D 55** (1997) 1771.
6. R. Bernabei et al., *Phys. Lett.* **B 424** (1998) 195.
7. J. Ellis, 1998, *this volume.*
8. L. Roszkowski, *Phys. Lett.* **B 262** (1991) 59.

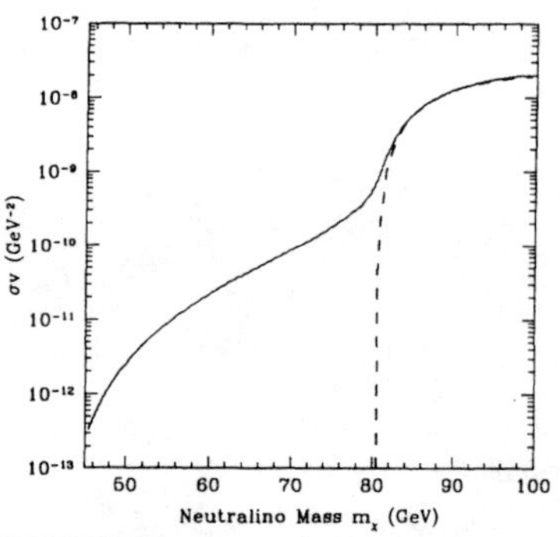

FIGURE 6. total annihilation cross section.

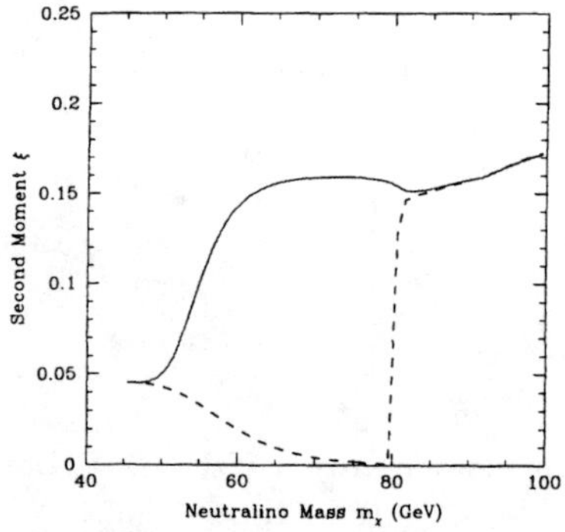

FIGURE 7. Detection Rate for neutrinos from the core of Earth.

Direct Dark Matter Searches:

WIMPS and Axions

Bernard Sadoulet

Center for Particle Astrophysics
Lawrence Berkeley National Laboratory
and Physics Department, University of California, Berkeley
Berkeley, California 94720 USA

Abstract. After briefly recalling the evidence which suggests that the dark matter pervading the universe is nonbaryonic, we review the present searches for the best motivated particle candidates: axions, light neutrinos and Weakly Interacting Massive Particles (WIMPs).

INTRODUCTION

In the last decade considerable additional evidence (1) has been gathered supporting the hypothesis that at least 90% of the mass in the universe is dark: it does not emit or absorb any form of electromagnetic radiation. Understanding this dark matter has become one of the more central problems in astronomy and cosmology. Once a subject of controversy among astronomers, its existence is now well established at a variety of scales.

The debate has shifted to measuring the amount of dark matter in the universe, studying its distribution and unraveling its nature. A central question is whether this dark matter is made of ordinary baryonic matter or is nonbaryonic. A number of cosmological observations indicate that it is probably nonbaryonic.

Searches for nonbaryonic dark matter are therefore well motivated and essential,. After a brief description of other nonbaryonic dark matter searches, we review the current direct detection efforts for Weakly Interacting Massive Particles (WIMPs).

THE CASE FOR NON BARYONIC DARK MATTER

Comparison of Ω and Ω_b

Figure 1 summarizes the current attempts of measuring the average density Ω of the universe in units of the critical density

$$\Omega = \frac{\rho}{\rho_c} \text{ with } \rho_c = 1.88 \times 10^{-26}\ h^2 \text{ kg m}^{-3}$$

CP478 , *COSMO—98*, edited by David O. Caldwell

where h is the Hubble expansion parameter in units of 100 km/s/Mpc (h = 0.65±0.1). Ω can be determined through an inventory of the masses of the various objects in the universe, for instance using the virial velocities in galaxy clusters. These techniques can give only a lower limit of Ω, since they only measure local density inhomogeneities. Dynamic methods attempt to relate the observed velocity deviations from the Hubble flow to the density concentrations and deduce an effective Ω, which unfortunately depends on how well the number density of galaxies tracks the mass density fluctuations. Cosmological tests can also be used to directly probe the geometry, but since very distant objects are used, it is difficult to correct the measured quantities for evolution. This fundamental difficulty, which foiled the earlier attempts (2), is still a cause for concern in the interpretation of high redshift supernovae(3). Taken at face value, these exciting observations indicate that the universe is accelerating. They provide an approximate measurement of the difference between the vacuum energy density and matter density, $\Omega_\Lambda - \Omega_m$. The sum of these quantities, $\Omega_\Lambda + \Omega_m$, can be obtained from the acoustic ("Doppler") peak in the microwave background power spectrum indicated by the Saskatoon and CAT data. Together these observations give $\Omega_m = 0.25$ (+0.18-0.12, 95%CL interval) (4).

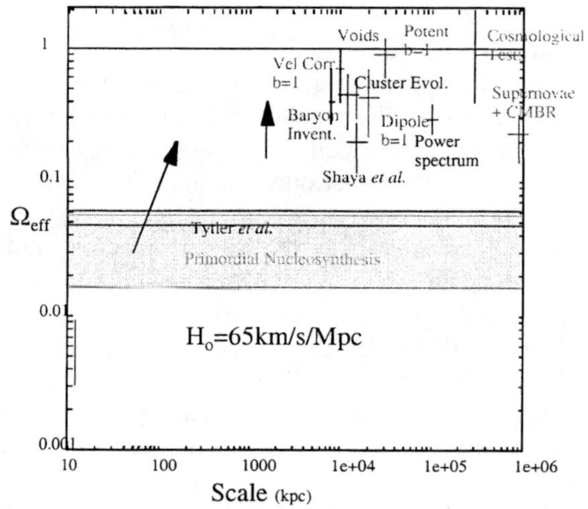

FIGURE 1. Effective Ω as a function of scale of measurement for H_0 = 65 km/s/Mpc. The bands give the Ω in baryons expected from primordial nucleosynthesis.

The combination of all observations makes a convincing case for the existence of dark matter, since the value obtained over large scales (≥ 0.3) is much greater than the contribution of stars (0.003-0.01).

The data also provides a convincing argument for the nonbaryonic nature of dark matter. The shaded band displays the narrow limits ($0.017 \leq \Omega_b \leq 0.056$) inferred from the observations of ^{4}He, D, ^{3}He and ^{7}Li in the very successful standard scenario of homogeneous primordial nucleosynthesis(5). It is definitely below most measurements of Ω at large scale. For the recent measurement of the D fraction in Lyman alpha systems by Tytler (6), Ω_b may be close to the upper boundary of this band but our conclusion remains solid.

Formation of the large-scale structure

A second argument for nonbaryonic dark matter is based on the fact that it provides the most natural explanation of the large-scale structure of the galaxies in the universe. These cold dark matter models are based on the growth through gravitational collapse of the initial density fluctuations. Such fluctuations can be inferred from the COBE measurement of the temperature fluctuations of the cosmic microwave background. The deduced power spectrum of the adiabatic mass fluctuations on very large scales connects smoothly with the galaxy power spectrum measured at smaller scales (7), giving strong evidence for the formation of the observed structure through gravitational collapse. The observed spectral shape is a natural consequence of models with cold nonbaryonic dark matter but cannot be explained with baryons only. The baryons are bound up with the photons until after recombination preventing sufficient growth of the initial fluctuations to form the structure that we see today.

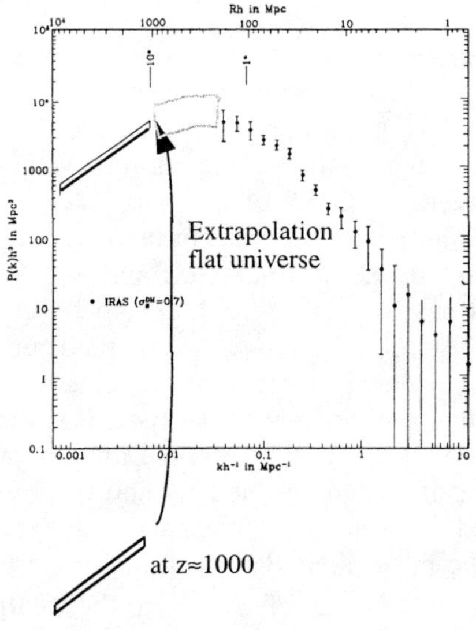

FIGURE 2. Measured power spectrum measured for IRAS galaxies and extrapolation of the COBE result assuming nonbaryonic dark matter and a flat universe (after Fisher et al., 1992). The contour in the middle gives an estimate of the power spectrum inferred from the measurement of the acoustic peak of the cosmic microwave background (after Scott et al., 1995).

Although the result is more model-dependent, the recent measurements of temperature fluctuations in the cosmic microwave background at the degree scale support this conclusion (8). These results smoothly bridge the gap between the COBE extrapolation and the large-scale structure.

Inefficiency of forming compact objects

A third general argument comes from the implausibility of hiding a large amount of baryonic matter in the form of MACHOs. For instance, since the ratio of the mass in gas and stars to the total mass in clusters is about 20%, 80% of the initial gas would have to condense into invisible MACHOs. Such a large fraction is very difficult to understand within the standard scenarios of star formation. The same argument applies to galactic halos.

Impact of the MACHO observations

The most important result from the microlensing searches for Massive Compact Halo Objects (MACHO) is the unambiguous exclusion of brown dwarfs between 10^{-7} and 0.08 solar masses as a significant fraction of our halo (See Figure 3). No short duration events are observed.

However a few long duration events are observed towards the Large Magellanic Cloud and these have sometimes been interpreted as evidence that our halo is composed mostly of MACHOs around 0.5 solar masses. This conclusion cannot unambiguously be reached because the lensing duration is a degenerate function of the mass, distance and transverse velocity of the lens. We simply do know where the lenses responsible for the observed events are. In fact, the measured location of the two observed double lenses are in the host galaxies, and the low event rate towards the Small Magellanic Cloud cast considerable doubt about the MACHOs forming the majority of our halo (9) (Figure 3). Moreover, a halo made up of 0.5 solar mass objects would not be dark unless they are primordial black holes.

Finally, if a cold nonbaryonic component exists, it is difficult to prevent it from accreting into galactic halos. Even in the presence of some MACHOs in the halos, it would constitute a significant portion of the halo and be present locally for detection. In fact, taking into account all kinematic information on the galaxy and the combined MACHO observations, the most likely density for a nonbaryonic component is close (10) to the canonical 0.3 GeV/cm^{3}as inferred from the rotation curves of our galaxy (11).

Conclusion

In conclusion, it is very difficult to construct a self-consistent cosmology without cold nonbaryonic dark matter, and the hypothesis that MACHOs constitutes a large fraction of our galactic halo is disfavored by the newest observations. Thus, searches for nonbaryonic dark matter have the highest scientific priority!

A large number of candidates have been proposed over the years for such a nonbaryonic component. They range from shadow universes existing in some string models, strange quark nuggets formed at a first-order quark-hadron phase transition (12), Charged Massive Particles (CHAMPs)(13), and a long list of usually massive particles with very weak interactions. We should probably search first for particles that would also solve major questions in particle physics. Using this criterion, three candidates appear particularly well motivated: axions, neutrinos and weakly interactive massive particles.

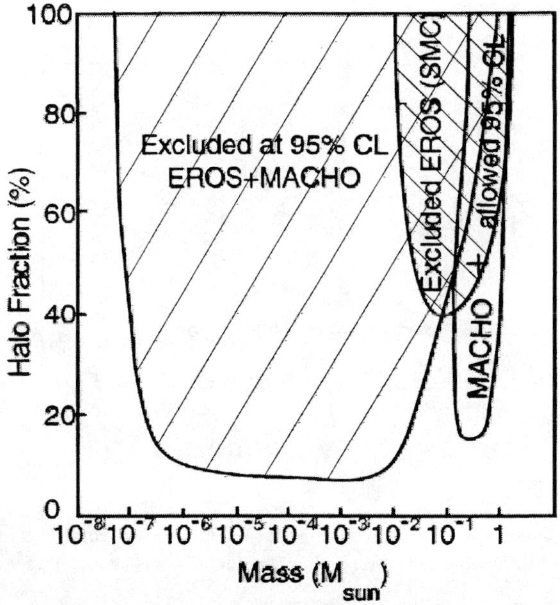

FIGURE 3. Excluded region (at 95% confidence level)of the halo fraction in MACHOs as a function of their mass in a standard halo model. The ellipse on the right is the 95% confidence level range allowed by the 2 year data of the MACHO collaboration. We display also the region excluded recently by the EROS observations of SMC (14). Note that large MACHO fractions are disfavored!

SEARCHES FOR AXIONS

Axions are an example of relic particles produced out of thermal equilibrium, a case where we depend totally on the specific model considered to predict their abundance. These particles have been postulated (15) in order to dynamically prevent the violation of CP in strong interactions in the otherwise extremely successful theory of quantum chromodynamics. Of course, there is no guarantee that such particles exist, but the present laboratory and astrophysical limits on their parameters are such that if they exist, they would form a significant portion of cold dark matter (16). Such low-mass cosmological axions could be detected by interactions with a magnetic field, which produce a faint microwave radiation detectable in a tunable cavity (17). The first two searches[18] for cosmological axions performed a decade ago were missing a factor of 1000 in sensitivity.

This is no longer the case; Livermore, MIT, Florida and Chicago are currently performing an experiment which has published preliminary limits (19) and will reach (Fig. 4) a cosmologically interesting sensitivity at least for one generic type of axion (hadronic models (20)). By replacing their HEMT-based amplifiers by SQUID amplifiers, the collaboration hopes to improve their sensitivity down to the lowest couplings currently predicted (DFZ model (21)). Matsuka and his collaborators in Kyoto are developing a more ambitious scheme using Rydberg atoms, which are very sensitive photon detectors and should be able to directly reach the DFZ limit.

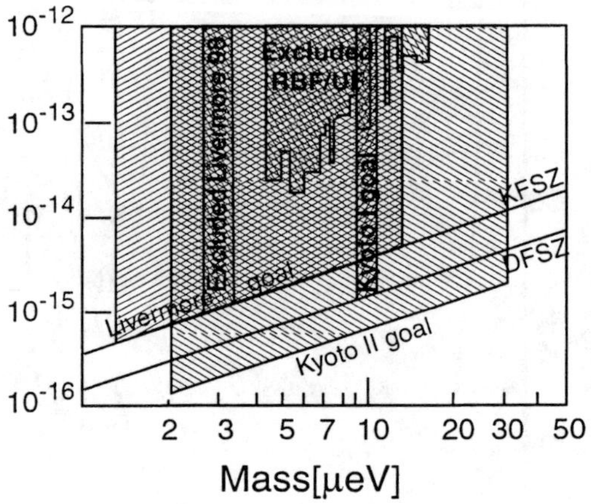

FIGURE 4. Expected sensitivity of the Livermore and Kyoto experiments. The lines labeled KSVZ and DFSZ refer to two generic species of axions. The shaded regions in the upper right are the previous experimental limits.

Although these experiments are very impressive, it should be noted that the decade of frequency (and therefore of mass) which can be explored with the present method is only one out of three which are presently allowed.

LIGHT NEUTRINOS

Neutrinos of mass much smaller than 2 MeV/c fall in the generic category of particles which have been in thermal equilibrium in the early universe and decoupled when they were relativistic. Their current number density is approximately equal to that of the photons in the universe. The relic particle density is therefore directly related to its mass, and a neutrino species of 25 eV would give an Ω of the order of unity (22). Note that neutrinos alone cannot lead to the observed large-scale structure, as fluctuations on scales greater than 40 h^{-1} Mpc are erased by neutrino streaming. They have to be mixed in with cold nonbaryonic dark matter (23) or seeded by topological defects. Moreover, because of phase space constraints, they cannot explain the dark matter halos observed around dwarf galaxies (24).

Unfortunately, no good ideas exist of possible ways to detect cosmological neutrinos (25), and one can only rely on the mass measurements of neutrinos in the laboratory through the study of beta spectra, neutrinoless double beta decay, and oscillation experiments.

The situation of the mass measurements may be summarized as follows: The direct mass measurement of the electron neutrino gives limits of 5 eV(26). Model dependent limits of the order of 1 eV on the mass of Majorana neutrinos are given by neutrinoless double beta decay searches. The claim by the LSND group (27) for muon to electron neutrino oscillation with relatively large $\Delta m^2 \approx 6$ eV^2 oscillation is now challenged by the Karmen experiment (28).

In June 1998, the SuperKamiokande group has presented statistically significant results (29) demonstrating the disappearance of atmospheric muon neutrinos which points to an oscillation with Δm^2 of a few 10^{-3} eV^2 and a large mixing angle. A muon to electron neutrino oscillation is disfavored, both by Chooz (30) and internally by SuperKamiokande.

At the date of this writing, the situation of solar neutrinos is less clear. The combination of the chlorine, water Cerenkov and gallium experiments have now indicated for some time a depletion of solar neutrinos with respect to the standard solar model (31). The most natural explanation is an MSW (32) or vacuum oscillation with Δm^2 of 10^{-6} eV^2 or 10^{-10} eV^2 respectively (33). However, the distortion of the energy distribution observed by SuperKamiokande, which would be a direct confirmation independently of the solar model, is not fully understood. The essential measurements of neutral current events by SNO and of the ^{7}Be neutrino flux depletion by Borexino will not be available before 2000 and 2001 respectively.

If the oscillation interpretation of both atmospheric and solar neutrinos is correct, we are led to nearly degenerate masses of the three types of neutrinos (unless we have a light sterile neutrino), without any direct indication of the mass which may well be in the electron volt range if we do not invoke the see-saw mechanism. It is therefore important for cosmology to keep pushing the resolution of direct electron neutrino mass measurement.

WEEKLY INTERACTING MASSIVE PARTICLES

A third generic class of candidates constitutes particles that were in thermal equilibrium in the early universe and decoupled when they were non-relativistic. In this case, it can be shown that their present density is inversely proportional to their annihilation rate (34). For these particles to have the critical density, this rate has to be roughly the value expected for weak interactions (if they have masses in the GeV/c^2 to TeV/c^2 range). This may be a precious hint that physics at the W and Z^o scale is important for the problem of dark matter. Inversely, physics at the W and Z^o scale leads naturally to particles whose relic density is close to the critical density. In order to stabilize the mass of the vector intermediate bosons, one is led to assume the existence of new families of particles such as supersymmetry in the 100 GeV mass range. In particular, the lightest supersymmetric particle could well constitute the dark matter. This class of particles is usually called Weakly Interactive Massive Particles (WIMPs).

The most direct method to detect these WIMPs is by elastic scattering on a suitable target in the laboratory (35). Elastic WIMP scattering would produce a roughly exponential spectrum with a mean energy dependent on their mass. The hope is to identify such a contribution in the differential energy spectrum measured by an ultra-low background detector, or at least to exclude cross sections that would lead to differential rates larger than observation.

Before reviewing the direct detection of WIMPs, let us note that several methods have been proposed for detecting WIMPs through their annihilation products(36).They of course assume dark matter exists in the form of both particles and antiparticles (or is self-conjugate) as otherwise no annihilation would occur. The detection of gamma ray lines from their annihilation into two photons(37) will require the resolution of the next generation of satellites and may be masked by the galactic background, especially if the dark matter density does not strongly peak at the galactic center. The first measurements of the energy spectra of antiprotons and anti-electrons offered tantalizing hints of dark matter particle annihilations(38), but they turned out to be inaccurate. The interpretation of such spectra would in any case be unclear because of the uncertainty in the confinement time of these antiparticles in the halo of our galaxy.

A much more promising method (39,40,41,42,43) is to search for high energy neutrinos coming from the centers of the earth and the sun. Since they can lose energy by elastic interactions, some dark matter particles would be captured by these objects, settle in their centers and annihilate with each other producing, among other products, high energy neutrinos which can then be detected by underground detectors, especially through the muons produced by their interactions in the rock. The current generation of such detectors (Baksan, MACRO and SuperKamiokande) of roughly 1000 m^2 area, put a limit of the order of 10^{-14} muon/cm^2/s above 3 GeV. Such results exclude any charge-symmetric Dirac neutrino or scalar sneutrino and put limits on supersymmetric models which are generally in agreement with but less restrictive than direct detection experiments. Fairly model-independent arguments (44) show that such an advantage of direct detection should be maintained for the next generation of detectors (cryogenic WIMP searches and 10^4 m^2 detectors such as AMANDA II), especially for scalar interactions. However, the very large neutrino detectors currently being studied (10^6 m^2) may be more sensitive than direct searches for large-mass WIMPs (45).

Direct searches for Weakly Interactive Massive Particles

We now review the status of the various efforts of direct detection of WIMP dark matter.

In specific models such as supersymmetry, the knowledge of the order of magnitude of the annihilation cross section allows an estimation of their elastic scattering, taking into account the coherence over the nucleus. Typically, if scalar (or "spin independent") couplings dominate, the interaction rate of WIMPs from the halo is expected to be of the order of a few events per kilogram of target per week for large nuclei like germanium. Note that these rates depend on the local dark matter density in the halo and are insensitive to the global value of Ω_m. We display in Figure 5, as the lower shaded region, the range of cross sections (rescaled to a proton target) expected (46) in grand unified theory inspired supersymmetric models, where scalar interactions usually dominate.

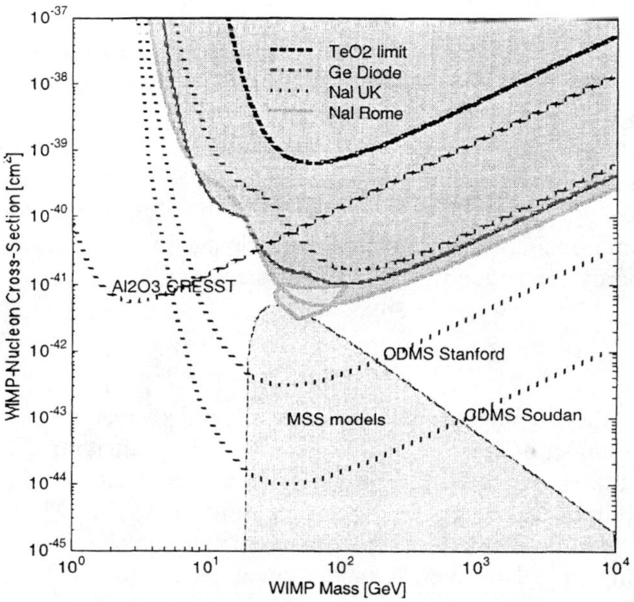

FIGURE 5. Current achieved limits for spin independent couplings as a function of the WIMP mass. This figure includes the results of the Rome (47) NaI, UK (48) NaI, Milan (49) TeO_2, Modane (50) Al_2O_3, and the Ge diode experiments: PNL-USC (51), Oroville (52), Neuchâtel-Caltech (53), Heidelberg-Moscow (54), and IGEX (55). All the results have been converted to WIMP-nucleon cross sections assuming scalar interactions scaling as the square of the atomic number. The shaded region at the top is excluded by these experiments. The heart shaped in the middle is the region corresponding to the modulation signal claimed by Bernabei et al (56). The shaded region at the bottom is the rate predicted by minimal supersymmetric models including the constraints from LEP and CDF. We also indicate the sensitivity projections of the current CRESST experiment, CDMS I at Stanford, and CDMS II in Soudan.

The upper shaded regions summarize the current limits achieved with state of the art techniques for low radioactivity background. These limits barely skirt the supersymmetric region, although relaxing the unification assumptions enlarges it somewhat (57).

Unfortunately, the expected rates can be very small for specific combinations of parameters where axial ("spin dependent") couplings dominate. In this case, the interaction takes place with the spin of the nucleus, which limits the number of possible targets, and the current limits are very far above the supersymmetry expectation (46).

It is therefore essential to construct experiments with very low radioactive backgrounds or, even better, with active background rejection. The main tool for this purpose is to use the fact that WIMP interactions produce nuclear recoils, while the radioactive background is dominated by electron recoils (if neutrons are eliminated).

A second challenge faced by the experimentalist comes from the fact that the energy deposition is quite small, typically 10 keV for the mass range of interest. For detectors based only on ionization or scintillation light, this difficulty is compounded by the fact that the nuclear recoils are much less efficient in ionizing or giving light than electrons of the same energy. This increases the recoil energy threshold of such detectors, and one should be careful to distinguish between true and "electron-equivalent" energy, which may differ by a factor three (Ge) to twelve (I).

A third challenge is to find convincing signatures linking detected events to particles in the halo of the galaxy. The best one would be the measurement of the direction of the scattered nucleus (58), a very difficult task. Short of this directionality signature, it is in principle possible to look for a change in the event rate and the spectrum of energy deposition with the time of the year (59).

Prominent direct search strategies

In spite of these experimental challenges, low expected rates and low energy depositions, a number of experimental teams are actively attempting to directly detect WIMPs. A number of interesting attempts have been made to use mica which integrates for billions of years (60), superheated microdots (61) which should be only sensitive to nuclear recoil, and low pressure time projection chambers which could give the directionality (62). However, the main developments occurred along three main experimental strategies.

1. A first approach is to attempt to decrease the radioactive background as much as possible. Germanium is the detector of choice as it is very pure, and the first limits (51,52,53) were obtained by decreasing the threshold of double beta experiments. The most impressive results have been obtained by the Heidelberg-Moscow group (54)with a background of 0.05 events/kg/day/electron-equivalent-keV around 20 keV (equivalent electron energy). This impressive performance comes from a careful screening of surrounding material, the large size of their crystal (2.5 kg), and signal shape discrimination. The IGEX and Baksan-USC-PNL (63) collaborations have achieved somewhat worse levels (0.25 events /kg/day/electron-equivalent-keV), but reached lower thresholds. The current combined exclusion plot is given in Fig. 5.

GENIUS, an ambitious proposal (64) to immerse one ton of germanium detectors in an ultra-pure liquid nitrogen bath, pushes this strategy to the extreme.

However, this approach is fundamentally limited by the absence of discrimination against the radioactive background. Not only can this background not be partially rejected, but it also cannot be measured independently of the signal (except by multiple scattering) and subtracted. Once the background level is measured with sufficient statistical accuracy, the sensitivity of the experiment does not improve with exposure. In contrast, the combination of an active background rejection and subtraction allows a sensitivity increase as the square root of the target mass and the running time, until the subtraction becomes limited by systematics (65). Note that with large setups like GENIUS, some discrimination will be obtained against gamma's through multiple scattering, but it is difficult to arrange self vetoing with a rejection efficiency greater than 95%.

2. A second approach has been to use large scintillators with pulse shape discrimination of nuclear and electronic recoils. The technique is simple and large masses can be assembled to search for modulation effects. The most impressive result so far has been obtained with NaI. The NaI groups (47,48) have published limits that claim to be slightly better than those obtained with conventional germanium detectors. However, these limits remain controversial, as they may not fully take into account systematics in the efficiency close to the threshold or in the rejection power from pulse shape discrimination. In any case, because sodium has a spin, these experiments so far give the best limits for spin dependent couplings. The Rome group has recently announced (56) a nearly three _ detection of a signal using the annual modulation expected for a WIMP spectrum (heart-shaped region in Fig. 5). This modulation signal represents less than 1% of the observed background and it is not yet clear that the systematics have been controlled at the required level. Overall, it is unlikely that NaI could make significant additional progress, as the small number of photoelectrons at the energies of interest and the lack of power of the pulse shape discrimination make it highly susceptible to systematics.

3. Therefore, more powerful discrimination methods need to be devised. Liquid xenon with simultaneous measurement of scintillation and ionization is a promising approach, albeit with relatively high thresholds, and not enough development so far to fully judge its potential. In contrast, the active development of novel "cryogenic" detectors based on the detection of phonons produced by particle interactions is beginning to bear fruit. In spite of the complexity of the very low temperature operation, four large setups are currently being routinely operated (Milano (49), CDMS (66) CRESST (67) and EDELWEISS (68)), with total detector mass ranging from 70 g to 7 kg.

For dark matter searches, this technology appears to have three advantages:

• It can lead to a much smaller threshold, as phonons measure the total energy of nuclear recoil without any loss. Already the performance of thermal phonon detectors in the laboratory exceeds that of ionization detectors. CDMS (69) are now routinely getting a resolution of better than 900 eV and 450 eV FWHM in phonons and ionization respectively with 165 g detectors. The CRESST group has also demonstrated a FWHM of 235 eV at 1.5 keV in a 250 g crystal of sapphire. Four of these detectors are now installed in the CRESST experiment, which hopes to obtain without discrimination the limits shown in Fig. 1.4. CDMS (70) have recently shown that it is even possible to detect athermal phonons after very few bounces on the surface and get similar baseline resolution (1 keV).

• With the simultaneous measurement (71) of ionization and phonons in crystals of germanium or silicon, it is possible to distinguish between nuclear recoils and electron recoils. This approach is used by both the CDMS and the EDELWEISS collaborations. CDMS has demonstrated greater than 99% rejection with thermal and athermal phonon plus ionization technology down to 20 keV recoil energy (Figure 6 a and b). This allows us to reach at a shallower site an effective gamma contamination better than Heidelberg-Moscow, and with much lower thresholds. Unfortunately, as often in such situations, a new background was uncovered: soft electrons incident on the surface suffer from ionization losses in a micron-thick dead layer and partially simulate nuclear recoils (Figure 6c). This dead layer is due to back diffusion of the carriers and can be decreased by suitable modification of the contacts. Combining these improvements with better shielding, the CDMS team have recently been able to drastically reduce this problem and hope to reach at our present Stanford site the limit displayed in Fig. 5. The lower line indicates the expected sensitivity of the CDMS II experiment in the Soudan mine.

• A third advantage of phonon-mediated detectors is the greater amount of information obtained about very rare events. Already the simultaneous measurement of phonons and ionization gives two pieces of information instead of one, and allows a more efficient rejection of microphonics and spurious instrumental effects. The detailed measurement of out-of-equilibrium phonons is even more promising. CDMS has recently demonstrated that geometrical fiducial cuts can be imposed using the phonon information and that the problematic surface electrons can be eliminated by a phonon rise time cut (Figure 5 d). In the long run, athermal phonons may allow a determination of the directionality for isotopically pure targets.

CONCLUSION

The case for nonbaryonic dark matter remains very strong and it is important to aggressively continue the current particle searches. An axion experiment is underway which should give us a definite answer about axions over a mass range of one order of magnitude (out of three that are still allowed).

Oscillation neutrino experiments are in progress, which cover the mass range of cosmological interest. The WIMPs search is very active with the installation of very large NaI scintillators, liquid xenon detectors, and phonon-mediated detectors that are beginning to be operational.

The combination of all these efforts may well give us the solution of a central puzzle of cosmology and astrophysics. It may even give us important information about particle physics, with perhaps the discovery of the long sought-for axions or supersymmetric particles.

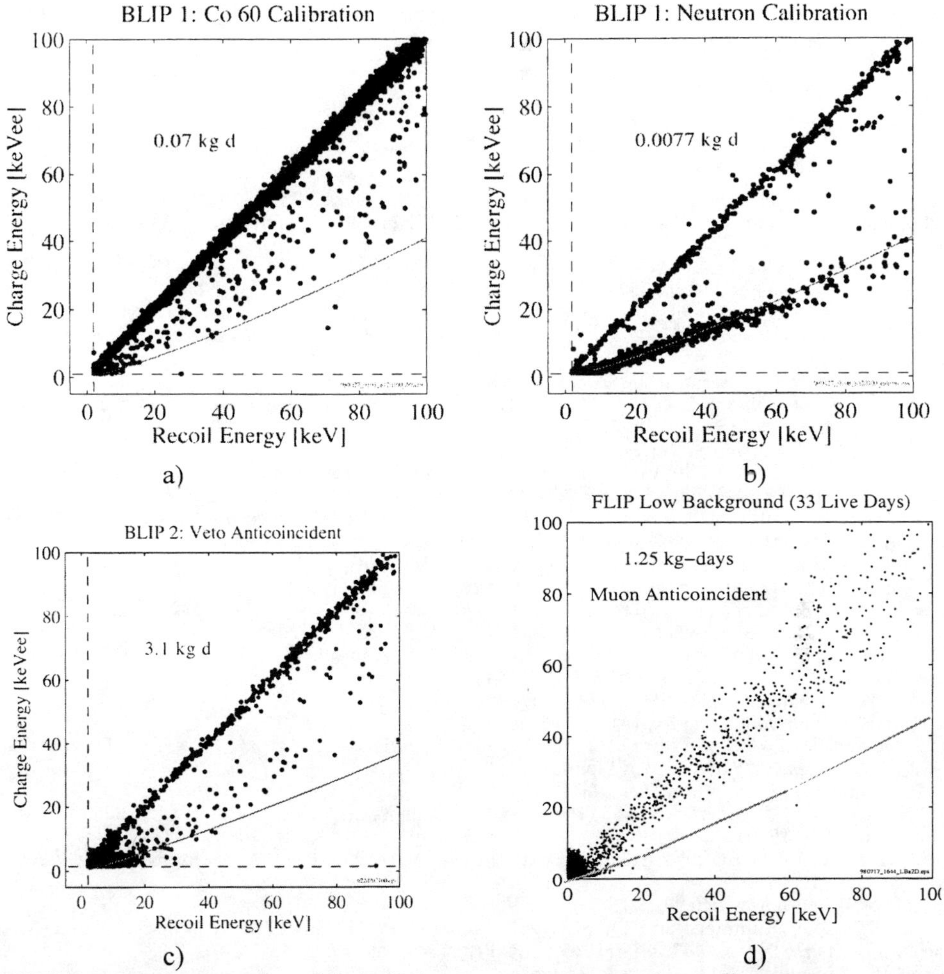

FIGURE 6. CDMS scatter plots of the ionization measurement versus the recoil energy measurement for (a, b, c) a 165 g Ge detector with thermal phonon readout (BLIP), and (d) a 100 g Si detector with athermal phonon sensing (FLIP) obtained at the Stanford Underground Facility Icebox. The ionization measurements are normalized to electron equivalent energy. Panels (a) and (b) show results of calibration runs with a ^{60}Co photon source (a) and a ^{252}Cf source producing neutrons (and photons). The line represents a fit to the region of nuclear recoil events. Panels (c) and (d) are obtained in low background running conditions. Note in (a) and (c) the soft electron component, intermediate between the diagonal photon line and the nuclear recoil line. In panel (d), after an athermal phonon signal rise time cut, only two events are left in the nuclear recoil region.

ACKNOWLEDGMENTS

This work was supported by the Center for Particle Astrophysics, a National Science Foundation Science and Technology Center operated by the University of

California under Cooperative Agreement no. AST-912005, and by the Department of Energy under the contract DE-AC03-76SF00098.

REFERENCES

1. See, e.g., the reviews by V. Trimble, *Ann. Rev. Astron. Astrophys.*, **25**, 425 (1987); J.R. Primack, D. Seckel and B. Sadoulet, *Ann. Rev. Nucl. Part. Sci.*, **38**, 751 (1988); S. Tremaine, *Physics Today* **45,** 28 (1992).
2. A. Sandage, *Physics Today*, **34** (1990); E. Loh and E. Spillar, *Astrophys. J.*, **303**, 154 (1986); E. Loh and E. Spillar, *Astrophys. J. Lett.*, **307**, L1 (1988); E. Loh, *Astrophys. J.*, **329**, 24 (1988).
3. S. Perlmutter *et al.*, *Astrophys. J.*, **483**, 565 (1997); S. Perlmutter *et al.*, *Nature*, **391**, 51 (1998); P.M. Garnavich *et al.*, *Astrophys. J. Lett.*, **493**, L53 (1998); S. Perlmutter *et al.*, Preprint 1998: Astro-ph 9812473.
4. G. Efstathiou *et al.*, Preprint 1998: Astro-ph 981226.
5. J. Yang et al., *Astrophys. J.*, **281**, 493 (1984); K.A. Olive, D.N. Schramm, G. Steigman and T. Walker, *Phys. Lett. B*, **426** (1990); D.N. Schramm and M. Turner, *Rev. of Modern Phys.*, **70**, 303 (1998).
6. D. Tytler, X.-M. Fan and S. Burles, *Nature*, **381**, 207 (1996).
7. C. Fisher, M. Davis, M.A. Strauss, A. Yahil *et al.*, *Astrophys. J.*, **389**, 188 (1992).
8. D. Scott, J. Silk and M. White, *Science*, **268**, 829 (1995).
9. M. Spiro, in Neutrino 98, *Proc. of the XVIII International Conference on Neutrino Physics and Astrophysics*, Takayama, Japan, 4–9 June 1998, edited by Y. Suzuki and Y. Totsuka.
10. E.I. Gates, G. Gyuk, M.S. Turner, *Phys. Rev. D*, **53**, 4138 (1996).
11. M. Fich and S. Tremaine, *Ann. Rev. Astron. Astrophys.*, **29**, 409 (1991).
12. E. Witten, *Phys. Rev. D*, **30**, 272 (1984); A. De Rujula and S. Glashow, *Nature*, **312**, 734 (1984).
13. A. de Rujula, S.L. Glashow and U. Sarid, *Nucl. Phys. B*, **333**, 173 (1990).
14. C. Afonso *et al.*, submitted to *Astron. and Astrophys. Lett.*, Astro-ph/98121173
15. R. Peccei and H. Quinn, *Phys. Rev. Lett.*, **38**, 1440 (1977).
16. M.S. Turner, *Phys. Reports*, **197**, 167 (1990).
17. P. Sikivie, *Phys. Rev. Lett.*, **51**, 1415 (1983).
18. S. DePanfilis *et al.*, *Phys. Rev. Lett.*, **59**, 839 (1987); S. DePanfilis *et al.*, *Phys. Rev. D.*, **40**, 3153 (1989); C.A. Hagmann, University of Florida thesis (1990).
19. C. Hagmann *et al.*, *Phys. Rev. Lett.*, **80**, 2043 (1998).
20. J.E. Kim, *Phys. Rev. Lett.*, **43**, 103 (1979); M.A. Shifman, A.I. Vainshtein and V.I. Zakharov, *Nucl. Phys. B*, **166**, 493 (1980).
21. M. Dine, W. Fischler and M. Srednicki, *Phys. Lett. B*, **104**, 199 (1981); A.P. Zhitniskii, *Sov. J. Nucl. Phys.*, **31**, 260 (1980).
22. R. Cowsik and J. McClelland, *Astrophys. J.*, **180**, 7 (1973).
23. M. Davis, F. Summers and D. Schlegel, *Nature*, **359**, 393 (1992); D. Pogosyan and A.A. Starobinsky, *Astrophys. J.*, **447**, 465 (1995); J.R. Primack et al., *Phys. Rev. Lett.*, **74**, 2160 (1995); A. Klypin, R. Nolthenius and J.R. Primack, *Astrophys. J.*, **474**, 533 (1997).
24. S.D. Tremaine and J.E. Gunn, *Phys. Rev. Lett.*, **42** , 407 (1979); J. Madsen, *Phys. Rev. D*, **44**, 999 (1991).
25. See, e.g., P.F. Smith and J.D. Lewin, *Physics Reports*, **187**, 203 (1990).
26. V.M. Lobashev, in Neutrino 98, *Proc. of the XVIII International Conference on Neutrino Physics and Astrophysics*, Takayama, Japan, 4–9 June 1998, edited by Y. Suzuki and Y. Totsuka.
27. C. Athanassopoulos *et al.*, *Phys. Rev. Lett.*, **75**, 2650 (1995); D.H. White et al., in Neutrino 98, *Proc. of the XVIII International Conference on Neutrino Physics and Astrophysics*, Takayama, Japan, 4–9 June 1998, edited by Y. Suzuki and Y. Totsuka.
28. B. Zeitnitz *et al.*, in Neutrino 98, *Proc. of the XVIII International Conference on Neutrino Physics and Astrophysics*, Takayama, Japan, 4–9 June 1998, edited by Y. Suzuki and Y. Totsuka.
29. T. Kajita *et al.*, in Neutrino 98, *Proc. of the XVIII International Conference on Neutrino Physics and Astrophysics*, Takayama, Japan, 4–9 June 1998, edited by Y. Suzuki and Y. Totsuka.
30. M. Apollonio *et al.*, *Phys. Lett. B*, **420**, 397 (1998).
31. J.N. Bahcall, S. Basu and M.H. Pinsonneault, 1998, astro-ph/9805135, to be published *in Phys. Lett. B*.
32. S.P. Mikheyev. and M.S. Smirnov, *Nuovo Cim.*, **9C**, 17 (1986); L. Wolfenstein, *Phys. Rev. D*, **20**, 2634 (1979).
33. N. Hata and P. Langacker, *Phys. Rev. D*, **56**, 6107 (1997).

34. B. Lee and S. Weinberg, *Phys. Rev. Lett.*, **39**, 165 (1977); J. Silk and M. Srednicki, *Phys. Rev. Lett.*, **53**, 624 (1984).
35. M.W. Goodman and E. Witten, *Phys. Rev. D*, **31**, 3059 (1985); J.R. Primack, D. Seckel and B. Sadoulet, *Ann. Rev. Nucl. Part. Sci.*, **38**, 751 (1988); J.D. Lewin and P.F. Smith, *Astropart. Phys.*, 6, **87** (1996).
36. J.R. Primack, D. Seckel and B. Sadoulet, *Ann. Rev. Nucl. Part. Sci.*, **38**, 751 (1988).
37. L. Bergstrom, *Nucl. Phys. B*, **325**, 647 (1989).
38. J.Silk and M. Srednicki, *Phys. Rev. Lett.* ,**53**, 624 (1984).
39. W.H.Press and D.N. Spergel, *Astrophys. J.*, **296**, 679 (1985).
40. J. Silk, K. Olive and M. Srednicki, *Phys. Rev. Lett.*, **55**, 257 (1985).
41. K. Freese, *Phys. Lett. B*, **187**, 295 (1986).
42. L. Krauss, M. Srednicki and F. Wilczek, *Phys. Rev. D*, **33**, 2079 (1986).
43. T. Gaisser, G. Steigman and S. Tilav, *Phys. Rev. D*, **34**, 2206 (1986).
44. M. Kamionkowski, K. Griest, G. Jungman and B. Sadoulet, *Phys. Rev. Lett.*, **74**, 5174(1995).
45. L. Bergstrom, J. Edsjo and P. Gondolo, hep-ph/9806293.; *Phys. Rev. D*, **55**, 1765 (1997).
46. G. Jungman, M. Kamionkowski and K. Griest, *Phys. Rep.*, **267**, 195 (1996).
47. R. Bernabei *et al.*, *Phys. Lett. B*, **389**, 757 (1996).
48. P.F. Smith *et al.*, *Phys. Lett. B*, **379**, 299 (1996); J.J. Quenby *et al.*, Astropart. Phys., 5, 249 (1996).
49. A. Alessandrello *et al.*, *Nucl. Instr. and Methods*, **A370**, 241 (1996).
50. A. de Bellefon *et al.*, *Nucl. Instr. and Methods*, **A370**, 230 (1996).
51. S.P. Ahlen *et al.*, *Phys. Lett. B*, **195**, 603 (1987).
52. D.O. Caldwell *et al.*, *Phys. Rev. Lett.*, **61**, 510 (1988).
53. D. Reusser *et al.*, *Phys. Lett. B*, **235**, 143 (1991).
54. M. Beck *et al.*, *Phys. Rev. Lett.*, **70**, 2853 (1993); L. Baudis *et al.*, Preprint 1998: hep-ex/9811045.
55. A. Morales, private communication (1997).
56. R. Bernabei *et al.*, Rome II preprints, 1998, INFN/AE-8/23 and ROM2F/98/34.
57. A. Bottino *et al.*, *Phys. Lett. B*, **402**, 113 (1997).
58. D.N. Spergel, *Phys. Rev. D*, **37**, 353 (1988).
59. A.K. Drukier, K. Freese and D.N. Spergel, *Phys. Rev. D*, **33**, 3495 (1986); K. Freese, J. Frieman and A. Gould, *Phys. Rev. D*, **37**, 3388 (1987).
60. D.P. Snowden-Ifft, E.S. Freeman and P.B. Price, *Phys. Rev. Lett.*, **74**, 4133 (1995).
61. J.I. Collar, *Phys. Rev. D*, **54**, R1247 (1996).
62. K.N. Buckland, M.J. Lehner, G.E. Masek and M. Mojaver, *Phys. Rev. Lett.*, **73**, 1067 (1994); M.J. Lehner, K.N. Buckland and G.E. Masek, *Astropart. Phys.*, **8**, 43 (1997).
63. A.A. Klimenko et al., *JETP Lett*,. **67**, 875 (1998).
64. J. Hellmig and H.V. Klapdor-Kleingrothaus, *Zeit. fur Phys. A*, **359**, 351 (1997).
65. R.J. Gaitskell *et al.*, *Nucl. Phys. B (Proc. Suppl.)*, **51B**, 279 (1996).
66. R.J. Gaitskell *et al.*, *Proc. of the VIIth International Workshop on Low Temperature Detectors,* Munich, 1997, published by the Max Planck Institute of Physics, p. 221, and Web site: http://avmp01.mppmu.mpg.de/ltd7.
67. M. Sisti *et al.*, *Proc. of the VIIth International Workshop on Low Temperature Detectors,* Munich, 1997, published by the Max Planck Institute of Physics, p. 232, and Web site: http://avmp01.mppmu.mpg.de/ltd7.
68. D. L'Hôte *et al.*, *Proc. of the VIIth International Workshop on Low Temperature Detectors,* Munich, 1997, published by the Max Planck Institute of Physics, p. 237, and Web site: http://avmp01.mppmu.mpg.de/ltd7.
69. S.W. Nam *et al.*, *Proc. of the VIIth International Workshop on Low Temperature Detectors,* Munich, 1997, published by the Max Planck Institute of Physics, p. 217, and Web site: http://avmp01.mppmu.mpg.de/ltd7.
70. R.M. Clarke *et al.*, *Proc. of the VIIth International Workshop on Low Temperature Detectors*, Munich, 1997, published by the Max Planck Institute of Physics, p. 229, and Web site: http://avmp01.mppmu.mpg.de/ltd7.
71. T. Shutt et al., *Phys. Rev. Lett.*, **29**, 3425 (1992). T. Shutt et al., *Phys. Rev. Lett.*, **29**, 3531 (1992).

Status of the CDMS Search for WIMPs.

Andrew Sonnenschein
University of California, Santa Barbara
(for the CDMS collaboration [1])

Abstract. The Cryogenic Dark Matter Search (CDMS) experiment uses germanium and silicon detectors cooled to 20 mK to attempt to detect weakly interacting dark matter particles in our galactic halo. Simultaneous measurement of phonons and ionization from interactions in these detectors allows a high degree of background rejection. We discuss recent results obtained with an array of two 165 g germanium detectors and a 100 g silicon detector after a 33 live-day exposure. Further development of this technology should allow us to explore a significant fraction of the MSSM parameter space for neutralino dark matter.

I THE CHALLENGE OF WIMP DETECTION.

One possible solution to the famous "dark matter problem" is that the dark matter is in the form of weakly interacting massive particles (WIMPs), which could be, for example, the neutralinos predicted by supersymmetry [1]. These particles should have masses in the range $10^1 - 10^4$ GeV. Large numbers of them could have been produced in thermal equilibrium in the early Universe. Today, they would be present in the dark halo of the Galaxy, which is known to have a local density

[1] The CDMS collaboration: D.S. Akerib[a], P.D. Barnes Jr.[b], D.A. Bauer[c], A. Bolozdynya[a], P.L. Brink[d], B. Cabrera[d], D.O. Caldwell[c], J.P. Castle[e], R.M. Clarke[d], P. Colling[d], M.B. Crisler[f], A. DaSilva[b], A.K. Davies[d], R. Dixon[f], S. Eichblatt[f],, R.J. Gaitskell[g], S.R. Golwala[g], E.E. Haller[h], J. Hellmig[g], M.E. Huber[i], K.D. Irwin[j], J. Jochum[g], F. Lipschultz[e], J. Martinis[j], S.W. Nam[d], B. Neuhauser[e], H. Nelson[c], T.A. Perera[a], R.R. Ross[h], T. Saab[d], B. Sadoulet[g], R.W. Schnee[a], T. Shutt[k], A. Smith[h], A. Sonnenschein[c], A.L. Spadafora[g], W.K. Stockwell[g], S. Yellin[c], B.A. Young[l] [a] *Case Western Reserve University, Cleveland, OH.* [b] *Lawrence Livermore National Laboratory, Livermore, CA.* [c] *University of California, Santa Barbara, CA.* [d] *Stanford University, Stanford, CA.* [e] *San Francisco State University, San Francisco, CA.* [f] *Fermi National Accelerator Laboratory, Batavia,IL.* [g] *Center for Particle Astrophysics and Department of Physics, University of California, Berkeley, CA.* [h] *Lawrence Berkeley National Laboratory, Berkeley, CA .* [i] *University of Colorado, Denver, CO.* [j] *National Institute of Standards and Technology, Boulder,CO.* [k] *Princeton University, Princeton, NJ.* [l] *Santa Clara University,Santa Clara, CA.* Address inquiries to Andrews@slac.stanford.edu.

CP478 , *COSMO—98*, edited by David O. Caldwell

FIGURE 1. Demonstration of nuclear recoil discrimination in a 165 g thermistor-instrumented Ge detector. The ratio of the ionization and thermal signals is plotted as a function of recoil energy for exposure to a Co-60 gamma ray source (left) and a Cf-252 neutron source (right). The Cf-252 source produces some gamma rays as well as neutrons. Dashed lines show the charge and thermal noise thresholds and the position of the nuclear recoil band, which is determined by fitting the neutron calibration data. Events between the gamma and nuclear recoil bands are due primarily to surface charge-loss effects.

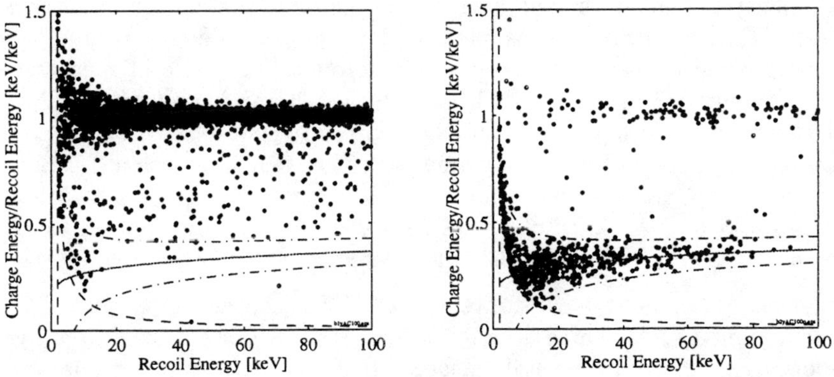

of $\simeq$ 0.3 Gev/cm^3. If the halo mass is dominated by WIMPs of mass m, they should have a local flux of (1 GeV/m)$\times 10^7$cm^{-2}s^{-1} and might be detected via interactions with detectors on Earth. Unfortunately, the small interaction cross sections on atomic nucleii imply very low event rates ($<$ 0.01 event/kg-day) and the energy transferred to the nucleus is also low ($\sim$ 10 keV).

The main challenge for WIMP detection is to achieve low background counting rates at low energies in a detector that incorporates at least a few kg of target material. One approach to solving this problem is to build detectors that can discriminate between WIMP scattering events and background events. Nuclear scattering produces less ionization in semiconductor targets per unit energy deposited than scattering on electrons. Since gamma rays, which are the dominant background source for dark matter searches, interact with electrons via compton and photoelectric scattering, while WIMPs make ionization only through the recoil of a target nucleus, a measurement of the ratio of ionization to deposited energy permits discrimination between the two types of interaction. At low temperatures. it is possible to measure both the heat deposited by an ionizing event and the amount of ionization, making this type of discrimination technologically feasible.

II THE CDMS DETECTORS AND SHIELDING.

The Cryogenic Dark Matter Search (CDMS) collaboration has built low-temperature germanium and silicon detectors capable of independent energy and ionization measurements. As illustrated in Figure 1, a high degree of discrimination between nuclear recoil events and gamma ray backgrounds has been achieved. Energy deposition is measured with two different technologies, (1) neutron-transmutation-doped germanium thermistors capable of measuring temperature changes of less than 1 μK in an attached 20 mK target crystal [2], and (2) tungsten superconducting transition-edge sensors (TES), which respond to the initial high-energy phonon burst associated with a particle interaction [3]. Both types of detector have a noise threshold of $\sim$3 keV in recoil energy. In the TES devices, the shapes and arrival times of pulses at a grid of 4 sensors contain information about the location of the event inside the detector.

Ionization is read out with conventional JFET charge amplifiers. There are charge-trapping effects which lead to a suppression of the ionization signal for events occurring near the crystal surfaces [4]. This leads to a degradation of the capability to discriminate nuclear scattering events from surface events, such as those caused by external β emitters. In the TES devices, this problem can be overcome by imposing a cut on the pulse rise times, which are faster for surface events.

The setup and operation of the CDMS experiment has been described before in e.g. Ref. [5]. The site of the CDMS I experiment is a tunnel on the campus of Stanford University, with a 17 meters water-equivalent overburden. A low-radioactivity cryostat installed in the tunnel is capable of holding 42 detectors, or about 20 kg of total detector mass. The cryostat is shielded from gamma rays by 16 cm of low radioactivity lead and from neutrons by 25 cm of polyethylene moderator. This shielding is sufficient to reduce the background event rate caused by the external gamma ray flux to below 3 events/(keV-kg-day) in the low energy (10-100 keV) region and the rate from the external neutron flux to below 0.01 events/(keV-kg-day). A set of plastic scintillator veto counters surrounds the apparatus and is used to tag events caused by neutrons made in the shielding by the fraction of cosmic ray muons which penetrate the overburden.

A new cryostat and shield are being prepared for an extension of our experiment called CDMS II, which will be located at the Soudan iron mine (2200 meters water-equivalent) in Tower, MN, USA. Detectors will be transferred to the new site when the irreducible neutron background at the shallow site from neutrons that leak through the shielding becomes the limiting factor in sensitivity. Monte Carlo studies indicate that the neutron background in the mine can be suppressed to about 0.0001 events/(keV-Kg-day) with a practical shield design.

FIGURE 2. Data from a 33 live-day exposure of two 165 g thermistor-instrumented Ge detectors (left) and a 100 g Si TES detector (right). Several cuts have been made on the data to remove noise, pileup, and muon-coincident events. In the case of the TES device, a cut was made to remove fast-rising events, which are more likely to be due to surface electrons. The Ge data is the equivalent of a 6.3 kg-day exposure after correcting for cut efficiencies and the Si data is the equivalent of 1.25 kg-days.

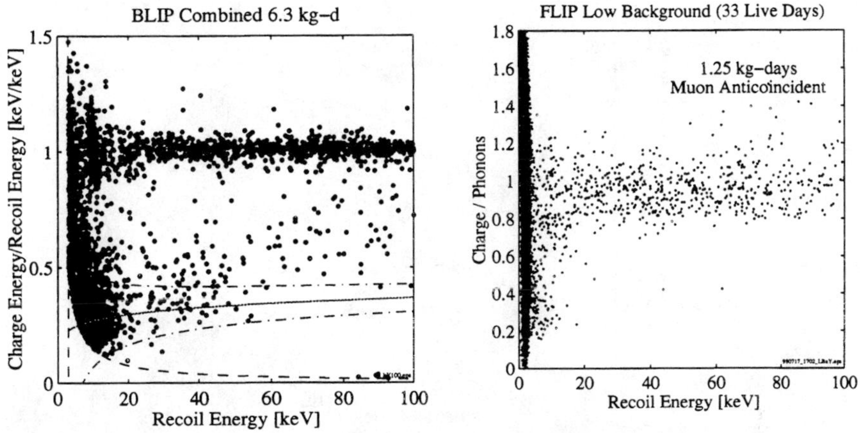

III RECENT RESULTS.

Figure 2 shows data from a recent 33 live-day exposure of an array of two 165 g thermistor-instrumented Ge detectors and a 100 g Si TES detector. The thermistor-instrumented devices are limited by contamination of the packaging materials with β emitters, including ^{3}H below the 18.6 keV endpoint and unidentified others at higher energy. The TES devices are nearly background-free above the $\sim$ 20 keV threshold where the risetime-based rejection of surface events is efficient.

Figure 3 shows our 90% CL limits on the masses and cross sections of WIMPs that might populate our galactic halo. In the future, we expect to improve these limits by reducing the background counting rates through reduction in β contamination, reduction in surface charge trapping, the use of self-shielding detector arrays, and further implementation of our TES technology. The TES sensors have recently been successfully transferred to Ge target crystals. Currently, we are collecting data from six thermistor-instrumented Ge detectors, with a total mass of 1 Kg. Four of these detectors are arranged in a self-shielding stack and have improvements to the charge electrodes which significantly increase the fraction of collected charge for surface events. An array of six Ge TES detectors is in preparation.

ACKNOWLEDGEMENTS.

This work is supported by the Center for Particle Astrophysics, an NSF Science and Technology Center operated by the University of California, Berkeley, under

FIGURE 3. Limits (90% CL) on masses and cross sections for coherent scattering of halo WIMPs assuming a halo density of 0.3 GeV/cm^3 and a velocity dispersion of 220 Km/s. Current results are shown for Si (dashed line) and Ge (solid line). Also shown are a set of MSSM predictions for the neutralino [1] and the expected ultimate CDMS I and II sensitivity (upper and lower dotted lines).

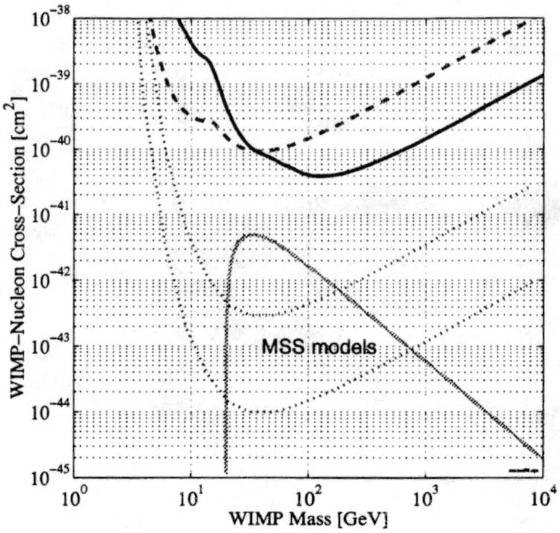

Cooperative Agreement No. AST-91-20005, and by the DOE under contracts DE-AC03-76SF00098, DE-FG03-91-ER40618 and DE-FG03-90ER40569.

REFERENCES

1. G. Jungman, M. Kamionkowski, K. Griest, *Phys. Rep.* **267**, 195 (1996).
2. N. Wang *et al.*, *Phys. Rev.* **B41** 3761 (1990). T. Shutt *et al.*, *Phys. Rev. Lett.* **69**, 3452 and 3531 (1992). A. Sonnenschein *et al.*, Proceedings of the International Conference on the Identification of Dark Matter (IDM98), 1998, Buxton, UK (ed. N. Spooner), in press.
3. K.D. Irwin *et al., Rev. Sci. Instrum.* **66**, 11 (1995). R.M. Clarke *et al.*, IDM98.
4. T. Shutt *et al.*, Proceedings of the VIIth International Workshop on Low Temperature Detectors, 1997, Munich, Germany (ed. S. Cooper) 221.
5. R.W. Schnee *et. al., Phys. Rep.* **307**, 283 (1996), D.S. Akerib *et al.*, Astroph/9712343.

Future of Double Beta Decay and Dark Matter Searches–GENIUS

H. V. Klapdor-Kleingrothaus, L. Baudis, G. Heusser, I. V. Krivosheina, S. Kolb, B. Majorovits, J.-U. Nabi,[1] H. Päs

Max-Planck-Institut für Kernphysik, Postfach 103980, 69029 Heidelberg, Germany.

Abstract. The recent results from the HEIDELBERG-MOSCOW experiment have demonstrated the large potential of double beta decay to search for new physics beyond the standard model. To increase further by a major step the present sensitivity for double beta decay and dark matter searches, we describe here a project, proposed recently [1], which would operate one tonne of 'naked' enriched germanium-detectors in liquid nitrogen as shielding in an underground set-up (GENIUS). It improves the sensitivity of neutrino masses to 0.01 eV. A 10 tonne version would probe neutrino masses even down to 10^{-3} eV.

INTRODUCTION

Two topics are of greatest interest in high-energy physics and astrophysics. These are test of the existence of supersymmetry (SUSY) and search for a dark matter candidate. A very close connection between both issues in addition to SUSY which could be responsible for the cold dark matter (neutralinos), could be established by non-zero neutrino masses as candidates for hot dark matter. Especially degenerate neutrino mass scenarios [2] explain the observations by the COBE satellite for dark matter [3]. Our new GErmanium in NItrogen Underground Set-up (GENIUS), first proposed and presented by [1], is an experiment which is optimized to address both issues. It is a large step beyond the HEIDELBERG-MOSCOW experiment [4], which is the most sensitive existing double beta decay experiment at present and for the next years and which has also given the most stringent limits on WIMP'S for several years [5]. The GENIUS project could represent the future of this field. It would not only be unique in probing *absolutely* the neutrino mass down to 10^{-2} or 10^{-3} eV (all running neutrino oscillation experiments probe only *differences* of masses!) but would also be more sensitive than all accelerator neutrino oscillation experiments running at present or being under construction for the future. It could

---

[1] Talk presented by Jameel-Un Nabi.

CP478, *COSMO—98*, edited by David O. Caldwell

further decisively contribute to the solution of the atmospheric and solar neutrino problems [6–8].

Its ultralow background allows GENIUS to also act as a very efficient dark matter detector. It would allow us, for the first time ever, to cover practically the complete MSSM neutralino parameter space (see Fig. 2). GENIUS thus should either confirm or rule out the present favorite particle physics candidate for CDM, the neutralino, and test the hypothesis of R-parity conservation to unanticipated levels. The infrastructure and space required for the experiment in an underground laboratory, could be provided by the INFN in the Gran Sasso Laboratory in Italy.

Below we discuss some of the physics potentials of GENIUS (for details see [1,6–8].

I POTENTIAL OF GENIUS

(1) *Sensitivity for neutrino masses:* GENIUS can test all (partially) degenerate neutrino-mass models. The GENIUS 1 tonne version can probe the neutrino masses, after one year of measurement, down to [8]

$$\langle m_\nu \rangle \leq 0.02\ eV \qquad (with\ \ 68\%\ \ CL)$$

whereas after ten years of measurement the limit increases to

$$\langle m_\nu \rangle \leq 0.006\ eV \qquad (with\ \ 68\%\ \ CL)$$

The ultimate experiment could test the $0\nu\nu\beta\beta$ half-life of ^{76}Ge up to a limit of 5.7 $\times$ 10^{29}y and the neutrino mass down to 2 $\times 10^{-3}$ eV using 10 tonnes of enriched germanium.

(2)*Left-right symmetry:* If GENIUS is able to reach down to $\langle m_\nu \rangle \leq$ 0.01 eV, it would at the same time be sensitive to right-handed W-boson masses up to $m_{W_R} \geq$ 8 TeV (for a heavy right-handed neutrino mass of 1 TeV) or $m_{W_R} \geq$ 5.3 TeV (at $\langle m_N \rangle = m_{W_R}$). A 10 tonne version of GENIUS could even reach a sensitivity of $m_{W_R} \geq$ 18 TeV (for a heavy right-handed neutrino mass of 1 TeV) or $m_{W_R} \geq$ 10.1 TeV (at $\langle m_N \rangle = m_{W_R}$).

(3) *R-parity violating SUSY:* If squarks are exceptionally heavy, $m_{\tilde{q}} \geq$ 1 TeV, LHC could not compete with GENIUS. However, for typical squark masses below 1 TeV, LHC can finally probe down to smaller couplings than the double beta decay experiment.

(4)*R-parity conserving SUSY:* Since limits on the Majorana sneutrino mass $\tilde{m}_M$ scale as $(T_{1/2})^{1/4}$ GENIUS with 1 tonne (with 10 tonnes) would test (B-L) violating 'Majorana' sneutrino masses lower by factors of about 7 (20), compared with the present constraints [9] (GENIUS 1 (10) ton(s) would probe the sneutrino mass down to $m^M_{\tilde{\nu}_e} < 3$ (1) MeV).

(5)*Leptoquarks:* For leptoquarks in the range of 200 GeV LQ-Higgs couplings down to (few) 10^{-8} could be explored. In other words, if leptoquarks interact with the Higgs boson with a coupling of order $\mathcal{O}(1)$ either $0\nu\nu\beta\beta$ decay must be found or LQs must be heavier than (several) 10 TeV(s).

(6)*Composite neutrinos:* GENIUS in the 1 tonne version (10 tonnes) would improve the previous limit on the mass of an excited composite Majorana neutrino [10] up to $m_N \geq 1.1(2.3)$ TeV, if $0\nu\beta\beta$ decay is not found.

(7) *Neutrino oscillations:* GENIUS 1 tonne probes LSND (for $m_1/m_2 > 0.02$), and can check the CHOOZ result for the $\nu_e - \nu_\mu$ solution of the atmospheric neutrino problem whereas GENIUS 10 tonnes can probe the large angle solution of the solar neutrino problem (even in the worst assumed case of strong hierarchical neutrino mass scenario), and even the small angle solution (for $m_1/m_2 \geq 0.5$) (see Fig. 1).

(8)*Cold dark matter:* GENIUS will be the only experiment, which could seriously test the MSSM predictions over almost the whole SUSY parameter space (see Fig. 2) even in a smaller version using only 100 kg of natural germanium . In this way, GENIUS could compete with LHC in the search for SUSY (see [12] for discussion).

It is interesting to note that if WIMP scattering is found by GENIUS this would extremely constrain the amount of R-parity violation within supersymmetric models. Due to the fact that neutralinos are bound in the galaxy even today, neutralino decays via R-parity violating operators would have to be highly suppressed. The details depend, of course, on the neutralino mass and composition. However, finding the neutralino with GENIUS would imply typical limits on R-parity violating couplings of the order of $10^{-(16-20)}$ for any of the $\lambda_{ijk}, \lambda'_{ijk}$ or λ''_{ijk} in the superpotential [19].

REFERENCES

1. Klapdor-Kleingrothaus H.V., *Proc. Int. Workshop on Physics Beyond the Desert: Accelerator and Non-accelerator Approaches, Castle Ringberg, Germany, June 1997*, ed H. V. Klapdor-Kleingrothaus and H. Päs, Bristol: IOP, 1998, pp. 485-532.
2. Caldwell D.O., and Mohapatra R.N., *Phys. Rev. D.* **50**, 3477 (1994).
3. Smoot G.F. et al., *Ap. J. Lett.* **396**, L1-L5 (1992).
4. HEIDELBERG-MOSCOW Collaboration, Günther M. et al., *Phys. Rev. D* **55**, 54 (1997), Baudis L. et al., *Phys. Lett. B* **407**, 219 (1997).
5. HEIDELBERG-MOSCOW Collaboration, Beck M. et al., *Phys. Lett. B* **336**, 141 (1994); HEIDELBERG-MOSCOW Collaboration, Baudis L. et al., *Phys. Rev. D* **59**, 022001-1 (1998).

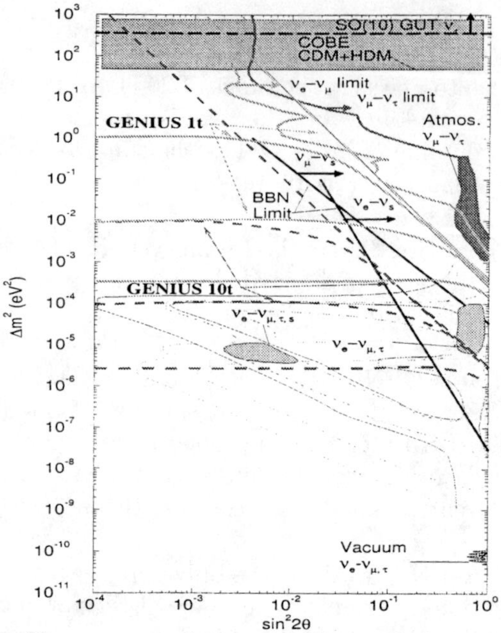

FIGURE 1. Summary of known constraints on neutrino oscillation parameters. Shown are the vacuum and MSW solutions (for two generation of neutrinos) for the solar neutrino problem, the parameter range which would solve the atmospheric neutrino problem and various reactor and accelerator limits on neutrino oscillations. In addition, the mass range in which neutrinos are good hot dark matter candidates is indicated, as well as limits on neutrino oscillations into sterile states from considerations of big bang nucleosynthesis. Finally, the heavy lines indicate the sensitivity of GENIUS (full lines 1 tonne, broken lines 10 tonne) to neutrino oscillation parameters for three values of neutrino mass ratios R = 0, 0.01 and 0.1 (from top to bottom). For GENIUS 10 ton also the contour line for R = 0.5 is shown. Non-observation of $0\nu\beta\beta$ decay would exclude the regions beyond and right to the curves. While already the 1 tonne GENIUS would be sufficient to constrain degenerate and quasi-degenerate neutrino mass models, and also would solve the atmospheric neutrino problem if it is due to $\nu_e \leftrightarrow \nu_\mu$ oscillations, the 10 tonne version of GENIUS could cover a significant new part of the parameter space, including the large angle MSW solution to the solar neutrino problem, even in the worst case of R = 0. For R > 0.5 it would even probe the small angle MSW solution (see [11]).

6. Klapdor-Kleingrothaus H.V., Hirsch M., *Z. Phys. A* **359**, 351 (1997).
7. Klapdor-Kleingrothaus H.V., *Int. J. Mod. Phys.* **A 13**, 3953 (1998).
8. Klapdor-Kleingrothaus H.V., Hellmig J., Hirsch M., *J. Phys. G* **24**, 483 (1998).
9. Hirsch M., Klapdor-Kleingrothaus H.V., and Kovalenko S.G., *Phys. Lett. B* **403**, 291 (1997).
10. Panella O. et al., *Phys. Rev. D* **56**, 5766 (1997).
11. Klapdor-Kleingrothaus H.V., *Proc. NEUTRINO'98, Takayama, Japan, June*

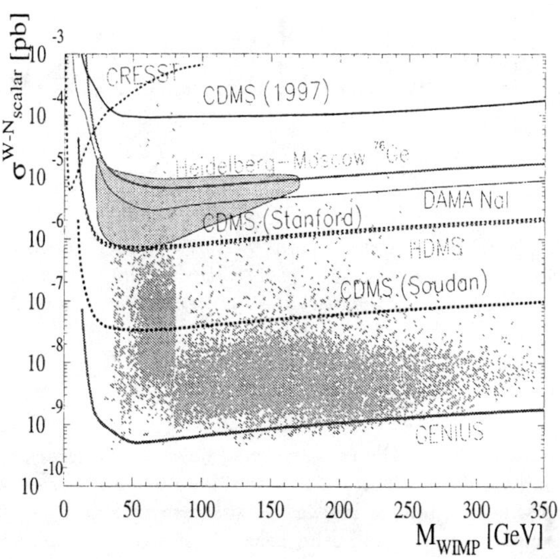

FIGURE 2. WIMP-nucleon cross section limits in pb for scalar interactions as function of the WIMP-mass in GeV. Hatched area excluded by the HEIDELBERG-MOSCOW experiment [5] and the DAMA experiment [13]. Dashed lines correspond to results from upcoming experiments (HDMS [14], CDMS [15] and GENIUS [1]). These experimental limits and perspectives are compared with expectations (scatter plot) for WIMP-neutralino cross-sections calculated in the MSSM framework with non-universal scalar mass unification [16]. The 90% allowed region according to [17] (light filled area), which is further restricted by indirect dark matter searches [18], could be easily tested with a 100 kg version of the GENIUS experiment.

1998, World Scientific, Singapore, 1999.

12. Klapdor-Kleingrothaus H.V., and Ramachers Y., *Proc. Int. Workshop on Physics Beyond the Desert: Accelerator and Non-accelerator Approaches, Castle Ringberg, Germany, June 1997* , ed H. V. Klapdor-Kleingrothaus and H. Päs, Bristol: IOP, 1998, pp. 786-802; *Eur. Phys. J.* **A 3**, 85 (1998).
13. Bernabei R. et al., *Phys. Lett. B* **389**, 757 (1997).
14. HEIDELBERG-MOSCOW Collaboration, Baudis L. et al., *Nucl. Instrum. Methods A* **385**, 265 (1997).
15. Barnes P.D. et al., *Nucl. Instrum. Methods A* **370**, 233 (1996).
16. Bednyakov V., Klapdor-Kleingrothaus H.V., and Kovalenko S.G., *Phys. Rev. D* **50**, 7128 (1994); *Z. Phys. A* **357**, 339 (1997).
17. Bernabei R. et al., *Phys. Lett B* **424**, 195 (1998).
18. Bottino A. et al., *Phys. Lett. B* **423**, 109 (1998).
19. Hirsch M., and Klapdor-Kleingrothaus H.V., *Proc. Int. Workshop on Dark Matter in Astro and Particle Physics, Heidelberg, Sept. 1996* ed H.V. Klapdor-Kleingrothaus and Y. Ramachers, Singapore: World Scientific, 1997, p. 640.

Dark Matter in Models of String Cosmology

Ram Brustein [1] and Merav Hadad [2]

[1] *Department of Physics, Ben-Gurion University, Beer-Sheva 84105, Israel*
[2] *School of Physics and Astronomy, Tel Aviv University, Tel Aviv 69978, Israel*
email: ramyb@bgumail.bgu.ac.il, meravv@post.tau.ac.il

Abstract.
The origin of dark matter in the universe may be weakly interacting scalar particles produced by amplification of quantum fluctuations during a period of dilaton-driven inflation. We present two interesting cases, the case of small fluctuations, and the resulting nonthermal spectrum and, the case of large fluctuations of a field with a periodic potential, the QCD axion.

We consider particle production in models of string cosmology which realize the pre-big-bang (PBB) scenario [1]. In this scenario the evolution of the universe starts from a state of very small curvature and coupling and then undergoes a long phase of dilaton-driven inflation (DDI) and at some later time joins smoothly standard radiation dominated (RD) cosmological evolution, thus giving rise to a singularity free inflationary cosmology. Particles are produced during DDI phase by the standard mechanism of amplification of quantum fluctuations [2]. Some debate about the naturalness of initial conditions and whether PBB models actually solve the cosmological problems has been taking place [3]. The smooth transition from DDI to RD is also not completely solved, for ideas about how this may come about see [4]. We will take a more phenomenological approach, and concentrate on interesting cosnsequences of the models which do not depend on the detailed resolution of these issues, assuming that a resolution exists.

In the simplified model of background evolution we adopt, the evolution of the universe is divided into four distinct phases with specific (conformal) time dependence of the scale factor of the universe $a(\eta)$ and the dilaton $\phi(\eta)$. We assume throughout an isotropic and homogeneous four dimensional flat universe, described by a FRW metric. All other scalar fields are assumed to have a trivial vacuum expectation value during the inflationary phase.

We have computed spectrums of produced particles for the models described above [5]. We have solved a linear perturbation equation, $\chi_k'' + (k^2 + M^2a^2 - s''/s)\chi_k = 0$, where $s(\eta) \equiv a(\eta)^m e^{l\phi(\eta)/2} = a_s^m e^{l\phi_s/2}(\eta/\eta_s)^{1/2-n_s}$,

CP478, *COSMO—98*, edited by David O. Caldwell

imposing initial conditions corresponding to normalized vacuum fluctuations. The parameter m depends on the spin of the particle and l depends on its coupling to the dilaton. Similar calculations have been performed by several groups and the results agree [6], whenever a comparison was possible.

We consider weakly interacting scalar particles, abundant in string theory and supergravity. For scalar fields $m = 1$, and we will consider for concreteness the values, $l = -1, 0, 1$ corresponding, respectively, to moduli, Ramond-Ramond axions, and Neveu-Schwartz axions. We will assume that the produced particles interact so weakly, that their interactions and decay are not sufficient to alter the primordial spectrum substantially. The particles we have in mind have typically gravitational strength interactions, which is definitely weak enough to satisfy our assumption, and their masses are below a fraction of an eV.

A typical spectrum of a light scalar may be divided, at a given time, into three physical momentum (PM) regions: *i*) The massless region, $\omega_S > \omega > M$, in this region particles are relativistic. *ii*) The "false" massive region, $M > \omega > \omega_m$, where $\omega_m = \omega_1 (M/M_s)^{1/2}$. In this region particles are NR, but have reentered the horizon as relativistic modes. *iii*) The "real" massive region $\omega_m > \omega$. In this region particles are non-relativistic (NR), and have reentered the horizon as NR modes. Note that PM redshift as the universe expands, and therefore boundaries of regions change in time,

$$\frac{d\Omega}{d\ln\omega} = \mathcal{N} g_1^2 \left(\frac{g_1}{g_S}\right)^{2l} \begin{cases} \left(\frac{\omega}{\omega_S}\right)^x & \omega_S > \omega > M \\ \frac{M}{\omega_S}\left(\frac{\omega}{\omega_S}\right)^{x-1} & M > \omega > \omega_m \\ \frac{M}{\omega_1}\sqrt{\frac{M}{M_s}}\left(\frac{\omega}{\omega_S}\right)^x & \omega_m > \omega, \end{cases} \tag{1}$$

where $x \equiv 2 + 2\alpha + l\beta$, and $\mathcal{N}$ is a numerical factor, estimated in [5], which we will set to unity in what follows. Parameters appearing in eq.(1) are, the string scale M_s, z_S the total redshift during the string phase, g_S and g_1, the string coupling at the beginning and end of the string phase, and ω_1, the PM (today), corresponding to the end of the string phase, estimated in [7] to be $\omega_1 \sim 10^{-5} eV$, and the PM $\omega_S = \omega_1 / z_S$, the PM (today) corresponding to the end of the DDI phase. In (1) we have assumed no substantial late entropy production has occured. In general [7], the effect of late entropy production is to further redshift the physical frequencies as $\sim (1 - \delta s)^{1/3}$ where δs is the fraction of produced entropy, and, more importantly, to dilute the contribution of modes which are already inside the horizon by a factor $(1 - \delta s)^{4/3}$. If a substantial amount of entropy is produced below $T \sim M_s / \sqrt{z_S}$, then spectrum (1) is no longer a good approximate spectrum.

The first example we look at is an example of small field fluctuations. The produced spectrum in this case is nonthermal, and may lead to an interesting case of cold and hot dark matter from the same source [8]. We look at generalized axions ($l = 1$) with masses below .1 eV in a cosmological model described in [1]. In this model, $d = 3$ spatial dimensions are expanding and $n = 6$ spatial dimensions are contracting, leading to $\alpha = -2/(d + n + 3) = -1/6$ and $\beta = -4d/(d + n + 3) =$

-1. For this specific model $x = 2/3$, $\Omega_{REL} \simeq g_1^2 \left(\frac{g_1}{g_S}\right)^2$, $\Omega_{NR} \simeq g_1^2 \left(\frac{g_1}{g_S}\right)^2 \frac{M}{\omega_S} \left(\frac{\omega_m}{\omega_S}\right)^{-1/3}$ Taking 10^{-10}eV $< M < 10^{-2}$eV, for which the above condition is comfortably satisfied, we observe that the ratio $\Omega_{REL} : \Omega_{NR}$ at the start of structure formation era can vary in a range from well above unity to well below unity, corrresponding to hot, mixed and cold dark matter. For example, choosing $g_1 = .1$ and $g_S = .01$ if the axion's mass is 10^{-10}eV, and for $z_S \sim 2 \times 10^4$ we get $\Omega_{REL} : \Omega_{NR} = 1 : 1$ with both energy densities being near critical, and if we choose $g_1 = .1$ and $g_S = .03$, making $\Omega_{REL} \simeq .1$, and if $z_S \sim 10^6$ we obtain $\Omega_{REL} : \Omega_{NR} = 1 : 10$, with $\Omega_{NR} \simeq 1$. Note that ratio depends on $z_S^{-2/3} M^{-5/6}$, so the previous examples correspond to a range of allowed values.

We now turn to the case of large field fluctuations. The spectrum in this case can be very different from that given by the naive result eq.(1). We look at the model independent axion [9], $l = 1$, assuming that it is the QCD axion [10] in a model of background evolution in which d=3 spatial dimensions expand, and $n = 6$ spatial dimensions are fixed, leading to $x = 3 - 2\sqrt{3} \sim -0.46$ [11].

Because of the negative exponent the spectrum is dominated by the lowest PM entering the horizon at a given time. The most interesting situation is when the axion potential turns on when the universe cools down to QCD temperatures. If we try to approximate the axion potential by a quadratic potential, leading to the result (1), we encounter a puzzle. The axion energy density becomes formally divergent as soon as the axion potential turns on! Once the potential is generated, all the low frequencies reenter the horizon at once, so to obtain the total energy density inside the horizon we need to integrate it from the minimal amplified frequency ω_{min}, which is either zero, if the duration of the dilaton-driven phase is infinite, or exponentially small if the duration is finite but large. The lower frequency part of the spectrum yields a divergent contribution, proportional to ${\omega_{min}}^{3-2\sqrt{3}}$. This result does not make sense.

The resolution of the puzzle depends crucially on the periodic nature of axion potential $V(\psi) = \frac{1}{2} V_0 \left(1 - \cos(\frac{\psi}{\psi_0})\right)$. This point was first understood by Kofman and Linde [12], and we have adopted their ideas to our particular situation. First, the total potential energy is limited to V_0 and does not continue to increase indefinitely as the axion field increases, providing a "topological cutoff" on the total axionic energy density and as important, large fluctuations in the axion field are also "topologically cutoff", producing exponentially small energy density perturbations. Large fluctuations lead to a uniform distribution of the axion field inside the horizon, with very small statistical fluctuations.

Using completely standard arguments [13], we may obtain a bound on the mass of the axion m_a (or equivalently on ψ_0) by requiring that the energy density in the coherent axion oscillations be subcritical at the beginning of matter domination epoch. This requirement leads to the standard bounds on the axion mass, $\Omega_a h^2 \sim \frac{10^{-6} eV}{m_a}$, where h is todays Hubble parameter in units of 100 km/Mpc/sec. Requiring subcritical Ω_a leads to the standard bound on ψ_0, $\psi_0 \lesssim 10^{12} GeV$ and $m_a \gtrsim 10^{-6} eV$.

In string theory, natural values of ψ_0 are approximately $10^{16}GeV$, which, if taken at face value, would lead to overclosure of the universe with axions many times over. Two possible resolutions have been suggested [14] to allow our universe to reach its old age of today.

If dark matter in the universe is indeed made of light particles with gravitational strength interactions its detection in current direct searches is extremely difficult, and will probably require new methods and ideas.

Acknowledgments

Work supported in part by the Israel Science Foundation.

REFERENCES

1. G. Veneziano, *Phys. Lett.* B265 (1991) 287; M. Gasperini and G. Veneziano, *Astropart. Phys.* 1 (1993) 317.
2. N. Birrell and P. Davies, *Quantum fields in curved space*, Cambridge University Press, 1984; V. Mukhanov, A. Feldman and R. Brandenberger, *Phys. Rep.* 215 (1992) 203.
3. G. Veneziano, *Phys. Lett.* B406 (1997) 297; E. Weinberg and M. Turner, *Phys. Rev.* D56 (1997) 4604; M. Maggiore and R. Sturani, *Phys. Lett.* B415 (1997) 335; A. Buonanno, et al, *Phys. Rev.* D57 (1998) 2543; A. Buonanno, T. Damour and G. Veneziano, hep-th/9806230.
4. R. Brustein and G. Veneziano, *Phys. Lett.* B329 (1994) 429; N. Kaloper, R. Madden and K. Olive, *Nucl. Phys.* B452 (1995) 677; M. Gasperini, M. Maggiore and G. Veneziano, *Nucl. Phys.* B494 (1997) 315; R. Brustein and R. Madden, *Phys. Lett.* B410 (1997) 110; *Phys. Rev.* D57 (1998) 712; hep-th/9901044; G. veneziano, hep-th/9902126.
5. R. Brustein and M. Hadad, *Phys. Rev.* D57 (1998) 725.
6. E. Copeland, et al, *Phys. Rev.* D56 (1997) 874; E. Copeland, J. Lidsey, D. Wands, *Nucl. Phys.* B506 (1997) 407; hep-th/9809105; A. Buonanno, et al, JHEP 01 (1998) 004; M. Giovannini, hep-th/9809185.
7. R. Brustein, M. Gasperini and G. Veneziano, *Phys. Rev.* D55 (1997) 3882.
8. R. Brustein and M. Hadad, hep-ph/9810526, to appear in *Phys. Rev. Lett.* .
9. E. Witten, *Phys. Lett.* B149 (1984) 351; *Nucl. Phys.* B268 (1986) 79; K. Choi and J. Kim, *Phys. Lett.* B154 (1985) 393; *Phys. Lett.* B165 (1985) 71.
10. R. Peccei and H. Quinn, *Phys. Rev. Lett.* 38 (1977) 1440; S. Weinberg, *Phys. Rev. Lett.* 40 (1978) 223; F. Wilczek, *Phys. Rev. Lett.* 40 (1978) 279.
11. R. Brustein and M. Hadad, *Phys. Lett.* B442 (1998) 74.
12. A. Linde,*Phys. Lett.* B158 (1985) 375; L. Kofman, *Phys. Lett.* B173 (1986) 400; L. Kofman and A. Linde, *Nucl. Phys.* B282 (1987) 555.
13. J. Preskill, M. Wise and F. Wilczek, *Phys. Lett.* B120 (1983) 127; L. Abbott and P. Sikivie, *Phys. Lett.* B120 (1983) 133; M. Dine and W. Fischler, *Phys. Lett.* B120 (1983) 137.
14. T. Banks and M. Dine, *Nucl. Phys.* B479 (1996) 173; T. Banks and M. Dine, *Nucl. Phys.* B505 (1997) 445; K. Choi, *Phys. Rev.* D56 (1997) 6588.

The Cosmological Consequences of the Preon Structure of Matter

Vladimir V. Burdyuzha,* Grigory M. Vereshkov,† Olga D. Lalakulich † and Yuri N. Ponomarev *

* Astro Space Center of Lebedev Physical Institute of Russian Academy of Sciences, Profsouznaya str. 84/32, 117810 Moscow, Russia
† Rostov State University, Stachki str. 194, 344104 Rostov on Don, Russia

Abstract. If the preon structure of quarks, leptons and gauge bosons will be proved then in the Universe during a relativistic phase transition the production of nonperturbative preon condensates has occured. Familons are collective excitations of these condensates. It is shown that the dark matter consisting of familon type pseudogoldstone bosons was undergone to two relativistic phase transitions temperatures of which were different. In the result of these phase transitions the structurization of dark matter and therefore the baryon subsystem had taken place. In the Universe two characteristic scales which have printed this phenomenon arise naturally.

The commonly accepted point of view on formation of Large Scale Structure (LSS) of the Universe is based on the assumption about development after inflation period of homogeneous and isotropic Gausian scalar perturbations (perturbations of density). The standard model predicts $\Omega_{total} = 1$ that is $\Omega_{CDM} + \Omega_{HDM} + \Omega_{baryon} + \Omega_{\Lambda} = 1$ in which large-mass objects are created at $z \sim 2-3$. For formation of LSS significantly earlier ($z \sim 6-8$) it is necessary to include in Ω_{total} the density of vacuum energy (Λ-term). Our hypothesis about the formation of seeds for LSS is connected with the late relativistic phase transitions (RPT) in the gas of familon type pseudo-Goldstone bosons (pseudo-GB) which are a probable candidate in dark matter (DM). The time of the beginning of these RPT depends on a mass of familons. If $m_{GB} > 3 \times 10^{-4}\ eV$ then the essential influence of these RPT may take place in no far last evolution of our Universe. If $m_{GB} < 3 \times 10^{-4}\ eV$ then similar influence of these RPT may take place in the nearest cosmological future when $T_{CMB} < 2.7\ K$.

We suppose that the theory of RPT in the gas and the vacuum of pseudo-GB allows more effectively to obtain $\delta\rho/\rho \sim 1$ on earlier stage of the Universe evolution ($z > 5$) than the standard model.

The cosmological consequenses of the simplest boson-fermion model of quarks and leptons [1] are investigated. Our interest to the preon model was induced by

CP478 , *COSMO—98*, edited by David O. Caldwell

the fact of possible leptoquarks resonance in the experiment HERA [2]. We have researched more detail the structure of preon nonperturbative vacuum arising in the result of the correlation of nonabelian fields on two scales ($\Lambda_{mc} \gg 1\ Tev$ is the confinement scale of metacolour and $\Lambda_c \sim 150\ Mev$ is QCD scale). We have detected that in spectrum of excitations of heterogenic nonperturbative preon vacuum pseudo-GB modes of familon type occurs. Familons are created in the result of spontaneous breaking symmetry of quark-lepton generations. Nonrezo masses of familons are the result of superweak interactions with quark condensates. We consider these particles as the basic constituent of cosmological DM. The distinguishing characteristic of these particles is the availability of the residual $U(1)$ symmetry and possibility of it spontaneous breaking for temperatures $\frac{\Lambda_c}{\Lambda_{mc}} \sim 10^{-3}\ eV$ in the result of a RPT.

We have proposed that this relativistic phase transition has direct relation to the production of primordial perturbations in DM the evolution of which leads to baryon large scale structure formation. The idea of RPT in the cosmological gas of pseudo-GB in connection with LSS problem was formulated by Frieman et al. [3]. Here we have investigated by quantitatively the preon-familon model of this RPT.

Observational data show that some baryon objects such as the quasar on $z \sim 4.9$, CO lines on $z \sim 4.69$ and galaxies on $z > 5$ [4] were produced as minimum on redshifts $z \sim 6 \div 8$. This is the difficulty for standard CDM and CHDM model to produce their (the best fit is $z \sim 2 \div 3$ and observations provide the support of this). If early baryon cosmological structures produced on $z > 10$ then the key role must play DM particles with nonstandard properties. In standard model DM consists of ideal gas particles with $m \approx 0$ practically noninteracting with usual matter (till now they do not detected because of their superweak interaction with baryons and leptons).

Certainly a characteristic moment of the most of cosmological structures formation finishing remains the same ($z \sim 2 \div 3$) and the appearence of baryon structures on high z ($z > 4$) will be a result of statistical outburst evolution of the spectrum of DM density perturbations. Early cosmological baryon structures are connected to statistical outbursts in sharp nonlinear physical system which is RPT (the production of inhomogeneities).

A new theory of DM must combain properties of superweak interaction of DM particles with baryons and leptons and intensive interaction of theses particles each other. Such interactions are provided by nonlinear properties of DM medium. This is the condition for realization of RPT.

The familon symmetry is experimentally observed (the different generations of quarks and leptons participate in gauge interactions the same way). Breaking of this symmetry gives a mass of particles in different generations. A hypothesis about spontaneous breaking of familon symmetry is natural and the origin of Goldstone bosons is inevitably. The properties of any pseudo-GB as and pseudo-GB of familon type depends on physical realization of Goldstone modes. These modes can be arisen from fundamental Higgs fields or from collective excitations of a heterogenic nonperturbative vacuum condensate more complex than quark-gluon one in QCD. The second possibility can realize the theory in which quarks and leptons are composite that is the preon model of elementary particles.

Thus, in frame of preon theory DM is interpretated as a system of familon collective excitations of a heterogenic nonperturbative vacuum. This system consists of 3 subsystems:1) familons of up-quark type; 2) familons of down-quark type; 3) familons of lepton type. On stages of cosmological evolution when $T \ll \Lambda_{mc}$ the heavy unstable familons are absent. Small masses of familons are the result of superweak interactions of Goldstone fields with nonperturbative vacuum condensates and therefore familons acquire status of pseudo-GB. The value of these masses is limited by astrophysical and laboratory magnitudes [5]:

$$m_{astrophysical} \sim 10^{-3} \div 10^{-5}\, eV, \quad m_{laboratory} \leq 10\, eV \tag{1}$$

The effect of familons mass production corresponds formally mathemathically the appearence of mass terms in the Lagrangian of Goldstone fields. From general considerations one can propose that mass-terms may arise as with "right" as and with "wrong" signs. The sign of the mass-terms predetermines the destiny of residual symmetry of Goldstone fields. In the case of "wrong" sign for low temperatures $T < T_c \sim m_{familons} \sim 0.1 \div 10^5\, K$ a Goldstone condensate produces and the symmetry of the familon gas breaks spontaneously.

As example we discuss a model only one familon subsystem corresponding to up- quarks of second and third generations. The chiral-familon group of the model is $SU_L(2) \times SU_R(2)$. The familon excitations are described by eight measure (on number of matrix components) reducible representation of this group factorized on two irreducible representations (F, f_a); (ψ, φ_a) which differ each other by a sign of space chirality. In this model the interaction of quark fields with familons occurs. However in all calculations quark fields are represented in the form of nonperturbative quark condensates. From QCD and a experiment the connection between quark and gluon condensates is known:

$$\langle 0 \mid \bar{q}q \mid 0 \rangle \approx \frac{1}{12 m_q} \qquad \langle 0 \mid \frac{\alpha_s}{\pi} G^n_{\mu\nu} G^{\mu\nu}_n \mid 0 \rangle \approx \frac{3\Lambda_c^4}{4 m_q} \tag{2}$$

Here: $q = t, c$; $m_c \sim 1.5\, Gev$; $m_t \sim 175\, Gev$; $\Lambda_c \sim 150\, Mev$.

The spontaneous breaking of symmetry $SU_L(2) \times SU_R(2) \to U(1)$ is produced by vacuum shifts $\langle \psi \rangle = v$; $\langle f_3 \rangle = u$. The numerical values $v, u \sim \Lambda_{mc}$ are unknown. They must be found by experimentally if our theory corresponds to reality. Parameters u and v together with the value of condensates (2) define numerical values of basic magnitudes characteristing the familon subsystem. After breaking of symmetry $SU_L(2) \times SU_R(2) \to U(1)$ light pseudo-GB fields contain the real pseudoscalar field with the mass:

$$m^2_{\varphi'} = \frac{1}{6(u^2 + v^2)} \langle 0 \mid \frac{\alpha_s}{\pi} G^n_{\mu\nu} G^{\mu\nu}_n \mid 0 \rangle \tag{3}$$

the complex pseudoscalar field with the mass:

$$m_\varphi^2 = \frac{1}{24v^2}\frac{m_t}{m_c}\langle 0 \mid \frac{\alpha_s}{\pi} G^n_{\mu\nu} G^{\mu\nu}_n \mid 0 \rangle \tag{4}$$

and complex scalar field the mass square of which is negative:

$$m_f^2 = -\frac{1}{24u^2}\frac{m_t}{m_c}\langle 0 \mid \frac{\alpha_s}{\pi} G^n_{\mu\nu} G^{\mu\nu}_n \mid 0 \rangle \tag{5}$$

Complex field with masses (4-5) are the nontrivial representation of residual symmetry of $U(1)$ group, but the real field (3) is the sole representation of this group. We propose that cosmological DM consists of particles with these masses and their analogies from the down- quark-familon and lepton-familon subsystems.

The negative square of mass of complex scalar field means that for

$$T < T_{c(up)} \sim \mid \bar{m}_f \mid \sim \frac{\Lambda_c^2}{\Lambda_{mc}} \sqrt{m_t/m_c} \tag{6}$$

pseudogoldstone vacuum is unstable that is when $T = T_{c(up)}$ in gas of pseudo-goldstone bosons should be the RPT in a state with spontaneous breaking $U(1)$ symmetry. Two other familon subsystems can be studied by the same methods. Therefore DM consisting of pseudo-GB of familon type is a many component heterogenic system evolving complex thermodynamical way.

We have detected that RPT in familon gas is RPT of first kind with wide region of phases coexistence. Therefore in epoch RPT or more exactly in region phase coexistence the Universe had block-phase structure containing domains of different phases. The numerical modelling of this RPT [8] has shown that average contrast of density in the block-phase structure is $\delta\epsilon/\epsilon \sim 1$. The size of domains and masses of baryon and dark matter inside domains are defined by distance to horison of events $L_{horiz.}$ at moment of RPT. As it is seen from (6-7) numerical values of these magnitudes which are important for LSS theory depend on values of unknown today parameter of preon confainment Λ_{mc}.

If inhomogeneties appearing during RPT in familon gas have the relation to observable scales of LSS (10 Mpc) then $\Lambda_{mc} \sim 10^5\ Tev$. More detail estimates today is premature but it is necessary to note that suggested theory contains two phase transitions and therefore two characteristic scales of LSS. Here also it is necessary to underline that a catastrophic phenomenon in familon gas could not influence on spectrum of relic radiation even if $m_f \geq 0.4\ eV$ due to superweak interaction familons with the usual matter but effects connected with the fragmentation of DM medium may be superimposed at the spectrum of CMB radiation.

Numerical estimates of inhomogeneities parameters arising as the result of strong interaction of domains LS and HS phases in region of their contact show that $\delta\epsilon/\epsilon \sim 1$ on scale $L \sim 0.1 L_{horison}$ at the moment of the phase transition.

We have prepared the more detail publication [8].

REFERENCES

1. Fritzsch H., Mandelbaum G. 1981,Phys.Lett.B 102,319.
2. Adloff C. et al., 1997, Z.Phys.C74, 191; Breitweg J., 1997, Z.Phys.C74, 207.
3. Frieman J.A., Hill C.T., Watkins R., 1991, Preprint Fermilab Pub-91/324-A; Hill C.T., Schramm D.N., Fry J., 1989, Comments of Nucl. and Particle Phys. 19, 25.
4. Afanas'ev V.L.,Dodonov S.N. (Private communication).
5. Barnett R.M. et al., 1996, Phys. Rev. D54.
6. Vereshkov G., Burdyuzha V., 1995, Intern. J. Modern. Phys. A. 10, 1343.
7. Burdyuzha V., Lalakulich O., Ponomarev Yu., Vereshkov G., 1997, Phys. Rev. D 55, 7340 R.
8. Burdyuzha V., Lalakulich O., Ponomarev Yu., Vereshkov G., 1998, Phys. Lett. B. (submitted).

NEUTRINOS

Atmospheric Neutrino Oscillations

David W. Casper

Department of Physics and Astronomy, University of California, Irvine, Irvine CA 92697

Abstract. Underground atmospheric neutrino measurements over a broad range of energies indicate a lower rate of ν_μ interactions than predicted from the calculated fluxes. In addition, the angular distribution of ν_μ events inside the detectors and entering from the surrounding rock is inconsistent with expectation. Neutrino oscillation ($\nu_\mu \rightarrow \nu_\tau$ or $\nu_\mu \rightarrow \nu_{sterile}$) with maximal mixing and $\Delta m^2 \sim$ few $\times 10^{-3}$ (eV)^2 appears to provide a consistent explanation of the data.

THE PHYSICS OF ATMOSPHERIC NEUTRINOS

Primary cosmic rays impinging on the Earth's atmosphere create showers of secondary particles, many of which decay into neutrinos (see Figure 1). The flux of these "atmospheric" neutrinos can be calculated from measured primary spectra, models of the atmospheric density profile and the terrestrial magnetic field, hadronic cross-sections measured at accelerators, and the well-known decay properties of pions, muons, and kaons. The primary cosmic-ray flux and modeling of hadronic interactions are the dominant uncertainties in predicting the rate of neutrino production in the atmosphere, leading to an estimated error of ±15-20% on the absolute fluxes (1,2).

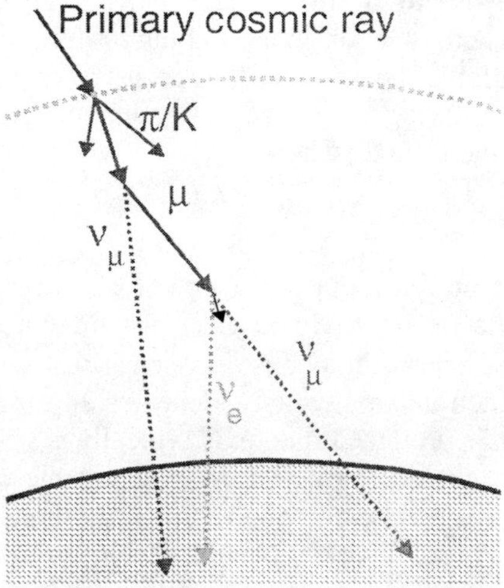

FIGURE 1. Schematic view of atmospheric neutrino production

CP478 , *COSMO—98*, edited by David O. Caldwell

The flavor composition of atmospheric neutrinos can be estimated from simple arguments. The relevant decay chains are:

$$\pi/K \rightarrow \mu\, \nu_\mu$$
$$\mu \rightarrow e\, \nu_\mu\, \nu_e$$

Thus, from flavor-counting, a ν_μ:ν_e ratio of roughly 2:1 is expected in the limit that all mesons decay before reaching the detector depth. In reality, the highest energy muons do not decay above the detector, and consequently this ratio grows with increasing energy. Nevertheless, calculations of the relative fluxes of ν_μ and ν_e are insensitive to uncertainties in the absolute primary cosmic ray flux and hadronic interaction models, and have quoted errors of roughly 5% (1,2,3).

A variety of independent estimates of the expected neutrino flux have been performed; differences in the predicted ν_μ/ν_e ratio are small and reasonably well-understood. Further, while the Earth's magnetic field deflects low-energy primaries from certain directions, it too has negligible effect on the flavor composition (4).

Atmospheric neutrinos are detected by two general methods. Neutrinos may interact inside the sensitive volume of an underground detector, allowing direct observation. These contained interactions allow the flavor of the neutrino to be identified based on the final-state lepton. Events with a single visible track (mainly the result of a quasi-elastic scattering) are the easiest to assign a flavor, however in principle a subset of multi-track events can also be identified as either ν_μ charged-current, ν_e charged-current, or neutral current-induced with reasonable efficiency. If the event is "fully-contained" in the detector, such that no particles escape, the visible energy provides an estimate of the initial neutrino energy. If the event is only "partially-contained", with one or more exiting tracks, only a lower-limit on the neutrino energy is available, however in a large detector, the vast majority of exiting tracks are muons, simplifying the flavor assignment. Contained neutrino data are usually reported using the ratio R, where

$$R \equiv \frac{(\mu / e)_{Data}}{(\mu / e)_{MC}}$$

Neutrinos may also interact outside a detector's sensitive volume, producing a lepton which enters the detector. Since electrons do not penetrate any appreciable distance in rock, this indirect technique is only sensitive to ν_μ interactions. Without the contained signature which unambiguously identifies an event as neutrino-induced, the acceptance for externally-produced muons is typically restricted to zenith angles at or below the horizon, to exclude downward-going cosmic-ray muons with sufficient energy to penetrate the overburden. Such "upward-going muons" may traverse the detector volume (and eventually exit) or stop inside the detector, but in neither case can the energy of the muon and any associated reaction products be measured. Even so, the through-going/stopping signature provides a statistical indication of the parent neutrino's energy: through-going upward-muons are typically produced by neutrinos of energy ~100 GeV, while stopping upward muons mainly result from neutrinos of energy ~10 GeV.

ATMOSPHERIC NEUTRINO OSCILLATIONS

Neutrino oscillation has been widely discussed as a solution to the solar neutrino problem, and is a quantum mechanical consequence of non-degenerate neutrino masses and mixing between neutrino mass and weak-interaction eigenstates. The well-known expression for a neutrino in a weak interaction eigenstate a to transform into a different eigenstate a' after traversing a length L is:

$$P(\upsilon_a \rightarrow \upsilon_{a'}) = \sin^2 2\theta \times \sin^2(1.27 \cdot \Delta(m^2 / 1\ eV^2) \cdot L(km) / E_\upsilon(GeV)),$$

where θ is the mixing angle between weak and mass eigenstates and $\Delta(m^2)$ is the difference in squared mass eigenvalues. Observation of neutrino oscillation would imply that neutrinos have finite mass, with possible implications for cosmology and grand-unification.

Atmospheric neutrinos provide a unique opportunity to search for neutrino oscillation, owing to the variation of both L and E_ν over three orders of magnitude. L varies with neutrino arrival direction at the detector, from ~15 km for neutrinos produced in the atmosphere directly above the detector to ~13000 km for neutrinos produced on the other side of the Earth. E_ν varies from a few hundred MeV for contained neutrino interactions to a few hundred GeV for through-going upward muons. Hence atmospheric neutrino measurements are sensitive to a wide range of $\Delta(m^2)$ and further allow independent checks by looking for variation with energy at constant L, and variation with L at constant energy. Oscillation between ν_e and ν_μ would distort the flavor composition of both observed electrons and muons, while oscillation into ν_τ manifests itself as a disappearance phenomenon, since only ν_τ neutral current reactions have significant cross-sections.

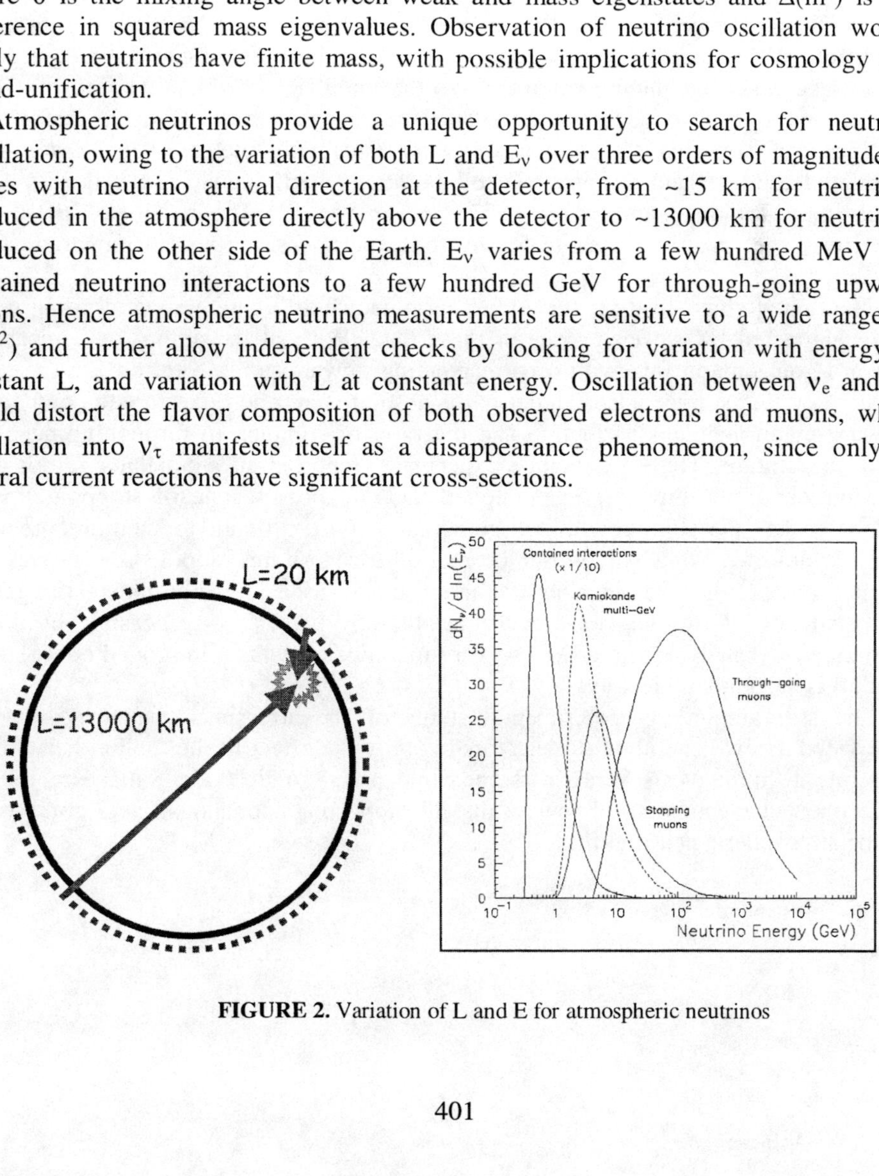

FIGURE 2. Variation of L and E for atmospheric neutrinos

THE ATMOSPHERIC NEUTRINO PROBLEM

The first high-statistics measurements of atmospheric neutrinos came from nucleon decay detectors in the 1980's. The IMB-1 collaboration noted a smaller than expected rate of events with a muon decay in their contained sample (5), however the detector's sensitivity was insufficient to perform direct particle identification on the data. The Kamiokande experiment, with smaller mass but considerably greater sensitivity, subsequently announced a ~30% deficit of muon-like events in its "sub-GeV" (E_{vis}<1.33 GeV) data sample, providing the first clear evidence for anomalous flavor composition of atmospheric neutrinos(6). The upgraded IMB-3 detector, with particle identification capability, later confirmed the Kamiokande result (7). Kamiokande also observed not only a muon deficit, but also an angular dependence of the depletion, in its higher energy "multi-GeV" (E_{vis}>1.33 GeV) sample (8).

At the same time, two smaller experiments, Fréjus(9) and NUSEX(10), found no evidence of the anomaly. Since the two experiments observing the effect used the water Cherenkov technique, while the two that didn't observe it were iron-based, the experimental status at the end of the 1980's was less than clear. To determine whether the particle identification abilities of water Cherenkov detectors had been over-estimated, the IMB and Kamiokande collaborations jointly conducted tests with muon, electron, and pion beams at the KEK accelerator(11). The results of these tests confirmed the sensitivities estimated by the experimental groups from Monte Carlo studies.

The first attempts to study the atmospheric neutrino flux with upward-going muons were hampered by the lower rate of these data. Without the ability to compare the upward-going muon flux with a corresponding upward-going electron flux, the only ways to evade the large uncertainties in the absolute neutrino fluxes are to compare the upward-muon angular distribution and the ratio of stopping to through-going events with expectation. The Kamiokande experiment reported an anomalous zenith angle distribution of its upward-going muon data(12). The low rate of stopping upward muons makes the stopping/through-going ratio more difficult to measure precisely. IMB-3 published limits on neutrino oscillation based on this ratio(13), but it was later pointed out(14) that the neutrino interaction model used probably underestimates the expected rate of stopping upward-muons, thereby masking any possible oscillation signature. A reanalysis of the IMB upward muon data (with an improved cross-section model) is currently underway.

The still ambiguous experimental status of the atmospheric neutrino problem motivated a new generation of detectors to study the effect further and either confirm or refute it. In the past several years, the Soudan-II, MACRO, and Super-Kamiokande experiments have announced new results, all supporting anomalous flavor composition of the atmospheric neutrino flux.

CONTAINED NEUTRINO INTERACTIONS

Super-Kamiokande is a very large (50 kt, 22.5 kt fiducial) water Cherenkov detector located at the site of the original Kamiokande experiment. The detector consists of a cylindrical tank of highly purified water, 34 m in diameter and 36 m high, optically divided into inner and outer detector regions. The outer detector comprises 1885 20-cm photo-multiplier tubes (PMTs) with wavelength-shifting plates, and is used to tag tracks entering or exiting the inner region. The inner detector's surface is instrumented with 11,146 50-cm PMTs which image the Cherenkov cone projected by relativistic ($\beta > 0.75$) charged particles. The position of an interaction is reconstructed to within about 25 cm, and angular resolution is a few degrees for single-ring events. Visible energy is measured by summing the corrected pulse-heights of firing PMTs. Tracks are identified as either e-like (showering) or μ-like (non-showering) based on the geometry of the Cherenkov pattern. Misidentification is estimated to be about 2%. Using the visible energy and particle identification, lepton momenta can be inferred.

Fully-contained events with $E_{vis} < 1.33$ GeV make up the "sub-GeV" data sample, while higher-energy fully-contained events, as well as all partially-contained events, make up the "multi-GeV" sample. For the principle oscillation analysis, events with a single Cherenkov ring are selected since these are 85% (95%) quasi-elastic scattering by ν_e (ν_μ) for the e-like and μ-like samples, respectively. In a 736 live-day (45.3 kt-yr) exposure, both sub-GeV(15) and multi-GeV(16) samples show a significant deficit of μ-like events, as summarized in Table 1.

Sub-GeV event Summary

$E_{vis} < 1.33$GeV
$P_e > 100$MeV/c
$P_\mu > 200$MeV/c

	DATA	MC(Honda)	MC(Gaisser)
1R	3224	3765.3	4177.9
e−like	1607	1502.7	1718.2
μ−like	1617	2262.6	2459.7
2R	1010	1214.6	1282.8
≧3R	261	416.4	428.8
TOTAL	4495	5396.3	5889.5

$$\frac{(\mu/e)_{DATA}}{(\mu/e)_{MC}} = 0.668 \pm^{0.024}_{0.023} \pm 0.007 \pm 0.052 \text{ (Honda)}$$

data stat. MC stat. sys.

$$\frac{(\mu/e)_{DATA}}{(\mu/e)_{MC}} = 0.703 \pm^{0.025}_{0.024} \pm 0.007 \pm 0.055 \text{ (Gaisser)}$$

data stat. MC stat. sys.

Multi-GeV event Summary

(1)FC($E_{vis} > 1.33$GeV)

	DATA	MC(Honda)	MC(Gaisser)
1R	687	771.0	792.6
e−like	386	351.9	370.4
μ−like	301	419.1	422.2
2R	388	506.7	521.2
≧3R	349	428.7	446.5
TOTAL	1424	1706.4	1760.2

(2)PC	DATA	MC(Honda)	MC(Gaisser)
TOTAL	374	508.9	539.0

*All events are assumed to be μ−like.
*Fraction of CC ν_μ, $\bar{\nu}_\mu$ events in the PC sample is estimated to be (97−98)%.

FC+PC

$$\frac{(\mu/e)_{DATA}}{(\mu/e)_{MC}} = 0.663 \pm^{0.044}_{0.041} \pm 0.013 \pm 0.078 \text{ (Honda)}$$

$$= 0.674 \pm^{0.045}_{0.042} \pm 0.012 \pm 0.079 \text{ (Gaisser)}$$

TABLE 1. Sub-GeV and multi-GeV event rates at Super-Kamiokande

The multi-GeV sample, as well as the sub-GeV sample for tracks with p > 400 MeV/c, also show clear evidence of an angular dependence of the muon depletion, consistent with the dependence on neutrino flight distance expected from oscillation (see Figure 3).

Super-Kamiokande has analyzed its data to determine whether known systematic uncertainties, when allowed to vary within allowed tolerances, can account for the anomalous flavor composition and angular distribution of its data. For no oscillations, the best fit χ^2=175/69 DoF. The best fit value oscillation hypothesis in the physical region is $\Delta(m^2)=3.5\times10^{-3}$ $(eV)^2$ and $\sin^2 2\theta$ =1, yielding χ^2=62/67 DoF(17).

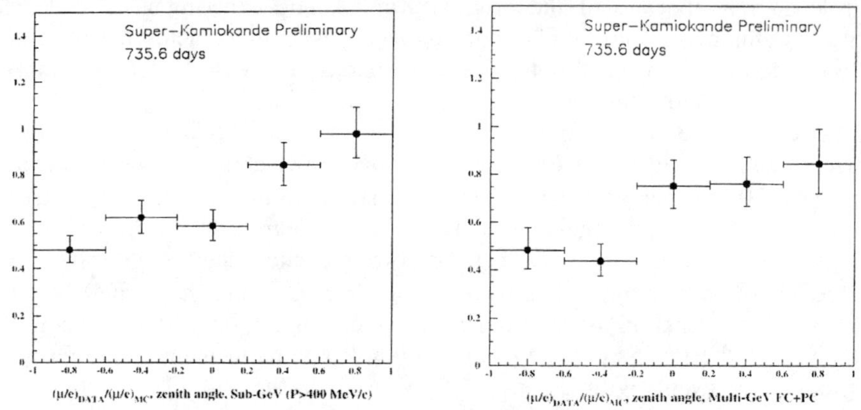

FIGURE 3. Zenith angle dependence of the Super-Kamiokande contained muon deficit

Soudan-II is a 960-ton fine-grained iron-based detector located in the Homestake Mine in Minnesota. The detector is comprised of 224 4.3-ton modules, surrounded by an active veto shield. Each module contains 7560 drift tubes layered between 1.6 mm steel plates. The highly segmented design provides excellent spatial resolution (~2 mm in x,y; ~1 cm in z). Particle identification is performed by distinguishing the long, straight geometry of a minimum-ionizing track from the diffuse, fuzzy pattern characteristic of an electromagnetic shower. Performance of the particle identification has been studied with Monte Carlo and in a test beam.

Results reported from a 3.89 kt-yr exposure are summarized in Table 2. Soudan-II observed an R value of 0.64 ± 0.11(stat) ± 0.06(sys), while 1.0 is expected. Soudan-II is thus the first iron detector to observe the atmospheric neutrino anomaly(18).

Soudan-II event sample	*"Track" (i.e. μ-like)*	*"Shower" (i.e. e-like)*
Background-subtracted data	76.9 ± 10.8	116.3 ± 12.8
Monte-Carlo prediction	137.4	133.8
Soudan-II: R=0.64 ± 0.11 ± 0.06		

TABLE 2. Summary of Soudan-II data sample

Using the high-resolution of their detector, Soudan-II has also identified a smaller sample of events with a clear recoiling proton track. Together with the neutrino-induced lepton and any pions produced, measurement of the recoiling proton allows a more accurate reconstruction of the initial neutrino energy and direction. The tighter requirements also decrease the sample size. This high-resolution sample is summarized in Table 2. The estimated resolution on the neutrino direction is ~25° and the neutrino energy is reconstructed to within 20%. Particle-identification for this sample is 93%. The measured value of R for this high-resolution sample is 0.52 ± 0.09 (stat) (preliminary) (19).

Soudan-II high-resolution sample	*"Track" (μ-like)*	*"Shower" (e-like)*	*Multi-Track (ν_μ)*	*Multi-Track (ν_e)*
Background-subtracted data	32.2	65.9	28.6	40.5
Monte-Carlo prediction	78.3	60.6	33.2	40.6
Soudan-II High-Resolution Sample: R=0.52 ± 0.09 (Preliminary)				

TABLE 3. Summary of Soudan-II high-resolution sample.

Soudan-II have analyzed their high-resolution data to determine its consistency with the oscillation hypothesis, and the best fit value of $\Delta(m^2)$. Using their electron sample for normalization, they find χ^2=24.1/18 DoF at $\Delta(m^2)=1.1\times10^{-2}$ $(eV)^2$. The allowed region is relatively broad, however, and encompasses the best fit point from Super-Kamiokande at 90% confidence level. To study the no-oscillation hypothesis, they have fit their data allowing only the overall normalization to float. The best fit is rather poor, with χ^2=43/20 DoF.

UPWARD-GOING MUONS

Super-Kamiokande has also collected data from ν_μ interactions outside their detector, using the upward-going muon signature. Muons arriving from below the horizon ($\theta_z > 90°$) are accepted if they traverse a total length of 7m or more in the inner detector region before exiting or stopping. Because the detector overburden is not uniform, a small amount of cosmic-ray muon contamination is measured in the azimuthal region where the overburden is shallow. This background (comprising roughly 1.5% of the total event rate, is subtracted from the $-0.1<\cos\theta_z < 0.0$ bin.

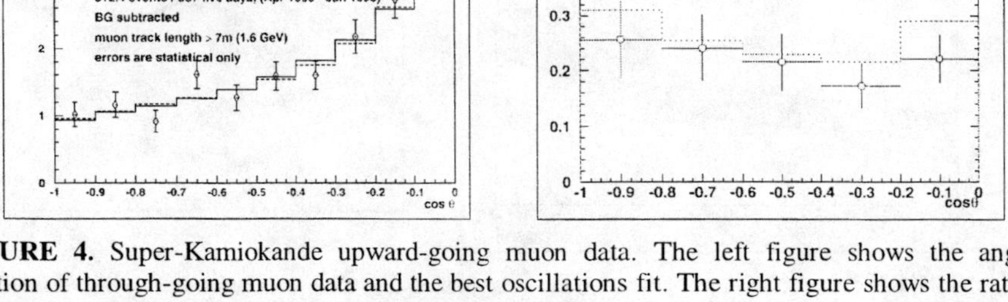

FIGURE 4. Super-Kamiokande upward-going muon data. The left figure shows the angular distribution of through-going muon data and the best oscillations fit. The right figure shows the ratio of stopping to through-going events with and without oscillation.

Due to uncertainties in the absolute atmospheric neutrino flux, the total rate of upward-going muons does not strongly constrain the oscillation hypothesis. The angular distribution of through-going upward muons and the ratio of stopping to through-going events are relatively insensitive to these uncertainties, however. Neither the angular distribution of through-going events nor the relative rates of stopping and through-going events agree with expectation in the absence of oscillation. Figure 4 shows the data, and the best fit oscillation hypothesis. For through-going muons, a fit without oscillation (allowing the normalization and other uncertain quantities to vary) gives a best-fit χ^2=19.2/10 DoF, compared to χ^2=7.5/8 DoF for oscillations at $\Delta(m^2)=6\times10^{-3}$ (eV)^2 (20). For stopping muons, the rate is roughly half that predicted relative to through-going events, yielding a similar value of $\Delta(m^2)$ (21).

MACRO is a large surface detector located in the Gran Sasso laboratory in Italy. Although designed as a magnetic monopole detector, it is also sensitive to upward-going muons. The completed detector is $76.7\times12\times9.3$ m^3, although data collection has proceeded while additional modules were added. A total of 4.7 effective live-years have been recorded. The components relevant for upward-muon detection are horizontal planes of streamer tubes and liquid scintillator – there is no acceptance for exactly horizontal tracks. Time-of-flight discrimination can distinguish upward and downward going muons. Three data-sets are identified: upward through-going muons, upward-stopping muons, and partially-contained (upward or downward) interactions.

The zenith angle distribution of the through-going MACRO data is shown in Figure 5. Allowing only the normalization to vary (without oscillations), the best fit to the angular distribution gives χ^2=26.1/8 DoF. With oscillation ($\Delta(m^2)$~few $\times$ 10^{-3} (eV)^2), the fit improves but is still rather poor. The conclusion of the MACRO collaboration is that "the data favor oscillations"(22). Preliminary results with the stopping muon and partially-contained samples also favor oscillations with $\Delta(m^2)=5\times10^{-3}$ (eV)^2. The rate of stopping upward-muons and partially-contained downward muons appears to be significantly suppressed, consistent with Super-Kamiokande measurements(23).

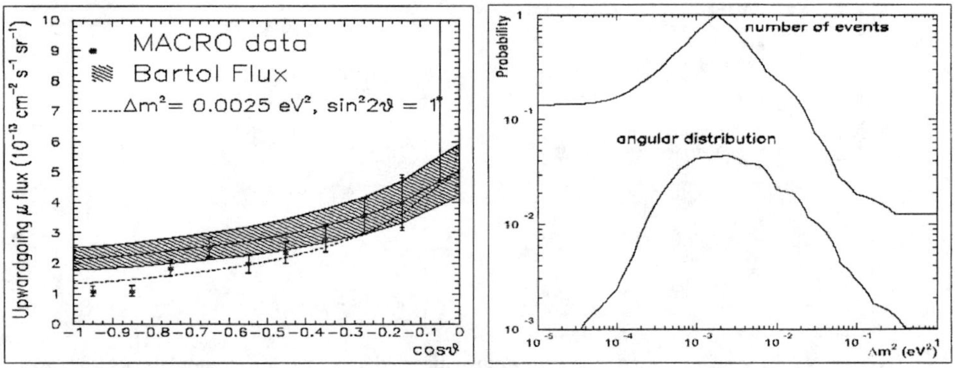

FIGURE 5. MACRO upward-muon data compared to expectation with and without oscillation (left). Likelihood vs. $\Delta(m^2)$ based on the angular distribution and number of events (right).

CONCLUSION

The atmospheric neutrino anomaly, first observed in the mid-1980's, has recently been established much more conclusively. Measurements by the present generation of detectors (Super-Kamiokande, Soudan-II, and MACRO), at a variety of energies and using a variety of techniques, are all consistent with $\nu_\mu \rightarrow \nu_\tau$ (or $\nu_\mu \rightarrow \nu_{sterile}$) oscillation with large mixing and $\Delta(m^2) \sim 5\times10^{-3}$ $(eV)^2$, although it must be emphasized that each experiment's allowed region still accommodates at least a factor 2 variation from the best-fit value). Earlier measurements at Kamiokande and IMB are also generally consistent with the presently favored value, if one accepts that an inadequate cross-section model may have skewed the original IMB stopping/through-going upward muon result.

Additional data from Super-Kamiokande and Soudan-II should remove whatever questions remain about underground measurements of the effect, but to definitively prove oscillation (and hence non-zero neutrino mass) will require confirmation using a more controlled neutrino source. The K2K experiment, to begin data-taking in Spring 1999, will be the first of several long-baseline experiments to attempt to verify the reality of neutrino oscillations and the value of $\Delta(m^2)$ inferred from atmospheric data.

REFERENCES

1. V. Agrawal, *et al., Phys Rev* .**D53**, 1313 (1996).
2. M. Honda, *et al., Phys. Lett.* **D52**, 4985 (1995).
3. T.K. Gaisser, *et al., Phys. Rev.* **D54**, 5578 (1996).
4. P. Lipari, T.K. Gaisser and T. Stanev, *Phys. Rev.* **D58**, 073003 (1998).
5. T.J. Haines, *et al., Phys. Rev. Lett.* **57**, 1986 (1986).
6. K.S. Hirata, *et al., Phys. Lett.* **B205**, 416 (1988).
   K.S. Hirata, *et al., Phys. Lett.* **B280**, 146 (1992).
7. D. Casper, *et al., Phys.Rev.Lett.* **66**, 2561 (1991).
   R. Becker-Szendy, *et al., Phys. Rev* **D46**, 3720 (1992).
8. Y. Fukuda, *et al., Phys. Lett.* **B335**, 237 (1994).
9. K. Daum, *et al., Z. Phys.* **C66**, 417 (1995).
10. M. Aglietta, *et al., Europhys. Lett.* **8**, 611 (1989).
11. S. Kasuga, *et al., Phys. Lett.* **B374**, 238 (1996).
12. S. Hatakayama, *et al., Phys. Rev. Lett.* **81**, 2016 (1998).
13. R. Becker-Szendy, *et al., Phys. Rev. Lett.* **69**, 1010 (1992).
14. P. Lipari, M. Lusignoli and F. Sartogo, *Phys. Rev. Lett.* **74**, 4384 (1995).
15. Y. Fukuda, *et al., Phys. Lett.* **B433**, 9 (1998).
16. Y. Fukuda, *et al., Phys. Lett.* **B436**, 33 (1998).
17. Y. Fukuda, *et al., Phys. Rev. Lett.* **81**, 1562 (1998).
18. W.M.W. Allison, *et al., Phys. Lett.* **B391**, 491 (1997).
19. H.Gallagher (for the Soudan-II collaboration), "Atmospheric Neutrinos in Soudan 2," presented at the International Conference on High-Energy Physics, Vancouver, British Columbia, July 1998.
20. Y.Fukuda, *et al.*, hep-ex/9812014 (accepted by *Physical Review Letters*).
21. Super-Kamiokande Collaboration, in preparation.
22. M. Ambrosio, *et al., Phys. Lett.* **B434**, 451 (1998).
23. M. Spurio (for the MACRO collaboration), "Low energy atmospheric muon neutrinos in MACRO," presented to the 16th ECRS, Madrid, July 1998 (hep-ex/9808001).

Atmospheric Neutrinos from Charm

L. Pasquali[1], M. H. Reno[1] and I. Sarcevic[2]

[1]*Department of Physics and Astronomy, University of Iowa, Iowa City, Iowa 52242*
[2]*Department of Physics, University of Arizona, Tucson, Arizona 85721*

Abstract. Using next-to-leading order perturbative QCD, we find that the charm contribution to the atmospheric lepton fluxes dominates over the conventional ones from π and K decays for energies higher than 10^5 GeV. We also discuss theoretical uncertainties involved in the calculations and compare our results with previous evaluations.

INTRODUCTION

Conventional lepton fluxes from π and K decays in the atmosphere are well known [1], but at energies of a few GeV charm-anticharm pairs can be produced and the leptons coming from charm decays, called prompt leptons, start to contribute to the atmospheric lepton fluxes. Results from several experiments [2] show a muon excess relative to the conventional muon flux in the 10 TeV energy range that could be an indication of a charm contribution. Large volume water and ice neutrinos experiments [3] are designed to search for muon neutrinos from extragalactic sources for which atmospheric muons and muon neutrinos represent the main background. Therefore it is important to determine the atmospheric lepton flux, including charm contributions, with better accuracy.

Atmospheric lepton fluxes from charm decays have been evaluated in the past using different models of charm production [4,5]. We present a new calculation [6] based on next-to-leading order (NLO) perturbative QCD. We also investigate the importance of the small-x behaviour of the parton distribution functions and study the uncertainties involved in the necessary extrapolation of cross sections and energy distributions beyond the experimentally measured regime.

LEPTON FLUX CALCULATION

The lepton fluxes can be calculated semi-analytically using approximate cascade equations which begin with the incident cosmic ray flux. Details of this solution method can be found, for example, in Refs. [7,6]. As in Ref. [4], we assume that the incident cosmic ray flux can be represented by protons and that at the top

CP478 , *COSMO—98*, edited by David O. Caldwell

of the atmosphere (X=0) the proton flux $\phi_p(E, X)$ scales as 1.7 $(E/\text{GeV})^{-2.7}$ for $E < 5 \cdot 10^6$ GeV and scales as 174 $(E/\text{GeV})^{-3}$ for $E \geq 5 \cdot 10^6$ GeV.

Production of charmed hadrons j of energy E by proton-air interactions enters into the calculation via the "Z-moment", expressed in terms of an integral over $x_E = E/E_p$, which is given by

$$Z_{pj} = 2f_j \int_0^1 \frac{dx_E}{x_E} \frac{\phi_p(E/x_E)}{\phi_p(E)} \frac{1}{\sigma_{pA}(E)} \frac{d\sigma_{pA\to c\bar{c}}(E/x_E)}{dx_E}, \tag{1}$$

where f_j is the fraction of charmed particles which emerge as hadron j and the factor of two is taking into account the fact that we implicitly sum over particles and antiparticles. The quantity $\sigma_{pA}(E)$ is the total proton-air cross section [8].

The charm production cross section and energy distribution depends on the value of the charm mass (m_c), the factorization scale (M) and renormalization scale (μ) as well as the choice of parton distribution functions. Comparing the NLO total charm cross section [9] with the experimental data from Ref. [10] led us to set $m_c = 1.3$ GeV. We use the CTEQ3 [11] and the MRSD- [12] parton distribution functions. We set our default factorization and renormalization scales to $M = 2\mu = 2m_c$. We also evaluate the lepton flux using the CTEQ3 parton distribution functions with $M = \mu = m_c$.

Because of the large computing time involved, we have used the NLO Monte Carlo program from Mangano, Nason and Ridolfi [13] to evaluate the appropriate rescaling factor K that, applied to the "LO" energy distribution, correctly accounts for the NLO corrections. Here "LO" means taking the leading order matrix element squared, but using the two-loop $\alpha_s(\mu^2)$ and the NLO parton distribution functions. The parametrization of $K(E, x_E)$ is given by the following expression

$$\begin{aligned} K(E, x_E) = {} & 1.36 + 0.42\ln(\ln(E/\text{GeV})) \\ & + \left(3.40 + 18.7(E/\text{GeV})^{-0.43} - 0.079\ln(E/\text{GeV})\right) \cdot x_E^{1.5}. \end{aligned} \tag{2}$$

The K-factor rescaled "LO" charm production distribution is used in Eq. (1) to obtain our lepton fluxes.

In Fig. 1 we show our results for the prompt atmospheric muon flux scaled by E^3 for two parton distributions and factorization scale choices. The highest flux at $E = 10^8$ GeV is with the MRSD- distribution and $M = 2\mu = 2m_c$ (dashed). The CTEQ3 distributions with the same choice of scale are represented by the solid line, while the dot-dashed line shows the result when $M = \mu = m_c$. For reference, we show the vertical conventional and prompt muon flux calculated and parameterized by Thunman *et al.* in Ref. [4]. The prompt flux is isotropic. The prompt lepton flux evaluated using perturbative QCD can be parameterized as

$$\log_{10}\left(E^3\phi_\ell(E)/(\text{GeV}^2/\text{cm}^2\,\text{s}\,\text{sr})\right) = -A + B\,x + C\,x^2 - D\,x^3, \tag{3}$$

where $x \equiv \log_{10}(E/\text{GeV})$. In Table I, we collect the constants for the MRSD- and CTEQ3 fluxes exhibited in Fig. 1. The prompt neutrino fluxes equal the prompt muon flux.

TABLE 1. Parameters for the prompt muon plus antimuon flux appearing in Fig. 1, parameterized by Eq. (3).

PDF	Scales	A	B	C	D
CTEQ3	$M = \mu = m_c$	5.37	0.0191	0.156	0.0153
CTEQ3	$M = 2\mu = 2m_c$	5.79	0.345	0.105	0.0127
MRSD-	$M = 2\mu = 2m_c$	5.91	0.290	0.143	0.0147

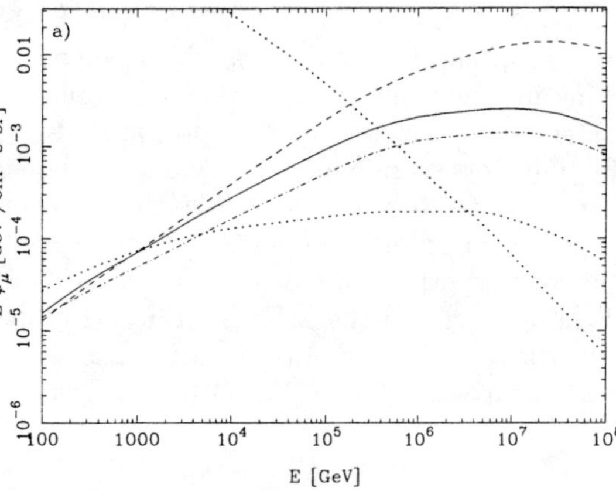

FIGURE 1. The prompt atmosheric muon flux scaled by E^3 versus muon energy for CTEQ3 (solid) and MRSD- (dashed) with $M = 2\mu = 2m_c$. Also shown is the scaled muon flux using CTEQ3 with $M = \mu = m_c$ (dot-dashed) and the Thunman *et al.* parametrization [4] of the prompt muon flux and the vertical conventional muon flux (dotted).

CONCLUSIONS

We find that the perturbative charm contributions to lepton fluxes are significantly larger than the recent Thunman *et al.* calculation [4]. The prompt muon flux becomes larger than the conventional muon flux from pion and kaon decays at energies above $\sim 10^5$ GeV. We have evaluated the energy distribution at NLO in QCD, and we find that the NLO corrections give a correction of more than a factor of two which is weakly energy and x_E dependent.

The main uncertainty in the perturbative calculation of the prompt flux, given fixed charm mass, factorization scale and renormalization scale, is the small-x behavior of the parton distribution functions. The spread in predictions at $E = 10^8$ GeV is indicative of this uncertainty.

We conclude that the prompt muon flux calculated in the context of perturbative QCD cannot explain the observed excess of muons in the TeV region [2], independent of the theoretical uncertainties associated with small parton x. Measurements of the atmospheric flux in the 100 TeV range would help pin down the charm cross section at energies above those currently accessible using accelerators and would provide valuable information about the small-x behavior of the gluon distribution function.

ACKNOWLEDGEMENTS

Work supported in part by National Science Foundation Grant Nos. PHY-9507688 and PHY-9802403 and D.O.E. Contract No. DE-FG02-95ER40906.

REFERENCES

1. Gaisser, T. K., *et al.*, *Phys. Rev. D* **54**, 5578 (1996) and references therein.
2. Rhode, W., *Nucl. Phys. (Proc. Suppl.)* **B35**, 250 (1994); Bakatanov, V. N., *et al.*, *Sov. J. Nucl. Phys.* **55**, 1169 (1992). For a summary of experimental results, see Bugaev, E. V., *et al.*, *Phys. Rev. D* **58**, 054001 (1998).
3. See, for example, Halzen, F., eprint archive astro-ph/9707289; Anassontzis, E. G., *et al.*, presented at the XVI Int. Workshop on Weak Interactions and Neutrino 97 Conference (Capri); Sokalski, I. (for the Baikal Collaboration), submitted to the Proceedings of the International Workshop XXXIInd Rencontres de Moriond - Very High Energy Phenomena (Les Arcs, Savoie, France, January 18-25, 1997).
4. Thunman, M., Gondolo, P., and Ingelman, G., *Astropart. Phys.* **5**, 309 (1996).
5. Volkova, L. V., *Yad. Fiz.* **31**, 1510 (1980) [English translation: *Sov. J. Nucl. Phys.* **31**, 784 (1980)]; Mitsui, K., Minorikawa, Y., and Komori, H., *Nuo. Cim.* **9c**, 995 (1986); Butkevich, A. V., Dedenko, L. G., and Zheleznykh, I. M., *Yad. Fiz.* **50**, 142 (1989) [English translation: *Sov. J. Nucl. Phys.* **50**, 90 (1990)]; Agrawal, V., Gaisser, T. K., Lipari, P., and Stanev, T., *Phys. Rev. D* **53**, 1314 (1996); Volkova, L. V., and Zatsepin, G. T., *Yad. Fiz.* **37**, 353 (1983); Inazawa, H., and Kobayakawa, K., *Prog. Theor. Phys.* **69**, 1195 (1983); Battistoni, G., *et al.*, *Astropart. Phys.* **4**, 351 (1996); Volkova, L. V., Fulgione, W., Galeotti, P., and Saavedra, O., *Nuo. Cim. C* **10**, 465 (1987); Bugaev, E. V., *et al.*, Ref. [2]; Bugaev, E. V., *et al.*, *Nuo. Cim.* **C12**, 41 (1989).
6. Pasquali, L., Reno, M. H., and Sarcevic, I., eprint archive hep-ph/9806428, to be published in *Phys. Rev. D.*
7. Lipari, P., *Astropart. Phys.* **1**, 195 (1993).
8. Mielke, H. H., Foller, M., Engler, J., and Knapp, J., *J. Phys. G* **20**, 637 (1994).
9. Nason, P., Dawson, S., and Ellis, R. K., *Nucl. Phys. B* **303**, 607 (1988); Beenakker, W., Kuijf, H., van Neerven, W. L., and Smith, J., *Nucl. Phys. B* **431**, 453 (1994); Matsuura, T., van der Marck, S. C., and van Neerven, W. L., *Nucl. Phys. B* **319**, 570 (1989).
10. Frixione, S., *et al.*, e-print archive hep-ph/9702287, to be published in *Heavy Flavour II*, edited by A.J. Buras and M. Lindner, Singapore: World Scientific.
11. Lai, H., *et al.* (CTEQ Collaboration), *Phys. Rev. D* **51**, 4763 (1995).
12. Martin, A. D., Stirling, W. J., and Roberts, R. G., *Phys. Lett.* **B306**, 145 (1993).
13. Mangano, M., Nason, P., and Ridolfi, G., *Nucl. Phys. B* **373**, 295 (1992).

SO(10) and Atmospheric ν Oscillations

Stephen M. Barr

Bartol Research Institute
University of Delaware, Newark, DE 19716

Abstract. The three neutrino angles are predicted in an SO(10) model.

In this talk I shall explain some work that I have done in collaboration with C.H. Albright and K.S. Babu (1), and a recent extension of that work (2).

This work suggests a possible simple answer to the following basic question: *Why is the mixing between the heavier two generations apparently so large for the leptons when it is so small for the quarks?* In other words, why is $\sin^2 2\theta_{\mu\tau} > 0.8$, when $V_{cb} \cong 0.04$? Actually, the question we set out to answer was whether in the context of supersymmetric $SO(10)$ it was possible with a simple Higgs content to reproduce the spectrum of the quarks and charged leptons. As it happens, we were led to a structure in which the mixing between ν_μ and ν_τ is large.

First, I will review some basic facts about $SU(5)$ and $SO(10)$. In (minimal) $SU(5)$, the quarks and leptons are contained in $\mathbf{10}_i + \overline{\mathbf{5}}_i + \mathbf{1}_i$. The subscript i is a family index. The light Higgs fields are contained in $\mathbf{5}_H + \overline{\mathbf{5}}_H$. The subscript H shall always stand for a Higgs multiplet. These Higgs multiplets each contain both a weak doublet field with weak-scale mass and a color-triplet field with GUT-scale mass. This is called the "doublet-triplet splitting", and explaining it is one of the central concerns of work in grand unification.

With such fields, the only quark and lepton mass terms that can be written down (except for the Majorana masses of the singlet neutrinos) are $(\mathbf{1}_i \cdot \overline{\mathbf{5}}_j)\mathbf{5}_H + (\mathbf{10}_i \cdot \mathbf{10}_j)\mathbf{5}_H + (\overline{\mathbf{5}}_i \cdot \mathbf{10}_j)\overline{\mathbf{5}}_H$. The first term leads to the Dirac mass matrix of the neutrinos (N_{ij}); the second to the mass matrix of the up quarks (U_{ij}); and the third to the mass matrices of the down quarks (D_{ij}) and the charged leptons (L_{ij}). Thus, one has in minimal $SU(5)$ the following features: (a) U and D are unrelated, so that the CKM angles are arbitrary and naturally of order 1. (b) D and L satisfy the relation $D = L^T$. This relation predicts $m_b^0/m_\tau^0 = 1$, $m_s^0/m_\mu^0 = 1$, and $m_d^0/m_e^0 = 1$. (Here superscript zeros refer to quantities evaluated at the GUT scale.) The first relation is well-known to work in SUSY GUTs, but the latter do not. In fact, empirically one finds that $m_s^0/m_\mu^0 \simeq 1/3$, and $m_d^0/m_e^0 \simeq 3$. These factors of three are often called the Georgi-Jarlskog factors (3). They can arise as Clebsch coefficients in various ways.

CP478 , *COSMO—98*, edited by David O. Caldwell

The fact that $D = L^T$ is vital, as we shall see. The reason for it is that in $SU(5)$ the $\overline{\mathbf{5}}_i$ contain the *right-handed* down quarks and *left-handed* leptons, while the $\mathbf{10}_i$ contain the *left-handed* down-quarks and *right-handed* leptons. Thus, in minimal $SU(5)$, D and L end up being equal *only up to a left-right transposition.*

In minimal $SO(10)$, all the known quarks and leptons are contained in three spinors, $\mathbf{16}_i$, and the light Higgs multiplets are contained in a single $\mathbf{10}_H$. Consequently, one has the relations $N = U \propto D = L$, where all the matrices are symmetric. The proportionality of U and D makes the CKM angles vanish, which is a good zeroth order approximation to reality. On the other hand, it also gives the notoriously bad prediction $m_c^0/m_t^0 = m_s^0/m_b^0$, which is wrong by about a factor of 20.

The moral is that one wants some minimal $SU(5)$ and $SO(10)$ relations satisfied (at least approximately) and others to be badly broken. Therefore everything depends on exactly how $SU(5)$ and $SO(10)$ break. The starting point of our work is to use the simplest set of Higgs fields to break $SO(10)$ down to the Standard Model group.

In $SO(10)$ it is necessary to have at least *two* kinds of Higgs fields to do the symmetry breaking at the unification scale. First, there must be some multiplets that "break the rank" of the group, from 5 (for $SO(10)$) down to 4 (for $SU(3) \times SU(2) \times U(1)$), and that give superheavy masses to the right-handed neutrinos. These tasks may be accomplished either by a spinor-antispinor pair, $\mathbf{16}_H + \overline{\mathbf{16}}_H$, or a pair of rank-5 tensors, $\mathbf{126}_H + \overline{\mathbf{126}}_H$. We consider the case of spinors, since that seems more economical, and because large representations, like **126**'s, can destroy the perturbativity of $SO(10)$ below the Planck scale. The vacuum expectation values of the $\mathbf{16}_H$ and $\overline{\mathbf{16}}_H$ must be in the direction that leaves $SU(5)$ unbroken (as, of course, any other possibility would also break the Standard Model group at the GUT scale).

Some other kind of Higgs field is required to complete the gauge breaking down to $SU(3) \times SU(2) \times U(1)$ and to produce the doublet-triplet splitting. This can be done either by adjoints (**45**) or by symmetric tensors (**54**) or both. It has been shown (4) that it is possible to do the breaking with only a single **45**, and so that is what we shall consider as the most economical possibility.

The assumption of only a single adjoint is a powerful one, since the fact that this adjoint is responsible for the doublet-triplet splitting determines the direction of its vacuum expectation value (VEV). The only mechanism for doing the doublet-triplet splitting naturally in $SO(10)$ is the Dimopoulos-Wilczek mechanism (also called the missing VEV mechanism). In its simplest form this involves an adjoint whose VEV is proportional to the $SO(10)$ generator $B - L$. The point is that the Higgs doublets which must remain light have $B - L = 0$, whereas the color-triplet Higgs that must be made superheavy have $B - L \neq 0$.

Thus, the assumption of economy and simplicity lead to a definite picture of how the unified groups are broken. The question we pose is whether such a picture can lead to realistic quark and lepton mass matrices.

The masses of the third family can come from the minimal term $(\mathbf{16}_3\mathbf{16}_3)\mathbf{10}_H$, giving the successful relation $m_\tau^0 \cong m_b^0$. However, for the second family the corresponding relation, $m_\mu^0 \cong m_s^0$ is not satisfied, which suggests that these masses feel the breaking of $SU(5)$. Since the GUT-scale breaking of $SU(5)$ is entirely done by the $\mathbf{45}_H$, one seeks an effective term containing that field. The lowest dimension term of that form is $(\mathbf{16}_2\mathbf{16}_3)\mathbf{10}_H\mathbf{45}_H/M_G$. (Such a term can arise simply from integrating out a spinor-antispinor pair.) It is easily shown that because $\langle \mathbf{45}_H \rangle \propto B - L$ this term gives a flavor-antisymmetric contribution to the mass matrices. These contributions ($\propto B - L$) will be denoted below by the letter ϵ.

For several reasons, which will become clear shortly, it is necessary to add one more contribution to get realistic mass matrices. The best way to do this is through a term of the form $[\mathbf{16}_2\mathbf{16}_H][\mathbf{16}_3\mathbf{16}'_H]$, where the products in the square brackets are contracted into $\mathbf{10}$'s of $SO(10)$. (Such a term can easily arise by integrating out $SO(10)$ vectors.) The field $\mathbf{16}'_H$ is supposed to acquire a VEV only in the $SU(2)_L \times U(1)_Y$-breaking direction. It is easily shown that this term contributes only to D and L, and in a lopsided way, *i.e.* to D_{23} and L_{32} but not to D_{32} and L_{23}. These entries are denoted below by the letter σ. The resulting matrices have the form

$$
\begin{aligned}
U^0 &= \begin{pmatrix} 0 & & \\ & 0 & \epsilon/3 \\ & -\epsilon/3 & 1 \end{pmatrix} m_U, \quad
D^0 = \begin{pmatrix} 0 & & \\ & 0 & \sigma + \epsilon/3 \\ & -\epsilon/3 & 1 \end{pmatrix} m_D, \\
N^0 &= \begin{pmatrix} 0 & & \\ & 0 & -\epsilon \\ & \epsilon & 1 \end{pmatrix} m_U, \qquad
L^0 = \begin{pmatrix} 0 & & \\ & 0 & -\epsilon \\ & \sigma + \epsilon & 1 \end{pmatrix} m_D.
\end{aligned}
\tag{1}
$$

The entries σ play several crucial roles which will now be explained. First, without them one has the see-saw relations $m_\mu^0/m_\tau^0 \cong \epsilon^2$ and $m_s^0/m_b^0 \cong \epsilon^2/9$. This gives a ratio of ratios of $1/9$ rather than the Georgi-Jarlskog value $1/3$. However, with the σ entries present, and $\sigma > \epsilon$, $m_\mu^0/m_\tau^0 \cong \epsilon\sigma$ and $m_s^0/m_b^0 \cong \epsilon\sigma/3$, giving the Georgi-Jarlskog result.

Second, without the σ entries $U^0 \propto D^0$, giving both the very bad relation $m_c^0/m_t^0 = m_s^0/m_b^0$ and $V_{cb}^0 = 0$. The σ entries break the proportionality of D and U and give good fits to both m_c/m_t and V_{cb}, as will now be seen.

To leading order in ϵ (which will turn out to be small), one can see that the matrices in Eq. (1) give the predictions

$$
\begin{aligned}
& m_s^0/m_b^0 \cong \tfrac{1}{3}\epsilon\tfrac{\sigma}{\sigma^2+1}, \quad m_\mu^0/m_\tau^0 \cong \epsilon\tfrac{\sigma}{\sigma^2+1}, \quad m_c^0/m_t^0 \cong \tfrac{1}{9}\epsilon^2, \\
& V_{cb}^0 \cong \tfrac{1}{3}\epsilon\tfrac{\sigma^2}{\sigma^2+1}, \qquad m_b^0/m_\tau^0 \cong 1.
\end{aligned}
\tag{2}
$$

A careful fit using more exact expressions gives

$$\epsilon \cong 0.136, \qquad \sigma \cong 1.81. \tag{3}$$

Basically, two parameters, ϵ and σ are able to account for five quantities relating to the charged fermions. Of the resulting three predictions, two are essentially built-in to the group theory of the model (namely the equality of m_b^0 and m_τ^0 and the Georgi-Jarlskog factor). The other is a prediction for m_c. This turns out to agree with the experimental value within the theoretical and experimental errors (1).

In addition, there is a prediction for the neutrinos, which is the most striking prediction of the model. It is that the mixing between ν_μ and ν_τ should be very large. This is a consequence of the element L_{32} in the charged-lepton mass matrix being large $(\sigma + \epsilon)$. The corresponding element of the down-quark mass matrix, D_{32} is small $(-\epsilon/3)$. Thus is explained the difference between V_{cb} and $\sin^2 2\theta_{\mu\tau}$ remarked in the beginning of this talk. The ultimate reason for it has to do with the fact that the σ entries come in transposed between L and D. This, in turn, is a consequence of the fact that these entries are $SU(5)$ invariant. (Recall that $D = L^T$ in minimal $SU(5)$.)

The key point is that $SU(5)$ unifies the left-handed down quarks with the right-handed charged leptons. Thus V_{cb} is related by symmetry not to the mixing of the left-handed leptons which is seen in oscillation experiments, but to the mixing of the right-handed leptons. Similarly, $\sin^2 2\theta_{\mu\tau}$ is related to the mixing of the right-handed down quarks.

In a recent preprint (2) Albright and I have extended this model to the first family. It is found that by introducing only two more parameters, one of which is complex, seven quantities related to the first family (mass ratios and mixing angles) can be very cleanly fit. Among the predictions are the following: (1) The Wolfenstein parameters ρ and η are predicted to satisfy the relation $\rho + i\eta \cong 0.47e^{i\theta} + i0.29$, where θ is unknown. This gives a circle in the (ρ, η) plane that slices neatly through the presently allowed region. (2) $\sin^2 2\theta_{e\mu} \cong 6.6 \times 10^{-3}$. (3) $\sin^2 2\theta_{e\tau} \cong 1.1 \tan^2 \theta_{\mu\tau} \sin^2 2\theta_{e\mu}$.

REFERENCES

1. Albright, C.H., Babu, K.S., and Barr, S.M., *Phys. Rev. Lett.* **81**, 1167-1170 (1998).
2. Albright, C.H. and Barr, S.M., hep-ph/9901318.
3. Georgi, H. and Glashow, S.L., *Phys. Lett.* **B86**, 297-303 (1979).
4. Barr, S.M. and Raby, S., *Phys. Rev. Lett.* **79**, 4748-4751 (1998).

A Comparison of the LSND and KARMEN $\bar{\nu}$ Oscillation Experiments

Steven J. Yellin

Univ. of Calif. at Santa Barbara, yellin@slac.stanford.edu

Abstract. The LSND and KARMEN ν oscillation experiments are compared in order to clarify the significance of the apparent disagreement between their results.

Introduction

The LSND experiment, performed at the Los Alamos National Laboratory, and the KARMEN experiment, performed at the ISIS spallation facility at Rutherford Appleton Laboratory, give results that appear to be in conflict. Both groups experimentally investigated the possibility of $\bar{\nu}_\mu \to \bar{\nu}_e$. They expressed their results in terms of a two-neutrino oscillation model in which only the muon and electron neutrinos are relevant, so that the probability of observing an electron neutrino of energy E_ν MeV a distance L meters away from formation of a muon neutrino is

$$P = sin^2 2\theta \, sin^2 \frac{1.27 \delta m^2 \, L}{E_\nu}, \tag{1}$$

with θ the mixing angle and δm^2 the difference in eV^2 between the squared masses. LSND has published evidence for a signal [1], and includes more recent data in figure 1. But recent KARMEN results, presented in conferences such as Neutrino '98 [2] and shown in the same figure, exclude almost all of the favored LSND region.

Both experiments have competent personnel (this author is in LSND), and both have been working a long enough time to eliminate serious mistakes. How, then, do the experiments differ, and are their results really inconsistent?

Comparison of Experimental Methods

Both experiments produce neutrinos with intense, 800 MeV proton beams. These beams hit targets and create, among other things, π^+ mesons, which decay mainly into $\mu^+\nu_\mu$, and the μ^+ mesons almost all decay at rest into $e^+\nu_e\bar{\nu}_\mu$. The experiments

CP478 , *COSMO—98*, edited by David O. Caldwell

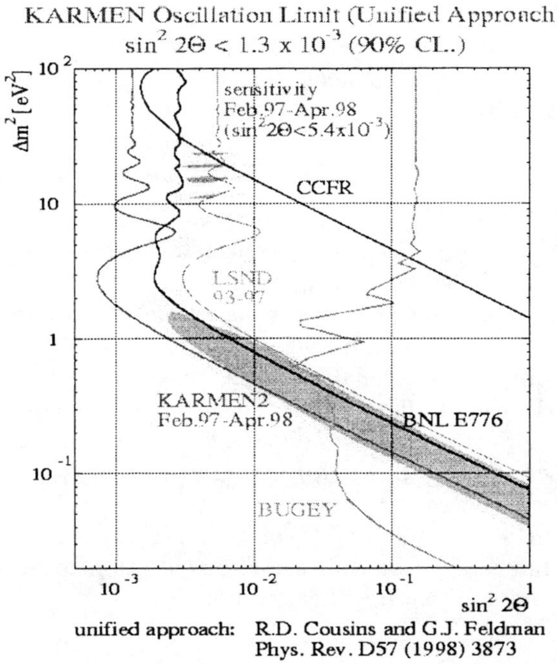

FIGURE 1. Comparison of preliminary LSND $\bar{\nu}_\mu \to \bar{\nu}_e$ results with those of KARMEN and other experiments. Everything to the right of the leftmost curve is excluded to 90% confidence level by KARMEN; the light and dark filled regions are ones favored by LSND.

investigate whether the resulting $\bar{\nu}_\mu$ oscillate into $\bar{\nu}_e$, which are detected by the reaction $\bar{\nu}_e p \to e^+ n$. Both the e^+ and the n are detected in a large, liquid scintillator detector, the n by delayed scintillation light caused by the γ from neutron capture in the detector. One might think that π^- would be produced in the target along with π^+, and that the antiparticle decay chain would lead to serious $\bar{\nu}_e$ background from $\mu^- \to e^- \bar{\nu}_e \nu_\mu$. But the beam and target consist of matter, rather than antimatter, and that makes a great difference between the π^+ and π^- decay chains. Production of π^- is about 1/8 of π^+ production, π^- decay is suppressed by a factor of 20 because these negative particles are attracted to positive nuclei in the target and are usually absorbed before they can decay, and μ^- decay is suppressed by a factor of 10 for the same reason. Thus $\bar{\nu}_e$ production in the target is more than three orders of magnitude below $\bar{\nu}_\mu$ production, allowing experiments to be sensitive to very small probability of $\bar{\nu}_\mu \to \bar{\nu}_e$ oscillation.

Both LSND and KARMEN suppressed cosmogenic neutron background by excluding events with a veto counter signal. LSND also excluded events with evidence of something other than a neutrino entering the detector by looking for even small veto signals and by using the e^+ direction and position to tell if it entered from outside. LSND used information from both before and after the e^+ to exclude events caused by $\mu \rightarrow e\nu\nu$, and KARMEN used only information from before the e^+ for that purpose.

Once events with an apparent e^+ were selected, both KARMEN and LSND selected events with a delayed γ from neutron capture. Both experiments required the γ to be near in both space and time to the e^+ and to have a reasonable energy. But KARMEN made its selection through cuts in position, time, and energy, while LSND needed a more elaborate analysis since it had poorer resolution in these parameters. For LSND, the product of the position, time, and energy distributions for correlated neutron capture events was divided by the similar product for accidental γ's to form a likelihood ratio, R. High R corresponds to a γ that appears more like a neutron capture correlated with the e^+ production, and low R corresponds to a γ that appears to be accidental. LSND could then either cut on high R in order to select clean candidates, or it could fit the R distribution to measure and subtract the contribution from accidental γ's. It did both.

KARMEN used plausible criteria to optimize cuts by computer, while some of LSND's cuts were chosen with a less systematic attempt to avoid human biases. Instead, LSND verified that its results were not very sensitive to variations of cuts.

In order to exclude $\nu_e C \rightarrow e^- X$, which almost always has measured e^- energy below 36 MeV, and which is the largest background with no correlated γ, LSND cleaned its sample by further requiring $E_e > 36$ MeV. It then required $R > 30$ (nearly unambiguous correlated γ), leaving 22 beam-on events and 36 beam-off events. Since LSND (beam-on time)/(beam-off time)=0.070, cosmic beam-on background is estimated to be $0.070 \times 36 = 2.5$ events. There was a beam-associated background of 0.4 events with a correlated photon from $\bar{\nu}_e$ produced in the beam dump, and the measured R distribution of accidentals was used to estimate 1.7 events with an accidental γ. This left LSND with a measured excess of 17.4 events, a signal for which conventional processes cannot account. The probability that this was due to a statistical fluctuation is 4.1×10^{-8}.

KARMEN, after its cuts, found zero beam-on candidates with an expected background of 2.88 ± 0.13. The frequentist "unified approach" of Feldman and Cousins [4], led to the 90% confidence level exclusion limit shown in figure 1.

LSND's contribution to figure 1 was done without the tighter E_e and R cuts used to establish the existence of apparent oscillation events. Instead the R distribution was used to measure the background from accidental γ's, and that background was used in a likelihood function of $\sin^2 2\theta$ and δm^2, based on the data's distribution in R, e^+ position, direction, and energy. The largest likelihood regions are the favored ones shown in the figure.

There are important differences between the beams of the two experiments. The LSND target was H_2O followed by a drift space, allowing $\pi^+ \to \mu^+\nu_\mu$ decay in flight at the 3.4% level, while KARMEN's target was $Ta + D_2O$ with no drift space; so only 0.1% of its π^+ mesons decayed in flight. Thus LSND could, and did [3], also measure the charge conjugate oscillation, $\nu_\mu \to \nu_e$, by detecting higher energy decay in flight neutrinos with $\nu_e C \to e^- X$. Other significant differences were that LSND had 1 ma proton beam intensity instead of KARMEN's 0.2 ma, and LSND had a much larger duty factor than KARMEN, 0.07 instead of 5×10^{-4}. KARMEN's tiny duty factor was a great advantage, because the two experiments suppressed and measured cosmic background by comparing data with and without beam.

There are also important differences between the two detectors. The LSND detector was centered 30 meters from the target, rather than the 17.6 meters of KARMEN. This greater distance of LSND lowered the neutrino flux, but increased the oscillation probability with low δm^2, and made the experiment sensitive to somewhat different δm^2. The LSND detector was over three times as large as KARMEN's, 167 tons of scintillator instead of 56 tons. LSND had a single tank whose walls were covered with photomultiplier tubes. KARMEN, on the other hand, was segmented into 512 modules. This segmentation gave KARMEN better position resolution, especially for the low energy γ produced upon capture of the neutron from $\bar{\nu}_e p \to e^+ n$. LSND used a much lower concentration of scintillator in the liquid. KARMEN therefore had a factor of four better energy resolution, which can be useful for distinguishing between neutron capture γ's and the lower energy accidental γ's from radioactivity. But KARMEN could not distinguish between neutrons and e^+ nearly as well as LSND. Because the LSND detector was one big tank with a low concentration of scintillator, the requirement of Cherenkov ring detection could exclude almost all high energy neutrons produced by cosmic muons, and the e^+ direction could be measured. Two other advantages of LSND were a) surrounding the detector was a superior veto against cosmic muons, and b) it had more shielding against neutrons from the target. The first of these two advantages was eliminated in 1996 by an upgrade of KARMEN's veto. The KARMEN data shown in figure 1 are based entirely on 2897 coulombs of integrated proton beam taken after their upgrade over a three month period. LSND's results shown in that figure are based on over 20000 coulombs collected between 1993 and 1997.

Comparison of the Analyses

LSND's analysis was more complex than what KARMEN used for its Neutrino '98 [2] presentation. LSND distinguished e^+ from high energy cosmogenic neutrons by requiring the showers to produce a good Cherenkov ring, to have a larger portion of fast light than would come from pure scintillation, and to be confined in position. The e^+ energy was required to be high enough to avoid $\nu_e\,^{12}C \to e^- n\,^{11}N$ contamination, but not too high to be from oscillation of $\bar{\nu}_\mu$ from decay at rest of μ^+: $20\ MeV < E_e < 60\ MeV$. KARMEN's energy cut was $20\ MeV < E_e < 50\ MeV$.

Critique

KARMEN's 90% confidence region is misleading. A 90% frequentist confidence region is one chosen by a **method** expected to give the truth 90% of the time it's used. That doesn't mean that **this time** the method has a 90% probability of giving the truth. For KARMEN's particular experimental result, frequentist methods can give a region with much less than 90% probability of giving the truth. For example, if KARMEN had used the conventional frequentist confidence region, instead of the "unified approach", it would have zero percent probability of giving the truth, for the region would be empty, excluding both $sin^2 2\theta > 0$ and $sin^2 2\theta = 0$! This happens because KARMEN found no events, a somewhat surprisingly low result even under the assumption that $sin^2 2\theta = 0$. While the "unified approach" cannot give an empty region, it can give one that is too small. In the absence of a priori assumptions one cannot prove that a non-empty region is too small. But if one takes the usual Bayesian a priori uniform distribution in the unknown parameters ($sin^2 2\theta$ and δm^2) then the region allowed by KARMEN with the "unified approach" covers less than half of that allowed by the 90% Bayesian probability.

Figure 1 shows LSND regions surrounded by contours of 0.1 and 0.01 in (likelihood)/(maximum likelihood). These favored regions can be misleading if interpreted as confidence regions. I therefore compared KARMEN's Bayesian 90% limit with LSND's data reanalyzed to give 90% Bayesian upper and lower limits. For simplicity, the fit ignored position and energy information. But unlike the figure 1 result, this fit itself used R to correct for accidentals, instead of getting the correction elsewhere. Figure 2 shows the result, based on the same events as figure 1, but with the effect of uncertain backgrounds lessened by requiring $E_e > 36$ MeV. Not only is there no inconsistency between LSND and KARMEN, but the figure also shows that LSND's $\nu_\mu \rightarrow \nu_e$ 90% frequentist confidence region [3] is consistent with this way of analyzing LSND's $\bar{\nu}_\mu \rightarrow \bar{\nu}_e$ data.

The LSND analysis of figure 2 can be replaced by one whose requirement of $E_e > 36$ MeV is replaced by $E_e > 20$ MeV. In that case, the lower LSND 90% limit shifts up almost a factor of four relative to that of figure 2. This probably happens because more high R events are seen than are expected from intermediate R. Since low E_e data have especially high background, high R, clean, events contribute more strongly to a fit with more low energy contamination. Fit results might then be increased by an upward statistical fluctuation in the number of high R events.

The alternative analyses shown in figure 2 are in some ways inferior to the ones LSND and KARMEN have reported elsewhere. It is better to include E_e and event position in the LSND analysis. Frequentist methods like the one KARMEN used do depend less on controversial a priori assumptions of experimentalists. The point of figure 2 is that statistical fluctuations, combined with the details of the method of experimental analysis, still allow too much freedom in the experimental results for us to conclude that the two experiments really disagree.

LSND at the time of this conference is just finishing up with the last of its data collection, and will not increase its statistics much over what has already been

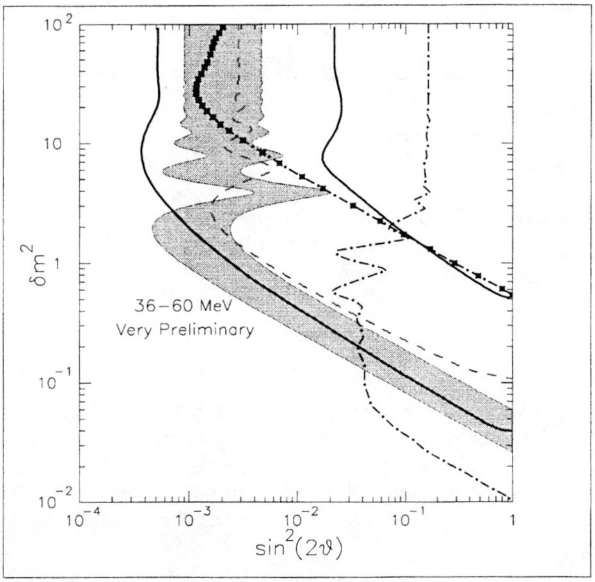

FIGURE 2. Comparison of preliminary LSND $\bar{\nu}_\mu \rightarrow \bar{\nu}_e$ results (filled region) with those of KARMEN (dashed curve), Bugey (dot-dashed), and NOMAD (dot-x-dash). KARMEN's exclusion curve is 90% Bayesian confidence level, while the LSND region is bounded by 90% Bayesian upper and lower limits with $36 < E_e < 60$ MeV. The pair of smooth curves surrounding the LSND region gives the LSND 90% confidence region for $\nu_\mu \rightarrow \nu_e$ oscillation.

reported. But LSND analysis is still underway. For the $\nu_\mu \rightarrow \nu_e$ analysis [3], LSND significantly improved particle identification, and position and energy resolution, all while using methods of cut selection that avoided human bias better than past $\bar{\nu}_\mu \rightarrow \bar{\nu}_e$ analyses. The $\nu_\mu \rightarrow \nu_e$ results were also displayed as a true frequentist confidence region, unlike past LSND $\bar{\nu}_\mu \rightarrow \bar{\nu}_e$ results. What was learned and invented for $\nu_\mu \rightarrow \nu_e$ is being applied to a joint analysis of both ν_μ and $\bar{\nu}_\mu$ data.

The upgraded KARMEN will eventually have much more data than were reported at Neutrino '98, and its analysis is also improving.

Perhaps the two experiments will eventually resolve their differences.

REFERENCES

1. Athanassopoulos, C., *et. al.* (LSND Collaboration), *Phys. Rev. C* **54**, 2685 (1996).
2. Eitel, K., and Zeitnitz, B., *Proceedings Contribution to Neutrino98 in Takayama, Japan, June 4-9, 1998*, hep-ex/9809007.
3. Athanassopoulos, C. et al. *Phys. Rev. C* **58**, 2489 (1998).
4. Feldman, G. J. and Cousins, R. D., *Phys. Rev. D* **57**, 3873 (1998).

Implications of Recent Neutrino Oscillation Data

Thomas J. Weiler[1]

Department of Physics & Astronomy
Vanderbilt University
Nashville, TN 37235
email: tom.weiler@vanderbilt.edu

Abstract. Implications of the oscillation interpretations of solar, atmospheric, and LSND neutrino data are discussed. We list many open questions, and point out the promise of future data to answer these. Two issues discussed in some detail are (i) the difficult nature of measuring CP-violation in oscillations, and (ii) the promise of reconstructing the flavor-basis neutrino mass matrix from oscillation data.

INTRODUCTION

The measured neutrino flux from the sun is about half of its predicted value. Moreover, the flux seems to show an energy–dependence relative to the expected spectrum. This deficit has been interpreted in terms of oscillation of the electron neutrino. The energy–dependence is explained as due to matter-effects in the sun (large– and small–angle "MSW" solutions) or to a vacuum solution with a long oscillation wavelength on the order of an AU. An anomalous flavor–ratio of atmospheric neutrinos is also apparent, interpretable in terms of muon–neutrino oscillation. And a terrestrial experiment at Los Alamos has presented data interpretable as oscillations between $\overline{\nu}_\mu$ and $\overline{\nu}_e$, and between ν_μ and ν_e.

In fact, present data raise more questions than they answer. Interesting open questions include:

• Are oscillations the correct explanation for the solar neutrino deficit? the atmospheric flavor anomaly? the LSND result? Alternative explanations may be possible, e.g. neutrino decay in the atmospheric context [2]. If oscillations are the answer to one or two but not three anomalies, then we must add neutrino mass and mixing to extend the Standard Model (SM). If oscillations are the answer to all three anomalies, then the collective data argue provocatively for at least four light neutrinos, because each of these three experiments requires an independent neutrino mass-squared difference δm^2 to explain their data. From measurements

[1]) Supported in part by the Department of Energy grant DE–FG05–85ER40226

CP478 , *COSMO—98*, edited by David O. Caldwell

of the invisible width of the Z at LEP, it is known that there are a only 3.00 ± 0.05 left-handed "active" neutrinos with mass below $M_Z/2$. Thus, the existence of four or more light neutrinos requires new right-chiral "sterile" neutrinos, coming from far beyond the SM.

• Might the solar deficit and/or atmospheric anomaly be due to neutrino oscillations into unknown right-handed "sterile" neutrinos? Some astrophysical constraints on this possiblity are already inferred from Big Bang Nucleosynthesis (BBN) and Supernova1987A dynamics. Applying these BBN rules conservatively, one finds that the SAM and VLW solar solutions are allowed to proceed via $\nu_e \to \nu_s$. It is a hotly debated BBN issue whether or not active-sterile oscillations are allowed when mass-squared splittings and mixing angles are larger. Future neutral current data from SNO and from SuperKamiokande (henceforth, "SuperK") will render an experimental opinion.

• Three solutions in amplitude-mass space are consistent with the solar neutrino deficit and its energy-dependence. These are the large-angle MSW (LAM), small-angle MSW (SAM), and vacuum long wavelength (VLW) solutions. Theorists tend to favor SAM and LAM, although the heartless data seem to favor VLW. How can these three solutions be discriminated? Day-night variations due to earth matter-effects are possible for SAM and LAM but not VLW. Seasonal variations are expected for VLW but not for SAM or LAM. A high statistics determination of the scattered-electron energy spectrum, particularly at the high and low energy ends, can also discriminate among the three proposed solutions. The SuperK and Borexino experiments will make these discriminating measurements.

• Is the tau neutrino involved in the oscillations so far inferred? Future long baseline appearance experiments or neutrino telescope experiments (not discussed in this talk) are needed to answer this question. At the moment, the elusive ν_τ hasn't been seen in any experiment, except indirectly through missing energy in final states that include a charged tau lepton.

• The evidence from the atmosphere, and the VLW solution for the sun, require a nearly maximal mixing of ν_μ and ν_e with their respective partners. Can we understand this maximality, which is so in contrast with the quark sector? Differences between charged Dirac fermions and Majorana neutrinos may be the key to the answer.

• The atmospheric data also suggest that the electron neutrino has almost no mixing with one neutrino mass eigenstate (call it ν_3), i.e., $U_{e3} \approx 0$. A matrix zero and the constraints from unitarity significantly reduce the free parameters in the mixing matrix. Future long baseline experiments looking for $\nu_e \to \nu_\mu$ or $\nu_e \to \nu_\tau$ oscillations will put even better upper limits on U_{e3}, or measure its value. Does the vanishing of U_{e3} hint at the fundamental flavor-dynamics underlying the neutrino mass matrix?

• Can CP- or T-violation be observed in neutrino oscillations? We will show that the answer is no for a three light-neutrino universe constrained by the solar disappearance experiment and the atmospheric flavor-anomaly. However, with more than three light neutrinos, CP- and T-violation may be observable.

• Can the "Maki-Nakagawa-Sakata" (MNS) mixing matrix [3], the neutrino analog of the quark sector's CKM matrix, be reconstructed from available data? from soon to be available data? As we will show, some progress has been made in this regard, but full reconstruction may require a difficult measurement of CP- or T-violation.
• Can the neutrino mass-matrix $M_{\alpha\beta}$ in the flavor-basis (i.e., the charged-lepton mass-basis) be reconstructed from data? This is an ultimate goal of neutrino physics, for $M_{\alpha\beta}$ is fundamental to the low-energy Lagrangian, and the first necessary term in the lagrangian from beyond the SM.
With over twenty neutrino experiments either actively taking data or proposed to do so in the near future, many of the open questions listed here will soon be answered.

In the next sections we analyze the data presently available, and begin a march toward reconstruction of $M_{\alpha\beta}$.

NEUTRINO OSCILLATION FORMULAE

Measurable quantities constructed from mixing matrix elements include the magnitudes $|U_{\alpha j}|^2 = U_{\alpha j}U^*_{\alpha j}$ ("U-quadratics") and certain U-quartics. All higher order observables may be expressed as combinations of these. We employ a shorthand notation for these obeservables, $P_{\alpha j} \equiv |U_{\alpha j}|^2$ for the quadratics, and ${}^{\alpha j}B_{\beta k} \equiv U_{\alpha j}U_{\beta k}U^*_{\alpha k}U^*_{\beta j}$ for the observable quartics. An attractive feature of working directly with the U-quadratics and quartics $P_{\alpha j}$ and ${}^{\alpha j}B_{\beta k}$ is that, being observables, they are invariant with respect to the convention-dependent parameterization chosen for the mixing matrix U. The quartic observables appear prominently in the probability amplitudes for neutrino oscillations [4]. The general formula for neutrino oscillation probabilities is

$$P(\nu_\alpha \to \nu_\beta) = \delta_{\alpha\beta} - 2\sum_{i=1}^{n}\sum_{j>i}\left[2\ {}^{\alpha i}R_{\beta j}\sin^2\Delta_{ij} - {}^{\alpha i}J_{\beta j}\sin 2\Delta_{ij}\right], \qquad (1)$$

where $\Delta_{jk} \equiv \delta m^2_{jk}\, L/4E = 1.27(\delta m^2_{jk}/\mathrm{eV}^2)(L/\mathrm{km})/(E/\mathrm{GeV})$, $\delta m^2_{jk} \equiv m^2_j - m^2_k$, and and the sum is over all j and k subject to $k < j$. We have defined ${}^{\alpha i}R_{\beta j} \equiv \mathrm{Re}\,(\,{}^{\alpha i}B_{\beta j}\,)$ and ${}^{\alpha i}J_{\beta j} \equiv \mathrm{Im}\,(\,{}^{\alpha i}B_{\beta j}\,)$. Examination of ${}^{\alpha j}B_{\beta k} \equiv U_{\alpha j}U_{\beta k}U^*_{\alpha k}U^*_{\beta j}$ reveals the symmetries in the indexing of the boxes: ${}^{\alpha i}B_{\beta j} = {}^{\beta j}B_{\alpha i} = {}^{\beta i}B^*_{\alpha j} = {}^{\alpha j}B^*_{\beta i}$. These symmetries imply that the Js are antisymmetric in both flavor indices and mass indices, while the Rs are symmetric in both. The symmetries have been applied to arrive at the summation limits in (1).

CP– and T–violation in neutrino oscillations

To observe T-violation in neutrino oscillations, one must measure a difference in the transition probability of some flavor-pair vs. the probability for the flavor-reversed channel. To observe CP-violation in neutrino oscillations, one must measure a difference in the flavor-transition probability for some channel vs. the charge-conjugate channel. Such a measurement requires discrimination of the charged-lepton's charge, as well as identification of the charged-lepton flavor. According to the CPT theorem, $P(\nu_\beta \to \nu_\alpha)$ must equal $P(\overline{\nu}_\alpha \to \overline{\nu}_\beta)$, and it does. Given the index symmetries of the R's and J's, probabilities for the CP- and T- reversed channels are obtained by reversing the sign of ${}^{\alpha i}J_{\beta j}$ in (1). Thus it is seen that the J's contribute to the T- and CP-violating probabilities, and the R's contribute to the T- and CP-conserving probabilities.

Unitarity of U places constraints on the R's and J's [4]. The unitarity constraints for the J's are simple: $\sum_{\alpha\neq\beta} {}^{\alpha j}J_{\beta k} = 0$ and $\sum_{j\neq k} {}^{\alpha j}J_{\beta k} = 0$. Summing the first equation in over β gives zero trivially since ${}^{\alpha j}J_{\beta k}$ is antisymmetric in (α, β). The number of independent Js after implementing these constraints is $\frac{1}{4}(n-1)^2(n-2)^2$. In three generations, these unitarity constraints lead to a common value for all the $|J|$'s, the "Jarlskog invariant". One consequence of common $|J|$ in three generations is that if any one of the nine $U_{\alpha j}$ is zero, then all J's vanish and there can be no CP-violation.

The unitarity constraints allow us to establish an important theorem governing the largest mass-squared scale in the neutrino spectrum.
Theorem: If the squared-masses of the neutrinos fall into two sets, call them heavy (H) and light (L), with the inter-set splitting δm^2_{HL} much larger than any intra-set splitting δm^2, then CP-violating effects in oscillation measurements made at the large inter-set mass-squared scale are suppressed to order $\delta m^2/\delta m^2_{HL}$ due to cancellations among the leading terms.
proof: At leading order in this largest mass-squared scale δm^2_{HL}, the CP-violating contribution is proportional to $J^{\alpha\beta}_{HL} \equiv 2\sum_{j\in H}\sum_{k\in L} J^{\alpha j}_{\beta k}$. But the unitarity constraint for the J's gives $\sum_{k\in L} J^{\alpha j}_{\beta k} = -\sum_{k\in H} J^{\alpha j}_{\beta k}$, so we come to $J^{\alpha\beta}_{HL} = -2\sum_{j\in H}\sum_{k\in H} J^{\alpha j}_{\beta k}$, which vanishes because the sums are symmetric in j and k whereas $J^{\alpha j}_{\beta k}$ is antisymmetric in j and k.

As an immediate application of this theorem, we may conclude that CP-violation is not observable in a three-neutrino universe where the smaller mass-squared splitting is the one inferred for the sun and the other mass-squared-splitting is the one inferred from the atmospheric or LSND anomalies. This is because the solar measurement is a disappearance measurement, which as we shall see cannot reveal CP-violation, and the larger mass-squared splitting satisfies the criterion of the theorem. A corollary of this remark is that any observation of CP-violation in neutrino oscillations at the atmospheric scale must implicate the existence of an even larger scale than $\delta m^2_{\rm atm}$, e.g. the LSND scale $\delta m^2_{\rm LSND}$. Another corollary is that if CP-violation is seen at the $\delta m^2_{\rm LSND}$ scale, then there must be an even larger mass-squared scale than $\delta m^2_{\rm LSND}$.

Channel and parameter counting

After neutrino oscillations are observed in as many flavor channels as possible, the next goal becomes reconstructing the mixing matrix from the data. If possible CP-violating phases in the mixing matrix are absent, then for n neutrino flavors the number of real parameters determining U is $N \equiv \frac{1}{2}n(n-1)$. Conveniently, the number of CP- and T-conserving transition probabilities $P(\nu_\alpha \to \nu_\beta) = P(\nu_\beta \to \nu_\alpha) = P(\overline{\nu}_\alpha \to \overline{\nu}_\beta) = P(\overline{\nu}_\beta \to \overline{\nu}_\alpha)$ is also N, and so form a convenient basis for determining U. Not all N transition probabilities will be needed to determine U if the same transition probability is measured at two or more different distances. Allowing for CP-violating phases in the mixing matrix, there are N real parameters and $\frac{1}{2}(n-1)(n-2)$ phases for Dirac neutrinos, for a total of $(n-1)^2$ parameters. With Majorana neutrinos, there are an additional $n-1$ possible phases for a total of $n(n-1)$ possible parameters, but these additional $n-1$ Majorana phases are not observable in neutrino oscillation measurements. With CP-violation, the number of independent transition probabilities $P(\nu_\alpha \to \nu_\beta) = P(\overline{\nu}_\beta \to \overline{\nu}_\alpha)$ is $2N = n(n-1)$, which exceeds the number of measurable parameters and so the transition probabilities again provide a convenient basis for fixing the MNS mixing matrix. (In reality, only the three flavor indices e, μ, τ are easily accessible in experiments, so a full reconstruction of U for large n may not be possible at all.) After examining the available oscillation data, we will attempt to construct the MNS matrix.

Oscillation amplitudes and disappearance probabilities

It proves useful to extract from eqn. (1) the amplitudes for oscillating transition-probabilities at a particular mass-squared scale or equivalently, at a value of L/E. $R^{\alpha\beta}_{\Delta} \equiv -4\sum_{[j>k]} {}^{\alpha j}R_{\beta k}$ is the CP-conserving amplitude, i.e. the coefficient of the $\sin^2\Delta$ oscillating factor, and $J^{\alpha\beta}_{\Delta} \equiv 2\sum_{[j>k]} {}^{\alpha j}J_{\beta k}$ is the CP-violating amplitude, i.e. the coefficient of the $\sin 2\Delta$ oscillating factor. The sum $\sum_{[j>k]}$ is over those mass states whose mass-squared differences δm^2_{jk} are discernible by the particular experiment and indistinguishable within the resolution of the particular experiment. Subscript labels on the amplitudes distinguish the various sums relevant to the different experiments. "LSND" will denote short-baseline experiments such as LSND, "atm" will denote atmospheric and long-baseline experiments, and "sun" will denote extraterrestrial experiments, especially those with solar neutrinos. Turning to the two-neutrino case as a simple and well-known example, U is just a 2x2 rotation matrix, there is only one mass-squared difference δm^2, and $R^{\alpha\beta}_{\delta m^2} = -4(\cos\theta)(\sin\theta)(\cos\theta)(-\sin\theta) = \sin^2(2\theta)$, $J^{\alpha\beta}_{\delta m^2} = 0$.

From their definition, boxes with $\alpha = \beta$ or $j = k$, are real, given by ${}^{\alpha j}B_{\alpha k} = P_{\alpha j}P_{\alpha k}$, ${}^{\alpha j}B_{\beta j} = P_{\alpha j}P_{\beta j}$, and ${}^{\alpha j}B_{\alpha j} = P^2_{\alpha j}$. From these, one may deduce the survival probabilities $P(\nu_\alpha \to \nu_\alpha) = 1 - \sum_{\beta\neq\alpha} P(\nu_\alpha \to \nu_\beta)$ to be

$$P(\nu_\alpha \rightarrow \nu_\alpha) = 1 - 4\sum_{i=1}^{n}\sum_{j>i} {}^{\alpha i}B_{\alpha j}\, \sin^2\Delta_{ij} = 1 - 4\sum_{i=1}^{n}\sum_{j>i} P_{\alpha i}P_{\alpha j}\sin^2\Delta_{ij}. \qquad (2)$$

Thus we have $R_\Delta^{\alpha\not\alpha} \equiv 1 - \sum_{\beta\neq\alpha} R_\Delta^{\alpha\beta} = 4\sum_{[k<j]} P_{\alpha j}\, P_{\alpha k}$, and $J_\Delta^{\alpha\not\alpha}$=0, for *any* scale Δ. The vanishing of $J_\Delta^{\alpha\not\alpha}$ means that CP- or T-violation cannot be directly probed in a disappearance experiment, such as the solar neutrino experiment.

THREE-NEUTRINO PHENOMENOLOGY

Confirmation of the LSND results awaits future experiments such as BOONE at Fermilab. Until the LSND results are independently validated, a cautious person might assume that oscillations need only account for the solar and atmospheric data. We will mainly work within this premise. In so doing, one avoids the need to introduce sterile neutrinos.

The three lightest neutrinos satisfy $m_\nu < 4.4$ eV

The present laboratory spectrum-endpoint measurements give only the upper limits $m_{\nu_e} < 4.4$ eV, $m_{\nu_\mu} < 170$ keV, and $m_{\mu_\tau} < 18.2$ MeV, which are not very restrictive in the ν_μ and ν_τ cases. In contrast, from oscillation measurements one infers tiny mass-squared splittings, but no absolute mass scale. The solar neutrino data measure the survival probability of ν_e at $\delta m^2_{\text{sun}}\lesssim 10^{-5}\text{eV}^2$ in the case of MSW-enhanced solutions, and at $\delta m^2_{\text{sun}}\lesssim 10^{-10}\text{eV}^2$ in the case of vacuum oscillations. The atmospheric data measure the survival probability of ν_μ and ν_e at $\delta m^2_{\text{atm}}\lesssim 10^{-3}\text{eV}^2$. Phenomenologically, we have the two inequalities $\delta m^2_{\text{sun}} \ll \delta m^2_{\text{atm}} \ll m^2_\beta$, where $m_\beta = 4.4$ eV is the upper limit on the effective ν_e mass from tritium beta decay. These constraints may be combined to place an upper bound of m_β on the ν_μ and ν_τ masses in a three-neutrino universe [5]: The fact that the neutron undergoes β–decay implies that ν_e is a linear combination of one or more mass eigenstates with mass below m_β. Then, the atmospheric and solar neutrino oscillation data tells us that both neutrino mass splittings in a three–neutrino universe are much smaller than m_β. We conclude therefore that all three neutrino mass eigenvalues satisfy $m_j \leq m_\beta$, and therefore that any linear combinations of these mass states, which includes ν_μ and ν_τ, are also $\leq m_\beta$.

Put another way, the neutrino masses allowed by the solar and atmospheric data are degenerate within the precision of the tritium experiment's energy resolution. Therefore, either tritium doesn't decay at all, or all three neutrino masses are below the tritium limit m_β.

The bound is even stronger for a hierarchical mass spectra $m_3 \gg m_1, m_2$ or $m_3 \simeq m_2 \gg m_1$, since here it must be that $m_3 \sim \sqrt{\delta m^2_{\text{atm}}} \sim 0.08$ eV and so $m_{\nu_\alpha} \lesssim 0.08$ eV for $\alpha = e, \mu, \tau$. These results are independent of the neutrino mixing matrix, and are valid for Majorana or Dirac neutrinos.

Similar considerations applied to a four–neutrino universe which includes the LSND data lead to an upper bound of $\sqrt{m_\beta^2 + \delta m_{LSND}^2} \sim 5.4$ eV on the four neutrino masses [5]. We remark that the mass bounds deduced from tritium decay are likely to remain stable even if more neutrinos exist (using an argument based on universality of various weak interaction processes).

Fitting the atmospheric and solar data

The values inferred for $\delta m^2_{\rm sun}$ and $\delta m^2_{\rm atm}$ are well separated. As a consequence, when analyzing atmospheric data, contributions from $\delta m^2_{\rm sun}$ may be ignored to order $\delta m^2_{\rm sun}/\delta m^2_{\rm atm}$, and when analyzing solar data, contributions from $\delta m^2_{\rm atm}$ may be averaged over the oscillation period. For definiteness, let us assume that $\delta m^2_{32} \sim \delta m^2_{\rm atm} \gg \delta m^2_{21} = \delta m^2_{\rm sun}$; the other possible case, $\delta m^2_{32} \ll \delta m^2_{21}$, is essentially the same after the label swap $3 \leftrightarrow 1$. From eqn. (1), the transition probabilities among the atmospheric neutrinos are one finds

$$P(\nu_\alpha \to \nu_\beta) = 4\, P_{\alpha 3}\, P_{\beta 3} \sin^2 \Delta_{\rm atm}\,. \tag{3}$$

These probabilities do not depend of the J's and so are not sensitive to possible CP-violation, in accord with the theorem derived earlier for the largest mass-squared splitting. The disappearance probabilities $P(\nu_e \to \nu_e)$ and $P(\nu_\mu \to \nu_\mu)$ are easily constructed from these flavor-transition probabilities.

In view of the unitarity constraints $\sum_\alpha P_{\alpha j} = 1$ and $\sum_j P_{\alpha j} = 1$, it is seen that these transition probabilities depend on just $\delta m^2_{\rm atm}$ and two $P_{\alpha 3}$, which may be taken to be P_{e3}, and $P_{\mu 3}$. Consequently, in a three light-neutrino universe, all predictions for atmospheric and long-baseline experiments are completely determined by the δm^2_{atm}, P_{e3} and $P_{\mu 3}$.

Best fit values [8] to the 535 day SuperK atmospheric neutrino data are $\delta m^2_{atm} = 2.8 \times 10^{-3}$ eV^2, $U_{e3} = 0.00$, and $|U_{\mu 3}| = 0.50$, with $\chi^2_{min} = 7.1$ for 12 degrees of freedom. This fit is consistent with maximal ν_μ-ν_τ mixing in the atmosphere, and close to the result of the SuperK simulation that assumed only two-neutrino oscillations, due to the (near) vanishing of U_{e3}. Although pure $\nu_\mu \to \nu_\tau$ oscillations of atmospheric neutrinos are found to be favored, there exist three-neutrino solutions with the $\nu_\mu \leftrightarrow \nu_e$ oscillation amplitude as large as 0.18, even after applying the constraints from the CHOOZ reactor experiment. Consequently $\nu_\mu \leftrightarrow \nu_e$ and $\nu_e \to \nu_\tau$ oscillations may be observable in future long-baseline experiments (e.g. MINOS, K2K, and ICARUS). In the more distant future, a long-baseline experiment utilizing a muon storage ring at Fermilab could improve the limit on U_{e3} even more, or measure it.

The ν_e survival probability for solar neutrinos is

$$P(\nu_e \to \nu_e) = 1 - 2P_{e3}(1 - P_{e3}) - 4P_{e1}\, P_{e2} \sin^2 \Delta_{sun}\,, \tag{4}$$

where the second term on the RHS is an oscillation-average over the contribution from the atmospheric scale. Thus, solar data can in principle be used to determine a third independent parameter of the mixing matrix, most obviously P_{e1} or P_{e2}, related by $P_{e1} + P_{e2} + P_{e3} = 1$. The 3x3 mixing matrix is determined by four independent parameters (for Dirac neutrinos), but there seems to be little hope of determining the fourth unless perhaps effects of order $\delta m^2_{\rm sun}/\delta m^2_{\rm atm}$ can be measured in the atmospheric data.

However, the vanishing best fit value for $|U_{e3}|$ implies that the Jarlskog invariant $|J|$ is zero or nearly so, which in turn implies that U is nearly real and well-parameterized by just $P_{e3} = 0$ and two additional independent $P_{\alpha j}$'s, or equivalently, by just two rotation angles. The general result for $U_{e3} = 0$ is then

$$U(\theta_1, \theta_2) = R_{23}(\theta_2)\, R_{12}(\theta_1) = \begin{pmatrix} c_1 & s_1 & 0 \\ -s_1 c_2 & c_1 c_2 & s_2 \\ s_1 s_2 & -c_1 s_2 & c_2 \end{pmatrix} . \tag{5}$$

Some attempts have been made to model this result with one rotation attributed to the charge-lepton sector, and the other attributed to the neutrino sector. The logic of these attempts will be seen in the next section.

The "bi-maximal model"

The ansatz of the bi-maximal model [6,7] is that both the observed ν_μ disappearance at the atmospheric scale and the ν_e disappearance at the solar scale are maximized in terms of vacuum parameters. For the sun this singles out the vacuum oscillation solution, now favored by recent SuperK data. From eqns. (3) and (4) it is determined that bi-maximization leads to $P_{e3} = 0$ and $P_{e1} = P_{e2} = P_{\mu 3} = P_{\tau 3} = 1/2$. Up to trivial sign changes of rows or columns, and modulo two possible nontrivial Majorana phases, The resulting bi-maximal mixing matrix is uniquely given as

$$U(\frac{\pi}{4}, \frac{\pi}{4}) = \begin{pmatrix} 1/\sqrt{2} & 1/\sqrt{2} & 0 \\ -1/2 & 1/2 & 1/\sqrt{2} \\ 1/2 & -1/2 & 1/\sqrt{2} \end{pmatrix} . \tag{6}$$

With $U_{e3} = 0$, the model necessarily conserves CP and automatically implies that there are no oscillations of atmospheric ν_e, consistent with indications from the SuperKamiokande experiment. The matrix further implies that atmospheric ν_μ disappearance is due to maximal mixing with ν_τ, and that solar ν_e disappearance is due to ν_e oscillations into equal numbers of ν_μ and ν_τ. If the solar and atmospheric oscillations are not quite maximal, then there must be $\nu_e \to \nu_\tau$ and $\nu_\mu \to \nu_e$ oscillations of nearly equal amplitude that may be visible in future long–baseline experiments.

With the unique bi-maximal mixing matrix U in hand, we may "undiagonalize" the mass-eigenvalues to obtain the fundamental neutrino mass matrix $M_{\alpha\beta}$ in the flavor basis, where the charged–lepton mass matrix is diagonal. We do this in the next section.

Remarks on four– and six–neutrino physics

A model-independent analysis of the oscillation data using four-neutrino unitarity constraints shows that a neutrino mass spectrum with one eigenstate separated from three other nearly-degenerate states (the "three of a kind" spectrum) is disfavored compared to a spectrum with two nearly degenerate mass pairs (the "pair of pairs" spectrum) [9,10]. With the pair of pairs spectrum, the ν_e and its partner $\tilde{\nu_e}$ with a mass-squared difference equal to $\delta m^2_{\rm sun}$ constitutes one pair, and the ν_μ and its partner $\tilde{\nu_\mu}$ with mass-squared difference $\delta m^2_{\rm atm}$ constitutes the other pair. Interestingly, the message from the small LSND amplitude $R^{e\mu}_{LSND} \sim 1\%$ is that there is little inter-pair mixing. The 4x4 mixing matrix in the (ν_e, $\tilde{\nu_e}$, ν_μ, $\tilde{\nu_\mu}$ basis that fits the data has the form

$$U = \begin{pmatrix} R(\theta_e) & \cdots \\ \cdots & R(\theta_\mu) \end{pmatrix}, \tag{7}$$

where the two rotation submatrices are 2x2, and the dots indicate smallness for the remaining elements. As required by the atmospheric data, θ_μ is nearly maximal; θ_e is determined by which of the three solar solutions nature chooses.

Two possible interpretations immediately emerge. The first is that the partners of ν_e and ν_μ are ν_s and ν_τ, or some linear combination of ν_s and ν_τ. Neutral current measurements (zero for ν_s) at SuperK and SNO, and earth matter-effects such as day/night differences (zero to leading order for ν_μ-ν_τ oscillations) each offers the potential to distinguish ν_s and ν_τ. The second possible interpretation is that $\tilde{\nu_e}$ and $\tilde{\nu_\mu}$ are both sterile neutrinos, the "mirror" partners of ν_e and ν_μ. In this type of model, there are six light neutrinos including the mirror $\tilde{\nu_\tau}$. Mirror models are attractive in that they may offer a natural reason for the large mixing and small mass-squared splitting between each neutrino and its mirror, while at the same time suppressing the mixing and increasing the splitting between neutrinos of differing flavor. Again, neutral current measurements and earth matter-effects hold the potential to distinguish sterile partners from the active partners.

Additional test of the model [10] include the prediction of (i) $\nu_e \to \nu_\tau$ and/or $\nu_e \to \nu_s$ oscillations in long-baseline experiments at $L/E \gg 1$ km/GeV, with amplitude determined by the LSND oscillation amplitude and argument given by the atmospheric δm^2; and (ii) the equality of the ν_e disappearance probability, the ν_μ disappearance probability, and the LSND $\nu_\mu \to \nu_e$ appearance probability in short-baseline experiments.

RECONSTRUCTING $M_{\alpha\beta}$ IN THE FLAVOR–BASIS

Other clues may reside in the mass-squared differences inferred directly from neutrino oscillations. These are the eigenvalues of the hermitian matrix $MM^\dagger$ diagonalized by U.

The unitary MNS matrix U diagonalizes $M^2 \equiv MM^\dagger$. The mixing matrix U is itself a product $U_{QL}^\dagger U_{NL}$, where U_{NL} and U_{NR} are the unitary matrices acting on the left- and right-chiral fields in the bi-unitary transformation which diagonalizes the neutrino mass-matrix; and U_{QL} and U_{QR} are the analogous matrices which diagonalize the charged-lepton mass-matrix. Thus, U measures the unitary rotation of the neutrinos *relative to the rotation of the charged leptons*, or equivalently, the rotation of the left-chiral neutrinos in the mass basis of the charged leptons. In general, the neutrino flavor-basis mass-matrix $M_{\alpha\beta}$ may be obtained from the diagonal mass-matrix with positive eigenvalues $\sqrt{M_{\rm diag}^2}$ by the inverse bi-unitary transformation $M_{\alpha\beta} = U\, M_{\rm diag}\, U_{NR}^\dagger\, U_{QR}$. The unitary matrices U_{QR} and U_{NR} cannot be determined from neutrino oscillations or charged current measurements. As a result, it appears that $M_{\alpha\beta}$ may not be obtainable from M^2 and U. However, in the theoretically-favored case of Majorana neutrinos, CP properties require that M be symmetric. As a consequence, $M_{\alpha\beta} = U\, M_{\rm diag}\, U^T$ with individual matrix elements $M_{\alpha\beta} = \sum_j m_j U_{\alpha j} U_{\beta j}$. I remind the reader that the matrix U will in general contain some Majorana phases that cannot be determined via neutrino oscillations (they are measurable in principle, e.g., in neutrinoless double beta-decay measurements). In the following discussion, we will ignore possible Majorana phases which are beyond the scope of oscillation measurements.

In the case of Dirac neutrinos, a simplification occurs *if* $M_{\alpha\beta}$ happens for some reason to be hermitian. Then it is given by $U\, M_{\rm diag}\, U^\dagger$, and the individual matrix elements are $M_{\alpha\beta} = \sum_j m_j U_{\alpha j} U_{\beta j}^*$.

As an example of reconstruction of a flavor-basis mass matrix, we return our focus to the bi-maximal model. The U matrix is specified in eqn. (6); it is real and so gives the same reconstructed $M_{\alpha\beta} = U\, M_{\rm diag}\, U^T$ for the Majorana and hermitian Dirac cases. The mass eigenvalues are determined from $\delta m_{\rm atm}^2$ and $\delta m_{\rm sun}^2$ once an overall mass-scale, say m_3, is specified. The result in the $(\nu_e, \nu_\mu, \nu_\tau)$ flavor basis is [6]

$$M = \frac{m_3}{2}\left[\begin{pmatrix} 0 & 0 & 0 \\ 0 & 1 & -1 \\ 0 & -1 & 1 \end{pmatrix} + \begin{pmatrix} 2\epsilon & \delta & \delta \\ \delta & \epsilon & \epsilon \\ \delta & \epsilon & \epsilon \end{pmatrix}\right], \tag{8}$$

where the parameters δ and ϵ are related to the mass eigenvalues as $\epsilon = (m_2 + m_1)/2m_3$ and $\delta = \sqrt{2}\,(m_1 - m_2)/2m_3$, and to the oscillation scales as $2\sqrt{2}\delta\epsilon m_3^2 = \delta m_{\rm sun}^2$ with $m_3^2 \simeq \delta m_{\rm atm}^2$.

From this result one may look for symmetry and broken symmetry hinting at the deep nature of flavor-dynamics. Thus one sees the potential of neutrino phenomenology to open new doors to new physics beyond the SM.

ACKNOWLEDGEMENTS

I acknowledge my good fortune to have V. Barger, J. Learned, S. Pakvasa, DJ Wagner, and K. Whisnant as collaborators on many of the topics included in this manuscript. I also acknowledge the huge body of work of others, from which I have benefitted, but have not referenced herein due to constraints on space, and, quite frankly, time as well.

REFERENCES

1. Due to the limited page length of these proceedings, many references are necessarily and blatantly neglected. Appropriate referencing may be found in the papers that are listed here.
2. V. Barger, J.G. Learned, S, Pakvasa, and T.J. Weiler, *Phys. Rev. Lett.*, to appear (1999), and astro-ph/9810121.
3. Z. Maki, M. Nakagawa, and S. Sakata, *Prog. Theo. Phys.* **28**, 870 (1962).
4. DJ Wagner and T.J. Weiler, *Phys. Rev.* **D**, to appear (1999), and hap-ph/9801327.
5. V. Barger, T.J. Weiler, and K. Whisnant, *Phys. Lett* **B442**, 255 (1998).
6. V. Barger, S. Pakvasa, T.J. Weiler, and K. Whisnant, *Phys. Lett* **B437**, 107 (1998).
7. A.J. Baltz, A.S. Goldhaber, and M. Goldhaber, *Phys. Rev. Lett.* **81**, 5730 (1998).
8. V. Barger, T.J. Weiler, and K. Whisnant, *Phys. Lett* **B440**, 1 (1998).
9. S.M. Bilenky, C. Giunti, and W. Grimus, *Eur. Phys. J.* **C1**, 247 (1998).
10. V. Barger, T.J. Weiler, and K. Whisnant, *Phys. Lett* **B427**, 97 (1998); V. Barger, S. Pakvasa, T.J. Weiler, and K. Whisnant, *Phys. Rev.* **D58**, 093016 (1998).

The Pattern of Neutrino Masses and How to Determine It

V. Barger

Physics Department, University of Wisconsin, Madison, WI 53706

I INTRODUCTION

Our knowledge of the neutrino sector of the Standard Model has recently undergone a revolution. Deficits of the atmospheric muon neutrino flux and the solar electron neutrino flux compared to their predicted values can be understood in terms of neutrino oscillations [1] and we can therefore infer that neutrinos have non-degenerate masses. Additional but somewhat less secure evidence for $\bar{\nu}_\mu \to \bar{\nu}_e$ and $\nu_\mu \to \nu_e$ oscillations has been found in the LSND accelerator experiment [1]. Because these experiments have widely different L/E_ν ranges (≈ 10 to 10^4 km/GeV for atmospheric, $\approx 10^{11}$ for solar, and ≈ 1 for LSND), the mass-squared differences required to explain the phenomena must be distinct. Given the observations, an important next step is to deduce the pattern of neutrino masses and mixings. Such studies depend on the number of neutrinos. The invisible width of the Z-boson measured in LEP experiments gives $N_\nu = 2.993 \pm 0.011$, consistent with the usual ν_e, ν_μ and ν_τ "active" neutrinos. But there may also be right-handed "sterile" neutrinos with no weak interactions. Only the observation of oscillations of the active neutrinos to sterile neutrinos can test for their existence. In the following, we first discuss the atmospheric and solar neutrino data in a 3-neutrino framework and then later generalize our considerations to include the LSND data with oscillations of four neutrinos. The proceedings of the Neutrino 98 conference [1], a current review by the organizer of COSMO 98 [2], and recent phenomenological analyses [3–9] may be consulted for references to the vast primary literature.

When neutrino flavor eigenstates ν_f are not the same as the mass eigenstates ν_i, e.g., for two neutrinos,

$$\nu_f = \cos\theta\nu_1 + \sin\theta\nu_2\,, \qquad \nu_{f'} = -\sin\theta\nu_1 + \cos\theta\nu_2\,, \tag{1}$$

then neutrinos oscillate. The vacuum oscillation probabilities are

$$P(\nu_f \to \nu_{f'}, L) = A\sin^2\left(\frac{\delta m^2 L}{4E}\right) \qquad \text{"appearance"}\,, \tag{2}$$

$$P(\nu_f \to \nu_f, L) = 1 - A\sin^2\left(\frac{\delta m^2 L}{4E}\right) \qquad \text{"survival"}\,, \tag{3}$$

CP478 , *COSMO—98*, edited by David O. Caldwell

where $A = \sin^2 2\theta$, $\delta m^2 = m_2^2 - m_1^2$; L is the path length and E is the neutrino energy. The neutrino anomalies can be explained by effective two-neutrino oscillations; vacuum oscillations can account for the solar [10], atmospheric [11] and LSND anomalies; matter-enhanced oscillations [12] are an alternative to explain the solar neutrino deficit.

II NEUTRINO ANOMALIES AND THEIR OSCILLATION INTERPRETATIONS

Atmospheric Cosmic ray interactions with the atmosphere produce π-mesons and the decays $\pi \to \mu\nu$ and $\mu \to \nu_e e \nu_\mu$ give ν_μ and ν_e fluxes in the approximate ratio $(\nu_\mu + \bar{\nu}_\mu)/(\nu_e + \bar{\nu}_e) \sim 2$ for $E_\nu \sim 1$ GeV. Measurements of $R = (N_\mu/N_e)_{\rm data}/(N_\mu/N_e)_{\rm MC}$ for $E_\nu \sim 1$ GeV find values of $R \sim 0.6$ [1]. In the water Cherenkov experiments the single rings from muons are fairly clean and sharp, while those from electrons are fuzzy due to electromagnetic showers. The separate distributions of μ-like and e-like events versus the zenith angle establish that the anomalous R-ratio is due to a deficit of upward μ-like events. As suggested long ago [11], the data are well described by $\nu_\mu \to \nu_\tau$ or $\nu_\mu \to \nu_s$ oscillations with $\delta m^2_{\rm ATM} \approx 3 \times 10^{-3}\,{\rm eV}^2$ and $A_{\rm ATM} \approx 1$. For sub-GeV neutrino energies, L/E is large at $\cos\theta < 0$ and the oscillations average, $P(\nu_\mu \to \nu_\mu) \approx 0.5$. At multi-GeV energies, L/E is large at $\cos\theta = -1$ and $P(\nu_\mu \to \nu_\mu) \approx 0.5$; also L/E is small at $\cos\theta = +1$ and $P(\nu_\mu \to \nu_\mu) \approx 1$.

Solar Three types of solar ν_e experiments, (i) ν_e capture in Cl [Homestake], (ii) $\nu_e e \to \nu_e e$ [Kamiokande and SuperKamiokande], (iii) ν_e capture in Ga, measure rates below standard model expectations. The different experiments are sensitive to different ranges of solar E_ν. Three regions of oscillation parameter space are found to accommodate all these observations [3,7,8]:

	$\delta m^2_{\rm SOL}$ (eV)^2	$A_{\rm SOL}$
Small Angle Matter (SAM)	$\sim 10^{-5}$	$\sim 10^{-2}$
Large Angle Matter (LAM)	$\sim 10^{-5}$	~ 0.6
Vacuum Long Wavelength (VLM)	$\sim 10^{-10}$	~ 1

LSND The Los Alamos experiment studied $\bar{\nu}_\mu \to \bar{\nu}_e$ oscillations from $\bar{\nu}_\mu$ of μ^+ decay at rest and $\nu_\mu \to \nu_e$ from ν_μ of π^+ decay in flight. The results, including restrictions from BNL, KARMEN and Bugey experiments, suggest $\nu_\mu \to \nu_e$ oscillations with

$$0.3\ {\rm eV}^2 < \delta m^2_{\rm LSND} < 2.0\ {\rm eV}^2\,, \qquad A_{\rm LSND} \approx 4 \times 10^{-2} \text{ to } 3 \times 10^{-3}\,. \tag{4}$$

Figure 1 illustrates the parameter regions for the solar, atmospheric and LSND oscillation interpretations. Since three distinct δm^2 are needed to explain the atmospheric, solar and LSND data, but there are only two independent δm^2 from

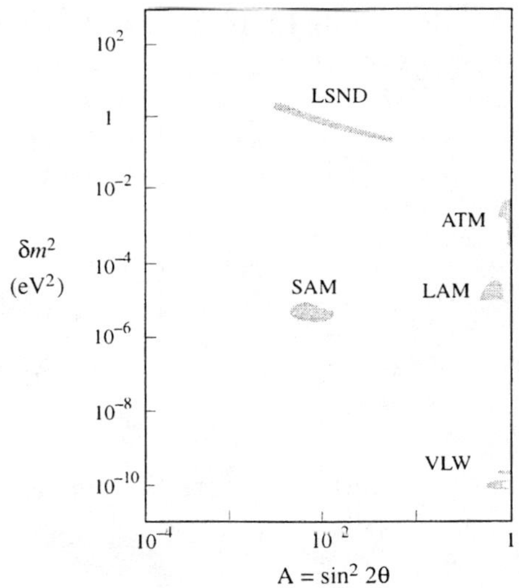

FIGURE 1. Regions of oscillation parameters $\delta m^2, \sin^2 2\theta$ that can explain the atmospheric, solar, and LSND anomalies.

ν_e, ν_μ, ν_τ neutrinos, there are two possible roads to follow: (i) put the LSND anomaly aside until it is confirmed by the KARMEN and mini-BooNE experiments, or (ii) explain all three anomalies by invoking oscillations to a sterile neutrino as well. We consider both routes in the following.

III NEW INFERRED LIMITS ON NEUTRINO MASS

The fact that the neutron undergoes β-decay implies that the electron-neutrino is a linear combination of one or more mass eigenstates with mass below the kinematic β-decay end-point limit of $m_\beta = 4.4$ eV. The interpretation of the atmospheric and solar neutrino anomalies in terms of oscillations tells us that the neutrino mass splittings in a three-neutrino universe are much smaller than m_β. We conclude therefore that all three neutrino mass eigenvalues satisfy $m_j \le m_\beta$ and that the linear combinations of these mass states which are ν_μ and ν_τ have effective masses $\le m_\beta$ as well [13]. These bounds represent a factor of 10^5 to 10^6 improvement over the current bounds $m_{\nu_\mu} < 170$ keV and $m_{\nu_\tau} < 18.2$ MeV. The largest mass eigenvalue is bounded below by $m_\beta \ge \sqrt{\delta m^2_{\rm atm}} \ge 0.002$ eV. Generalizing to include $\nu_\mu \to \nu_e$ oscillations in the LSND experiment with one sterile and three active neutrinos, we obtain an upper bound of 5.4 eV on all four neutrino masses and a lower bound on the largest mass eigenstate of $m_4 \ge \sqrt{\delta m^2_{\rm LSND}} \gtrsim 0.5$ eV.

IV FROM EFFECTIVE 2-GENERATION TO 3-GENERATION OSCILLATIONS

The neutrino MNS mixing matrix [14], for either Dirac or Majorana neutrinos, can be parametrized by three angles θ_i and a CP-violating phase as

$$\begin{pmatrix} \nu_e \\ \nu_\mu \\ \nu_\tau \end{pmatrix} = U \begin{pmatrix} \nu_1 \\ \nu_2 \\ \nu_3 \end{pmatrix} = \begin{pmatrix} c_1c_3 & c_1s_3 & s_1e^{-i\delta} \\ -c_2s_3 - s_1s_2c_3e^{i\delta} & c_2c_3 - s_1s_2s_3e^{i\delta} & c_1s_2 \\ s_2s_3 - s_1c_2c_3e^{i\delta} & -s_2c_3 - s_1c_2s_3e^{i\delta} & c_1c_2 \end{pmatrix} \begin{pmatrix} \nu_1 \\ \nu_2 \\ \nu_3 \end{pmatrix}, \tag{5}$$

where $c_j \equiv \cos\theta_j$, $s_j \equiv \sin\theta_j$, and ν_1, ν_2, ν_3 are the mass eigenstates. The vacuum oscillation probabilities of interest are

Atmospheric:

$$P(\nu_\mu \to \nu_\mu) = 1 - (c_1^4 \sin^2 2\theta_2 + s_2^2 \sin^2 2\theta_1)\sin^2 \Delta_{atm} \tag{6}$$

$$P(\nu_e \to \nu_e) = 1 - \sin^2 2\theta_1 \sin^2 \Delta_{atm} \tag{7}$$

$$P(\nu_e \leftrightarrow \nu_\mu) = s_2^2 \sin^2 2\theta_1 \sin^2 \Delta_{atm} \tag{8}$$

$$P(\nu_e \leftrightarrow \nu_\tau) = c_2^2 \sin^2 2\theta_1 \sin^2 \Delta_{atm} \tag{9}$$

$$P(\nu_\mu \leftrightarrow \nu_\tau) = c_1^4 \sin^2 2\theta_2 \sin^2 \Delta_{atm} \tag{10}$$

Solar

$$P(\nu_e \to \nu_e) = 1 - \tfrac{1}{2}\sin^2 2\theta_1 - c_1^4 \sin^2 2\theta_3 \sin^2 \Delta_{sun} \tag{11}$$

When $\theta_1 = 0$, the atmospheric and solar oscillations decouple and ν_e does not oscillate in atmospheric and long-baseline experiments. Best fits [13] to the SuperK data are obtained with $\theta_1 = 0$, $\theta_2 = \pi/4$ with 2σ bounds of $\theta_1 < 17°$ and $|\theta_2 - 45°| < 13°$. The angle θ_3 is determined by the solar data (for the just-so [10] or the MSW solutions [12], whichever is chosen by experiment). Thus we already have achieved the partial reconstruction of the neutrino MNS matrix!

V BI-MAXIMAL MIXING MODEL

If the atmospheric and solar data are both described by maximal mixing, then there exists a unique mixing matrix for three flavors [15],

$$U = \begin{pmatrix} \frac{1}{\sqrt{2}} & -\frac{1}{\sqrt{2}} & 0 \\ \frac{1}{2} & \frac{1}{2} & -\frac{1}{\sqrt{2}} \\ \frac{1}{2} & \frac{1}{2} & \frac{1}{\sqrt{2}} \end{pmatrix} \tag{12}$$

The off-diagonal oscillation probabilites in this model are

$$P(\nu_\mu \to \nu_\tau) = \sin^2 \Delta_{\text{ATM}} - \tfrac{1}{4}\sin^2 \Delta_{\text{SUN}} \tag{13}$$

$$P(\nu_e \to \nu_\mu) = \tfrac{1}{2}\sin^2 \Delta_{\text{SUN}} \tag{14}$$

$$P(\nu_e \to \nu_\tau) = \tfrac{1}{2}\sin^2 \Delta_{\text{SUN}} \tag{15}$$

where $\Delta \equiv 1.27\delta m^2 L/E$. A variety of models with nearly bi-maximal mixing have also been considered [15].

VI UNIFIED MODELS

In unified models based on SO(10) [16], SU(5) [17], flipped SU(5) [18], or anomalous U(1) [19], large $\nu_\mu \to \nu_\tau$ oscillations are accommodated but the small-angle MSW solar solution is required. This prediction is a clear distinction from the bi-maximal mixing model.

VII DISTINGUISHING SOLAR ν-OSCILLATION SCENARIOS

The solar ν oscillation solutions will eventually be distinguished by use of all the following measurements: (i) time-averaged total flux, (ii) day-night dependence (earth-matter effects), (iii) recoil electron energy spectra in $\nu e \to \nu e$ events), (iv) seasonal variation, and (v) the neutral-current to charged-current event ratio (SNO experiment). The non-observation of a day-night effect has already ruled out substantial regions of the $\delta m^2, \sin 2\theta$ parameter space [13]. The electron energy distribution from recent SuperKamiokande data (708 days) now favor the vacuum long-wavelength interpretation and nearly exclude the MSW solutions [20] unless an enhanced hep flux component is involved [21]. The day-night dependence due to the Earth is possible only for MSW solutions and has already ruled out substantial regions of otherwise allowed $\delta m^2, \sin^2 2\theta$ values. A seasonal variation beyond the $1/r^2$ dependence of the flux is another characteristic of only long-wavelength vacuum oscillations and the 708-day SuperK data seem to show a seasonal effect above the eccentricity correction. The neutral current measurements of the SNO experiment will distinguish active from sterile oscillations.

VIII FOUR-NEUTRINO OPTIONS

With four neutrinos it is possible to account for the LSND data as well as the atmospheric and solar data. The preferred mass spectrum is two nearly degenerate mass pairs separated by the LSND scale [4,5]; see Fig. 2. The figure also shows the options for oscillation solutions to all data. The alternative of a $1+3$ mass hierarchy with one heavier mass scale separated from three lighter, nearly degenerate states is disfavored when the null results of reactor and accelerator disappearance experiments are taken into account.

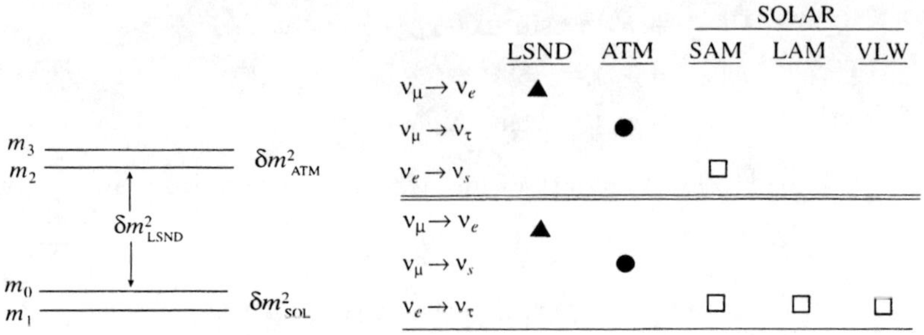

FIGURE 2. Neutrino mass spectrum showing which mass splittings are responsible for the LSND, atmospheric, and solar oscillations, and four-neutrino oscillation possibilities.

Neutrino mass matrices have been proposed [22] that can account for the observations.

IX LONG-BASELINE EXPERIMENTS

Long-baseline experiments with $L/E \approx 10$–10^2 km/GeV will measure $P(\nu_\mu \leftrightarrow \nu_\tau) \simeq \sin^2 \Delta_{\rm ATM}$ and test the atmospheric oscillation result. In addition, the existence of $\nu_\mu \to \nu_e$ and $\nu_e \to \nu_\tau$ oscillations may be tested. For this purpose intense neutrino beams are required. The MINOS experiment (Fermilab to Soudan) could confirm the SuperK $\nu_\mu \to \nu_\tau$ parameter region with 4σ sensitivty, provided that $\delta m^2_{\rm ATM} > 2 \times 10^{-3}\,{\rm eV}^2$. The K2K experiment is sensitive to $\delta m^2 \gtrsim 3 \times 10^{-3}\,{\rm eV}^2$.

In the future, a special purpose muon storage ring could provide high intensity neutrino beams with well-determined fluxes that could be directed towards any detector on the earth [23,24]. It could be possible to store $\sim 10^{21}$ μ^+ or μ^- per year and obtain $\sim 10^{20}$ neutrinos from the muon decays. Oscillations give "wrong sign" leptons from those produced by the beam. For example, μ^- decays give $\bar\nu_e$ and ν_μ fluxes so detection of $\mu^+, e^-, \tau^\pm$ leptons tests for $\bar\nu_e \to \bar\nu_\mu(\bar\nu_\tau)$ and $\nu_\mu \to \nu_e(\nu_\tau)$ oscillations. Taus can be detected via their $\tau \to \mu$ decays and the τ-charges so determined to distinguish $\nu_\mu \to \nu_\tau$ and $\bar\nu_e \to \bar\nu_\tau$ oscillations.

REFERENCES

1. Proceedings of Neutrino 98, Takayama, Japan [to be published], and references therein, http://www-sk.icrr.u-tokyo.ac.jp/nu98/scan/index.html
2. D.O. Caldwell, IJMP **A13**, 4409 (1998).
3. J. Bahcall, P. Krastev, and A. Smirnov, Phys. Rev. **D58**, 096016 (1998).
4. V. Barger, S. Pakvasa, T.J. Weiler, and K. Whisnant, Phys. Rev. **D58**, 093016 (1998).
5. S.M. Bilenky, C. Guinti, and W. Grimus, Eur. Phys. J. **C1**, 247 (1998); C. Guinti, C.W. Kim, and M. Monteno, Nucl. Phys. **B521**, 3 (1998).

6. D.O. Caldwell and R. Mohapatra, Phys. Rev. **D48**, 3259 (1993); S.C. Gibbons, R.N. Mohapatra, S. Nandi, and A. Raychoudhuri, Phys. Lett. **B430**, 296 (1998); S. Mohanty, D.P. Roy, and U. Sarkar, hep-ph/9810309.
7. G.L. Fogli, E. Lisi, A. Marrone, and G. Scioscia, Phys. Rev. **D59**, 033001 (1999).
8. N. Hata and P. Langacker, Phys. Rev. **D56**, 6107 (1997).
9. M.C. Gonzalez-Garcia, H. Nunokawa, O. Peres, T. Stanev, and J.W.F. Valle, Phys. Rev. **D58**, 033004 (1998).
10. V. Barger, R.J.N. Phillips, and K. Whisnant, Phys. Rev. **D24**, 538 (1981); S.L. Glashow and L.M. Krauss, Phys. Lett. **B190**, 199 (1987).
11. J.G. Learned, S. Pakvasa, and T.J. Weiler, Phys. Lett. **B207**, 79 (1988); V. Barger and K. Whisnant, Phys. Lett. **B209**, 365 (1988); K. Hidaka, M. Honda, and S. Midorikawa, Phys. Rev. Lett. **61**, 1537 (1988).
12. L. Wolfenstein, Phys. Rev. **D 17**, 2369 (1978); S.P. Mikheyev and A. Smirnov, Yad. Fiz. **42**, 1441 (1985); Nuovo Cim. **9C**, 17 (1986); S.P. Rosen and J.M. Gelb, Phys. Rev. **D 34**, 969 (1986); S.J. Parke, Phys. Rev. Lett. **57**, 1275 (1986); W.C. Haxton, Phys. Rev. Lett. **57**, 1271 (1986);
13. V. Barger, T.J. Weiler and K. Whisnant, Phys. Lett. **B442**, 255 (1998).
14. Z. Maki, M. Nakagawa and S. Sakata, Prog. Theor. Phys. **28**, 870 (1962).
15. V. Barger, S. Pakvasa, T.J. Weiler, and K. Whisnant, Phys. Lett. **B437** 107 (1998); A.J. Baltz, A.S. Goldhaber, and M. Goldhaber, Phys. Rev. Lett. **81**, 5730 (1998); M. Jezabek and Y. Sumino, Phys. Lett. **B440**, 327 (1998); Y. Nomura and T. Yanagida, Phys. Rev. **D59**, 017303 (1999); G. Altarelli and F. Feruglio, Phys. Lett. **B439**, 112 (1998); JHEP **9811**, 021 (1998); H. Georgi and S. Glashow, hep-ph/9808293; S. Davidson and S.F. King, Phys. Lett. **B445**, 191 (1998). R.N. Mohapatra and S. Nussinov, Phys. Lett. **B441**, 299 (1998); hep-ph/9809415; S.K. Kang and C.S. Kim, hep-ph/9811379; H. Fritzsch and Z. Xing, Phys. Lett. **B372**, 265 (1996); hep-ph/9807234; Phys. Lett. **B440**, 313 (1998); E. Torrente-Lujan, Phys. Lett. **B389**, 557 (1996); M. Fukugita, M. Tanimoto, and T. Yanagida, Phys. Rev. **D57**, 4429 (1998); M. Tanimoto, Phys. Rev. **D59**, 017304 (1999); Y.-L. Wu, Academia Sinica report AS-ITP-99-04 [hep-ph/9901230].
16. J.K. Elwood, N. Irges, and P. Ramond, Phys. Rev. Lett. **81**, 5064 (1998); C.H. Albright, K.S. Babu, and S.M. Barr, Phys. Rev. Lett. **81**, 1167 (1998); K.S. Babu, J.C. Pati, and F. Wilczek, report OSU-HEP-98-11 [hep-ph/9812538].
17. K. Hagiwara and N. Okamura, report KEK-TH-605 [hep-ph/9811495].
18. J. Ellis, G.K. Leontaris, S. Lola, and D.V. Nanopouos, report ACT-7-98 [hep-ph/9808251].
19. J. Harvey, D.B. Reiss, and P. Ramond, Nucl. Phys. **B199**, 223 (1982); .
20. M.B. Smy, talk presented at the DPF 99 meeting, UCLA, 1999; G. Sullivan, talk presented at the Aspen Winter Conference on Particle Physics, Aspen, CO, Jan. 1999.
21. J.N. Bahcall and P. Krastev, Phys. Lett. **B436**, 243 (1998).
22. V. Barger, Y.-B. Dai, K. Whisnant, and B.-L. Young, University of Wisconsin-Madison report MADPH-99-1096 [hep-ph/9901388], and references therein.
23. S. Geer, Phys. Rev. **D57**, 6989 (1998).
24. V. Barger, T.J. Weiler, and K. Whisnant, Phys. Lett. **B427**, 97 (1998).

Sterile Neutrinos: Phenomenology and Theory

Rabindra N. Mohapatra

Department of Physics, University of Maryland
College Park, MD 20742 USA
E-mail: rmohapat@physics.umd.edu

ABSTRACT

A simultaneous understanding of the results of the LSND experiment indicating $\nu_\mu - \nu_e$ oscillation together with other evidences for neutrino oscillations from solar and atmospheric neutrino data seems to require the existence of at least one sterile neutrino. One can also give other plausible astrophysical arguments that seem to require light sterile neutrinos. If such sterile neutrinos exist, a major theoretical challenge is to understand why they are so light. A scenario for the light sterile neutrino is presented where one assumes a parity doubling of the standard model with identical matter and gauge content. The neutrinos of the parity doubled (mirror) sector are light for the same reason that the known neutrinos are light and since they do not couple to the known W and Z bosons, they can be identified with the sterile neutrinos. Some of the implications and possible tests of this hypothesis are mentioned.

1. Why we may need a sterile neutrino?

The announcement by the Super-Kamiokande collaboration[2] of the evidence for neutrino oscillation (and hence nonzero neutrino mass) in their atmospheric neutrino data is a major milestone in the search for new physics beyond the standard model. In addition to the Super-Kamiokande atmospheric neutrino data, there are now strong indications for neutrino oscillations[1] from the five solar neutrino experiments (Kamiokande, Homestake, Gallex, Sage and Super-Kamiokande[3,4]), other atmospheric neutrino observations [5,6] and the direct laboratory observation in the LSND experiment[7]. To explain all three evidences for neutrino oscillations, three different mass differences (Δm^2) are needed. The atmospheric neutrino data requires $\Delta m^2_{\mu-X} \sim 3\times10^{-4} - 7\times10^{-3}$ eV^2 whereas the solar neutrino data prefers either $3\times10^{-6} - 7\times10^{-6}$eV^2 or $\sim 10^{-10}$ eV^2 depending on whether the solution arises via the MSW mechanism or via oscillation in vacuum[8]. The LSND data on the other hand prefers $0.2\text{eV}^2 \leq \Delta m^2_{e\mu} \leq 2.0$ eV^2 with Δm^2 as high as 10 eV^2 also in the allowed range. The LSND observations need to be corroborated by either KARMEN[9] or the proposed BooNE experiment. If however, we take the LSND experiment seriously, we have a crisis for the three neutrino picture since with the three known neutrinos one can get only two independent Δm^2's. A simple solution to this crisis is to adopt the proposal[10] that an ultralight fourth neutrino, to be called the sterile neutrino exists. For the uninitiated a sterile neutrino is defined as one whose interaction strength with standard model particles (such as the W, Z etc) is much weaker than that of the usual weak interaction. The reason for this is the discovery at LEP and SLC that

CP478, *COSMO—98*, edited by David O. Caldwell

only three neutrino species couple to the Z-boson. It is the goal of this talk to discuss other motivations for sterile neutrinos and then discuss possible theoretical scenarios for the sterile neutrinos after a brief discussion of how its introduction solves the neutrino puzzles.

2. Scenarios for solving the neutrino puzzles

In the presence of this extra neutrino species (ν_s), one can construct several scenarios for solving the three neutrino puzzles[10,11,12,13]. In the original paper introducing the sterile neutrino to solve the neutrino puzzles[10], it was proposed that the solar neutrino puzzle is solved via the oscillation of the ν_e to ν_s using the MSW mechanism[14] and the atmospheric neutrino puzzle is solved via the $\nu_\mu - \nu_\tau$ oscillation with maximal mixing. The solar neutrino puzzle fixes the $\Delta m^2_{e-s} \simeq (0.35 - .75) \times 10^{-5}$ eV^2, whereas the atmospheric neutrino puzzle implies that $\Delta m^2_{\mu-\tau} \simeq 10^{-3}$ eV^2. This gives a picture where the ν_s and ν_e are close by in mass with nearly zero mass and the ν_μ and ν_τ are nearly degenerate in mass. The masses of the $\nu_\mu - \nu_\tau$ system is determined by the LSND experiment $m_{\nu_2} \simeq m_{\nu_3} \simeq \sqrt{\Delta m^2_{LSND}}$. If the universe has a hot component in its dark matter, as some recent analyses suggest[15], then this requires the ν_μ and ν_τ masses to be each 2-3 eV implying that in LSND and KARMEN[9] one should observe a $\Delta m^2 \simeq 4 - 9$ eV^2. This point about hot dark matter has however become controversial[16] in view of recent indications that the total mass density of the universe may be considerably less than critical[17], an assumption that was used in concluding that the universe has 20% HDM.

This scenario is testable in the SNO[18] experiment once they measure the solar neutrino flux (Φ^{NC}_ν) in their neutral current data and compare with the corresponding charged current value (Φ^{CC}_ν). If the solar neutrinos convert to active neutrinos, then one would expect $\Phi^{CC}_\nu / \Phi^{NC}_\nu \simeq .5$, whereas in the case of conversion to sterile neutrinos, the above ratio would be nearly $\simeq 1$.

A second scenario advocated in Ref.[12] and in [13] suggests that it is the atmospheric ν_μ's that oscillate into the sterile neutrinos whereas the solar neutrino oscillation could be involving either active or sterile ones depending on how many sterile neutrinos one postulates. The present atmospheric neutrino data cannot distinguish between the ν_τ and ν_s as the final oscillation products. There is however an interesting suggestion[19] that monitoring the pion production via the neutral current reaction $\nu_\tau + N \rightarrow \nu_\tau + \pi^0 + N$ (which is absent in the case of sterile neutrinos) can help in distinguishing between these two possibilities. Another possible way to distinguish the $\nu_\mu - \nu_\tau$ oscillation possibility from the $\nu_\mu - \nu_s$ one for the atmospheric case may be to observe dips in the zenith angle distribution of the data due to matter oscillation in the earth for higher energy neutrinos[20].

There is yet a third scenario according to which[23] both solar and atmospheric neutrino oscillations involve the active neutrinos whereas the LSND data is an indirect oscillation that goes via the sterile neutrino which may have a mass of $\sqrt{\Delta m^2_{LSND}}$.

A possible mass matrix for the first case is[22]:

$$M = \begin{pmatrix} \mu_1 & \mu_3 & 0 & 0 \\ \mu_3 & 0 & 0 & \epsilon \\ 0 & 0 & \delta & m \\ 0 & \epsilon & m & \pm\delta \end{pmatrix}. \quad (1)$$

Solar neutrino data requires $\mu_3 \ll \mu_1 \simeq 2 \times 10^{-3}\ eV$ and $\epsilon \simeq .05m$. In the case with the negative sign in the 44-entry, the Δm^2 in the atmospheric data as well as the mixing in the LSND oscillation are linked to one another.

Finally, one has to ensure that all the new light particles introduced to explain the sterile neutrino do not spoil the success of big bang nucleosynthesis which cannot tolerate more than 1.5 extra neutrinos[27]. The contribution of a sterile neutrino is governed by its mass difference -squared and mixing with the normal neutrinos. For instance, the contribution of a sterile neutrino is suppressed[28] as long as the following inequality is satisfied:

$$\Delta m^2 sin^4 2\theta \leq 3 \times 10^{-6} \quad eV^2 \quad (2)$$

Any theoretical model must respect this constraint.

3. Other motivations for the sterile neutrino

There are several astrophysical observations whose understanding becomes easier if one assumes the existence of a sterile neutrino. The first example is a proper understanding of the heavy element abundance in the universe. A very plausible site for the production of heavy nuclei in the universe appears to be the hot neutron rich environment surrounding type II supernovae. The first road block that this proposal runs into arises from the fact that there are also hot neutrinos (ν_e's) surrounding the supernova core (indigenous as well as oscillation generated from ν_μ's) which have the effect of suppressing this process by reducing the neutron fraction via the reaction $\nu_e + n \rightarrow e^- + p$. It has recently been shown[23] that the introduction of the sterile neutrino seems to alleviate the problem by providing a "sucking" mechanism for the "bad" neutrinos into sterile neutrinos.

A second. albeit speculative motivation for the sterile neutrinos comes from the observed pulsar velocities (of about 500 to 100 km/sec.), for which there seems to be no convincing astrophysical explanation. One suggestion[24] based on the idea of neutrino oscillation is that in the presence of large magnetic fields, matter induced oscillations of the neutrinos will become asymmetric due to the presence of **k.B** terms in the resonance condition. Thus the neutrinos emerging on one side would have a larger momentum than those on the other side, thereby giving a momentum to the pulsar at its birth. However, for this mechanism to work one needs a neutrino with mass in the 100 to 300 eV range. Such a large mass would be unacceptable for the $\nu_{\mu,\tau}$ since that would overclose the universe. The sterile neutrinos on the other hand could have such masses without giving $\Omega_m \geq 1$ since their interaction with known matter is ultraweak and that makes them decouple above $T = 200$ MeV (where T is the temperature of the universe); the subsequent annihilation of known particles heats up the universe of the known particles thereby reducing the concentration of the sterile neutrinos at the present time. It is not very difficult to reduce their present density by 10-20, thus making it possible to have their mass be higher. Thus if matter enhanced neutrino oscillation is identified as the only way to account for observed pulsar velocities, that would provide indeed a very strong motivation for the sterile neutrinos.

A third motivation of similar speculative nature has to do with an understanding of the observed diffuse ionization in the milky way and other galaxies. A possible understanding of the various properties of the observed ionization (such as its large scale height $\sim$ 700 pc and uniform distribution etc.) suggests that they could be caused by the radiative decay of some species of heavy neutrinos[25]. But energetics of the problem require that the decaying neutrino must have mass of 27.4 eV. Any of the active neutrinos with a mass of this type would lead to $\Omega_m = 1$ in contradiction with recent indications. On the other hand, the sterile neutrino is free of such mass constraints and it has been recently shown[26] how a sterile 27.4 eV neutrino could provide a combined resolution of both the diffuse ionization problem along with the other neutrino puzzles.

4. Mirror universe theory of the sterile neutrino

If the existence of the sterile neutrino becomes confirmed say, by a confirmation of the LSND observation of $\nu_\mu - \nu_e$ oscillation or directly by SNO neutral current data to come in the early part of the next century, a key theoretical challenge will be to construct an underlying theory that embeds the sterile neutrino along with the others with appropriate mixing pattern, while naturally explaining its ultralightness.

It is clear that if a sterile neutrino was introduced into the standard model, the gauge symmetry does not forbid a bare mass for it implying that there is no reason for the mass to be small. It is a common experience in physics that if a particle has mass lighter than normally expected on the basis of known symmetries, then it is an indication for the existence of new symmetries. This line of reasoning has been pursued in recent literature to understand the ultralightness of the sterile neutrino by using new symmetries beyond the standard model.

We will focus on the recent suggestion that the ultralightness of the ν_s may be related to the existence of a parallel standard model[29,12,30] which is an exact copy of the known standard model (i.e. all matter and all gauge forces identical). The mirror sector of the model will then have three light neutrinos which will not couple to the Z-boson and would not therefore have been seen at LEP. We will call the ν'_i as the sterile neutrinos of which we now have three. The lightness of ν'_i is dictated by the mirror $B' - L'$ symmetry in a manner parallel to what happens in the standard model. The two "universes" communicate only via gravity or other forces that are equally weak. This leads to a mixing between the neutrinos of the two universes and can cause neutrino oscillation between ν_e of our universe to ν'_e of the parallel one in order to explain for example the solar neutrino deficit.

At an overall level, such a picture emerges quite naturally in superstring theories which lead to $E_8 \times E'_8$ gauge theories below the Planck scale with both E_8s connected by gravity. We hasten to emphasize that despite this apparent promising connection, no vacuum state that leads to the details needed for our neutrino model has been discussed to date. In this paper, we will assume the sub-Planck GUT group to be a subgroup of $E_8 \times E'_8$ in the hope of possible future string embedding of our model. For alternative theoretical models for the sterile neutrino, see Ref.[32].

As suggested in Ref.[29], we will assume that the process of spontaneous symmetry breaking introduces asymmetry between the two universes e.g. the weak scale v'_{wk} in the mirror universe is larger than the weak scale $v_{wk} = 246$ GeV in our universe. The ratio of the two weak scales $\frac{v'_{wk}}{v_{wk}} \equiv \zeta$ is the only parameter that enters the fit to the solar neutrino data. It was shown in Ref.[29] that with $\zeta \simeq 20 - 30$, the gravitationally generated neutrino masses[31] can provide a resolution of the solar neutrino puzzle (i.e. one parameter generates both the required Δm^2_{e-s} and the mixing angle $sin^2 2\theta_{e-s} \simeq 10^{-2}$).

There are other ways to connect the visible sector with the mirror sector using for instance a bilinear term involving the righthanded neutrinos from the mirror and the visible sector. An $SO(10) \times SO(10)$ realization of this idea was studied in detail in a recent paper[33], where a complete realistic model for known particles and forces including a fit to the fermion masses and mixings was done and the resulting predictions for the masses and mixings for the normal and mirror neutrinos were presented. In this model, the fermions of each generation are assigned to the $(\mathbf{16}, \mathbf{1}) \oplus (\mathbf{1}, \mathbf{16'})$ representation of the gauge group. The $SO(10)$ symmetry is broken down to the left-right symmetric model by the combination of $\mathbf{45} \oplus \mathbf{54}$ representations in each sector. The $SU(2)_R \times U(1)_{B-L}$ gauge symmetry in turn is broken by the $\mathbf{126} \oplus \overline{\mathbf{126}}$ representations. These latter fields serve two purposes: first, they guarantee automatic R-parity conservation and second, they lead to the see-saw suppression for the neutrino masses. The standard model symmetry is then broken by several $\mathbf{10}$-dim. Higgs fields. Using the charged fermion masses and mixings as inputs, we obtain the following absolute values of the neutrino masses (in eV's): $m_{\nu'_\tau} = 90.56, m_{\nu'_\mu} = -90.56, m_{\nu'_e} = 0.0034$ $m_{\nu_\tau} = 1.51, m_{\nu_\mu} = -1.509, m_{\nu_e} = 0.001$. The squared mass diffences (in eV^2) are $\Delta m^2_{e-s} = 9.9 \times 10^{-6}$, $\Delta m^2_{\mu-\tau} = 0.003$ and $\Delta m^2_{e-\mu} = 2.27$, where the numbers are given in eV^2. The neutrino mixing matrix O^ν

of the six neutrinos in the basis $(\nu_e, \nu_\mu, \nu_\tau, \nu'_e, \nu'_\mu, \nu'_\tau)$ is approximately given as,

$$O^\nu = \begin{pmatrix} -0.99 & 0.037 & 0 & 0.039 & -0.00025 & 0 \\ -0.031 & -0.85 & -0.52 & -0.00072 & 0 & 0 \\ 0.019 & 0.525 & -0.85 & -0.00043 & 0 & 0 \\ -0.042 & 0.0014 & 0.00071 & -0.999 & 0.0062 & 0 \\ 0 & 0 & 0 & -0.0053 & -0.850 & -0.525 \\ 0 & 0 & 0 & 0.0032 & 0.52 & -0.85 \end{pmatrix} \qquad (3)$$

Combining this with the mixing angle for the charged leptons, we obtain the final mixing matrix among the four neutrinos which serves the desired purpose of fitting all the neutrino oscillation data.

Turning now to the consistency of our model with big bang nucleosynthesis (BBN), we recall that present observations of Helium and deuterium abundance can allow for as many as 4.53 neutrino species[27] if the baryon to photon ratio is small. In our model, since the neutrinos decouple above 200 MeV or so, their contribution at the time of nucleosynthesis is negligible (i.e. they contribute about 0.3 to ΔN_ν.) On the other hand the mirror photon could be completely in equilibrium at $T = 1$ MeV so that it will contribute $\Delta N_\nu = 1.11$. All together the total contribution to δN_ν is less than 1.5.

There may be another potentially very interesting application of the idea of the mirror universe. It appears that there may be a crisis in understanding the microlensing observations[34]. It has to do with the fact that the best fit mass for the 14 observed microlensing events by the MACHO and the EROS group is $0.5 M_\odot$ and it appears difficult to use normal baryonic objects of similar mass such as red dwarfs or white dwarfs to explain these events, since they lead to a number of cosmological and astrophysical problems[35]. Speculations have been advanced tha this crisis may also be resolved by the postulate that the MACHOs with $0.5 M_\odot$ may be mirror stars[36,37] which would then have none of the difficulties that arise from the white dwarf or other interpretations.

In conclusion, if the LSND result stands the test of time, an ultralight sterile neutrino would be required to understand all neutrino data. Combined with other astrophysical arguments based on supernova r-process nucleosynthesis, it would appear that future developements in particle physics may very well require that we build models beyond standard unification framework that incorporate light sterile neutrinos. An interesting class of models that lead to a naturally light sterile neutrino postulate mirror matter and mirror forces. The existing MACHO and EROS data on microlensing events may also find a more satisfactory explanation in terms of compact near solar mass objects built from mirror matter rather than those from ordinary matter. On the theoretical front, it is possible to construct fully grand unified models for mirror matter using the $SO(10) \times SO(10)$ or $SU(5) \times SU(5)$ group.

I am grateful to D. Caldwell, Z. Berezhiani, B. Brahmachari, D. Sciama and V. L. Teplitz for collaborations on different aspects of the sterile neutrino idea. I am also grateful to G. Fuller, R. Volkas and A. Smirnov for discussions. This work is supported by the National Science Foundation under grant no. PHY-9802551.

References

1. For two recent excellent reviews of the neutrino oscillations and their implications, see S. Bilenky, C. Giunti and W. Grimus, hep-ph/9812360 and K. Zuber, hep-ph/9811267 (to appear in Phys. Reports).
2. T. Kajita, Invited talk at the Neutrino98, Takayama, Japan (to appear in the proceedings); Super-Kamiokande Collaboration, UMD-PP-98- 117 and ICRR Report-411-98-7.
3. B. T. Cleveland et al. Nucl. Phys. B(Proc. Suppl.) **38**, 47 (1995); K.S. Hirata et al., Phys. Rev. **44**, 2241 (1991); GALLEX Collaboration, Phys. Lett. **B388**, 384 (1996); J. N. Abdurashitov et al., Phys. Rev. Lett. **77**, 4708 (1996).
4. Super-Kamiokande collaboration, Phys. Rev. Lett. **81**, 1562 (1998).
5. K.S. Hirata et al., Phys. Lett. **B280**, 146 (1992); R. Becker-Szendy et al., Phys. Rev. **D 46**, 3720 (1992); W. W. M. Allison et al., Phys. Lett. **B 391**, 491 (1997); Y. Fukuda *et al*, Phys. Lett. **B 335**, 237 (1994).
6. T. Montaruli, talk given at TAUP 97, Gran Sasso, Italy.
7. C. Athanassopoulos et al. Phys. Rev. Lett. **75**, 2650 (1995); C. Athanassopoulos et al. Nucl-ex/9706006.
8. For a recent review and update, see J. Bahcall, P. Krastev and A. Smirnov, hep-ph/9807216 and P. Langacker, hep-ph/9811460.
9. B. Armbruster et al., hep-ex/9801007.
10. D.O. Caldwell and R.N. Mohapatra, Phys. Rev. **D 48**, 3259 (1993); J. Peltoniemi and J. W. F. Valle, Nucl. Phys. **B 406**, 409 (1993).
11. S. Bilenky, C. Giunti and W. Grimus, Eur. Phys. J. **C 1**, 247 (1998); hep-ph/9805368; S. Goswami, Phys. Rev. **D 55**, 2931 (1997); N. Okada and O. Yasuda, hep-ph/9606411.
12. R. Foot and R. Volkas, Phys. Rev. **D52**, 6595 (1995).
13. Q. Liu and A. Smirnov, hep-ph/9712493.
14. S. P. Mikheyev and A. Smirnov, Sov. J. Nucl. Phys. **42**, 913 (1985); L. Wolfenstein, Phys. Rev. **D 17**, 2369 (1978).
15. E. Gawiser and J. Silk, astro-ph/9806197; Science, **280**, 1405 (1998).
16. J. A. Primack and M. A. K. Gross, Astro-ph/9810204.
17. S. Perlmutter et al. Nature **391**, 51 (1998); V. Eke, S. Cole, C. S. Frenk and J. P. Henry, Astro-ph/9802350.
18. SNO collaboration, E. Norman *et al.*, "*Proceedings of The Fermilab Conference: DPF 92* ed. by C. Albright, P. H. Kasper, R. Raja and J. Yoh (World Scientific), p.1450.
19. A. Smirnov and F. Vissani, hep-ph/9710565; J. G. Learned, S. Pakvasa and J. Stone, hep-ph/9805343.

20. Q. Liu, S. P. Mikheyev and A. Smirnov, hep-ph/9803415.
21. See G. Fuller, these proceedings.
22. V. Barger, K. Whisnant and T. Weiler, hep-ph/9712495; S. Gibbons, R. N. Mohapatra, S. Nandi and A. Raichoudhuri, hep-ph/9803299, Phys. Lett. **B 430**, 296 (1998).
23. D. Caldwell, G. Fuller and Y. Qian, to appear.
24. A. Kusenko and G. Segre, Phys. Rev. Lett. **77**, 4872 (1996).
25. For a review of the problem as well as the radiative neutrino decay solution, see D. W. Sciama, *Modern Cosmology and the Dark Matter Problem*, Cambridge University Press, 1993.
26. R. N. Mohapatra and D. Sciama, hep-ph/9811446.
27. S. Sarkar, Rep. Prog. Phys. **59**, 1493 (1996); P. Kernan and S. Sarkar, Phys. Rev. **D 54**, 3681 (1996).
28. R. Barbieri and A. Dolgov, Phys. Lett. **B 237**, 440 (1990); K. Enquist, K. Kainulainen and J. Maalampi, Phys. Lett. **B 249**, 531 (1992); D. P. Kirilova and M. Chizov, hep-ph/9707282.
29. Z. Berezhiani and R. N. Mohapatra, Phys. Rev. **D 52**, 6607 (1995).
30. S. I. Blinikov and M. Yu Khlopov, Sov. Astron. **27**, 371 (1983); Z. Silagadze, hep-ph/9503481.
31. R. Barbieri, J. Ellis and M. K. Gaillard, Phys. Lett. **90 B**, 249 (1980); E. Akhmedov, Z. Berezhiani and G. Senjanović, Phys. Rev. Lett. **69**, 3013 (1992).
32. K. Benakli and A. Smirnov, Phys. Rev. Lett. **79**, 4314 (1997); P. Langacker, hep-ph/9805281; D. Suematsu, Phys. Lett. **B392**, 413 (1997); E. J. Chun, A. Joshipura and A. Smirnov, Phys. Rev. **D 54**, 4654 (1996); E. Ma and P. Roy, Phys. Rev. **D 52**, 4342 (1995); M. Bando and K. Yoshioka, hep-ph/9806400; N. Arkani-Hamed and Y. Grossman, hep-ph/9806223.
33. B. Brahmachari and R. N. Mohapatra, hep-ph/9805429.
34. C. Alcock et al. Ap. J. **486**, 697 (1997); R. Ansari et al. A & A **314**, 94 (1996); for a recent review, see the talk by K. Griest, COSMO98 (to appear in the proceedings).
35. K. Freese, B. Fields and D. Graaf, Astro-ph/9901178.
36. Z. Berezhiani, A. Dolgov and R. N. Mohapatra, Phys. Lett. **B 375**, 26 (1996); S. Blinnikov, Astro-ph/9801015.
37. R. N. Mohapatra and V. L. Teplitz, Astro-ph/9902085; R. Foot, Astro-ph/9902065.

Neutrino Masses and Mixings: Big Bang and Supernova Nucleosynthesis and Neutrino Dark Matter

George M. Fuller

Department of Physics
University of California, San Diego
La Jolla, CA 92093-0319

Abstract. The existence of small mixings between light active and sterile neutrino species could have implications for Big Bang and Supernova Heavy Element Nucleosynthesis. As well, such mixing would force us to abandon cherished constraints on light neutrino Dark Matter. Two proposed 4-neutrino mass and mixing schemes, for example, can both accomodate existing experimental results and lead to elegant solutions to the neutron-deficit problem for r-Process nucleosynthesis from neutrino-heated supernova ejecta. Each of these solutions is based on matter-enhanced (MSW) active-sterile neutrino transformation. In plausible extensions of these schemes to the early universe, Shi & Fuller have shown that relatively light mass ($\sim 200\,\mathrm{eV}$ to $\sim 10\,\mathrm{keV}$) sterile neutrinos produced via active-sterile MSW conversion can have a "cold" energy spectrum. Neutrinos produced in this way circumvent the principal problem of light neutrino dark matter and would be, essentially, Cold Dark Matter.

INTRODUCTION

In a shocking affront to the sensibilities of many theoretical physicists the current laboratory and observational neutrino data point to the requirement for at least one light sterile neutrino species [1,2]. In this paper we explore the implications of this conclusion for astrophysics and cosmology. Namely, we examine what nontrivial mixings between light active and sterile neutrino species might imply for: (1) the synthesis of the light elements in the early universe; (2) the synthesis of the heaviest elements in supernova or neutron star merger-based neutrino-heated ejecta; and (3) neutrino Dark Matter.

Might we be spared this exercise if we could argue that one or more of the interpretations of the recent neutrino experiment's results in terms of neutrino oscillations is false? Yes, possibly, but this is increasingly difficult to do. The Superkamiokande (SuperK) atmospheric neutrino data [3] gives convincing evidence

CP478 , *COSMO—98*, edited by David O. Caldwell

of ν_μ disappearance, most likely in the $\nu_\mu \rightleftharpoons \nu_\tau$ channel, with a mass-squared difference $\delta m^2 \sim 5 \times 10^{-3}\,\mathrm{eV}^2$ and maximal or near-maximal vacuum mixing angle. We will argue below that active-sterile vacuum mixing with such a large mixing angle is ruled out as a potential solution channel as a result of Big Bang Nucleosynthesis (BBN) considerations. There is a well established deficit of solar ν_e's [3,4]. With the recent helioseismology data, the solar neutrino problem appears to be incompatible with solar model alterations as a solution, but there are several neutrino mixing solutions which remain viable [5]. Either active-active ($\nu_e \rightleftharpoons \nu_{\mu,\tau}$) or active-sterile ($\nu_e \rightleftharpoons \nu_s$) matter-enhanced MSW (Mikheyev, Smirnov, Wolfenstein [6]) channels have been proposed with vacuum mass-squared differences $\delta m^2 \sim 10^{-5}\,\mathrm{eV}^2$, or vacuum oscillation in the active-active channel with $\delta m^2 \sim 10^{-10}\,\mathrm{eV}^2$. Finally, the Los Alamos Liquid Scintillator Neutrino Detector (LSND) finds an excess of $\bar{\nu}_e$ events which can be interpreted as arising from vacuum oscillations in the antineutrino channel $\bar{\nu}_\mu \rightleftharpoons \bar{\nu}_e$ with mass-squared difference $\delta m^2 \approx 0.2\,\mathrm{eV}^2$ to $8\,\mathrm{eV}^2$ [7].

The entire data set taken at face value then requires three very different and disparate mass-squared differences for neutrino transformation solutions. However, the three active neutrino species can produce only two independent δm^2 values. Attempts to explain the data by employing one mass-squared difference for two of the three experimental results have failed, either because of the well established zenith angle dependence of the SuperK atmospheric data, or because of the apparent energy dependence of the solar neutrino deficit [8]. We are then forced to either discount one of the experimental/observational results, or else introduce a fourth light neutrino species (which must necessarily be sterile on account of the Z^0-width limits). Here we explore the astrophysical consequences of accepting *all* the data, and therefore adopting the latter viewpoint.

NEUTRINOS AND BBN

Big Bang Nucleosynthesis considerations argue strongly against a significant population of sterile neutrino seas in the early universe, unless accompanied by fine tuning of the ν_e and $\bar{\nu}_e$ energy spectra and number densities. This conclusion follows on a simple tour through the basic physics of BBN and weak decoupling and freeze-out.

Primordial nucleosynthesis is a freeze-out from Nuclear Statistical Equilibrium (NSE). At high enough temperature in the early universe all nuclear reactions that both build up and tear down nuclei are in equilibrium. As the universe expands and the temperature drops, reactions will become slow compared to the expansion rate, and the universe will experience a succession of "freeze-out" events as various nuclear species and groups of nuclei cease to be in equilibrium with each other and the background plasma.

Computation of nucleosynthesis yields from this freeze-out then depends on specifying five parameters (not all of which are independent!): (1) the expansion rate; (2) the average entropy per baryon $s/k \approx 2.53 \times 10^8 \Omega_b{}^{-1} h^{-2}$ (where Ω_b is the fractional contribution of baryon rest mass to the closure density and h is the Hubble parameter at the current epoch in units of $100\,\mathrm{km\,s^{-1}\,Mpc}$), or, equivalently the baryon-to-photon ratio $\eta \approx 2.79 \times 10^{-8}\Omega_b h^2$; (3) the *spatial* distribution of entropy; (4) the net lepton numbers, neutrino energy spectra, and spatial distribution of lepton number; (5) the neutron-to-proton ratio. The neutron-to-proton (n/p) ratio can be computed from the lepton parameters of point (4) by evolving the following reactions through the Weak Freeze-out Epoch:

$$\nu_e + n \rightleftharpoons p + e^-, \tag{1}$$

$$\bar{\nu}_e + p \rightleftharpoons n + e^+, \tag{2}$$

$$n \rightleftharpoons p + e^- + \bar{\nu}_e. \tag{3}$$

Weak Freeze-out occurs when these reactions become slow compared to the expansion rate. Subsequently, n/p drops only slowly with expansion - free neutron decay, Eq. (3), being the only significant weak process. The Weak Freeze-out temperature T_{WFO}, where the weak charge changing processes drop out of equilibrium, is a crucial determinant of the neutron abundance since $n/p \approx exp(-\delta m_{\mathrm{np}}/T_{\mathrm{WFO}})$, where the neutron-proton mass difference is $\delta m_{\mathrm{np}} \approx 1.293\,\mathrm{MeV}$.

In practice, two assumptions are usually made for parameters (3) and (4): homogeneous distribution of baryons and entropy; and homogeneous and small net lepton numbers. We can define the net lepton number contributed by any neutrino species α in terms of proper neutrino and antineutrino number densities as $L_{\nu_\alpha} \equiv (n_{\nu_\alpha} - n_{\bar{\nu}_\alpha})/n_\gamma$, where $n_\gamma = (2\zeta(3)/\pi^2)T^3 \approx 3.17 \times 10^{31}\,\mathrm{cm^{-3}}(T/\mathrm{MeV})^3$ is the photon proper number density at an epoch with temperature T. By small lepton numbers we mean $L_{\nu_\alpha} \sim \eta$ for all relevant neutrino flavors α.

Adoption of homogeneity and isotropy as spacetime symmetries in the early universe leads to the Friedmann, Lemaitre, Robertson, Walker (FLRW) metric solution for the Einstein field equations and, in turn, this symmetry allows extention of local energy conservation to spherically symmetric regions of finite 3-volume. This latter condition means that the bigger the energy density, the faster will be the expansion rate. Radiation domination in the early universe implies that we can compute the expansion rate $H \equiv \dot{a}/a$ (where a is the FLRW scale factor) from the number of relativistic boson and fermion degrees of freedom $g = \Sigma g_b + 7/8\Sigma g_f$,

$$H = \left(\frac{8\pi^3}{90}\right)^{1/3} g^{1/2}\frac{T^2}{m_{\mathrm{PL}}}, \tag{4}$$

where $m_{\mathrm{PL}} \approx 1.22 \times 10^{22}\,\mathrm{MeV}$ is the Planck Mass. It is common practice to parameterize the contributions to g beyond those from photons and electron-positron pairs by an effective number of thermal neutrino species N_ν.

Ultimately, very nearly all the neutrons that are extant at $T \sim 100\,\text{keV}$ are incorporated into alpha particles. This makes the expansion rate the crucial determinant of the helium yield. Conversely, the deuterium yield depends principally on the baryon density, as this determines the temperature at which alpha particles form, neutrons disappear, and, hence, the deuteron falls out of NSE. Observations of the deuterium isotope shift in hydrogen absorption lines in Lyman limit systems along lines of sight to distant QSO's then determine what is probably a good estimate of the primordial deuterium abundance. These observations imply $\Omega_b h^2 \approx 0.02$ [9]. The primordial helium abundance is, in my opinion, much less securely determined from the observations, though it is clear that the He mass fraction cannot exceed about 25% [10]. This implies a conservative limit on the expansion rate at the BBN epoch corresponding to $N_\nu \lesssim 3.3$ (cf. Ref. [11]).

The three active neutrino species alone would give us $N_\nu = 3$ if their masses are $\lesssim 10\,\text{MeV}$. What then about the sterile neutrino species, if any?

Active neutrino species will thermally decouple from the background plasma when the rates of neutrino-electron scattering (i.e, the principal energy transfer process) fall below the expansion rate. This Weak Decoupling Epoch for active neutrinos occurs when the temperature is $T \sim 1\,\text{MeV}$ to $3\,\text{MeV}$. Sterile neutrinos have much smaller cross sections and scattering rates so presumably they would have decoupled at a considerably earlier epoch. If they decouple *prior* to the QCD Epoch ($T \sim 150\,\text{MeV}$), they will not partake of the entropy transfer associated with the incorporation of quark degrees of freedom into color singlets; whereas, the active neutrinos will [12]. Since the change in g at the QCD epoch is of order a factor of ~ 3, this is a major effect. Decoupled sterile neutrinos which were as abundant as the active species prior to this event will be considerably diluted, and vitiated as far as their contribution to g at the BBN epoch is concerned.

However, after the QCD epoch but prior to or during the BBN epoch, neutrino oscillation processes and matter-enhanced MSW neutrino transformation in the channels $\nu_{\tau,\mu,e} \rightleftharpoons \nu_s$ could act to significantly populate one or more sterile neutrino seas (cf. Ref. [13]). Therefore the BBN ^{4}He limit becomes a major constraint on proposed active-sterile neutrino mixing schemes.

Is that the end of the story? No, because although an accelerated expansion rate from a sterile neutrino sea will lead to more neutrons (and , hence, more ^{4}He), this effect could be countered by an increased (decreased) number density of (or high energy tail for) ν_e ($\bar{\nu}_e$). This would act to decrease the number of neutrons at T_{WFO} by influencing the competition between the processes in Eq.s (1) and (2) in favor of fewer neutrons and more protons.

Active-sterile plus active-active neutrino mixing channels together can produce a rich and complicated lepton number generation mechanism, which is inherently chaotic [14]. In these processes, active-sterile transformation could result in a large net excess in one active neutrino species, even if initial acive populations have $L_{\nu_\alpha} \sim \eta$. This lepton number then could be transferred between active species via active-active transformation.

Foot and Volkas have pointed out a very clever means by which this might occur [15]. They envision a matter-enhanced level crossing $\nu_{\mu,\tau} \rightleftharpoons \nu_s$, followed by $\nu_{\mu,\tau} \rightleftharpoons \nu_e$ to build up an enhanced ν_e sea. Ultimately, they find that choosing the $\nu_\mu \rightleftharpoons \nu_s$ parameters necessary to explain the SuperK atmospheric neutrino data will give an acceptable ^{4}He BBN yield, so long as the mass of the tau neutrino is $\gtrsim 4\,\mathrm{eV}$. However, Shi & Fuller [11] find in a different calculation of the same processes that the required tau neutrino mass must be $\gtrsim 15\,\mathrm{eV}$, clearly unacceptable from a cosmological (large scale structure) standpoint. It should be pointed out that this process is difficult to model as it involves nonlinear MSW, out-of-equilibrium neutrino production/destruction/scattering processes, all coupled with expansion and nuclear reactions. Even if it turns out that the Foot & Volkas scheme cannot allow enough neutron suppression to allow $\nu_\mu \rightleftharpoons \nu_s$ for SuperK atmospheric neutrinos, it still represents a means by which active-sterile plus active-active neutrino mixing can generate lepton number.

SUPERNOVA CONSIDERATIONS

Ironically, the nucleosynthesis of the heaviest nuclei could bear a striking similarity to BBN. The astrophysical production site of the rapid neutron capture (r-Process) nuclei like iodine, platinum, and uranium remains a mystery. However, wherever these nuclides are made it must be in neutron-rich environments, as the nuclei themselves possess considerable neutron excesses. That essentially narrows the candidate production sites to core collapse supernovae and neutron star merger events. Like BBN, both of these sites are dominated by neutrinos. Some have argued that neutral current neutrino post-processing has left its "finger prints" on the solar system r-Process abundance distribution [16].

There is enough circumstantial evidence to fuel speculation. Recent meteoritic abundance arguments suggest that nearly every supernova makes at least some r-Process material [17]. Since huge reserves of neutrino energy are the common denominator of all core collapse supernovae, it is then likely that some r-Process nuclides are made in material which is ejected to space by neutrino heating. Establishment of this scenario is what we would require to be able to employ known r-Process abundance systematics to probe neutrino physics [18].

Perhaps our best model for r-Process nucleosynthesis involves the neutron-rich wind which could accompany the hot neutron star some several seconds after a supernova explosion [19]. Figure 1 shows the geometry and landmark events in r-Process nucleosynthesis in this model. The idea is that the supernova explosion has blown away nearly all the overlying material. What is left is a very tenuous, near hydrostatic "envelope," with total mass $\sim 10^{-4}\,\mathrm{M}_\odot$ sitting on top of the neutron star. Though basically transparent to neutrinos, the intense neutrino fluxes from the star cause ablation and heating, principally through the forward reactions in Eq.s (1) and (2), causing the envelope material to expand and, hopefully, be ejected to space. As a heated fluid element flows away from the neutron star it experiences weak (n/p) freeze-out and NSE freeze-out just as in the BBN case. In fact, we can

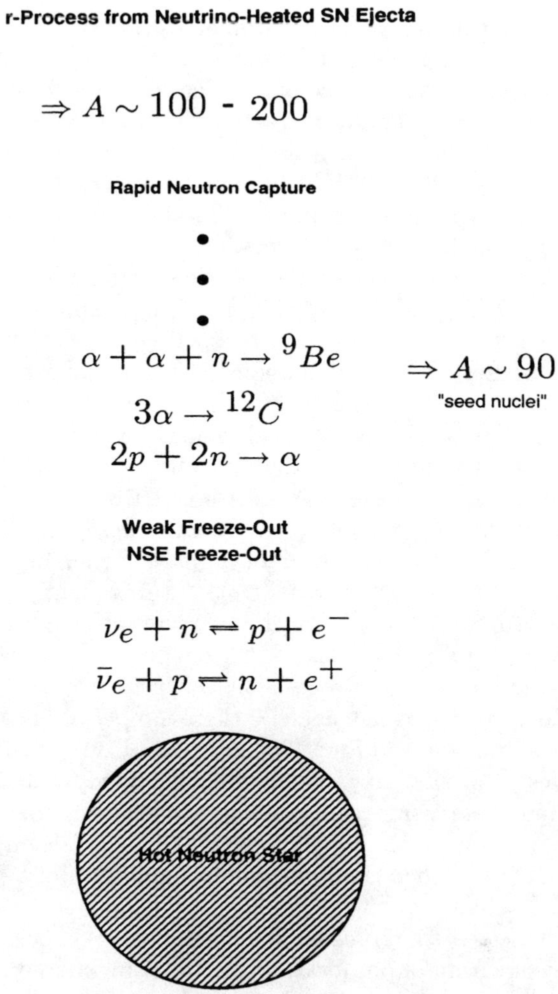

FIGURE 1. Flow chart for r-Process nucleosynthesis in a neutrino-heated wind from a hot neutron star. In this figure, the sense of the material flow is from the neutron star surface toward the top of the page. The flow will be self similar for a time, so that every fluid element will experience the same temperature-density history, crudely. Therefore, we can make an analogy to the expansion of the universe and Big Bang Nucleosynthesis. The radius of the outflowing fluid element here will be the analog of the time in BBN. A fluid element will cool as it moves out, encountering temperature $T_9 \approx 22.5/(S_{100}r_6)$ when it reaches radius $r_6 = r/(10^6\,\text{cm})$. The entropy per baryon in units of 100 times Boltzmann's constant is $S_{100} \sim 1$ and can be thought of as very roughly constant in the wind over the timescale of relevance, 3 s to 10 s after the supernova explosion. An outflowing fluid element will experience first weak freeze-out ($T_9 \approx 10$), then alpha particle formation and the production of seed nuclei with masses A$\sim$ 50 to 100. When the fluid element reaches a radius where the temperature has fallen to $T_9 \sim 3$, rapid neutron capture on the seed nuclei will ensue. The neutron-to-seed nucleus ratio must be large enough that neutron-rich nuclei with masses of order 200 or so can be built-up in this process.

think of the radius in this figure as a parametrization of time and temperature. As in BBN, the nucleosynthesis can be computed from a specification of the 5 parameters noted above for NSE freeze-out nucleosynthesis. From a nuclear physics standpoint, the entropy per baryon characteristic of the wind, $s/k \sim 100$, is also similar to that in the BBN case, i.e. high.The relationship between radius r_6 in units of 10^6 cm, temperature in 10^9 K, and entropy in units of $100k$ at any point in the history of an outflowing fluid element is, $T_9 \approx 22.5/(r_6 S_{100})$. We can think of the entropy as being roughly constant over the timescale of interest.

Unlike the BBN case, here we expect a generic hierarchy of average neutrino energies $\langle E_{\nu_\tau} \rangle \approx \langle E_{\bar{\nu}_\tau} \rangle \approx \langle E_{\nu_\mu} \rangle \approx \langle E_{\bar{\nu}_\mu} \rangle > \langle E_{\bar{\nu}_e} \rangle > \langle E_{\nu_e} \rangle$. Above the neutrino sphere, we expect the neutrino energy spectra for all species to be low chemical potential, Fermi-Dirac black bodies. The neutrino sphere is, roughly, the neutron star surface, where neutrinos effectively transport-decouple from the star. In the absence of neutrino oscillations/transformations, this generic energy hierarchy should guarantee that $n/p > 1$ in regions beyond the weak freeze-out radius. Nucleosyntheis in this environment is nearly the isospin mirror of BBN.

As the material moves out it cools to the point where, just like in BBN, every nucleon that can be is incorporated into an alpha particle. In neutron-rich conditions this will leave a sea of left-over neutrons. Subsequently, after a few of the alpha particles react to form "seed" nuclei as shown in the figure, the neutrons are captured on these seeds to produce the r-Process nuclei.

The numerical simulations of r-Process nucleosynthesis in this model fail to give a satisfactory yield of the heaviest nuclei - there simply are not enough neutrons (the neutron to seed ratio is too low) [20]. This neutron deficit can be traced to a number of causes: too slow expansion; too low entropy; and/or too low n/p. However, even fixing these issues with a different outflow model will not overcome the biggest problem involving the neutron deficit. This is the "alpha effect," where the process in Eq. (1) destroys the neutrons isolated by alpha particle formation [21].

The alpha effect seems to represent a paradox for r-Process models sited in neutrino-heated ejecta from supernovae or neutron star mergers. This is because these models require an intense enough flux of neutrinos that heating through the processes in Eq.s (1) and (2) as well as neutrino-antineutrino annihilation can impart some 100 MeV to each baryon: the amount by which baryons are bound in the gravitational potential well of the neutron star. This implies that each ejected baryon interacts with neutrinos ~ 10 times. But if the ν_e and $\bar{\nu}_e$ fluxes are high enough to accomplish this, they will also be high enough to produce an intense neutron-destroying alpha effect later in the flow when the alpha particles form ($T \lesssim 0.75$ MeV).

All hope may not be lost. Most of the baryon heating occurs quite close to the neutron star surface, while the ν_e captures of the alpha effect occur at the point of alpha particle formation, just at or beyond the weak freeze-out radius. Suppose we let the ν_e flux remain high near the neutron star surface to give efficient heating, but engineer its disappearance at some radius below where the alphas form?

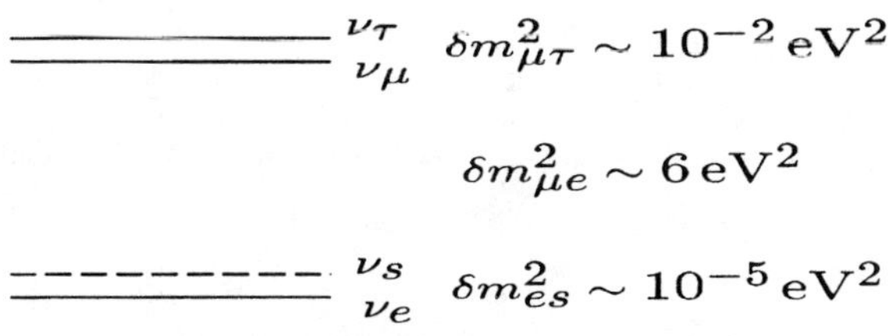

FIGURE 2. Caldwell, Fuller, & Qian (1999) neutrino mass/mixing scheme.

This could be done with matter-enhanced active-sterile neutrino transformation. Though active-sterile neutrino transformations have been discussed before in the context of supernova dynamics and nucleosynthesis [22] and four-neutrino arangements have been previously examined [23], two recent works have picked out neutrino mass and mixing schemes designed specifically to negate the alpha effect [24,25] while fitting other neutrino data constraints.

The CFQ [24] scheme for neutrino mass splittings is shown schematically in Figure 2. This scheme is quite similar to the Caldwell & Mohapatra suggestion [23], though here all parameters are chosen to aid r-Process nucleosynthesis. In this model, a lower mass doublet of ν_s split from a ν_e with the indicated δm^2 gives a matter-enhanced active-sterile solution for solar neutrinos (this would have a small vacuum mixing and would therefore evade BBN bounds). A heavier maximally-mixed ν_τ-ν_μ doublet would give the SuperK atmospheric neutrino result. The doublets would be split by $\delta m^2 \approx 0.2\,\mathrm{eV}^2$ to $8\,\mathrm{eV}^2$ to yield a vacuum neutrino oscillation solution $\bar{\nu}_\mu \rightleftharpoons \bar{\nu}_e$ for LSND. Note that nucleosynthesis as well as neutrino oscillation experiments pick out δm^2 values, not absolute masses, so that we are free to slide the scale shown in the figure up or down in mass. Even placing the lower doublet at zero mass however could imply a significant active neutrino dark matter component in this model.

In the supernova this scheme works beautifully at the late epoch of the neutrino-driven wind. High energy mu and tau neutrinos originating on the neutrino sphere undergo a mass level crossing (MSW resonance) and efficiently are changed to sterile neutrinos for vacuum mixing angles $\sin^2 2\theta \gtrsim 10^{-3}$. Matter-suppression effects argue for a low to vanishing flux of sterile neutrinos from the core. Subsequent to the $\nu_{\mu,\tau} \rightleftharpoons \nu_s$ resonance, at a larger radius neutrinos will encounter an active-active MSW resonance in the channel $\nu_e \rightleftharpoons \nu_{\mu,\tau}$. If neurinos evolve through this resonance adiabatically, then we will efficiently convert the electron neutrinos to mu/tau neutrinos and there will be none of those to convert back. Therefore, we will have succeeded in removing the ν_e flux at a sufficiently large radius to allow for heating but small enough to kill the alpha effect and produce a modest increase in n/p. There are many potential pitfalls here: (1) we require rapid outflow and small neutrino background to avoid reversal of the order of the resonances (probably meaning we must do this at quite late times $\sim 5 - 10\,\mathrm{s}$ after core bounce); and (2) we must be careful that this scenario does not wreck shock re-heating and

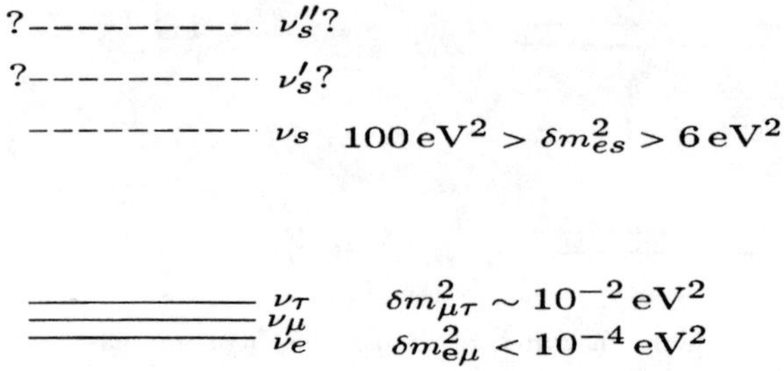

FIGURE 3. McLaughlin, Fetter, Balantekin, & Fuller (1999) neutrino mass/mixing scheme.

drive material neutron-rich at a few tenths of a second after core bounce.

McLaughlin, Balantekin, Fetter, & Fuller [25] have come up with a different four-neutrino mass and mixing scheme as depicted in Figure 3. Here we envision all the active neutrinos bunched together at lower mass with the indicated splittings. This lower mass scale can be chosen at will to be either near zero mass or $\sim 1\,\mathrm{eV}$ to accomodate a hot dark matter component. In this model we require at least one heavier sterile neutrino, split from the actives with $200\,\mathrm{eV}^2 > \delta m^2_{\mathrm{es}} > 0.2\,\mathrm{eV}^2$, depending on the neutrino background and luminosities and outflow entropy (a somewhat broader range than is indicated on the figure). Again we will have maximal $\nu_\mu \rightleftharpoons \nu_\tau$ mixing for SuperK, and any active-active channel for the solar neutrino problem. Here the LSND result would have to arise from a so-called indirect neutrino oscillation $\bar{\nu}_\mu \to \bar{\nu}_s \to \bar{\nu}_e$. This is still (barely) compatible with the neutrino oscillation constraint data.

In the supernova this scheme acts to evade the alpha effect in a different manner than the CFQ scheme. Here the matter-enhanced channels $\nu_e \rightleftharpoons \nu_s$ and $\bar{\nu}_e \rightleftharpoons \bar{\nu}_s$ both can come into play. The key point which McLaughlin et al. discovered was that the rapid outflow acted to keeep these two different resonance processes from reaching equilibrium at $n/p = 2$. In fact because of the rapid outflow, n/p can be even larger and, more importantly, most of the ν_e's are converted to steriles while few convert back and few $\bar{\nu}_e$'s are converted. Again, the effect can be to remove the ν_e flux at sufficiently large radius to kill the alpha effect but allow for baryon ejection.

NEUTRINO DARK MATTER

A cherished tenet of cosmology has been that light neutrinos, be they active or sterile, cannot be the bulk of the Dark Matter. For if they were, it would be impossible to understand how smaller structures like galaxies would exist at all and how clusters of galaxies would form at high redshift as observed. However, this conclusion presupposes that the neutrinos all have thermal energy spectra - a dangerous and incorrect assumption if MSW resonances are the production mechanism for sterile neutrinos.

If neutrinos have energy spectra like that of photons with comparable number densities (though colder by the factor $(4/11)^{1/3}$ from $e^{\pm}$ annihilation entropy transfer to the photons), then a single flavor with rest mass m_ν would contribute to closure a fraction $\Omega_\nu h^2 \approx m_\nu/(92\,\text{eV})$. Therefore, to be more important dynamically than baryons in the universe, the sum of the light neutrino masses need only be $\Sigma m_\nu \gtrsim 1.5\,\text{eV}$.

Could neutrinos be all of the Dark Matter? The conventional answer is no. Large scale structure simulations suggest that there will be constraints from observed structure unless $\Omega_\nu \lesssim 0.2$ [26]. These same simulations, however, also point to the salutory effect of a small admixture of neutrino hot Dark Matter. In any case, with normal "hot," near-zero chemical potential Fermi-Dirac energy spectra, the sum of all the light neutrino masses must be $\Sigma m_\nu \lesssim 8\,\text{eV}$.

The generic disease that hot light neutrino dark matter possesses is that the neutrinos will revert from relativistic to nonrelativistic kinematics only when the temperature scale of the universe has dropped below their rest masses. Until that epoch, neutrinos will freely stream, and therefore damp the growth of density fluctuations below the horizon scale. This free streaming (damping) scale co-moving to the present epoch is $\lambda_{\text{f.s.}} \sim 40\,\text{Mpc}(30\,\text{eV}/m_\nu)$.

However, Shi and Fuller in Ref. [27] have suggested a way in which relatively light sterile neutrinos could be produced in the early universe via MSW active-sterile matter-enhanced transformation. These sterile neutrinos would have masses between about 200 eV and 10 keV. (This mass range for sterile neutrinos has been discussed before [28] as Warm Dark Matter - with a very different production mechanism: vacuum mixing.)

In our model the production of the sterile neutrinos via MSW resonance would require a driving potential from a pre-existing net lepton number in any or several of the three active species in the range $L \sim 10^{-1}$ for a sterile mass of 200 eV to $L \sim 10^{-3}$ for $m_{\nu_s} \approx 10\,\text{keV}$. Note that this range of net lepton numbers is benign with respect to BBN and will be consistent with observed light element abundance constraints.

The beauty of this model is that MSW production here favors adiabatic evolution through resonances (and therefore efficient conversion of active to heavier sterile neutrinos) only for **lower energy** neutrinos. This is partly because lower energy neutrinos have larger ratios of resonance width to oscillation length. Also, the lepton number tends to be reduced by the transformation process as the resonance sweeps from lower energy to higher energy with the expansion of the universe. The net result will be a spectrum of sterile neutrinos which is grossly skewed toward lower energies: roughly resembling a Fermi-Dirac spectrum up to a cut-off energy $E_{\text{c.o.}}$ which is low compared to the temperature.

Therefore, these sterile neutrinos are born "cold." They will have much reduced free streaming lengths, $\lambda_{\text{f.s.}} \sim 40\,\text{Mpc}(30\,\text{eV}/m_\nu)((E_{\text{c.o.}}/T)/3.15)$. In turn, this would allow these neutrinos to behave like Cold Dark Matter, but with the twist of a small scale cut-off in structure. Their fractional contribution to closure would be

$$\Omega_{\nu_s} \approx \left(\frac{m_{\nu_s}}{350\,\text{eV}}\right)\left(\frac{0.5}{h}\right)^2\left(\frac{L_{\text{pot}}}{0.1}\right), \tag{5}$$

where $L_{\text{pot}} \approx 2L_{\nu_\alpha} + \sum_{\beta\neq\alpha} L_{\nu_\beta}$ with $\alpha, \beta = e, \mu, \tau$ is the actual driving MSW potential. For $\Omega_{\nu_s} = 0.4$ and $h = 0.5$, sterile neutrinos with $m_{\nu_s} = 200\,\text{eV}$ would have a structure cut-off $\sim 10^{12}\,\text{M}_\odot$, like Warm Dark Matter; whereas, $m_{\nu_s} = 10\,\text{keV}$ would have damping only for structures below $\sim 10^5\,\text{M}_\odot$, truly CDM.

First note that the sterile neutrino mass range in this model is larger than that required for aiding the r-Process. These would have to be other light steriles, e.g., perhaps those suggested in Figure 3. Second, though allowed by BBN constraints, the required lepton asymmetry still appears to be large. Note, however, that the true net lepton number could be $L_{\text{net}} \sim \eta$ as a result of cancellations, while L_{pot} is much larger.

Where would this net lepton number come from? Except possibly in the lepton number cancellation case, it could not arise from processes associated with Electroweak Baryogenesis, because in these models baryon and *net* lepton number are equal. It could possibly result from a Foot & Volkas lepton number generation scenario [15], or else something like Affleck & Dine baryogenesis [29] must occur (see also Ref. [30]).

Light sterile neutrinos represent a challenge to elementary particle physics and astrophysics. Verification of LSND's result and the solar neutrino deficit's interpretation in terms of neutrino mixing is pivotal for physics and cosmology.

I would like to acknowledge partial support from NSF Grant PHY98-00980 at UCSD and input from X. Shi, D. Caldwell, Y.-Z. Qian, A. B. Balantekin, J. Fetter, and G. McLaughlin.

REFERENCES

1. V. Barger, these proceedings.
2. T. Weiler, these proceedings.
3. Y. Fukuda, *et al.*, Phys. Rev. Lett. **81**, 1562 (1998); Phys. Lett. B **436**, 33 (1998).
4. B.T. Cleveland, T. Daily, R. Davis, J.R. Distel, K. Lande, C.K. Lee, P.S. Wildenhain, and J. Ullman, Astrophys. J. **496**, 505 (1998); W. Hampel, *et al.*, Phys. Lett. B **388**, 384 (1996), **420**, 114 (1998), *ibid.* **436**, 158 (1998); J.N. Abdurashitov, *et al.*, Nucl. Phys. Proc. Suppl. **48**, 299 (1996), hep-ph/9803418.
5. J.N. Bahcall, P.I. Krastev, and A.Yu. Smirnov, hep-ph/9807216; V. Barger, S. Pakvasa, T. J. Weiler, and K. Whisnant hep-ph/9806328; N. Hata and P. Langacker, Phys. Rev. D **56**, 6107 (1997).
6. Wolfenstein, L., *Phys. Rev. D* **17**, 2369 (1978); **20**, 2634 (1979); Mikeyev, S. P., and Smirnov, A. Yu., *Sov. Phys. JETP*, **69**, 4 (1986).
7. C. Athanassopoulos, *et al.*, Phys. Rev. Lett. **75**, 2650 (1996), **77**, 3082 (1996), Phys. Rev. C **54**, 2685 (1996), Phys. Rev. C **58**, 2489 (1998).

8. C. Y. Cardall and G. M. Fuller, Phys. Rev. D**53**, 4421 (1996); G. L. Fogli, E. Lisi, and G. Scioscia, Phys. Rev. D**52**, 5334 (1995); A. Acker and S. Pakvasa, Phys. Lett. **B397**, 209 (1997); P. Harrison, D. Perkins, and W.Scott, Phys. Rev. **B349**, 137 (1995); V. Barger, T. J. Weiler, and K. Whisnant hep-ph/9807319; T. Teshima and T. Sakai hep-ph/9805386 and hep-ph/9801276.
9. S. Burles and D. Tytler, in *Proceedings of the Second Oak Ridge Symposium on Atomic and Nuclear Astrophysics*, edited by A. Mezzacappa (Institute of Physics, Bristol, 1998); and references therein.
10. K. A. Olive, E. Skillman, and G. Steigman, Astrophys. J., **483**, 788 (1997); Y. I. Izotov and T. X. Thun, Astrophys, J., **500**, 188 (1998).
11. X. Shi and G. M. Fuller, Phys. Rev. D, in press (1999).
12. G. M. Fuller and R. A. Malaney, Phys. Rev. D, **43**, 3136 (1991).
13. R. Barbieri and A. Dolgov, Phys. Lett., **B 349**, 743 (1991); K. Enqvist, K. Kainulainen, and M. Thomson, Nucl. Phys. **B 373**, 498 (1992).
14. X. Shi, Phys. Rev. D **54**, 2753 (1996); R. Foot, M. J. Thomson, and R. R. Volkas, Phys. Rev. D **53**, 5349 (1996); R. Foot and R. R. Volkas, Phys. Rev. D **56**, 6653 (1997).
15. R. Foot and R. R. Volkas, Phys. Rev. D **55**, 5147 (1997).
16. Y.-Z. Qian, W. C. Haxton, K. Langanke, and P. Vogel, Phys. Rev. C **55**, 1532 (1997).
17. Y.-Z. Qian, P. Vogel, and G. J. Wasserburg, Astrophys. J., **494**, 285 (1998).
18. G. M. Fuller, R. Mayle, J. R. Wilson, and D. N. Schramm, Astrophys. J., **322**, 795 (1987); G. M. Fuller, R. Mayle, B. S. Meyer, and J. R. Wilson, Astrophys. J., **389**, 517 (1992); Y.-Z. Qian, *et el.*, Phys. Rev. Lett. **71**, 1965 (1993); Y.-Z. Qian and G.M. Fuller, Phys. Rev. D **51**, 1479 (1995), **52**, 656 (1995).
19. B. S. Meyer, G. J. Mathews, W. M. Howard, S. E. Woosley, and R. D. Hoffman, Astrophys. J., **399**, 656 (1992); S. E. Woosley, G. J. Mathews, J. R. Wilson, R. D. Hoffman and B. S. Meyer, Astrophys. J., **433**, 229 (1994); K. Takahashi, J. Witti and H.-T. Janka, Astron. Astrophys. **286**, 857 (1994).
20. B. S. Meyer and J. S. Brown, Astrophys. J. Suppl. **112**, 199 (1997); R. D. Hoffman, S. E. Woosley, and Y.-Z. Qian, Astrophys. J. **482**, 951 (1996).
21. G.M. Fuller and B.S. Meyer, Astrophys. J., **453**, 792 (1995); B. S. Meyer, G. C. McLaughlin, and G. M. Fuller, Phys. Rev. C, **58**, 3696 (1998).
22. J. T. Peltoniemi, Astron. Astrophys., **254**, 121 (1992); J. T. Peltoniemi, preprint hep-ph/9511323, unpublished (1995).
23. D. O. Caldwell and R. N. Mohapatra, Phys. Rev. D**48**, 3259 (1993); D. O. Caldwell and R. N. Mohapatra, Phys. Lett. B **354**, 371 (1995); G. M. Fuller, J. R. Primack, and Y.-Z. Qian, Phys. Rev. D**52**, 1288 (1995).
24. D. O. Caldwell, G. M. Fuller, and Y.-Z. Qian, Phys. Rev. D, *submitted* (1999).
25. G. C. McLaughlin, J. Fetter, A. B. Balantekin, and G. M. Fuller, Phys. Rev. C, in press (1999).
26. A. Klypin, J. Holtzman, J. Primack, and E. Regos, Astrophys, J., **416**, 1 (1993).
27. X. Shi and G. M. Fuller, Phys. Rev. Lett. , in press (1999).
28. S. Dodelson and L. M. Widrow, Phys. Rev. Lett. **72**, 17 (1994).
29. L. Affleck and M. Dine, Nucl. Phys. **B249**, 361 (1985).
30. J. Cline, these proceedings.

A Relic Neutrino Detector

C. Hagmann

Lawrence Livermore National Laboratory
7000 East Avenue, Livermore, CA 94550

Abstract. Probably the most promising way of detecting cosmic neutrinos is measuring the mechanical force exerted by elastic scattering of cosmic neutrinos from macroscopic targets. The expected acceleration is $\sim 10^{-23}$cm/s^2 for Dirac neutrinos of mass ~ 10 eV and local density $\sim 10^7/\text{cm}^3$. A novel torsion balance design is presented, which addresses the sensitivity-limiting factors of existing balances, such as seismic and thermal noise, and angular readout resolution and stability.

INTRODUCTION

The standard Big Bang model predicts a universal background of relic neutrinos, with an average density of $\sim 100/\text{cm}^3$ per flavor. Relic neutrinos of mass $\gg 10^{-3}$ eV would be nonrelativistic today, and contribute to the cosmological energy density an amount $\Omega_\nu \sim \sum_i m_{\nu_i}/(90h^2\text{eV})$, where $h \simeq 0.65$ is the Hubble expansion rate in units of 100 km/(s Mpc). Massive neutrinos are a natural candidate for the hot component in currently favored "Mixed Hot+Cold Dark Matter (HCDM)" [1] models of galaxy formation. In this scenario, neutrinos would contribute ~ 20 %, and CDM (e.g. Wimps and axions) the remainder of the dark matter. Nonrelativistic neutrinos would be clustered around galaxies and move with a typical velocity $v \sim 300\,\text{km/s}$. The Pauli principle [2] restricts the local neutrino number density to $n_\nu \lesssim 2 \times 10^6\text{cm}^{-3}\ (v_{\text{max}}/10^{-3}c)^3\ \sum_i(m_{\nu_i}/10\text{eV})^3$. The detection of relic neutrinos is hindered by the extremely small cross sections and energy deposits expected from interactions with electrons and nucleons. Past proposals have focused on detecting the mechanical force on macroscopic targets due to the "neutrino wind". Here, spatial coherence increases the cross section of targets smaller than the neutrino wavelength $\lambda_\nu \sim 100\,\mu\text{m}(10\,\text{eV}/m_\nu)$. In the nonrelativistic limit, one must distinguish between Majorana and Dirac neutrinos. For Dirac μ or τ neutrinos, the cross section is dominated by the vector neutral current contribution $\sigma_D = (G_F^2 m_\nu^2/8\pi)N_n^2 = 2 \times 10^{-55}\text{cm}^2(m_\nu/10\,\text{eV})^2 N_n^2$, where N_n is the number of neutrons in the target of size $\lesssim \lambda_\nu/2\pi$. For Majorana neutrinos, the vector contribution is suppressed by a factor $(v/c)^2$. The Sun's peculiar motion through the galactic halo will produce a wind force, whose direction is mod-

CP478 , *COSMO—98*, edited by David O. Caldwell

ulated by the Earth's rotation. For Dirac neutrinos, the acceleration of a target of density ρ and radius $\lambda_\nu/2\pi$ is [3] $a = 8 \times 10^{-24}\mathrm{cm/s^2}((A-Z)/A)^2\ (v_{\mathrm{sun}}/10^{-3}c)^2\ (n_\nu/10^7\mathrm{cm^{-3}})\ (\rho/20\,\mathrm{gcm^{-3}})$ and is independent of m_ν. A harmonic oscillator driven by the neutrino wind on resonance would experience a displacement amplitude of $\Delta x \simeq 10^{-15}\mathrm{cm}(\tau/\mathrm{day})$. A target of size $\gg \lambda_\nu$ can be assembled, while avoiding destructive interference, by using foam-like or laminated materials [3]. Alternatively, grains of size $\sim \lambda_\nu$ could be randomly embedded in a low density host material.

PROPOSED DETECTOR

At present, the most sensitive detector of small forces is the "Cavendish" torsion balance, which has been widely used for measurements of G, searches for new forces, and tests of the equivalence principle. A typical arrangement consists of a dumbbell-shaped test mass suspended by a tungsten fiber. The angular deflection is read out with an optical lever. The most serious noise backgrounds are 1. thermal noise, 2. seismic noise, 3. time-varying gravity gradients. The smallest measurable acceleration is $\sim 10^{-12}\mathrm{cm/s^2}$ [4]. Several improvements seem possible: Thermal noise can be decreased by lowering the temperature and by employing a low dissipation (high-Q) suspension, as seen from the expression for the thermal noise acceleration [5] $a_{\mathrm{th}} = 2 \times 10^{-23}\mathrm{cm/s^2}\ (T/\mathrm{K})^{1/2}\ (1\mathrm{day}/\tau_0)^{1/2}\ (10^6\mathrm{s}/\tau)^{1/2}\ (10^{16}/Q)^{1/2}$, where τ is the measurement time, τ_0 is the oscillator period, and T is the operating temperature. A promising low-dissipation suspension method uses the Meissner effect. Niobium or NbTi based suspensions have been employed in gravimeters, gyros, and gravitational wave antennas. Generally, the magnetic field applied to the superconductor is limited to $\lesssim 0.2\mathrm{T}$ to avoid flux penetration or loss of superconductivity. A remaining problem is flux creep noise and dissipation because of incomplete flux expulsion. An alternative method would employ a passive persistent-mode superconducting magnet floating above a fixed suspension magnet as shown in Fig. 1. This allows a much higher lifting force because the critical field of NbTi wire is several T. Moreover, the flux lines are strongly pinned by artificial wire defects, leading to a small field decay rate of $\dot{B}/B \sim 10^{-8}/\mathrm{hour}$. The cylindrical symmetry of the suspension magnets allows a very long rotational oscillation period, which can be matched to the diurnally varying neutrino wind by applying a suitable restoring force. The effects of seismic noise and gravity gradients can be reduced with a highly symmetric target (see Fig. 1). With the c.m. of the target centered below the suspension support, the leading order gravitational torques arise from the dipole and quadrupole moments of the target which need to be minimized by balancing. Braginsky et.al. [6] have given estimates of the seismic power spectra for vertical and horizontal, and rotational seismic modes. For example, the horizontal acceleration at $\omega \sim 10^{-4}\mathrm{s^{-1}}$ is $a \sim 10^{-12}\mathrm{cm/s^2}(\Lambda/100\,\mathrm{km})(10^6/\tau)^{-1/2}$, where Λ is the seismic wavelength. Hence, the coupling of this mode to the wanted rotational mode of the torsion oscillator must be made very small. Especially worrisome is rotational seismic noise which will directly mask the signal and needs to be

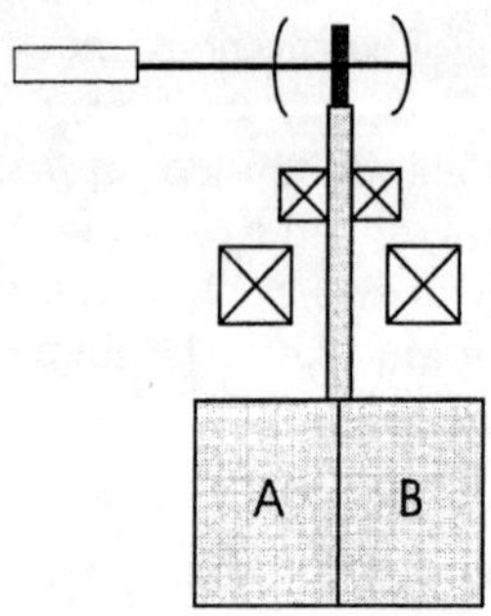

FIGURE 1. Schematic diagram of the torsion oscillator. The target consists of two hemicylindrical masses with similar densities but different neutrino cross sections. The mass is suspended by a "magnetic hook" consisting of a superconducting magnet in persistent mode floating above a stationary magnet. The rotation angle is read out by measuring the resonance frequency of a tunable Fabry-Perot cavity with a laser.

compensated. The proposed angular rotation readout is composed of a parametric transducer, which converts the angle to an optical frequency. As shown in Fig. 2, the transducer consists of a high-Q optical cavity of length l, tuned by a Brewster-angled low loss dielectric plate of thickness d. A cavity finesse $F \simeq 10^5$ should be obtainable with dielectric mirrors for the Gaussian TEM_{00p} modes. The frequency tuning sensitivity is $\Delta f/f \sim (d/l)\Delta\theta$ yielding a resolution of $\sim 10^{-14}$rad/Hz. The angular measurement precision depends on the number of photons N and laser wavelength λ via $\Delta\theta \sim \lambda/(dF\sqrt{N})$. This is a factor $\simeq F$ better than the optical lever for the same laser power. Cryogenic optical resonators have excellent long-term stability and have been proposed as secondary frequency standards. The measured frequency drifts range from ~ 1 Hz over minutes to ~ 100 Hz over days [7]. For the measurement of the rotation angle, a stable reference frequency will be required. This can be implemented using a laser locked to a second (untuned) cavity. Because of its symmetry, the described angle readout has high immunity against lateral, tilt, and vertical seismic noise, but couples to rotational noise. A possible solution would be to suspend the target as well as the optical cavity in order to suppress the common rotation mode. Additional background forces will arise from gas collisions, cosmic ray hits, and radioactivity, etc., resulting in a Brownian motion of the target. The equivalent acceleration is $a \sim (\bar{p}/m)(\Gamma/\tau)^{1/2}$, where $\bar{p}$ is the average momentum transfer, Γ is the collision or decay rate, and m is the test mass. The residual gas pressure must hence be kept very low by cryopumping and the use of getters. The cosmic muon flux at sea level is $\simeq 100/(\mathrm{m}^2\mathrm{s})$ and can be reduced by going underground. Further disturbing forces may be caused by time-varying electric and magnetic background fields, which can be shielded with superconducters. Thermal radiation and radiometric effects are much reduced by a temperature controlled cryogenic environment. Finally, there is a fundamental

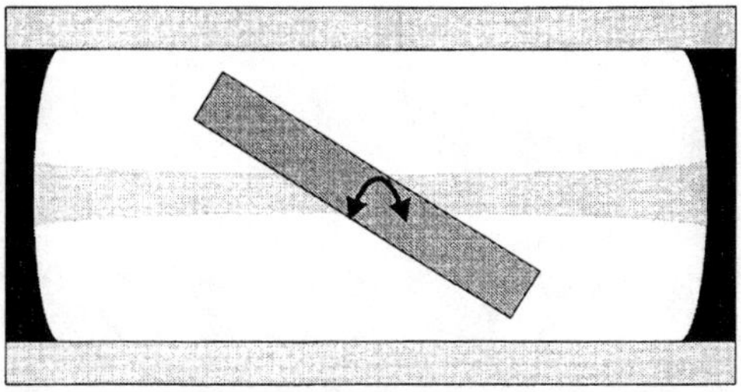

FIGURE 2. Topview diagram of the tunable high-Q optical cavity. The cavity is operated in the Gaussian TEM_{00p} mode. A narrow linewidth $\sim$ 100 kHz is achieved with high reflectivity mirrors and a low-loss dielectric Brewster-angled plate.

limit imposed by the uncertainty principle, with a minimum measurable acceleration given by $a_{SQL} = 5 \times 10^{-24} \text{cm/s}^2 (10\text{kg}/m)^{1/2}$ $(1\text{day}/\tau_0)^{1/2}$ $(10^6\text{s}/\tau)$. In our proposed position readout, the disturbing back action force arises from spatial fluctuations in the photon flux passing through the central tuning plate. There is hope that relic neutrinos will be detected in the laboratory early in the next century, especially if they have masses in the eV range and are of Dirac type. The most viable way seems to be a much improved torsion balance operating underground. Naturally, a slightly modified balance could be used to test the equivalence principle [8] and to search for new macroscopic forces [4].

ACKNOWLEDGEMENTS

This research was supported by DOE grant W-7405-ENG-048. I thank R. Newman for helpful discussions.

REFERENCES

1. Primack J., *Science* **280**, 1398 (1998); Gawiser E., and Silk J., *ibid.* , 1405 (1998).
2. Tremaine S., and Gunn J.E., *Phys. Rev. Lett.* **42**, 407 (1979).
3. Shvartsman V.F., Braginski V.B., Gershtein S.S., Zel'dovich Y.B., and Khlopov M.Y., *JETP Lett.* **36**, 277 (1982); Smith P.F., and Lewin J.D., *Phys. Lett.* **B127**, 185 (1983).
4. Adelberger E., Heckel B., Stubbs C., and Rogers W., *Annu. Rev. Nucl. Part. Sci.* **41**, 269 (1991).
5. Braginsky V., and Manukin A., *Measurements of Weak Forces in Physics Experiments*, Chicago: The University of Chicago Press, 1977.
6. V. Braginsky, C. Caves, and K. Thorne, Phys. Rev. **D15**, 2047 (1977).
7. Seel S., Storz R., Ruoso G., Mlynek J., and Schiller S., *Phys. Rev. Lett.* **78**, 4741 (1997).
8. Roll P., Krotkov R., and Dicke R., *Ann. Phys. (N.Y.)* **26**, 442 (1964); Braginsky V., and Panov V., *JETP Lett.* **34**, 463 (1972).

See-saw and supersymmetry

Goran Senjanović

International Center for Theoretical Physics, 34014 Trieste, Italy

Abstract. We review recent results on the connection between the see-saw mechanism and supersymmetry. It is shown that spontaneously broken B−L symmetry, in theories with a renormalizable see-saw, determines completely the low energy structure of the supersymmetric Standard Model: it implies exact R-parity. We then discuss how the violation of the survival principle in supersymmetric SO(10) naturally leads to an intermediate see-saw scale and the lowering of the unification scale. This affects the predictions for proton decay: $p \to \pi^o e^+$ may be the dominant mode.

INTRODUCTION

The see-saw mechanism [1] has become a canonical way of accounting naturally for a small neutrino mass. It says that the neutrino mass takes the form $m_\nu = m_D^2/M_R$, where m_D is the neutrino Dirac mass term and M_R is the mass of the right-handed neutrino. The solar and atmospheric neutrino data suggest a scale M_R way above the weak scale, and maybe even close to the grand unification scale M_X. In what follows, we assume this to be the case.

The see-saw mechanism can be naturally implemented in theories based on gauging the B−L anomaly-free accidental symmetry of the SM, such as in the Left-Right (LR) symmetric model [2] and their grand unified extensions such as $SO(10)$. This talk reviews the status of these theories.

First, we wish to address the issue of the see-saw mechanism in the context of supersymmetry. This is an extremely important question for at least two reasons. First, a central issue of the supersymmetric standard model is the fate of baryon and lepton numbers. The conservation of B and L is normally connected to R-parity, and it would be of great use to have a more fundamental, underlying principle settling this question. On the other hand, broken R-parity implies nonvanishing neutrino masses and hence obscures the see-saw predictions. It is here where the LR symmetry (or B−L gauge invariance) provides an important role: it implies R-parity in the underlying theory [3]. The results can be summarized in the following theorem [4]:

The renormalizable supersymmetric see-saw mechanism in theories with B−L symmetry implies an exact R-parity even in the low energy effective theory.

CP478, *COSMO—98*, edited by David O. Caldwell

SEE-SAW AND EXACT R-PARITY

Let us assume that the original theory possesses a $U(1)_{B-L}$ symmetry. Then M_R must be induced through the vev $\langle S\rangle$, where S is a B$-$L= 2 superfield (again, we assume that the see-saw has a renormalizable form). The B$-$L anomaly cancellation requires the existence of a B$-$L=-2 superfield $\bar{S}$. The superpotential is then given by

$$W_R = \frac{1}{2} f\, N^2\, S + g(S\bar{S}) \tag{1}$$

where N is the B$-$L=-1 singlet superfield which contains a right-handed neutrino $\nu^c \equiv C\bar{\nu}_R^T$ and $g(S\bar{S})$ is some function of the B$-$L invariant combination $S\bar{S}$, which gives $\langle S\rangle \neq 0$ at a scale M_R. Then from $F_N = f\,N\,S = 0$, it is clear that $\langle N\rangle = \langle\tilde{\nu}^c\rangle = 0$ in our vacuum. Since S is invariant under R-parity, at the scale M_R it remains unbroken.

Now what happens as one descends in energies all the way to M_W? Well, the see-saw and supersymmetry guarantee a large mass for $\tilde{\nu}^c$: $m_{\tilde{\nu}^c} = m_{\nu^c} = fM_R$ and thus $\tilde{\nu}^c$ is stable against any perturbation due to soft supersymmetry breaking terms. It remains firmly at the origin: $\langle\tilde{\nu}^c\rangle = 0$. However, the same is not true of $\tilde{\nu}$, since it is massless at the scale M_R. To see what happens, let us recall first the situation with the MSSM.

We define the MSSM as the SSM without R-parity breaking terms. Then, a nonvanishing vev for $\langle\tilde{\nu}\rangle$ implies the existence of a "doublet" [5] Majoron [6], the Goldstone boson associated with the spontaneous breaking of the continuous lepton number. Such a Majoron is ruled out experimentally. The point is that $\langle\tilde{\nu}\rangle$ must be small. In the most conservative case when $\tilde{\nu}$ is $\tilde{\nu}_\tau$, one still gets an upper limit [7] $\langle\tilde{\nu}_\tau\rangle \leq 10\,GeV$.

Now, the scalar partner R of the (pseudoscalar) Majoron, J, has a mass of the order of $\langle\tilde{\nu}_\tau\rangle$ and thus $m_R < M_Z$. This implies a Z decay $Z \longrightarrow J + R$, which, of course, is ruled out experimentally from the Z-decay width.

Let us see now what happens in the see-saw case. Since all the effects at low energies must be suppressed by a large scale $1/M_R$, simply the would-have-been Majoron of the MSSM remains light with $m_J \sim 1/M_R \ll M_Z$. Again we have the forbidden Z decay discussed above, and again this implies that R-parity remains exact, just as in the MSSM.

As we have seen, in order to have a theory of R-parity, one needs to assume B$-$L symmetry. This happens automatically in any theory based on LR symmetry, which offers a clue for the understanding of parity violation in weak interactions. The LR symmetry breaking in supersymmetric models with $M_R \gg M_W$ has been studied extensively in [8], and there already has been noticed that R-parity remains an exact symmetry of the low energy effective theory.

SO(10) AND INTERMEDIATE SEE-SAW SCALE

What is the value of the see-saw scale M_R? The solar and atmospheric neutrino data strongly indicate non-vanishing neutrino masses in the range 10^{-5} to 1 eV. In the context of see-saw, this suggests M_R somewhere above 10^{10} GeV and below the GUT scale. On the other hand, successful gauge coupling unification in the minimal supersymmetric Standard Model (MSSM) [9] discourages the existence of intermediate mass scales.

The way out is to have additional light states below the unification scale M_X, which are often postulated in an *ad hoc* manner. However, in supersymmetry these new light states arise naturally. The reason this tends to be overlooked is that normally one subscribes to the so-called survival principle [10], according to which both fermion and scalar masses take values at the highest scale allowed by the symmetries in question. Although this works perfectly well in ordinary field theories, it fails dramatically in their supersymmetric counterparts.

The argument is very simple: in many cases symmetry arguments in supersymmetry forbid the existence of trilinear couplings in the superpotential, which amounts to no renormalizable quartic couplings in the potential (other than those coming from the D-terms). Of course there will be in general non-renormalizable such couplings, induced when the heavier (with mass M) particles are integrated out, but they will be suppressed by the cut-off scale M. Namely, when a chiral supermultiplet Φ takes a vev $v \ll M$ the Higgs mechanism implies that the particle masses in Φ are of order λv, where λ stands for quartic couplings in general. The only exceptions are the states that belong to the superHiggs multiplets and have masses equal to those of the massive gauge bosons. When λ is generated through non-renormalizable interactions of order v/M, the particle masses associated with the scale v become suppressed. Instead of being of order v as the survival principle would imply, they become only of order v^2/M or smaller. This is an extremely important result of general nature and it can modify radically the running of gauge couplings and the unification predictions. It offers a serious hope for the existence of intermediate scales in supersymmetric GUTs.

In [11] we show that such light states are naturally present in the minimal supersymmetric SO(10) with a see-saw mechanism. We find that the right-handed scale (M_R) can lie between 10^{13-14} GeV, which fits nicely with the Kamiokande results. This can only be achieved at the expense of lowering the unification scale so that the lower limit on M_R results from proton stability. Thus the existence of an intermediate scale M_R predicts observable proton decay $p \to \pi^0\ e^+$, which is normally assumed to be strongly suppressed in favor of neutrino and kaon final states. This is a surprising result which establishes an important link between neutrino mass and the nature of proton decay, and at the same time sheds a new light on supersymmetric grand unification.

SUMMARY

As we stressed in the Introduction, the fate of R-parity is probably the central issue of the MSSM. We showed how *the renormalizable see-saw mechanism through the spontaneous breaking of B−L symmetry implies exact R-parity at all energies.*

When incorporated in SO(10), this leads to the possibility of an intermediate scale for M_R as suggested by the solar and atmospheric neutrino data. At the same time, the unification scale gets lowered providing an important connection between the proton decay and neutrino masses.

ACKNOWLEDGEMENTS

I wish to thank the organizers of COSMO98, and particularly David Caldwell, for a great conference and warm hospitality. I am deeply indebted to my collaborators Charan Aulakh, Borut Bajc, Alejandra Melfo and Andrija Rašin.

REFERENCES

1. M. Gell-Mann, P. Ramond, and R. Slansky in *Supergravity* (P. van Niewenhuizen and D. Freedman, eds.), (Amsterdam), North Holland, 1979; R. N. Mohapatra and G. Senjanović *Phys. Rev. Lett.* **44** (1980) 912; T. Yanagida in *Workshop on Unified Theory and Baryon number in the Universe* (O. Sawada and A. Sugamoto, eds.), (Japan), KEK, 1979.
2. J. Pati and A. Salam *Phys. Rev.* **D10** (1974) 275; R. Mohapatra and J. Pati *Phys. Rev.* **D11** (1975) 2558; G. Senjanović and R. Mohapatra *Phys. Rev.* **D12** (1975) 1502; G. Senjanović *Nucl. Phys.* **B153** (1979) 334.
3. R. Mohapatra *Phys. Rev.* **D 34** (1986) 3457; A. Font, L. Ibáñez, and F. Quevedo *Phys. Lett.* **B228** (1989) 79; S. P. Martin, *Phys. Rev.* **D46** (1992) 2769.
4. C. S. Aulakh, A. Melfo, A. Rašin, and G. Senjanović, hep-ph/9902409.
5. C. S. Aulakh and R. N. Mohapatra, *Phys. Lett.* **119B** (1982) 136.
6. Y. Chikashige, R. N. Mohapatra, and R. D. Peccei, *Phys. Lett.* **98B** (1981) 265.
7. C. S. Aulakh and R. N. Mohapatra, *Phys. Lett.* **121B** (1982) 147; D. E. Brahm, L. J. Hall, and S. D. H. Hsu, *Phys. Rev.* **D42** (1990) 1860.
8. C. S. Aulakh, A. Melfo, A. Rašin, and G. Senjanović, *Phys. Rev.* **D58** (1998) 115007.
9. M.B. Einhorn and D.R. Jones, Nucl. Phys. **B196**, 475 (1982); W.J. Marciano and G. Senjanović, Phys. Rev. **D25**, 3092 (1982).
10. H. Georgi, Nucl. Phys. **B156**, 126 (1979). For the case of scalars, see R.N. Mohapatra and G. Senjanović, Phys. Rev. **D27**, 1601 (1983).
11. C. S. Aulakh, B. Bajc, A. Melfo, A. Rašin and G. Senjanović, to appear.

LIST OF PARTICIPANTS

Andreas Albrecht	U.C. Davis	albrecht@physics.ucdavis.edu
Roland Allen	Texas A&M Univ.	allen@phys.physics.tamu.edu
Richard Arnowitt	Texas A&M Univ.	arnowitt@phys.physics.tamu.edu
Francisco Astorga	Michoacan Univ., Mexico	fastorga@zeus.ccu.umich.mx
Stephen Asztalos	MIT	Asztalos1@llnl.gov
Barut Bajc	Ljublyana Univ.	borut.bajc@ijs.si
Vernon Barger	Univ. Wisconsin	barger@goshawk.physics.wisc.edu
Steve Barr	Univ. Delaware	smbarr@bartol.bartol.udel.edu
Mar Bastero-Gil	Southampton Univ.	mbg@hep1.phys.soton.ac.uk
Dan Bauer	U.C. Santa Barbara	dan_bauer@SLAC.Stanford.EDU
Arjun Berera	Vanderbilt Univ.	berera@vuhep.phy.vanderbilt.edu
Lars Bergstrom	Stockholm Univ.	lbe@physto.se
Andrew Blain	Cambridge Univ.	a.w.blain@mrao.cam.ac.uk
Elliott Bloom	SLAC	Elliott@slac.stanford.edu
Norma Borstnik	Ljublyana Univ.	norma.mankoc@ijs.si
Raphael Bousso	Stanford Univ.	bousso1@leland.stanford.edu
Biswayjoy Brahmachari	Univ. Maryland	biswayjoy@katherine.physics.umd.edu
Ram Brustein	Ben Gurion Univ.	ramyb@bgumail.BGU.AC.IL
James Bullock	U.C. Santa Cruz	bullock@maxwell.ucsc.edu
Vladimir Burdyuzha	Lebedev, Moscow	burdyuzh@dpc.asc.rssi.ru
Chris Burgoyne	Cambridge Univ.	C.Burgoyne@damtp.cam.ac.uk
David Caldwell	U.C. Santa Barbara	caldwell@SLAC.Stanford.EDU
Sean Carroll	U.C. Santa Barbara	carroll@itp.ucsb.edu
David Casper	U.C. Irvine	casper@master.ps.uci.edu
Swapan Chattopadhyay	LBL	s_chattopadhyay@lbl.gov
Xuelei Chen	Columbia Univ.	xuelei@cuphy3.phys.columbia.edu
Jim Cline	McGill Univ.	jcline@physics.mcgill.ca
Joanne Cohn	Univ. Illinois, Chicago	jdc@astro.uiuc.edu
K.C. Cole	L.A. Times	KC.Cole@latimes.com
France Cordova	U.C. Santa Barbara	cordova@omni.ucsb.edu
Laura Covi	Lancaster Univ.	l.covi@lancaster.ac.uk
Hank Crawford	LBL	hjcrawford@lbl.gov
Owen Dando	Durham Univ.	O.R.Dando@durham.ac.uk
Marc Davis	U.C. Berkeley	marc@bkyast.berkeley.edu
Savas Dimopoulos	Stanford Univ.	savas@leland.stanford.edu
Alexandre Dolgov	TAC, Copenhagen	dolgov@x4.tac.dk
Richard Easther	Brown Univ.	easther@het.brown.edu
John Ellis	CERN	John.Ellis@cern.ch
Gary Felder	Stanford Univ.	gfelder@leland.stanford.edu
Nicolao Fornengo	Univ. Turin	fornengo@to.infn.it
Paul Frampton	Univ. North Carolina	frampton@physics.unc.edu
Josh Frieman	FNAL	frieman@fnal.gov
George Fuller	U.C. San Diego	gfuller@willy.ucsd.edu
Marcelo Gleiser	Dartmouth Univ.	gleiser@dartmouth.edu

Gary Godfrey	SLAC	godfrey@slac.stanford.edu
Jackie Grant	Univ. Sussex	kapt8@pcss.maps.susx.ac.uk
Patrick Greene	Toronto Univ.	greene@cita.utoronto.ca
Kim Griest	U.C. San Diego	griest@astrophys.ucsd.edu
Christopher Hagmann	LLNL	Hagmann1@llnl.gov
Stephen Hawking	Cambridge Univ.	Hawking@damtp.cam.ac.uk
Pat Henry	Univ. Hawaii	henry@ifa.hawaii.edu
Richard Holman	Carnegie Mellon	holman@cmuhep2.phys.cmu.edu
Dragan Huterer	Univ. Chicago	dhuterer@aether.uchicago.edu
Namanja Kaloper	Stanford Univ.	kaloper@leland.stanford.edu
Gordy Kane	Univ. Michigan	gkane@umich.edu
Shinta Kasuya	Tokyo Univ.	kasuya@icrr.u-tokyo.ac.jp
Jihn Kim	Seoul Univ.	JEKIM@phyp.snu.ac.kr
William Kinney	Univ. Florida	kinney@phys.ufl.edu
Joe Kiskis	U.C. Davis	jekiskis@ucdavis.edu
Lev Kofman	Toronto Univ.	kofman@cita.utoronto.ca
Rocky Kolb	FNAL	rocky@rigoletto.fnal.gov
Alexander Kusenko	UCLA	Kusenko@physics.ucla.edu
Andrei Linde	Stanford Univ.	linde@physics.stanford.edu
Phil Lubin	U.C. Santa Barbara	lubin@cfi.ucsb.edu
Joe Lykken	FNAL	lykken@fnal.gov
David Lyth	Lancaster Univ.	d.lyth@lancaster.ac.uk
Richard MacKenzie	Univ. British Columbia	rbmack@physics.ubc.ca
Joao Magueijo	Imperial College	j.magueijo@ic.ac.uk
Julie McEnery	LANL	jmcenery@flossie.lanl.gov
Joel McKibben		MCinCA@aol.com
Sanjib Mishra	Harvard Univ.	mishra@huhepl.harvard.edu
Rabindra Mohapatra	Univ. Maryland	rmohapatra@umdhep.umd.edu
Carmen Molina-Paris	LANL	carmen@T6-serv.lanl.gov
Jameal Nabi	MPI-Heidelberg	Jameel.Nabi@mpi-hd.mpg.de
Priyamvada Natarajan	Toronto Univ.	priya@cita.utoronto.ca
Al Odian	SLAC	odian@slac.stanford.edu
Lara Pasquali	Univ. Iowa	pasquali@hepsun1.physics.uiowa.edu
Saul Perlmutter	LBNL	saul@lbl.gov
Serguey Petcov	SISSA-Trieste	petcov@susy.he.sissa.it
Joel Primack	U.C. Santa Cruz	joel@ucolick.org
Harvey Reall	Cambridge Univ.	hsr1000@cam.ac.uk
Adam Riess	U.C. Berkeley	ariess@magicbean.berkeley.edu
Leszek Roszkowski	Lancaster Univ.	lr@virgo.lancs.ac.uk
Bernard Sadoulet	U.C. Berkeley	sadoulet@cfpa.berkeley.edu
David Salopek	Univ. British Columbia	dsscosmos@yahoo.com
John Schwarz	CalTech	jhs@theory.caltech.edu
Goran Senjanovic	ICTP-Trieste	goran@ictp.trieste.it
Joseph Silk	U.C. Berkeley	silk@bkyast.berkeley.edu
Andrew Sonnenschein	U.C. Santa Barbara	andrews@slac.stanford.edu
Glenn Starkman	Case Western Reserve Univ.	gds6@po.cwru.edu

Charles Steidel	CalTech	ccs@phobos.caltech.edu
Igor Tkachev	Purdue Univ.	tkachev@london.physics.purdue.edu
Mark Trodden	Case Western Reserve Univ.	trodden@theory1.phys.cwru.edu
Michael Turner	Univ. Chicago	mturner@oddjob.uchicago.edu
Tanmay Vachaspati	Case Western Reserve Univ.	txv7@po.cwru.edu
Gabriele Veneziano	CERN	gabriele.veneziano@cern.ch
Alexander Vilenkin	Tufts Univ.	vilenkin@cosmos2.phy.tufts.edu
Yun Wang	Princeton Univ.	ywang@astro.princeton.edu
Stefan Westerhoff	U.C. Santa Cruz	stefan@milagro.lanl.gov
Tom Weiler	Vanderbilt Univ.	tom.weiler@vanderbilt.edu
Martin White	Univ. Illinois, Chicago	white@physics.uiuc.edu
Serge Winitzki	Cambridge Univ.	S.Winitzki@damtp.cam.ac.uk
Stanford Woosley	U.C. Santa Cruz	woosley@helios.ucsc.edu
Steve Yellin	U.C. Santa Barbara	yellin@slac.stanford.edu
Gaurang Yodh	Univ. Calif., Irvine	yodh@master.ps.uci.edu
Junichi Yokoyama	Kyoto University	yokoyama@yukawa.kyoto-u.ac.jp
Hong-Sheng Zhao	Leiden University	hsz@strw.leidenuniv.nl
Ariel Zhitnitsky	Univ. of British Columbia	arz@physics.ubc.ca

COSMO-98 Program

Sunday, Nov. 15

Arrival and Registration from 3 pm onwards.

Dinner: 6:00-7:00 pm

Monday, November 16

Breakfast: 7:00-9:00 am

Morning: 9:00 am – 12:30 pm (Chair: F. CORDOVA)
- Opening Remarks (D. CALDWELL, F. CORDOVA) – 15 min.
- Inflation and Particle Physics (D. LYTH) – 30 min.
- Preheating and Reheating (L. KOFMAN) – 25 min.
- Matter Creation in the Early Universe (I. TKACHEV) – 20 min.
- Baryogenesis and Leptogenesis (J. CLINE) – 30 min.
- New and Improved Superstring Phenomenology (J. LYKKEN) – 30 min.

Lunch: 12:30-1:30 pm

Afternoon: 1:30 pm – 6:00 pm (Chair: G. KANE)
- Cosmological Parameters (M. TURNER) – 30 min.
- Current Issues in Structure Formation (J. SILK) – 30 min.
- Cosmic Microwave Background (M. WHITE) – 30 min.
- High-z Supernovae – experiment (S. PERLMUTTER) – 30 min.
- High-z Supernovae – experiment (A. RIESS) – 15 min.
- High-z Supernovae – theory (S. WOOSLEY) – 15 min.
- Dynamical Vacuum (J. FRIEMAN) – 30 min.
- Gamma Ray Telescopes, AGNs, GRBs (S. WESTERHOFF) – 30 min.

Dinner: 6:00-7:00 pm

Evening: 7:00 pm – 9:00 pm (Chair: E. KOLB)
- Debate on Quantum Cosmology (S. HAWKING, A. VILENKIN, A. LINDE)

Tuesday, Nov. 17

Breakfast: 7:00-8:30 am

Morning: 8:30 am – 12:00 Noon (Chair: B. SADOULET)
- Large Scale Structure (M. DAVIS) – 30 min.
- Galaxy Clusters and Cosmology (P. HENRY) – 30 min.
- High-z Universe (C. STEIDEL) – 30 min.
- Infrared Sources (A. BLAIN) – 30 min.
- Early Galaxy Formation (J. PRIMACK) – 30 min.

Lunch: 12 Noon-1:00 pm

Afternoon: 1:00 pm – 6:00 pm

Parallel Sessions

----

20 minute talks + 5 minute discussion, unless otherwise noted

Session A – INFLATION AND QUANTUM COSMOLOGY (Chair: A. LINDE)

Creation and Structure of Baby Universes (T. VACHASPATI)
Global Structure of the Inflationary Universe (R. BOUSSO)
Inflation with Quantum Corrections (L. COVI)
Non-equilibrium Dynamics of Inflationary Models (R. HOLMAN)
Formation of Topological Defects During Preheating (S. KASUYA)
Reheating after Inflation (P. GREENE)
Chaotic New Inflation (J. YOKOYAMA)
Inhomogeneity and the Post-Inflationary Universe (R. EASTHER)
Update on Open Universes (J. COHN)
Inflation after Preheating (G. FELDER)
Density Perturbations and CMB Anisotropy in
Open Inflation (A. LINDE) – 15 + 5 minutes

Session B – OBSERVATIONAL COSMOLOGY (Chair: J. PRIMACK)

Cosmology of Strongly Coupled Gravitational Systems (D. SALOPEK)
CMB Polarization Constraints on Inflation (W. KINNEY)
Position of the First Doppler Peak (P. FRAMPTON)
Evidence for Non-Gaussianity in COBE 4-year Sky Map (J. MAGUEIJO)
Model-Independent Measurement of the Primodial Power Spectrum (Y. WANG)
Lyman Break Galaxies as Collision-Driven Starbursts (J. BULLOCK)
Evidence for Dark Matter from Cluster Lensing Studies (P. NATARAJAN)
Gravitational Waves from Collapsing Vacuum Domains (M. GLEISER)
Instanton Cosmology (R. ALLEN)
Multi-wavelength Observations of TeV Gamma Ray Emitting AGNs (J. MCENERY)
GLAST (G. GODFREY) 15 + 5 minutes

Session C – NEUTRINOS (Chair: J. KIM)

Weak Interaction Rates at High Temperatures and Densities (J. NABI) 15 + 5 minutes
Atmospheric Neutrinos from Charm (L. PASQUALI)
See-Saw and Supersymmetry: the Fate of R-Parity (G. SENJONOVIC)
Large Mixing of Mu and Tau Neutrinos and SO(10) (S. BARR)
Neutrinos and Nucleosynthesis (A. DOLGOV)
Future of Double Beta Decay and Dark Matter
Searches – GENIUS (J. NABI) 15 + 5 minutes
A Mechanical Detector for Cosmic Neutrinos (C. HAGMANN)
A Comparison of the LSND and KARMEN Experiments (S. YELLIN)
Discussion of the Pattern of Neutrino Masses and
How to Determine H (V. BARGER) - 45 minutes

Dinner: 6:00-7:00 pm

WEDNESDAY, Nov. 18

Breakfast: 7:00-8:30 am

Morning: 8:30 am – 12:30 pm

Parallel Sessions

----

20 minute talks + 5 minutes discussion, unless otherwise noted

Session D – PARTICLE PHYSICS AND COSMOLOGY (Chair, G. STARKMAN)

Baryogenesis, Domain Walls and CP-Odd Bubbles in QCD (A. ZHITNITSKY)
A Next-to-Minimal Model of Hybrid Inflation in SUGRA Theories (M. BASTERO-GIL)
Sphalerons from CP-violating Higgs Potentials (J. GRANT)
SUSY and Symetry -Breaking at High Temperature (B. BAJC)
Cosmology with a Time-Varying Speed of Light (A. ALBRECHT)
Dirichlet Solitons in Field Theory (M. TRODDEN) - 15 + 5 minutes
Energy-Momentum of Particles Created in an
Expanding Universe (C. MOLINA-PARIS) - 15 + 5 minutes
Warm Inflation (A. BERERA) - 15 + 5 minutes
Unifications of Spins and Charges (N. BORSTNIK) - 15 + 5 minutes
Unification at Finite Temperature (F. ASTORGA) - 15 + 5 minutes

Session E – DARK MATTER (Chair, S. BARR)

Cryogenic Dark Matter Search (A> SONNENSCHEIN)
Relic Neutralinos and Dark Matter (N. FORNENGO)
Cosmological Consequences of Supergravity Models (R. ARNOWITT)
Three-Body Annihilation of Neutralinos Below Two-Body Thresholds (X. CHEN)
Quintessence and the Rest of the World (S. CARROLL)
SUSY Q-balls in the Early Universe (A. KUSENKO)
Lifetime Limit of the LSP from Cosmological Light Elements (J. KIM)
Dark Matter in String Cosmology Models (R. BRUSTEIN)
The Cosmological Consequences of the Preon Structure of Matter (V. BURDYUZHA)

Lunch: 12:30-1:30 pm

Excursion: 1:30-5:30 pm

Dinner: 6:00-7:00 pm

Evening: 7:00 pm- 9:00 pm (Chair, P. FRAMPTON)
From Superstrings to M-Theory (J. SCHWARZ) – 40 minutes
Testing String Theory by Probing the Pre-Big-Bang
Universe (G. VENEZIANO) - 30 minutes
Problems with Initial Conditions in Stringy Cosmology (N. KALOPER) – 20 min

THURSDAY, Nov. 19
Breakfast: 7:00-8:30 am

Morning: 8:30 am – 12:30 pm (Chair, G. YODH)
Theoretical Overview of Non-Baryonic
Dark Matter (L. ROSZKOWSKI) – 30 min
Limits on Sparticle Dark Matter (J. ELLIS) – 30 min
Direct DM Searches – WIMPS and Axions (B. SADOULET) – 30 min
Indirect Detection of Dark Matter (L. BERGSTROM) – 30 min
Microlensing of Baryonic Dark Matter (K. GRIEST) – 30 min
What MACHOs Could Be (H.-S. ZHAO) – 15 min

Lunch: 12:30-1:30 pm

Afternoon: 1:30 pm – 6:00 pm (Chair, A. DOLGOV)
Solar Neutrinos (S. PETCOV) – 30 min
Atmospheric Neutrinos (D. CASPER) – 30 min
Accelerator and Reactor Neutrino Oscillations (S. MISHRA) – 30 min
Neutrino Masses and Mixing (T. WEILER) – 30 min
Theoretical Views on Sterile Neutrinos (R. MOHAPATRA) – 30 min
Sterile Neutrinos: Big Bang and Supernova
Nucleosythesis (G. FULLER) – 30 min

Banquet: Buses depart at 6:45 pm

FRIDAY, Nov. 20
Breakfast: 7:00-9:00 am

Morning: 9:00 am – 12:00 pm (Chair, G. SENJANOVIC)
Higher Dimensions (S. DIMOPOULOS) – 30 min
Concluding Remarks (E. KOLB) – 45 min
Roundtable Discussion of the Near Future of the Field (1 hour)

Lunch: 12:00-1:00 pm

Conference Ends!

Author Index

A

B

C

D

E

F

G

H

J

K

L

M

N

P

R

S

T

V

W

Y